Sustainable Deve s

The **Sustainable Development Goals Series** is Springer Nature's inaugural cross-imprint book series that addresses and supports the United Nations' seventeen Sustainable Development Goals. The series fosters comprehensive research focused on these global targets and endeavours to address some of society's greatest grand challenges. The SDGs are inherently multidisciplinary, and they bring people working across different fields together and working towards a common goal. In this spirit, the Sustainable Development Goals series is the first at Springer Nature to publish books under both the Springer and Palgrave Macmillan imprints, bringing the strengths of our imprints together.

The Sustainable Development Goals Series is organized into eighteen subseries: one subseries based around each of the seventeen respective Sustainable Development Goals, and an eighteenth subseries, "Connecting the Goals," which serves as a home for volumes addressing multiple goals or studying the SDGs as a whole. Each subseries is guided by an expert Subseries Advisor with years or decades of experience studying and addressing core components of their respective Goal.

The SDG Series has a remit as broad as the SDGs themselves, and contributions are welcome from scientists, academics, policymakers, and researchers working in fields related to any of the seventeen goals. If you are interested in contributing a monograph or curated volume to the series, please contact the Publishers: Zachary Romano [Springer; zachary.romano@springer.com] and Rachael Ballard [Palgrave Macmillan; rachael.ballard@palgrave.com].

More information about this series at http://www.springer.com/series/15486

Xingxing Zhang
Editor

Data-driven Analytics for Sustainable Buildings and Cities

From Theory to Application

Springer

Editor
Xingxing Zhang
Dalarna University
Falun, Sweden

ISSN 2523-3084 ISSN 2523-3092 (electronic)
Sustainable Development Goals Series
ISBN 978-981-16-2780-4 ISBN 978-981-16-2778-1 (eBook)
https://doi.org/10.1007/978-981-16-2778-1

This Springer imprint is published by the registered company Springer Nature Singapore Pte Ltd.
The registered company address is: 152 Beach Road, #21-01/04 Gateway East, Singapore 189721, Singapore

Preface

The integration of data-driven analytics can be a solution to the challenges of energy transition, sustainable economic growth and mitigated climate change in built environment and city context. Buildings, communities and cities can be more vibrant, efficient and resilient if they are analysed and optimized as a complex multi-physics system based on big data sets.

This book explores the interdisciplinary and transdisciplinary fields of energy systems, occupant behaviour, thermal comfort, air quality and economic/business modelling across levels of building, communities and cities, through various data analytical approaches. It highlights the complex interplay of heating/cooling, ventilation and power systems in different processes, such as design, renovation and operation, for buildings, communities and cities. Methods from classical statistics, machine learning and artificial intelligence (e.g. regression/correlation, reinforcement learning, neural networks, genetic algorithm, clustering, agent-based modelling) are applied into the analyses for different building/urban components and systems. Knowledge from this book will assist to accelerate sustainability of the society, which would contribute to a prospective improvement through data analysis in the liveability of both built and urban environment.

This book targets at a broad readership with specific experience and knowledge in data analysis, energy system, built environment and urban planning. As such, it will appeal to researchers, graduate students, data scientists, engineers, consultants, urban scientists, investors and policymakers, with interests in energy flexibility, building/city resilience and climate neutrality.

Stockholm, Sweden
December 2020

Xingxing Zhang

Contents

Part II Thermal Comfort and Air Quality in Buildings

Part III Sustainability in Communities and Cities

The Evolving of Data-Driven Analytics for Buildings and Cities Towards Sustainability

1

Xingxing Zhang

Abstract

Buildings, communities and cities are now undergoing an accelerated transition in order to achieve goals of sustainability, security and resilience. Smart buildings and cities are generating a great amount of data by a very wide variety of sources. Data from these sources can be used to understand occupancy behaviour, evaluate energy performance, improve RES market competitiveness, enhance overall resources efficiency and so on. The emergence of the internet of things, improved data standards, big data analytical technologies and visualisation techniques are increasingly enabling the comprehensive applications in building and cities, allowing decision makers to understand and interrogate complex data from a variety of sources. The integration of data-driven analytics in building and cities could be a solution to the achievement of Sustainable Development Goals (SDGs). This chapter introduces background, motivation and structure for the whole book.

X. Zhang (✉)
Department of Energy and Community Buildings, Dalarna University, 79188 Falun, Sweden
e-mail: xza@du.se

1.1 Introduction

1.1.1 Background

The global contribution from buildings towards energy use has steadily increased, reaching figures between 20 and 40% and about 1/3 of greenhouse gas emission. For instance, in Sweden, buildings account for roughly 30% of the total energy and for about 7% of its total emissions of greenhouse gases. The Swedish government has set the target for enable energy use in buildings to be reduced by 20% by 2020 and 50% by 2050 compared with 1995 (Swedish Sustainable Building 2017). Progress in energy and buildings represent the biggest challenge or have insufficient steps so far. In addition, cities are the largest and most complex systems, and also the most resource-intensive and waste-producing ones.

It is no doubt that buildings, communities and cities are now undergoing an accelerated transition in order to achieve goals of sustainability, security and resilience. The main drivers for these transitions are the emergence of climate change, pandemic as Covid-19, renewable-energy-source solutions, building renovation requirements, digitalization, automation, smart mobility, circular economic models, carbon emission reduction, and policy/regulation change. These facts are significantly motivating the increase of comfort/health needs, energy and

X. Zhang (ed.), *Data-driven Analytics for Sustainable Buildings and Cities*, Sustainable Development Goals Series, https://doi.org/10.1007/978-981-16-2778-1_1

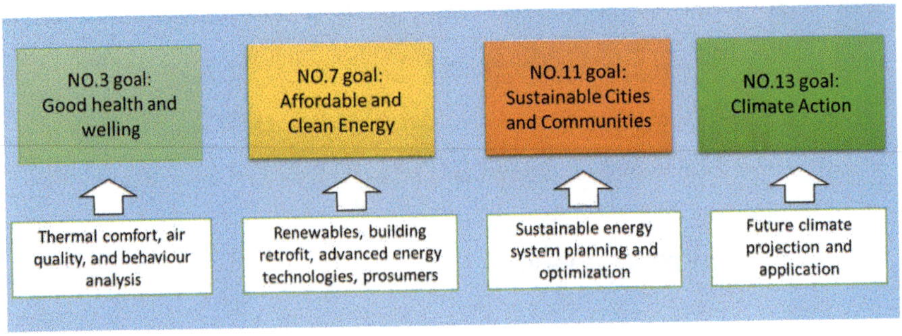

Fig. 1.1 Connection of UN's sustainable development goals

recourse efficiency at both building and city levels, as well as decarbonized economy for the aid in transition to sustainable economic growth and social welfare (International Energy Agency 2013; UNEP 2017). In September 2015, the United Nations (UN) General Assembly adopted the 2030 Agenda for Sustainable Development that includes 17 Sustainable Development Goals (SDGs) (The United Nations 2021), where health, energy, buildings/cities and climates are the main issues for us to endeavour. This book thus focuses on different ways to proceed towards several SDGs, such as *Good Health and Well-being*, *Affordable and Clean Energy*, *Sustainable Cities and Communities*, and *Climate Action*. Figure 1.1 illustrates total SDGs and the focused goals by this book.

In the framework of buildings: the goal of '*Good Health and Well-being*' are influenced greatly by occupanty behaviour and indoor environmental quality (IEQ), such as thermal comfort and air quality. Different occupancy condition causes various issues in air stagnation that may concentrate airborne viruses or dust indoor, which will decrease the indoor air quality and influence occupants' health. There are usually two ways to investigate IEQ and occupant's health: (i) analysis of IEQ from indoor air pollutant factors and indoor climate factors; and (ii) analysis of the occupant's health for both mental health and physical health (Kim et al. 2020). In the first approach, indoor air pollutant factors and indoor climate factors are further divided into different subcategories. Toxic chemical, soil gases, CO_2 concentration, suspended particles and microbes are the primary factors of indoor air pollution in our homes. While humidity, ventilation, and temperature are the main factors of indoor climate. In the second theory, occupants' mental health can be evaluated by psychological responses, such as verbal scale of 'comfort' feeling, which usually include the main indices of thermal comfort vote, thermal satisfaction, thermal sensation vote and thermal preference. Their physical health can be represented by the physiological responses (i.e., blood pressure) to check if homeostasis is maintained (Budd and Warhaft 1996). Both physiological and psychological impacts on occupants by different IEQ index. Therefore, buildings play an important role in supporting our health and well-being, especially in response to global health challenges, such as the COVID-19. During the normal periods, individuals spend more than 85% of their total time in indoor environments for home stay and work. In the special period of COVID-19 or similar crisis, this percentage is even higher up to 100% for a fully lock-off scenario. In addition, future climate change, particularly in temperature and humidity, will exacerbate the uncertainty in IEQ and the related impact on human's health and well-being. It is critical to investigate the impact and the effective means to improve IEQ in response to this SDG.

During the progress of achieving '*Affordable and Clean Energy*' goal, the emergence of renewable-energy-source (RES) envelope solutions, building retrofit requirements and advanced energy technologies are bringing challenges to the existing paradigm of energy systems at both building and city scales. This growth is predominantly due to the success and popularity of adaptive building envelope solutions, such as building integrated photovoltaics (BIPV) or building integrated photovoltaics/thermal, solar thermal façade, heat pump components and their accompanying power storage or thermal storage systems (Zhang et al. 2018). The emergence of these RES envelope solutions not only indicates a shift in the energy landscape towards more sustainable and resilient practices, but also entails an evolution in city energy planning, modelling techniques, operation/control intelligence and management schemes for matching of energy supply and demand across various system scales. Buildings are becoming prosumers, rather than purely standalone energy consuming units of the grid. They are increasingly turning into active elements of the energy network by consuming, producing, storing and supplying energy. Thus, they transform the energy market characterized by centralized, fossil-fuel based national systems to a decentralized, renewable, interconnected and viable system.

Above transitions then accelerate the achievements in the goal of '*Sustainable Cities and Communities*'. They lead to the development of novel approaches that address buildings and their energy systems at different scales: from single buildings to cluster, district and city levels. It is envisioned that sustainable energy strategy should combine energy efficiency retrofit and local RES supply, through the enhancement of district energy systems and decentralized energy supply. As a result, sustainable energy planning could foster the economic effectiveness and the operation feasibility to maximize the distributed RES harvesting and match with the respective energy demand and supply. It is essential to determine which RES solutions are synergic, and where the they should be implemented for operation in order to fully utilize the potential of distributed RES harvesting, storage, distribution, load aggregation and demand side management. The shift from the single building to the city is crucial for the improvement of local energy resource efficiency, through the interaction between the buildings and the energy infrastructure domain.

Buildings nowadays are able to erect for more than 100 years, and cities are even lasted for centuries. But they are vulnerable to the climate challenges, which can be observed from rise in the average environmental temperature and humidity levels to extreme and severe events (such as strong wind, floods, sea level rise and wildfires). When climate changes, the IEQ and energy performance will be different. Although many strategies are now implemented to mitigate the tread of climate change, it is changing now and will change in the future. Therefore, future climate projections need to be considered when studying building and cities by integrating climate change measures into design strategies, modelling and planning, which then leads to the contribution to '*Climate Action*' goal.

1.1.2 Data-Driven Analytics for Sustainability Goals

The 'data' represents discrete facts. Data can be used to deliver information, analysed to produce knowledge and applied to give guidance in built and urban environment, such as planning energy infrastructure (Designing Buildings Ltd. 2021). When computational power and data storage capabilities increase, and the cost of sensors reduces, the amount of data has risen significantly in recent decades.

Nowadays, smart buildings and smart cities are generating a great amount of data by a very wide variety of sources, such as design and construction, post occupancy evaluation, utilities, building services, operational/building management systems, energy infrastructure and transport systems (e.g. electrical vehicle), maintenance and replacement systems etc. Data from these sources can be used to understand occupancy behaviour, evaluate energy performance, improve RES market competitiveness, enhance overall resources efficiency and so on. The emergence of the internet of things, improved data standards, big data analytical technologies and visualisation techniques are increasingly enabling the comprehensive applications in building and cities, allowing decision makers to understand and interrogate complex data from a variety of sources.

The integration of data-driven analytics in building and cities could be a solution to the achievement of SDGs. The concept of data-driven analytics for buildings and cities refers to the development of different analytical models to simulate, predict and optimize the performance of buildings and cities as a complex multi-physics system based on big data sets. These approaches usually integrate traditional physics and engineering solutions with internet of things, artificial intelligence, machine learning and software analytical tools to discover the potentials and unknowns from data point of view. With the assistance by data-driven analytics, we could tackle the challenges of energy transition, sustainable economic growth and mitigated climate change in built environment and city context. Buildings, communities and cities can be more vibrant, efficient and resilient if they are analysed and optimized as a complex multi-physics system based on big data sets.

Data-driven analytics for buildings and cities enables a revolutionary way to accelerate sustainability goals of the society. For more and more buildings and cities, sensors will be set up to collect all kind of information towards digital twins, such as occupancy (mobility), temperature, moisture, energy consumption, renewable production, CO_2 concentration, costs, waste, carbon footprint, etc. With such big data sets, various data-analytical models can be developed and used to assess energy demand/supply/storage, indoor air quality, CO_2 emissions, expenses for operating & maintenance, building renovation/replacement needs (including recycle of waste construction material), carbon emissions and payback periods of energy saving measures over lifetimes of buildings and cities.

As a result, data-driven analytics are becoming more and more important in achieving the United Nations' SDGs. They could highghly enables the realization of high-value buildings and districts/cities over life span, where people live and work to facilitate social interactions, foster communities, and could thus improve individual outcomes, drive loyalty, build a brand, and create healthier, happier, and more resilient society. So buildings and districts/cities are able to maximize the wellbeing of their inhabitants, and achieve real sustainability.

1.2 Aim and Objectives

This book aims to explore the interdisciplinary and transdisciplinary fields of energy systems, occupant behaviour, thermal comfort, air quality, and economic/business modelling across levels of building, communities and cities, through various data analytical approaches. It highlights the complex interplay of heating/cooling, ventilation, and power systems in different processes, such as design, renovation and operation, for buildings, communities and cities.

The book has comprehensive objectives as improving thermal comfort and air quality, maximizing benefits of local renewable energy system, developing new business models, optimizing energy system at different scales, and mitigating the impact of future climate change on buildings and cities. All these objectives are connected further with the focused SDGs as explained in above section.

Methods from classical statistics, machine learning and artificial intelligence (e.g. regression/correlation, reinforcement learning, neural networks, genetic algorithm, clustering, agent-based modelling) are applied into analyses for different building/urban components and systems.

Knowledge from this book will assist to accelerate sustainability of the society, which would contribute to a prospective improvement through data analysis in the liveability of both built and urban environment.

1.3 Motivations and Novelties

Buildings and cities are complex, stochastic and multi-disciplinary under diverse cultural and geographical contexts. It is a common challenge to precisely describe the performance of buildings and cities through traditional modelling engines since it depends on a wide range of factors, i.e. weather condition, building envelope, system components' performance, operating schedules, energy sharing, and economic figures etc. It thus requires novel approaches to have in-depth quantitative analysis by integrating physical/engineering, statistical, machine learning, business models, to facilitate the transition in design, construction, operation, and retrofit of buildings and cities.

A few books have started to touch the related topics, e.g., '*Energy Simulation in Building Design*' (Clarke 2001), '*Building Performance Simulation for Design and Operation*' (Henson and Lambers 2011), '*Building Performance Analysis*' (Wilde 2018), '*Data Mining and Machine Learning in Building Energy Analysis*' (Magoules and Zhao 2016) and '*Data-driven Multivalence in the Built Environment*' (Biloria 2019). These books mostly highlight forward modelling methods, design strategies or energy performance of individual building, while very few on data-driven methods for predicting occupant schedules and behaviours, nor on community and urban scales.

There are also other existing studies that addresses the sustainability issues on data and cities. For instance, '*Data Cities*' (Jackson 2019) explains how Satellite data are transforming architecture and design. Similarly, '*ArcGIS and the Digital City*' (Huxhold et al. 2004) shows how a city goes digital and use geographic information systems (GIS) to store and access information. '*Data and the City*' (Kitchin et al. 2018) considers the social and political ramifications of data-driven urbanism. '*The Responsive City*' (Goldsmith and Crawford 2014) provides a guide to civic engagement and governance in the digital age that will help leaders link important breakthroughs in technology and data analytics with age-old lessons of small-group community input to create more agile, competitive, and economically resilient cities. In the book of '*Big Data Science and Analytics for Smart Sustainable Urbanism*' (Bibri 2019), the author gives the systemic implications for smart sustainable urbanism in light of big data science and analytics. Moreover, '*Smart and Sustainable Cities and Buildings*' (Roggema and Roggema 2020) collects a few conference papers and it is prominently featured in the design and planning of buildings and cities. In '*Urban Energy Systems for Low Carbon Cities*' (Eicker 2018), indicators to evaluate urban energy performance are introduced and the status quo of monitoring and efficiency valuation schemes are discussed. The author focuses on a bottom-up modelling approach used for the simulation of energy consumption, energy conversion systems and distribution networks using engineering methods. It is obvious that these books are mainly from the perspective of sociology, policy building design, urban planning and traditional engineering modelling, while few evidence on data analytical approaches and their applications in energy systems, occupant behaviour, thermal comfort, air quality, and economic/business modelling across different levels of building, communities and cities. Thus, it is desired to have a book to further cover the suitable methods and case studies in this area.

This book focuses on data-driven modelling and analysing methods of both individual building and buildings in community and urban scales. Due to the complex, dynamic and non-linear characteristics of buildings, communities and cities, data-driven methods (such as reinforcement learning, artificial neural networks, support vector machines, genetic algorithm, clustering, regressing etc.) can be applied into analyses of energy use, thermal comfort, indoor environment and grid interaction of buildings.

This book presents the up-to-date data-driven modelling methods and their application in sustainable buildings and cities, concerning not only the energy performance but also the thermal comfort and air quality of buildings. It will offset the gap of interdisciplinary fields of building, occupant behaviour, thermal comfort, air quality, energy system etc., using various data analytical methods.

1.4 Structure and Contents

Apart from the first chapter, this book consists of three main parts. Part I covers the data analytics (i.e. artificial neural networks, clustering, statistics) for energy systems in buildings, while Part II highlights the application of data analysis for thermal comfort and air quality in buildings using regression, reinforcement learning, support vector machines and clustering methods. Part III discusses machine learning and artificial intelligence (AI) methods (genetic algorithm, agent-based modelling and digital mapping etc.) for sensitivity analysis and energy performance in community/urban/city scale.

In Part I, there are six chapters. Chapter 2 starts with a comprehensive review about data-driven approaches for prediction and classification of building energy consumption. In Chap. 3, it presents a case study in prediction of occupancy level and energy consumption in office building using blind system identification and neural networks. The next chapter shows the cluster analysis for occupant-behaviour based electricity load pat-terns in buildings. In Chap. 5, a data-driven model predictive control for lighting system is studied based on historical occupancy in an office building. Chapter 6 considers the statistical analysis in future climate for building energy simulation. A solar photovoltaic/thermal (PV/T) concentrator for building application is covered in Chap. 7 using Monte Carlo method, in order to address the potential of solar energy application for buildings.

Another six chapters are considered in Part II. Chapter 8 estimates the influencing factors for occupants' window-opening behaviour in an office building through logistic regression and Pearson correlation approaches. Reinforcement learning method is reviewed in Chap. 9 and applied for improving occupant comfort via window opening and closing in Chap. 10. An adaptation table is developed in Chap. 11 to enhance the accuracy of the predicted mean vote model. In addition, a prediction accuracy weighted voting ensemble method is proposed in Chap. 12 for thermal sensation evaluation. In Chap. 13, it carries an analysis and interpretation of the particulate matter concentrations at the subway stations by general linear model (GLM) and correlation analysis.

Part III contains final eight chapters. Genetic Algorithm is applied for transforming a residential building cluster into electricity prosumers in Chap. 14, which is also studied in Chap. 15 for a coordinated control to improve performance for a building cluster with energy storage, electric vehicles, and energy sharing. Chapter 16 displays how to use Genetic Algorithm and Mont Carlo method for global sensitivity analysis of key parameters identification of net zero energy buildings towards power grid interaction optimization. In Chap. 17, the agent-based modelling is investigated in a peer-to-peer (P2P) business model for individual PV prosumers in a local electricity market. Chapters 18 and 19 tackle the building energy consumption at district/city level based on auto-encoder algorithm and digital mapping respectively. Chapter 20 gives an overview about digital twin approach to accelerate sustainability in positive energy districts. In Chap. 21, digital mapping is applied again for the techno-economic analysis of a solar photovoltaic/thermal collector over large geographical cities.

All the chapters are interlinked and well categorised to tackle the challenges in different aspects within built environment, communities and cities. The overall contents are derived from the techno-economic point of view, which are supplementary to the existing books regarding the sustainable transition of buildings and cities.

References

Bibri SE (2019) Big data science and analytics for smart sustainable urbanism: unprecedented paradigmatic shifts and practical advancements. Springer. ISBN 9783030173128

Biloria N (2019) Data-driven multivalence in the built environment. Published by Springer, ISBN, p 9783030121808

Budd GM, Warhaft N (1996) Body temperature, shivering, blood pressure and heart rate during a standard cold stress in Australia and Antarctica. J Physiol 186:216

Clarke J (2001) Energy simulation in building design, 2nd edn. Published by Elsevier. ISBN 978-0-7506-5082-3

De Wilde P (2018) Building performance analysis. Published by Wiley-Blackwell, ISBN, p 9781119341925

Designing Buildings Ltd. (2021) Big data for buildings. Accessed on 14 May 2021. https://www.designingbuildings.co.uk/wiki/Big_data_for_buildings

Eicker U (2018) Urban energy systems for low-carbon cities. Academic Press. ISBN 978012811554

Goldsmith S, Crawford S (2014) The responsive city: engaging communities through data-smart governance. Wiley. ISBN 9781118910931.

Henson J, Lambers R (2011) Building performance simulation for design and operation, 2nd edn. Published by Routledge. ISBN 9781138392199

Huxhold WE, Fowler EM, Parr B (2004) ArcGIS and the digital city: a hands-on approach for local government. Published by ESRI, Inc. ISBN 9781589480742

International Energy Agency (2013) Transition to Sustainable Buildings: Strategies and Opportunities to 2050

Jackson D (2019) Data cities: how satellites are transforming architecture and design. Published by Lund Humphries. ISBN 978-1848222748

Kim J, Hong T, Kong M, Jeong K (2020) Building occupants' psycho-physiological response to indoor climate and CO_2 concentration changes in office buildings. Build Environ 169:106596

Kitchin R, Lauriault TP, McArdle G (2018) Data and the city. Published by Routledge, ISBN, p 9781138222632

Magoules F, Zhao H-X (2016) Data mining and machine learning in building energy analysis. Published by ISTE Ltd. and John Wiley & Sons Inc., ISBN, p 9781848214224

Roggema R, Roggema A (2020) Smart and sustainable cities and buildings. Springer Nature. ISBN 9783030376352

Swedish Sustainable Building (2017) Accessed on 17 Aug 2017. Swedish Research Council

The United Nations (2021) Department of economic and social affairs sustainable development, THE 17 GOALS, Accessed on 14 May 2021. https://sdgs.un.org/goals

UNEP (2013) Energy efficiency for buildings. Accessed on 16 Aug 2017. http://www.studiocollantin.eu/pdf/UNEP%20Info%20sheet%20-%20EE%20Buildings.pdf

Zhang X, Lovati M, Vigna I, Widén J, Han M, Gal C, Feng T (2018) A review of urban energy systems at building cluster level incorporating renewable-energy-source (RES) envelope solutions. Appl Energy 230:1034–1056

Part I

Energy in Buildings

2 Data-Driven Approaches for Prediction and Classification of Building Energy Consumption

Yixuan Wei, Xingxing Zhang, and Yong Shi

Abstract

A recent surge of interest in the building energy consumption has generated a tremendous amount of energy data, which boosts the data-driven algorithms for broad application throughout industry. This chapter reviews the prevailing data-driven approaches used in building energy analysis under different archetypes and granularities including those for prediction (artificial neural networks, support vector machines, statistical regression, decision tree and genetic algorithm) and those for classification (K-mean clustering, self-organizing map and hierarchy clustering). To be specific, we introduce the fundamental concepts and major technical features of each approach, together summarizing its current R&D status and practical applications while pointing out existing challenges in their development for prediction and classification of building energy consumption. The review results demonstrate that the data-driven approaches, although they are constructed based on less physical information, have well addressed a large variety of building energy related applications, such as load forecasting and prediction, energy pattern profiling, regional energy-consumption mapping, benchmarking for building stocks, global retrofit strategies and guideline making etc. Significantly, this review refines a few key tasks for modification of the data-driven approaches in the contexts of application to building energy analysis. The conclusions drawn in this review could facilitate future micro-scale changes of energy use for a particular dwelling through appropriate retrofit in building envelop and inclusion of renewable energy technologies. They also pave an avenue to explore potential in macro-scale energy-reduction with consideration of customer demands. All these will be useful to establish a better long-term strategy for urban sustainability.

Keywords

Data driven approach · Building · Energy consumption · Prediction · Classification

Y. Wei (✉)
School of Civil Engineering and Resources, University of Science & Technology Beijing, Beijing, China
e-mail: weiyixuan@ustb.edu.cn

X. Zhang
Department of Energy and Community Buildings, Dalarna University, 79188 Falun, Sweden
e-mail: xza@du.se

Y. Shi
Research Centre for Fluids and Thermal Engineering, University of Nottingham Ningbo China, Ningbo, China
e-mail: yong.shi@nottingham.edu.cn

X. Zhang (ed.), *Data-driven Analytics for Sustainable Buildings and Cities*, Sustainable Development Goals Series, https://doi.org/10.1007/978-981-16-2778-1_2

2.1 Introduction

2.1.1 The Need for Energy Consumption Analysis

The global contribution from buildings towards energy consumption has steadily increased reaching figures between 20 and 40% in developed countries and about 1/3 of greenhouse gas emission. The case of China is particularly striking-the country only takes two decades to double its building energy consumption at an average growing rate of 3.7% (Pérez-Lombard et al. 2008; UNEP 2013). These facts demonstrate that to facilitate energy efficiency of building is a cost-effective resource for reducing energy consumption and carbon emission from building (Mathew et al. 2015). Also, large potential saving in economy has been anticipated by a large variety of previous studies. For instance, Nikolaidis et al. have shown that among various energy saving measures for common building types, isolation of roof constitutes the most superiority nearly €5000 economic benefit during 30 years (Nikolaidis et al. 2009). As the central approaches transmitting to energy efficiency, prediction and classification of energy consumption in building are significantly necessary with the aim to improve building performance, reduce environmental impact, and estimate economical potential for further energy conservation and renewable energy program (Zhao and Magoulès 2012).

Energy consumption in building has been heavily analyzed by substantial studies during the entire building lifecycle, with different focuses on identifying the sub-component energy use at the building level (Kang and Jin 2014; Bojić and Lukić 2000) or measuring energy performance in a nationwide analysis (Farahbakhsh et al. 1998; Huang 2000; Shimoda et al. 2004). This comprehensive set of analyses on different levels could help us not only optimize the energy use of a particular dwelling through appropriate retrofit in building envelop or inclusion of state-of-the-art renewable energy technologies (at the microscale), but also explore possible energy reduction opportunities and establish better urban-sustainability strategies (at the macroscale).

2.1.2 Advantage and Motivation

However, it is recognized that realization of a precise energy consumption analysis is a formidable task at the current stage. As an alternative, great efforts have been paid to developing models to predict and classify approximately energy consumption. Generally, these models possess some prominent functions. They includes (1) taking measures for energy conservation based on accurate prediction, (2) implementing demand-side management (DSM) after profiling electricity consumption, (3) outlining/mapping energy on the urban level, (4) establishing benchmark database of multi-scale building communities, and (5) integrating the processes of design, operation, retrofit of contemporary building (Perino et al. 2015; Hong et al. 2014). The results simulated by these models can not only offer essential information about energy footprint in regional building stocks, but also facilitate estimations of financial return on investment. It is thus not surprising energy simulation has become a favourable tool for stakeholders throughout building industry including policymakers, building owners, investors, operators and engineers (Mathew et al. 2015).

2.1.3 Usage of Building Energy and Performance Data

Management and optimization of building energy consumption call for a full understanding of building performance, which should first identify energy resources and major end-uses of a building. Energy resources in a building usually refer to electricity, natural gas and district heating supply. The corresponding major end-uses include heating, ventilation and air-conditioning (HVAC) system, domestic hot water, lighting, plug-loads, elevators, kitchen equipment, ancillary equipment and appliances. Figure 2.1 illustrates a representative classification of building energy use adopted in ISO Standard 12655:2013 (ISO 2013). Note that on top of the above building energy resources and major end-uses, HVAC operation schedule and indoor/outdoor conditions are also two important contributing factors to be considered in a building performance analysis.

Generally, reliability of a building performance analysis relies heavily on the datasets in use, which should contain sufficient energy consumption information of the buildings under investigation. Utility bills for electricity and natural gas from power supply companies are the common type of databases of building energy consumption. Facility managers or research institutes also collect information via survey and questionnaire for large-scale buildings, such as the residential sector (Residential Energy Consumption Survey (RECS), EIA 2009) and commercial buildings (Commercial Building Energy Consumption Survey (CBECS), EIA 2012) (Hong et al. 2014). In addition, in today's building performance analyses, virtual building database (VBD) developed from simulation software (e.g. TRNSYS and EnergyPlus) and energy disclosure laws (Mathew et al. 2015; Nikolaou et al. 2012) (e.g., US Energy Information Administration database) are the other two possible data resources. It is particularly worth mentioning that the empirical datasets taking advantage of smart meters and building energy system have emerged in recent years. These databases substantially improved accuracy and reliability of the related analyses (Mathew et al. 2015) despite their expensive costs and technical complexity involved for many practical commercial-uses.

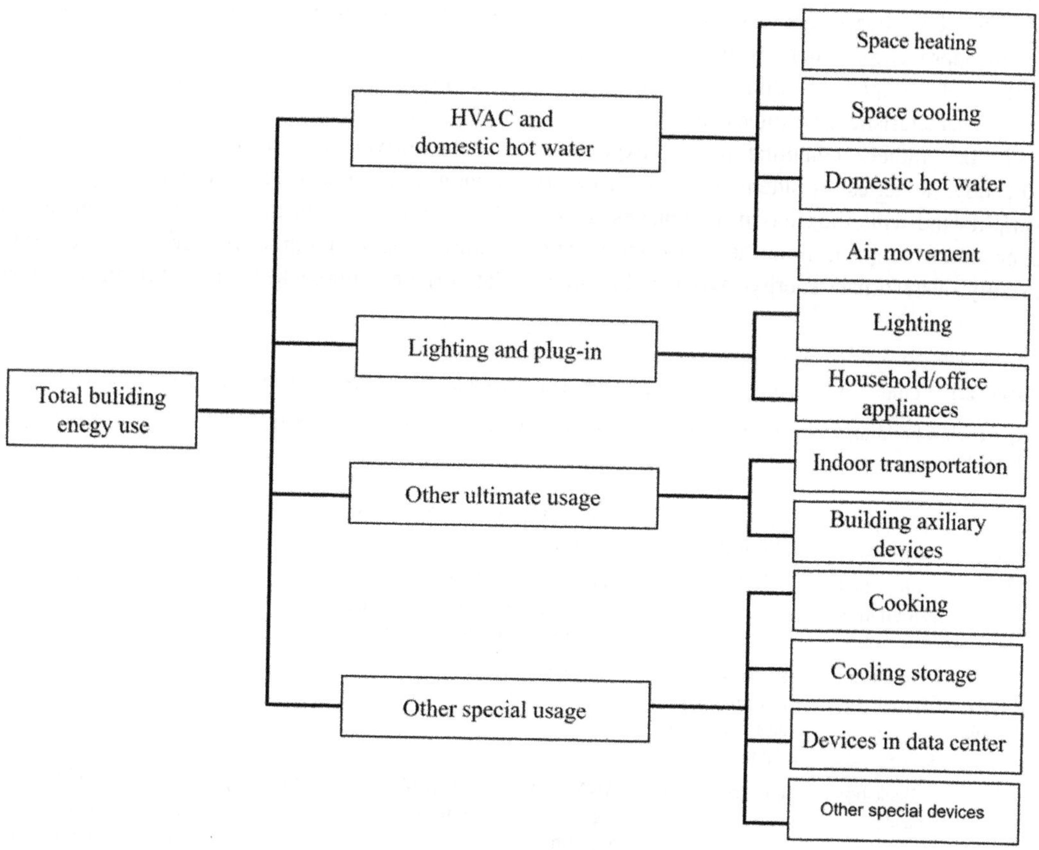

Fig. 2.1 The usage of energy in buildings (ISO 2013)

2.1.4 Proposed Methodologies for Building Energy Consumption

It is a challenging task to precisely describe energy consumption in a building as such an energy performance depends on a wide range of factors, such as weather condition, thermal properties of building envelope, occupancy behaviour, sub-level components' (lighting, HVAC and plug equipment) performance and schedules (Zhao and Magoulès 2012). A large number of efforts have been paid in the literature to ascertain the complexity pertinent to building energy consumption and strive to a precise depiction of building energy performance. Currently, these approaches used for building energy simulation are categorized roughly as: (1) white-box based approaches, (2) grey-box based approaches and (3) black-box based approaches, whose main features are summarized in Table 2.1.

White-box based approaches are physical-based approaches, which require detailed information of complex building phenomena. This basic characteristic determines their simulations will be rather computationally expensive. Recently, a series of attempts have made to simplify the white-box based approaches. However, these simplifications are error-prone and usually overestimate energy-saving of buildings (Al-Homoud 2001; Barnaby and Spitler 2005). Grey-box based approaches are a modification of these white-box based approaches through use of statistical methods combining the simplified physical information with historical data to simulate building energy. One primary issue in current grey-box version is computational inefficiency as the approaches involve uncertain inputs and complex interactions among elements and stochastic occupant behaviours (Paudel et al. 2015; Li et al. 2014). To circumvent the above shortfalls of white- and grey box based approaches, black-box based approaches are developed which are able to conduct a building energy consumption analysis only based on historical data without the detailed knowledge of on-site physical information. This essential change enable black-box based approaches fast calculations in high accuracy in comparison to their white- and grey-box counterparts (Zhao and Magoulès 2012). In many practical scenarios, the black-box based approaches are also called as data-driven approaches due to the statistical algorithm structures and a large amount of data in use. We will follow this convention and use the data-driven approaches throughout the following discussion in this review.

The remainder in this review is organized as follows: we will introduce various mainstream data-driven approaches and summarize their

Table 2.1 Comparison among white-box, grey-box and black-box approaches for building energy consumption

Approaches building simulation		Inputs needed	Typical software methods	Easy to use	Running speed	Accuracy
White-box based	Elaborated simulation	Detailed physical information	DOE-2, EnergyPlus, TRYSYS, ESP-r	No	Low	High
	Simplified simulation		Degree day method, temperature frequency method, residential load factor method	Yes	High	Fairly high
Data-driven models	Grey-box based	Physical information and historical data	RC network	No	Low	Fairly high
	Black-box based	Historical data	ANNs, SVMs, statistical regression, GA, cluster algorithms	No	High except SVM	High except regression

applications in prediction and classification of building energy consumption in Sects. 11.2.2 and 11.2.3, respectively. In Sect. 11.2.4, a few promising future directions in data-driven approaches with applications to building energy will be proposed. Finally, we draw our salient conclusions in Sect. 11.2.5.

2.1.5 Data-Driven Approaches

Data-driven models are constructed based on a group of datasets consisting of historical data records. These historical data will be used as benchmarks to justify the model's performance and guide its algorithm design. To be specific, all the parameters in a data-driven model will be carefully selected and modified through systematical comparisons between the model outputs and the historical data. This is the so-called learning process and only when the output errors fall within the required threshold, the corresponding data-driven models are deemed to be qualified for practical applications with fresh input data. Currently, the data-driven models is very prevailing in medical diagnosis (Kuo et al. 2001), political campaigns (Sides 2014) and commerce (Alhamazani et al. 2015) because of their low costs with no need of expensive equipment and audit activity. As to the building energy consumption studies, data-driven models are widely applied to either estimate the building energy demands (i.e., data-driven prediction models) or profile the energy consumption patterns (i.e., data-driven classification models), which are grouped in Fig. 2.2.

2.1.6 Data-Driven Prediction Models

Among the most popular data-driven prediction models are artificial neural networks (ANNs), support vector machine (SVM), statistical regression, decision tree (DT) and genetic algorithm (GA). This subsection will introduce each of these models.

2.1.6.1 Artificial Neural Networks

ANNs are designed mimicking the basic architecture of human brain, whose basic element is called as processing unit modelling a biological neuron. The network consists of a large number of these process units arrayed in layers, and process units in different layers are connected with one another via connections, shown in Fig. 2.3.

Each process unit, say l, will deal with signals, $x_{il}(i = 1, 2, \ldots, m)$, from units connected with it in the other layers. These signals are input through the incoming connections with a weight $w_{il}(i = 1, 2, \ldots, m)$. The process unit then takes two basic operations on the input signals: summation and activation, and delivers an output y_l (Magoules and Zhao 2016).

$$y_l = f\left(\sum_{i=1}^{m} w_{il}x_i + b_l\right) \quad (2.1)$$

where b_l is a bias set specifically for each process unit and f is the activation function, commonly defined as the sigmoid function (Magoules and Zhao 2016).

$$f(x) = \frac{1}{1 + e^{-x}} \quad (2.2)$$

The output y_l will be used as an input signal for the process units in the next layer connecting to the process unit l.

As we discussed, all the process units in ANNs are arranged in a layer-structure and process units in different layers are interconnected based on a designed architecture. Figure 2.3b shows a simple example: feed-forward ANNs where process units are arrayed in the input, hidden and output layers and the information flows in one direction throughout these layers. In today's ANNs studies, ANNs models also take other architectures to more effectively approximate human brain activities. Two representative are back-propagation neutral network (BPNN) and recurrent neutral network (RNN), see Fig. 2.4. The former computes the error of output

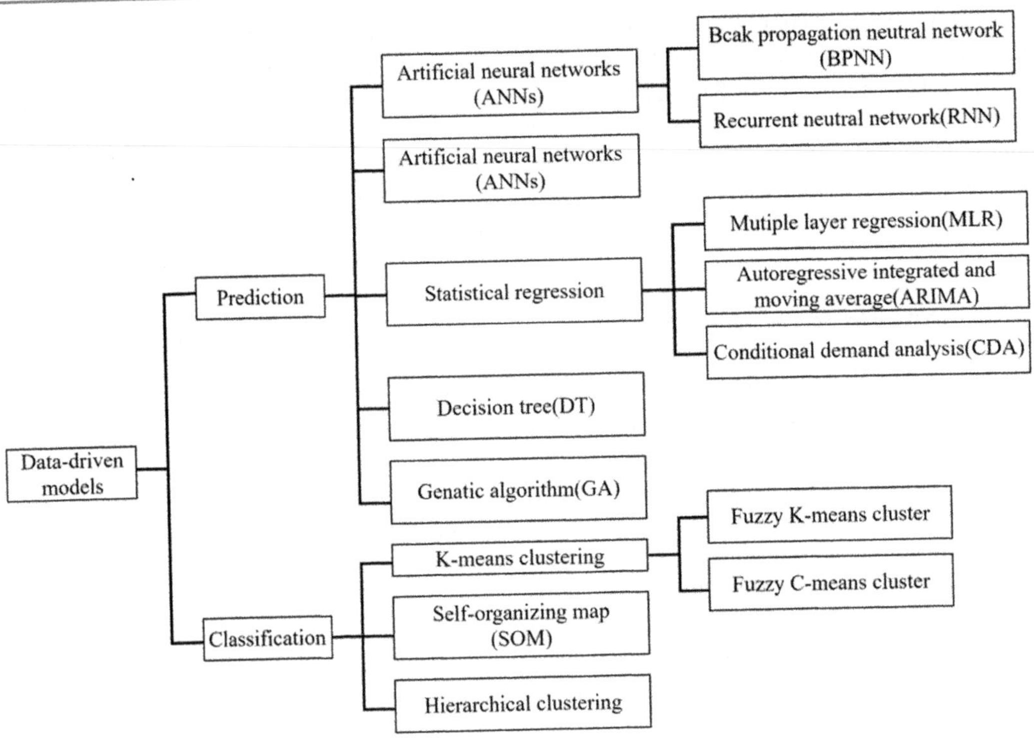

Fig. 2.2 Different data-driven models for building energy consumption

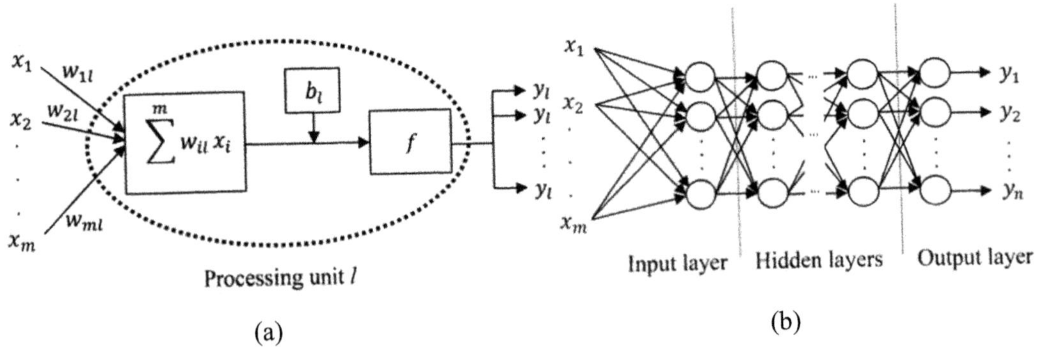

Fig. 2.3 Schematic of ANN. **a** A single process unit; **b** Artificial neural networks

every time, and then propagates this information as a negative feedback to tune the incoming connection weight and bias. This manipulation offers flexibility to modify the output error to a minimum, and thus improving accuracy of ANN calculation. As to RNN, it involves the backward connections feeding back the outputs themselves as the inputs to the process units in the former-layer or even the current unit to capture tempore behaviours. Such a recurrent design makes RNN deal with time series datasets without random data, which leads it to being particularly welcome for sequence events (Kalogirou and Bojic 2000).

No matter what kind of network architecture is in use, an ANNs model must experience a training (learning) process to specify all needed connection weights and biases before real

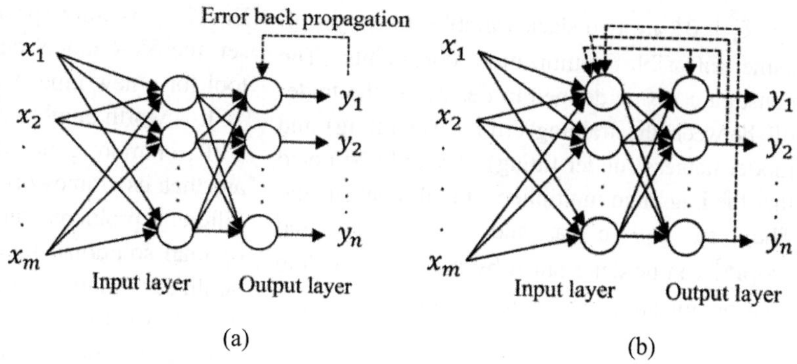

Fig. 2.4 Schematic of **a** two-layer BPNN and **b** two-layer RNN. White cycles: process units in different layers. Solid arrows: connections; dashed arrows: feedbacks

applications. This training process will take advantage of available historical data records, which will used as benchmarks to cultivate the proper response of the ANNs model for given inputs. Therefore, ANNs are capable of learning the relationship among input signals, and capturing key information through a training process based on historical data records. On top of that, it also possesses a number of other advantages, such as fault tolerance, robustness and noise immunity. Thanks to these favourable features, ANNs have achieved great success in solving non-linear problems so far. On the other hand, meanwhile, it should be also pointed out that the architecture choice and learning-rate optimization in the current ANNs are still developed on an ad hoc base. This implies ANNs applications are usually case-dependent nonetheless. They have to be designed and validated for every and each time for different applications (Kalogirou 2001).

2.1.6.2 Support Vector Machine

Supported vector machine (SVM) is another popular artificial intelligent method (Vapnik et al. 1996), which deals with n data records, i.e., $\{(x_i, Y_i)\}_{i=1}^{n}$, with the input $x_i \in R^N$ and the target $Y_i \in R$. (Note that Y_i could also be in binary for some applications (Zhao and Magoulès 2010)). Nowadays, this method has been widely applied to solve regression problems to estimate an underlying relationship between the nonlinear inputs to the continuous real-valued target. The SVM used for regression is called as support vector regression (SVR), which has become a particularly important data-driven approach for predicting building energy consumption.

The core task in SVR is to construct a decision function, $F(x_i)$, by use of a training process based on historical data. It is required that for a given input x_i, the result estimated by this function should not deviate from the actual target Y_i larger than the predefined threshold. In SVR, such a function is usually assumed in the form of

$$F(x_i) = \langle w, \varphi(x_i)\rangle + b \tag{2.3}$$

where the bias $b \in R$. $\langle \cdot, \cdot \rangle$ and w represent the dot product and weight defined in R^N. $\varphi(x_i)$ is a non-linear mapping of the input space to a high-dimensional feature space (Dong et al. 2005). w and b are two unknown in Eq. (2.3), and need to be estimated through minimizing the regularized risk function (Dong et al. 2005). In SVM theory, the latter is easily solved in its dual formulation by an introduction of a Lagranian L (Magoules and Zhao 2016),

$$\begin{aligned} L := &\frac{1}{2}\|w^2\| + c\sum_{i=1}^{n}(\xi_i + \xi_i^*) - \sum_{i=1}^{n}(\eta_i\xi_i + \eta_i^*\xi_i^*) \\ &- \sum_{i=1}^{n} a_i(\varepsilon + \xi_i - y_i - \langle w, \varphi(x_i)\rangle - b) \\ &- \sum_{i=1}^{n} a_i^*(\varepsilon + \xi_i^* - y_i - \langle w, \varphi(x_i)\rangle - b) \end{aligned} \tag{2.4}$$

where $\{a_i, a_i^*, \eta_i, \eta_i^* \geq 0\}$ are the Lagrange multiplier. $\|w\|$ is the Euclidean norm.

$\{\xi_i, \xi_i^* \geq 0\}$ are two slack variables to copy with some infeasible optimization constraints. The constant $c > 0$ is defined to determine the trade-off between the training error (over-fitting) and model flatness (under-fitting). It should be noted that the Lagrange multipliers are all independent. They are $\eta_i = c - a_i$ and $\eta_i^* = c - a_i^*$, and $\{a_i, a_i^*\}$ can be determined by the corresponding dual optimization (Dong et al. 2005),

$$\text{Maximize } W(a_i, a_i^*) = -\frac{1}{2}\sum_{i=1}^{n}\sum_{j=1}^{n}(a_i - a_i^*)(a_j - a_j^*)(\varphi(x_i)\cdot\varphi(x_j)) + \sum_{j=1}^{n}(a_i - a_i^*)y_i - \varepsilon\sum_{j=1}^{n}(a_i + a_i^*)$$
$$\text{subject to}\begin{cases}\sum_{j=1}^{n}(a_i - a_i^*) = 0\\ a_i, a_i^* \in [0, c]\end{cases} \tag{2.5}$$

With the computed a_i, a_i^*, the weight w can be written a function of $\{a_i, a_i^*, x_i\}_{i=1}^{n}$. This gives rise to the decision function in SVR

$$F(x) = \sum_{x_i \in SV}(a_i - a_i^*)K(x, x_i) + b \tag{2.6}$$

where $K(x, x_i) = \varphi(x)\cdot\varphi(x_i)$. In SVR, this is called as the kernel function, having different formulas for various applications in the literature, e.g., $K(x, x_i) = \exp(-\gamma\|x - x_i^2\|)$. It should be pointed out the sum in Eq. (2.6) does not cover all inputs. Instead, only those (i.e., support vectors $x_i \in SV$) corresponding to $(a_i - a_i^*) \neq 0$ are included. Moreover, the bias b in Eq. (2.6) is also computed by these support vectors

$$b = \frac{1}{N_1}\left\{\sum_{a_i \in (0,C)}\left[Y_i - \sum_{x_j \in SV}(a_j - a_j^*)E(x_i, x_j) - \varepsilon\right] + \sum_{a_j^* \in (0,C)}\left[Y_i - \sum_{x_j \in SV}(a_j - a_j^*)K(x_i, x_j) + \varepsilon\right]\right\} \tag{2.7}$$

Here, N_1 is the number of support vectors with either $\{a_i \in (0, c), a_i^* = 0\}$ or $\{a_i = 0,\ a_i^* \in (0, c)\}$. Once the decision function, i.e., Eq. (2.6), is fully specified by the training dataset, the SVR model can be used as a predicting tool for a new input x.

It is worth emphasizing that the superiority of SVR, or more generally SVM, to other models are that its framework is easily generalized for different problems and it can obtain globally optimal solutions. Its capability of dealing with nonlinear relations by transferring them into high-dimensional linear problem is also impressive for practical applications. Nonetheless, the method is rather time-consuming for large-scale problems (Zhao and Magoulès 2010; Li et al. 2009a). Recently, immerse efforts has been paid to developing possible ways to optimize its computational efficiency.

2.1.6.3 Statistical Regression

Prediction of building energy-consumption relies on a regression analysis to devise a relationship linking an output (i.e. response, Y_i, $i = 1, 2\ldots n$) to the contributing inputs (i.e., predictors, $x_{i,j}$, $i = 1, 2\ldots n, j = 1, 2\ldots m$). In the previous section, we have discussed a regression process based on the SVM theory-SVR. On top of that, there still exist other regression models, e.g., statistical regression, used for predicting building energy consumption. Statistical regression investigates the relationship among different variables in a probabilistic framework, which formulate the output as

$$\text{Multiple}: Y_i = \alpha_i + \beta_1 x_{i,1} + \beta_2 x_{i,2} + \ldots + \beta_m x_{i,m} + \varepsilon_i \tag{2.8}$$

or

$$\text{Polynomial}: Y_i = \tilde{\alpha}_i + \tilde{\beta}_1 x_{i,1} + \tilde{\beta}_2 x_{i,2}^2 + \ldots + \tilde{\beta}_m x_{i,m}^m + \varepsilon_i \tag{2.9}$$

where ε_i represents a random error assumed to be normally distributed, and α_i, $\tilde{\alpha}_i$, β_j and $\tilde{\beta}_j$ $(j = 1, \ldots\ldots m)$ are the parameters to be estimated. Note that both Eqs. (2.6) and (2.7) are linear with respect to these parameters whilst they are not necessarily linear with respect to the contributing predictors, as seen as Eq. (2.7). Like other data-driven approach for prediction, the statistical regression equations make use of the finite number of historical data to estimate the involved parameters. For demonstration, we choose the multiple linear regression Eq. (2.6) as an example, in which the estimates of all parameters will derived using the least squares (LS). To be specific, the sum of squared errors (SSE) is first defined

$$SSE = \sum_{i=1}^{n} \left(y_i - A_i - B_1 x_{i,1} - B_2 x_{i,2} - \ldots - B_m x_{i,m}\right)_i^2 \tag{2.10}$$

In Eq. (2.8) A_i and $B_j (j = 1, \ldots\ldots m)$ are the corresponding LS estimates of α_i, $\beta_j (j = 1, \ldots\ldots m)$ in Eq. (2.6). SSE is then minimized which gives rise to $m + 1$ equations. Each of these equations includes one of partial derivatives of SSE with respect to A_i and B_j $(j = 1, \ldots\ldots m)$, to be set zero, respectively. It is these equations that are used to solve A_i and $B_j (j = 1, \ldots\ldots m)$ directly subject to the given historical dataset $\{ x_{i,j}, Y_i,\ i = 1, 2 \ldots n,\ j = 1, 2 \ldots m \}$. Finally, the prediction equation with the estimated parameters in multiple linear regression is specified as

$$y_i = A_i + B_1 x_{i,1} + B_2 x_{i,2} + \cdots + B_m x_{i,m} \tag{2.11}$$

In statistical regression, there is another variable introduced to quantify the goodness of fit of the regression line by Eq. (2.9), that is the coefficient of determination R^2,

$$R^2 = 1 - \frac{SSE}{SS_{tot}} \tag{2.12}$$

where $SS_{tot} = \sum_{i=1}^{n} \left(Y_i - \overline{Y}\right)^2$, with the mean value $\overline{Y} = \sum_{i=1}^{n} Y_i$. Generally, a regress equation with a larger R^2 indicates it can better fit the original data.

Based on the above discussion, it is seen that statistical regression is an easy-to-use approach for predicting building energy consumption. In particular, it was popular to predict average consumption over a long period in the early studies. However, the regress models require a large number of historical data for training, and the resulting accuracy of a short-term prediction is yet poorer than that of other data-driven approaches, such as ANN or SVM. It is also challenging for statistical regression to select a set of plausible predictors and an appropriate time scale to well fit energy consumption for buildings under a wide range of environment and weather conditions. Worse, the selected predictors in some cases may not be literally independent. The unforeseen correlations among them would result in uncertain inaccuracy in the regression outputs (Swan and Ugursal 2009).

2.1.6.4 Decision Tree

Decision tree (DT) is a technique to partition data into groups using a tree-like flowchart. In this sense, a DT model manifest itself as a graph consisting of a root node and a couple of branch nodes. A DT starts from the root node where the input data are split into different groups based on some predictor variables predefined as splitting criteria. These split data are then disseminated to sub-nodes as branches emanating from the root node. The data on sub-nodes will undergo either further or no splits. The former are the internal nodes where the subsequent data split is conducted to form new subgroups as son-branches emanated graphically at the next level. Whereas the latter are leaf nodes which treat the corresponding data group at the current level as their final outputs. Figure 2.5 illustrates a DT representation used for medium annual source energy consumption per unit floor ($kWh/m^2/yr$) of a commercial building. In this case, the gross floor area and building use ratio are chosen as predictor variables in the root node and internal node, respectively, and a mixture of data about energy consumption has been purified into a hierarchy of groups.

Significantly, in a DT analysis the information entropy is an important concept used to quantify data group homogeneity. It is defined by

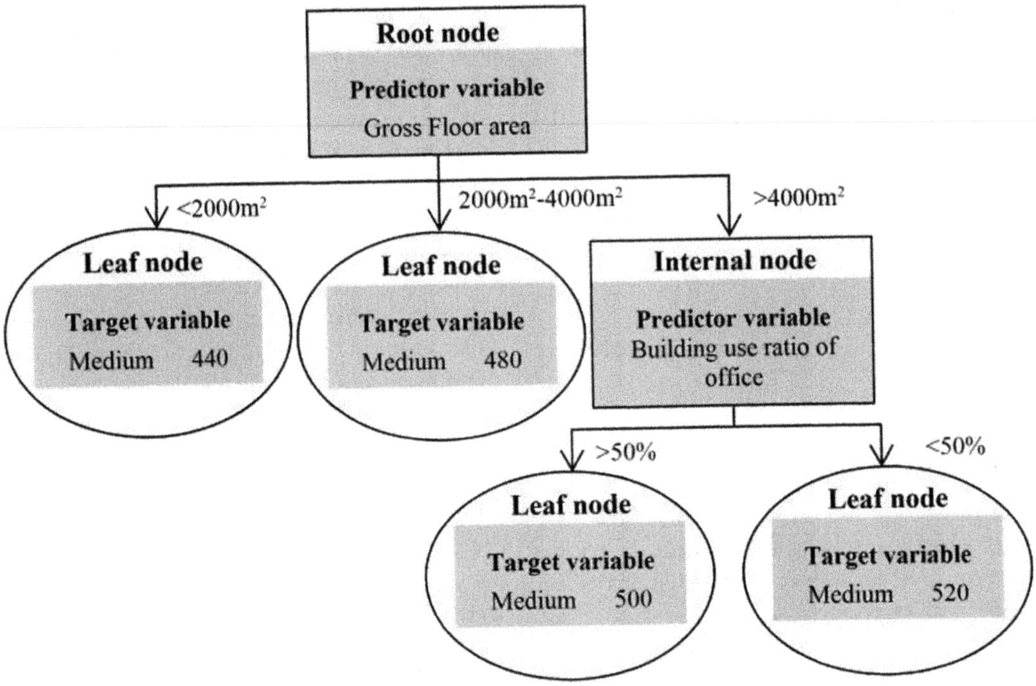

Fig. 2.5 Decision tree illustration of a medium annual source energy consumption per unit floor of a commercial building

$$E = \sum_{i=1}^{n} -P_i \log_2 P_i \tag{2.13}$$

where E is the information entropy. n and P_i are the number of different target values and the probability of a dataset taking the ith target value, respectively. This entropy is used to calculate the information grain or gain ratio, based on which a DT structure linking the top root node to each branch node is specified. Readers can refer to Quinlan (1986) for detailed splitting procedure using the gain ratio or information gain.

In comparison to other data-driven approaches, DT's tree-like structure is easy to understand and its implementation does not involve complex computation knowledge. However, its deficiency is also evident—the targets used in a DT are primarily based on expectations. This usually leads to significant deviations of its predictions from the real results. The DT architecture is also a restriction so as the method is unable to deal well with time-series and non-linear data.

2.1.6.5 Genetic Algorithms

Genetic algorithms (GAs) are stochastic optimization inspired by natural evolution based on the idea of "survival of the fittest" (Goldberg 1986). Many GAs in building energy prediction formulate three kinds of algebraic equations to compute the output (as solution) according to the given inputs:

$$\text{Linear}: y = w_1 x_1 + \cdots + w_m x_m, \tag{2.14}$$

$$\begin{aligned}\text{Quadratic}: y = {} & w_1^{(1)} x_1 + \cdots + w_m^{(1)} x_m + w_{1,2} x_1 x_2 \\ & + \cdots + w_{1,m} x_1 x_m + w_{2,3} x_2 x_3 + \cdots \\ & + w_{m-1,m} x_{m-1} x_m + w_1^{(2)} x_1^2 + \cdots + w_m^{(2)} x_m^2,\end{aligned} \tag{2.15}$$

$$\text{Exponential}: y = w_0 + w_1 x_1^{\widetilde{w}_1} + w_2 x_2^{\widetilde{w}_2} + \ldots + w_m x_m^{\widetilde{w}_m}, \quad (2.16)$$

where $(x_1, x_2, \ldots, x_m)$ are m independent inputs contributing to the output, y, and w_i, $w_i^{(1,2)}$ and $\tilde{w}_i$ are the real-valued weights. In GAs, different sets of weights compose a search space where a point represents a feasible solution to the problem under investigation. The core task of a GA is to model an evolution process to identify the best among all feasible solutions in this space. In implementation, a GA first randomly chooses n sets of weights and encode each weight as a l bit binary string, e.g. $w_i = \overbrace{100\ldots01}^{l}$. In so doing, a set of weights is then represented as a chromosome $X_j = \overbrace{100\ldots01}^{w_1}\overbrace{000\ldots11}^{w_2}\ldots\overbrace{100\ldots10}^{w_m}$, and the n chromosomes form an initial population r. Importantly, every chromosome X_j in the population r is mapped to a fitness $h(X_j)$ (a real value) and assigned a probability P_j. In most cases, these two variables are defined by

$$h(X_j) = \left(y\left(x_1, x_2, \ldots, x_m | X_j\right) - Y\right) \quad (2.17)$$

and

$$P_j = \frac{h(X_j)}{\sum_r h(X_k)} \quad (2.18)$$

where Y is the targeted output from historical datasets and the Greek letter "Σ" denotes a sum of the fitness of all chromosomes in the population r. Next, pairs of chromosomes are selected as parents to reproduce the offspring (still chromosomes). Generally, the better fitness the chromosomes have, the more possible they are selected. The chosen parents then proceed crossover and mutation. One simple crossover operation is to randomly choose a crossover point and exchange the alleles up to this point of the two parent chromosomes. As to mutation, a few of bits in the chromosome after crossover, again chosen randomly, are switched between 0 and 1 (e.g. 10$\underline{0}$01 → 10$\underline{1}$11). Selection, crossover and mutation will be repeated to generate sufficient new offspring to form a new population, r′, at the next level. It should be pointed out that the fitness of all offspring chromosomes in this new generated population will be computed and compared with the user's requirements. Generally, a GA will continue further runs of the above evolution process unless a chromosome (i.e., a set of weights) with satisfactory fitness is reproduced.

The aforementioned introduction of GAs indicates this method is a powerful optimization tool in dealing with complex multi-modal problems (Beyer 2000). The algorithms can obtain suitable solutions based on either the objective functions or subjective judgements when large and sophisticated input data are given. Meanwhile, two major deficiencies in the current GAs are also noted—non-unique results and large computation time. In the literature, attempts to combine a GA with other data-driven approaches (e.g. ANN) have been made to mitigate the negative impacts arisen from the deficiencies.

2.1.7 Data-Driven Classification Approaches

Besides great success in predicting building energy consumption, data-driven approaches have been extensively used to attack building energy classification over the last several decades, among which K-means algorithm, self-organizing map (SOM), hierarchical clustering and regression are the most popular choices.

2.1.7.1 K-Means Cluster

The K-means clustering algorithm is a classification approach quite popular in building load analysis. Technically, this algorithm partitions a set of data into a number of non-hierarchical groups of similar data points, i.e., clusters. The similarity among data points is quantified by the Euclidean distance, based on which a K-mean clustering procedure includes the following steps. A data set $(x_i,\ i = 1,2...n)$ is first input with the cluster centers $(\mu_j,\ j = 1,2...K)$ being specified randomly. The Euclidean distances between each data point and each cluster center are then computed. A datum x_i is set to belong to a cluster C_j if its distance to the cluster center μ_j is shorter than those to any other center. As a consequence, this classification forms K clusters in the input dataset, and the center of each cluster is re-calculated as a mean based on new data grouping. The K mean clustering algorithm will repeat the above distance computation, data classification and center relocation till all the K cluster centers do not move their locations with further iterations (Magoules and Zhao 2016). In many cases, a squared error function J is introduced to characterize this convergence,

$$J = \sum_{j=1}^{K} \sum_{x_i^{(j)} \in C_j} \left(x_i^{(j)} - \mu_j\right)^2 \quad (2.19)$$

where $x_i^{(j)}$ represents a data point belonging to the cluster C_j (Panapakidis et al. 2014). In the K mean clustering algorithm, a priori specifications of the cluster number K and initial positions of the cluster centers are required. This results in the algorithm has to be conducted several times in practice with these parameters with different values. Only the best results after comparison will be deemed as the algorithm's ultimate outcomes.

It is worth mentioning to improve its feasibility, the K-means clustering algorithm has been modified using the fuzzy methods. The modified version, in contrast to the aforementioned discussion, allows soft clustering, i.e., every data point can potentially belong to multiple clusters and a degree of membership is defined to characterize such relationships (Dunn 1973). Nikolaou et al. (2012) discusses one widely-used fuzzing cluster approach in building energy projects, i.e., fuzzy C-means (FCM) cluster. Interested readers can refer to it for more details.

2.1.7.2 Self-organizing Map

Self-organizing map (SOM) is developed from ANNs which transfers an incoming signal pattern in arbitrary dimensions into a one- or two- or multi-dimensional topographic map (Magoules and Zhao 2016). The method is trained by an unsupervised learning process and capable of classifying new inputs into clusters with different features in a neurobiological-like manner. Figure 2.6 illustrates a frequently-used network architecture of SOM consisting of a one-dimensional input layer and a two-dimensional computational layer. In this computational layer, a number of process units, i.e., neurons $(j = 1,2...m)$, are arranged in rows and columns, each of which connects all input signals $(x_i,\ i = 1,2...n)$ with connection weights w_{ij}. The output of the neuron j is sometimes given by $y_j = \sum_{i=1}^{n} w_{ij}x_i$.

In SOM, a squared Euclidean distance between all the input signals and connection weights pertinent to every neuron is computed

$$d_j = \sum_{i=1}^{n} (x_i - w_{ij})^2 \quad (j = 1,2...m) \quad (2.20)$$

This distance is termed as the discriminant function, and the neuron with the smallest discriminant function is designated as the winner for a given set of input signals. Typically, a SOM iteration starts from initializing all correction weights with small random numbers and choosing a set of input signals from historical database at random to form the input layer. Computation of the discriminant function for each neuron in the computational layer is then performed. Only the neuron with the smallest discriminant function is identified as the winner at this iterative level. Immediate to this, a topological neighborhood centered at the selected winner is defined, in which the connection weights linking every neuron to the input signals are adjusted subject to

$$w_{ij}(n+1) = w_{ij}(n) + g_j\left[x_i - w_{ij}(n)\right] \quad (2.21)$$

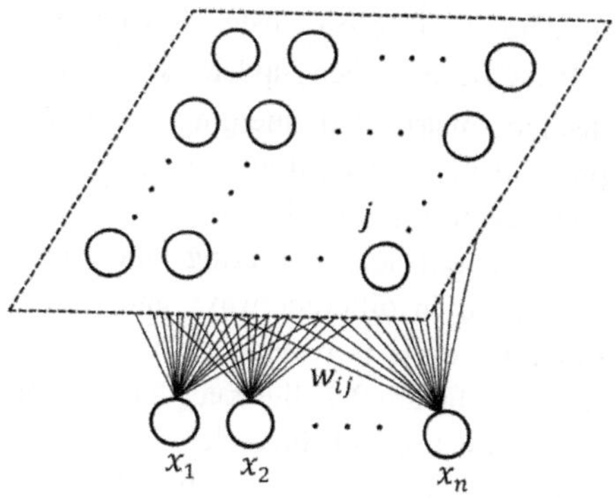

Fig. 2.6 Schematic of SOM. White cycles: process units; Solid lines: connections

where n represents the current iterative level, g_j is the learning rate depending on n and the distance between the winner and the neighboring neuron j. The next iteration at $n+1$ will be conducted with these adjusted correction weights and the new randomly-chosen input signals. Note that while the SOM iteration proceeds, both the learning rate and the size of the winner's neighborhood will decrease. The whole iteration will terminate once a threshold is met, e.g., $g_j \le g_{j,min}$ or only the winner itself or none being included in the neighborhood. After training, a particular neuron (i.e., winner) in SOM will be activated the most for a particular type of input signals. This correspondence ensures SOM to be effective means used for clustering new input signals.

In sum, SOM can effectively reduce the dimensions of a high-dimensional signal pattern to a feature map in which the similarities and differences among input objects are easily discerned. Moreover, its outputs can be directly followed by further classification using other clustering algorithms. This will lead to more mutually exclusive and well-separated groups. On the other hand, it is also noted that SOM clustering suffers from oscillation if a rambling dataset without any pretreatments is used as the input. Importantly, its computational cost will dramatically increase with the increasing dimension of the data. Therefore, a good SOM should be equipped with a well-designed tuning process and a clear parametric analysis on the impacts of different parameters. These parameters usually include the learning rate, neighborhood function, number of process units, and et al.

2.1.7.3 Hierarchical Clustering

Hierarchical clustering in building energy consumption commonly uses the bottom-up fashion to organize data points into a tree-like hierarchy of clusters (Nikolaou et al. 2012). Such clustering is known as the agglomerative algorithm starting with n data points. $(x_i,\ i = 1, 2, \ldots\ldots n)$, each of which is treated as a singleton cluster. To characterize the inter-cluster similarity, the distances among different clusters are computed, and form a $n \times n$ matrix

$$H = \begin{bmatrix} 0 & \cdots & D(C_n, C_1) \\ \vdots & \ddots & \vdots \\ D(C_1, C_n) & \cdots & 0 \end{bmatrix} \quad (2.22)$$

In the above matrix, the distance between two clusters $D(C_i, C_j)$ is defined by

$$D(C_i, C_j) = \min d(x_i,\ x_j), \text{with } x_i \in C_i \text{ and } x_j \in C_j,$$

where $d(x_i,\ x_j)$ is the distance (i.e., Euclidean distance) between two data points in these two cluster and $D(C_i, C_j) = 0$ when $i = j$ (Nikolaou et al. 2012). In the literature, there are the other ways to define the distance between two clusters. Interested readers can refer to Vesanto and Alhoniemi (2000) for more details. After computing the inter-cluster distances, the next step is to merge two closest clusters having the minimal $D(C_i, C_j)$, and then update the corresponding distance matrix. This merging manipulation will proceed iteratively till all data points have been included in a single cluster.

In hierarchical clustering, merging can be conducted in different ways and terminated at different levels provided the similarity criterion requires. Figure 2.7 illustrates an example where two distinct sets of three clusters are obtained in different merging routes based on different merging criteria. In building energy studies, hierarchical clustering has been proven that it can reveal the data internal structure and generate useful knowledge about energy consumption in a building (Magoules and Zhao 2016).

2.2 Practical Application of Data-Driven Approaches

2.2.1 R & D Works and Practical Applications

All the aforementioned data-driven approaches are widely applied to a large variety of prediction or classification applications of load prediction, energy pattern profile of specific use-cases, regional energy consumption mapping, energy benchmark for building stock, retrofit strategies and guideline making, see a summary in Table 2.2. This broad range of applications covers micro-scale and macro-scale studies that provide useful information and instructive suggestions for different stakeholders, including government, investors, engineers and occupants throughout the building life cycle from the early planning/design stage to later operation/retrofit stage.

2.2.1.1 Prediction

Originally, many data-driven approaches were established to predict the energy consumption of building, in particular electricity usage. It is well recognized that estimations of energy usage in the long-, medium- and short-term (i.e., annual, monthly and daily) are of importance for energy market planning and investments. Especially, a very short-term (hours or minutes ahead) estimation of electricity usage can exert a vital influence on the final dispatch for national electricity market (Setiawan et al. 2009). Therefore, a precise prediction in these scenarios would lead to more efficient energy management and direct to considerable reduction in operational cost for both energy suppliers and end-users in buildings (Setiawan et al. 2009; Mathieu et al. 2011; Neto and Fiorelli 2008). At the current stage, ANN and SVM are the two favourable data-driven approaches used for prediction of building energy consumption.

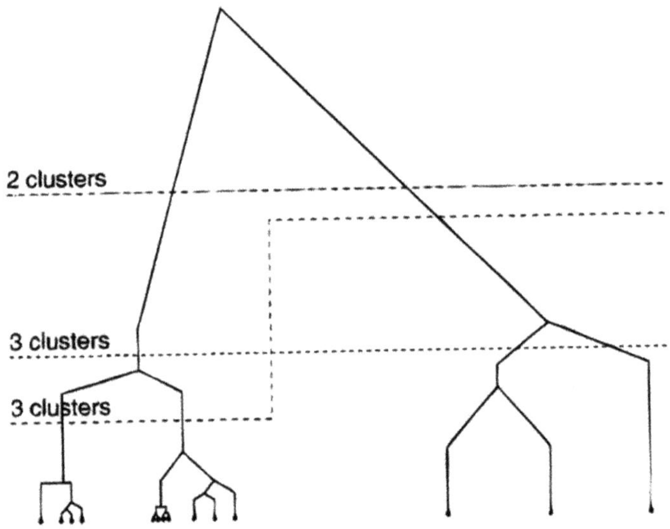

Fig. 2.7 Schematic of hierarchical clustering algorithm. The partitive clusters can be obtains at different levels of similarity (Vesanto and Alhoniemi 2000)

Table 2.2 Summary of data-driven approach for applications in building energy consumption

Data-driven approaches	Applications prediction classification	
ANN	(Kalogirou 2001; Kalogirou and Bojic 2000; Neto and Fiorelli 2008; Sözen and Arcaklioglu 2007; Yang et al. 2005; Canyurt et al. 2005; An et al. 2013; Yezioro et al. 2008; Yan and Yao 2010; Yokoyama et al. 2009; Olofsson and Andersson 2002; Aydinalp-Koksal and Ugursal 2008; Aydinalp et al. 2002; Yalcintas 2008; Asadi et al. 2014)	(Yalcintas 2006; Yalcintas and Ozturk 2006)
SVM	(Zhao and Magoulès 2012, 2010; Magoules and Zhao 2016; Dong et al. 2005; Li et al. 2009a, 2010, 2009b; Setiawan et al. 2009)	N/A
Regression	(Zhao and Magoulès 2012; Amjady 2001; Mejri et al. 2011; Wauman et al. 2013; Tiedemann 2007; Aydinalp-Koksal and Ugursal 2008; Larivière and Lafrance 1999; Mastrucci et al. 2014; Howard et al. 2011; Nikolaou et al. 2011; Chung et al. 2005; Yalcintas and Ozturk 2006)	N/A
DT	(Tso and Yau 2007; Yu et al. 2010)	(Park et al. 2016)
GA	(Canyurt et al. 2005; Sadeghi et al. 2011; Azadeh et al. 2007; Li and Su 2010; Juan, et al. 2009; Asadi et al. 2014)	N/A
K-means cluster	N/A	(Nikolaou et al. 2012; Panapakidis et al. 2014; Tsekouras et al. 2007; Heidarinejad et al. 2014; Arambula Lara et al. 2014; Jones et al. 2007; Fonseca and Schlueter 2015; Santamouris et al. 2007; Wang 2015; Lannon et al. 2050)
SOM	N/A	(Nikolaou et al. 2012; Panapakidis et al. 2014)
Hierarchical cluster	N/A	(Tsekouras et al. 2007; Xiao et al. 2012; Wang 2015)

Prediction Application of ANNs

ANNs have been extensively used as a prediction means in diverse areas (Kalogirou 2001). In building sector, ANNs excels in predicting building energy consumption, electricity demand, heating/cooling loads, important energy parameters and even assessment of software etc. Table 2.3 has centrally summarized these applications of ANNs in the literature.

In terms of energy consumption, ANNs are the popular candidate for both the short-term and long-term prediction. Kalogirou and Bojic (2000) used ANNs to predict energy consumption in a holiday passive solar building, where engineers working in the HVAC field were not included. In their study, the RNN model based on the back-propagation architect was applied for the training process. In so doing, such a model could detect features in the raw data of previous knowledge, e.g., the changing rules of operating conditions along different time epochs. In addition, Sözen and Arcaklioglu (2007) even derived an ANN model to shed light on causality link behind economic indicators, population and net energy consumption. Their study suggested economic indicators (e.g. gross national product (GNP) and gross domestic product (GDP) etc.), rather than conventional energy indicators (e.g. gross generation, installed capacity and years), are playing a more important role for an accurate prediction of energy consumption.

Table 2.3 Summary of ANNs in predicting building energy consumption

Type of house	Scale	Inputs	Output	Data source	Measure length	Algorithm
Holiday passive house (Kalogirou and Bojic 2000)	Single	Season, insulation function, wall thickness, heat transfer coefficient, time of day	Energy consumption	Measured data: ZigBee Input Device (ZID)	Two seasons	RNN combined with BPNN
Multiple (Sözen and Arcaklioglu 2007)	National	Economic indicators (GNP and GDP), population	Net energy consumption	World Energy Council	37 years (1968–2005)	BPNN
Office building (Yang et al. 2005)	Single	Outdoor dry-bulb temperature, outdoor humility, water temperature of chiller, compressor status etc	Dynamic chiller electric demand	Simulated data (DOE 2.1E) and measured data	1 year	Sliding window ANN and accumulative ANN
Office building (Canyurt et al. 2005)	Single	Previous load, temperatures of previous day, occupancy condition, sin and cosine of the hour	One day ahead electric power consumption	Great building energy predictor shootout I and measured data	1 year and a half	BPNN
Multiple An et al. 2013)	Reginal	Previous electricity consumption	Half-hour ahead electricity demand	Australian Energy Market Operator	9 weeks	Multi-output BPNN
Residential (Yan and Yao 2010)	7 builidngs	18 building envelope parameters, heating degree day, cooling degree day	Heating and cooling energy consumption	Simulated data (DeST)	1 year	BPNN
Commercial (Yokoyama et al. 2009)	Single	Previous cooling demand, air temperature and relative humidity	Cooling demand	Measured data	45 weekdays	BPNN

(continued)

Table 2.3 (continued)

Type of house	Scale	Inputs	Output	Data source	Measure length	Algorithm
Residential (Olofsson and Andersson 2002)	7 single family-building	Supplied heating demand, electricity domestic demand, flag parameter	Indoor-outdoor temperature difference	Measured data	2 years	BPNN
Solar house (Yezioro et al. 2008)	Single	Outdoor temperature, relative humility, set point temperature, occupancy schedule	Heating/cooling consumption	Measured data	2 days	BPNN
Office building (Neto and Fiorelli 2008)	Single	Outdoor dry-bulb temperature, day type (working day or weekend)	Daily total consumption	Measured data: energy demand measurement system	54 days	BPNN

As to electricity demand, the majority of ANN models focus on dynamic and short-term predictions, which require careful selection and pre-treatment of input data. One example is Yang et al. (2005) where an on-line chiller electricity prediction model was established through use of both the simulated data and measured data. Their results recommends the sliding-window ANN, which constantly drops the oldest data and adds new measurements during training process, showed better performance than the accumulative ANN based on measured data. Besides, Karatasou et al. has reported one-day ahead prediction of electricity consumption, called a 24-steps predictor, in Canyurt et al. (2005). The predictor used previous energy consumption data records with time delays larger than 24 h as inputs to train the network to perform next day's prediction. Interestingly, An et al. (2013) further developed an (EMD)-based signal filtering which is able to forecast half-hour electricity demand ahead. Such an EMD-based signal filtering can decompose an incoming signal into a series of pure modes and residues. The results revealed that the EMD-based filter a critically-functioned component in the ANNs prediction model. In fact, ANNs also play an important role in prediction of heating/cooling loads. In this particular type of applications, the ANN models usually require to input detailed climate information, envelop parameters and occupancy schedules (Yezioro et al. 2008; Yan and Yao 2010). Besides reliable input data, algorithm optimization is the other way to promote the prediction accuracy. To minimize the drawback of BPNN (e.g. local optimization of model parameters in training process), a global optimization called "Modal Trimming Method" was proposed by Yokoyama et al. (2009). This method was composed of two steps, shown as Fig. 2.8: (1) search for local optimal solution of input variables in an objective function $\left(x_o^{fs} \rightarrow x_1^{lo}\right)$, Normally, the objective function is defined as calculation error between predicted and measured values; (2) search for another feasible solution of the same objective function value with previous local optimization $\left(x_1^{lo} \rightarrow x_1^{fs}\right)$. These two steps were repeated $\left(x_1^{fs} \rightarrow x_2^{lo} \rightarrow x_2^{fs}\right)$ until tentative global optimal one x_3^{lo} is found. They validated

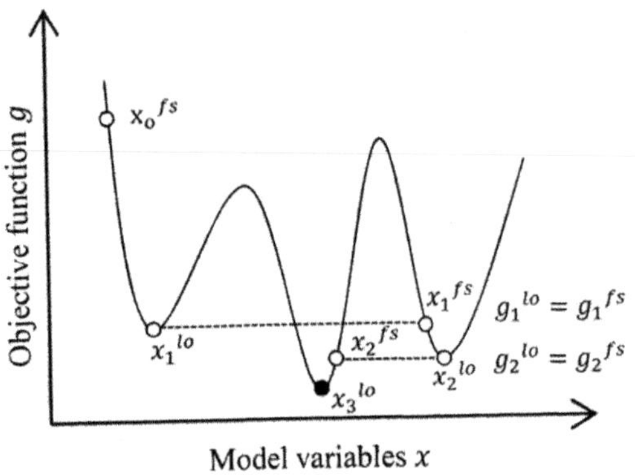

Fig. 2.8 Concept of modal trimming method

this method and concluded that significant error of predicted cooling demand from measured data was reduced compared to traditional local optimization method.

On top of the above applications, ANNs' application is also extended to predicting the key parameters of energy performance of building. For instance, Olofsson and Andersson (2002) proposed a use of the BPNN model to estimate the total heat loss coefficient (HLC) and domestic energy gain factor of inhabited single-family buildings. Here, the total HLC characterizes heat loss resulted from transmission and air-flow while the domestic energy gain factor focuses on the gain of heating or cooling from inside sources. In this kind of ANN model, flag parameter of each measured case was introduced to distinguish non-linear dependences among various predictors, instead of average dependency from previous experience.

It is worth mentioning that ANNs are sometimes used as tools to assess simulation software for building performance. Neto and Fiorelli (2008) compared the BPNN against EnergyPlus by using both to predict building energy consumption. The latter is recognized a mainstream simulator in building sector which can deliver much more accurate results than Energy_10, Green Building Studio web tool, and eQuest (Yezioro et al. 2008). Interestingly, Neto et al. found that when building and climate data were just briefly described, the used BPNN model works much better in daily energy demands prediction than EnergyPlus does. Importantly, especially for hourly prediction, all these simulation tools in current market give rather poor results in comparison to ANNs. This finding equips ANNs a new function as a benchmark to test accuracy of commercial software for estimating building energy performance.

Prediction Application of SVM

Prediction is also a primary function of SVM use in building energy simulation. Table 2.4 lists the up-to-data studies on SVM-prediction applications. Generally speaking, SVM works in high accuracy in the medium-term (Dong et al. 2005) and short-term (Setiawan et al. 2009) prediction. Significantly, the method only requires a few model-parameters to implement its calculation. On the other hand, however, computing speed of SVM is slower than that of other approaches, such as linear regression and the ANNs. Currently, how to optimize SVM algorithm is regarded as the core task for its future development.

Many efforts were actually made on SVM optimization in recent years. To save the computer memory and expedite the time-consuming training process, Zhao and Magoulès (2010) proposed targeted solutions for dual optimization process (see Eq. (2.5)) and Kernel function calculation. The main idea was to divide the entire dual optimization problem into sub-problems and calculate them in parallel. Then, the Kernel function matrix would be updated for each sub-problem calculation. This parallelized training process could be stopped until convergence. The modified SVM gains a capability of dealing with a large amount of data to predict energy consumption of multiple buildings. Another possible optimization solution is to develop a hybrid SVM. For example, Li et al. (2010) presented a hybrid approach combining SVM and FCM clustering algorithm to forecast building cooling loads. In this research, FCM was first employed to extract valid data records from the pool of raw data, and then the SVM followed with a training procedure based on the extracted valid data records. Clearly, such a pretreatment of data records effectively reduce the noise of inputs for SVM calculation. It should be pointed out the SVM is compatible with diverse input

Table 2.4 Summary of SVM in predicting building energy consumption

Type of house	Scale	Inputs	Output	Data source	Measure length	Algorithm
Commercial (Dong et al. 2005)	4 single buildings	Outdoor temperature, relative humility, global solar radiation, previous electricity consumption	Building energy consumption per month	Survey: monthly utility bill, National Environment Agency	3 years	SVM
Multiple (Setiawan et al. 2009)	Regional	Historical electricity consumption data	5-min ahead electricity load	Australian electricity operator	3 years	SVM, statistical regression, and BPNN
Office buildings (Zhao and Magoulès 2010)	100 buildings	Heating consumption, electrical consumption	Heating demand, electrical load	Simulated data (EnergyPlus)	5 months	Parallel SVM
Campus building (Li et al. 2010)	Single	Cooling load	Cooling load	Measured data	4 months	Fuzzy SVM combined FCM clustering
Office buildings (Zhao and Magoulès 2012)	Single	Previous energy load, building dynamic characteristics, outdoor temperature, occupancy schedule	Building energy demand	Measured data: data acquisition system	7 months	SVM with pseudo dynamic approach
Office buildings (Li et al. 2009b)	Single	Previous cooling load, air temperature, relative humidity, solar radiation intensity	Hourly cooling load	Simulated data (DeST)	Half year	SVM, BPNN

information. Besides the conventionally-used energy loads and climate conditions, Zhao and Magoulès (2012) also used energy load characteristics and hidden inertial effects of building as their SVM inputs. The energy load characteristics were described as operation level of HVAC system and the occupancy profile, the hidden inertial effects were provided as fluctuation of internal temperature. The block diagram of such a model for predicting building energy consumption is shown in Fig. 2.9. As we can see, partial selection of input data called dynamic time warping (DTW) was adopted during prediction process, which measures on the outdoor temperature difference between training days and prediction days. The minimal difference between two time series was chosen as optimal path for solution. Similarly, the previous energy load database was also partially selected by DTW as inputs to consider the most recent data rather than whole data. The result showed that the designed training leads to higher accuracy and better computational efficiency in comparison to that based on the whole input data.

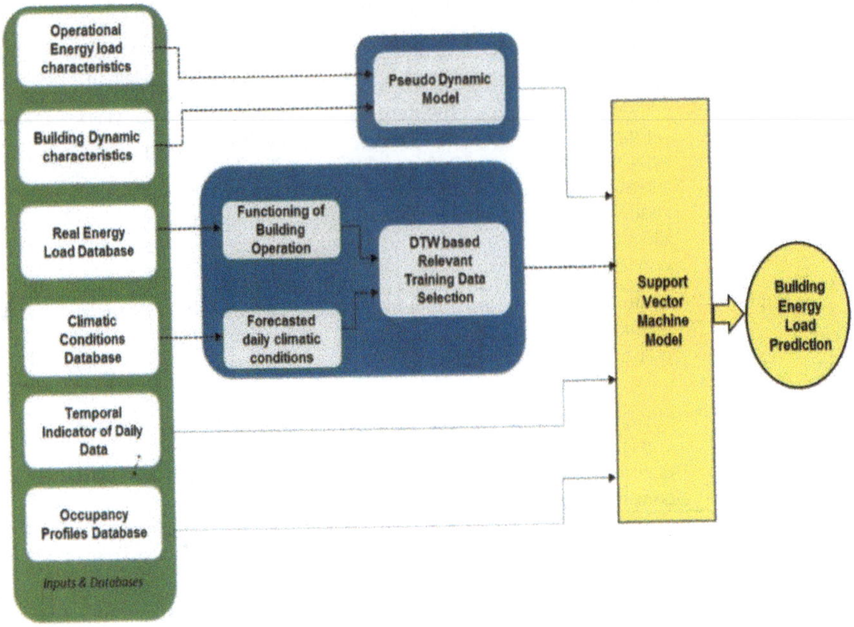

Fig. 2.9 Block diagram of SVM in prediction of energy demand using pseudo dynamic approach (Zhao and Magoulès 2012)

Prediction Application of Statistical Regression

Statistical regression is well treated as a simple tool for prediction for a long time (Zhao and Magoulès 2012). However, this approach suffers from low-accuracy in its prediction results, and such a deficiency has greatly limited its applications in building energy consumption analysis. This motivates a great deal of modification and optimization in statistical regression, which are briefly illustrated in Table 2.5. Among various modifications are multiple linear regression (MLR) proposed by Li and Huang (2013) for short-term prediction. This model utilized not only climate data, room temperature set point, but also the cooling loads of previous four hours as its inputs. The obtained prediction results achieved very impressive accuracy higher than that of conventional ANN models. Moreover, autoregressive, integrated and moving average (ARIMA) model under the statistical regression framework was designed to correlate time-series data. Amjady's study (2001) has well examined the exactitude of ARIMA model for predicting daily peak and hourly load based on national power net. He further extended ARIMA model with use of the estimated electricity load as an extra input. The accuracy of his model reach a higher level even compared to original ARIMA and ANNs.

In most cases, statistical regression models are adopted to estimate important parameters characterizing energy performance. For instance, Mejri's et al. (2011) investigated statistical regression modelling for predicting indoor air temperature. In their study, they analyzed the similarity in dynamic behaviours among different thermal zones for HVAC system design. Another example goes to Wauman's et al. (2013), where they used statistical regression to explore correlation between heat balance ratio and heat gain factor of some school buildings exemplified in their research. These obtained correlations are regarded of crucial significance for designing, tracing and analyzing building thermal behaviours. They are also important supportive materials for drafting heating control strategy for energy saving.

Table 2.5 Summary of statistic regression, DT and GA in predicting building energy consumption

Type of house	Scale	Inputs	Output	Data source	Measure length	Algorithm
Office building (Li and Huang 2013)	Single	Dry bulb outdoor air temperature, solar horizontal radiation, and room temperature set point, cooling load of previous 4 h	Cooling load	Simulated data from TRNSYS	60 measured case	MLR, ANN, grey-box approach
Multiple (Amjady 2001)	National	Previous load, estimated current load, temperature	Hourly electricity load and daily peak	National dispatching center	1 year	ARIMA
Office building (Mejri et al. 2011)	Single	Indoor temperature of four rooms	Room temperature	Measured data	2 months	Statistical regression
Multiple (Tso and Yau 2007)	Groups	Power rating of appliance, consumption time	Average weekly electricity consumption	Survey	Two seasons	BPNN, Least-squares regression, DT
Residential (Yu et al. 2010)	80 buildings	Outdoor temperature, building characteristics, appliance energy source and usage (10 inputs)	Energy use intensity	Survey and research committee	3 years	DT
residential-commercial (Canyurt et al. 2005)	National	GDP, population, import, export, house production, basic house appliance consumption figures	Future energy demand (2003–2030)	World Energy Council and State Statistics Institute	8 years	GA
Residential (Sadeghi et al. 2011)	National	GDP, real price of electricity and natural gas in residential sector	Future per-capita consumption of electricity (2009–2025)	Iran Statistics Center, Central Bank of Iran	39 years	GA
Hotel (Li and Su 2010)	Single	Outdoor temperature of past 2 days, air conditioning consumption of past 3 days	Daily air conditioning consumption	Measured	7 months	GA-HANFIS

Prediction Application of Decision Tree

In the large family of data-driven approaches for building energy consumption prediction, DT is a relatively new member, but involves much simple techniques. Tso and Yau (2007) compared statistic regression method, BPNN and DT by predicting the electricity consumption in summer and winter periods. Results showed that DT used in their study performed as well as BPNN, both of which deliver accurate results than statistical regression did. Yu et al. (2010) also applied the DT approach to predict energy use intensity (EUI) of residential buildings. They designed ten predictor variables concerning indoor temperature, building envelop, appliance types and occupant number in the DT framework. Their result clearly demonstrated that DT is able to well predict building energy consumption level as high/medium/low. The significances of these predictor variables were ranked in terms of degree of closeness to the outdoor temperature (predictor variable of root node), which is the most important determinant of EUI. The results showed that several building parameters, e.g. heat loss coefficient and equivalent leak area, deserve more attention at early design stage and benefit energy conservation in retrofit.

Prediction Application of Genetic Algorithms

GA has been regarded as a powerful prediction approach in building energy consumption. As shown in Table 2.5, most applications of GA models are national analysis. One typical example is prediction model of energy consumption for residential-commercial building section in Canyurt et al. (2005). Three different scenarios were proposed in order to find out the best fit solution. The result showed that GA model, which considers residential housing production, house appliances of washing machine, television, vacuum cleaner and refrigerator as the input parameters, can obtain the most accurate quadratic prediction model of energy consumption. Sadeghi et al. (2011) developed prediction model of electricity consumption using GA on national level. It was found out that exponential equation had the more accurate results compared to linear and quadratic forms.

Hybrid methods of GA and ANNs are widely used in electricity prediction application (Azadeh et al. 2007). Li and Su (2010) predicted the daily air-conditioning consumption by using the genetic algorithm-hierarchical adaptive network-based fuzzy inference system (GA-HANFIS). Before developing prediction model, clustering algorithm was applied to identify the nature groups and qualities of a large data set, and GA was used to optimize the unknown cluster-parameters through minimizing the error of predicting result. Figure 2.10 shows the architecture of GA-HANFIS, in which the outdoor temperature of predicted day $T(k)$, the air-conditioning consumption of past two days $y(k-1)$ and $y(k-2)$ were identified as more significant inputs of network layer 1. These less significant variables $T(k-1)$, $y(k-3)$, $T(k-2)$ and $T(k-3)$ were selected as inputs of network layer 2 and layer 3. Output $y(k)$ was air-conditioning consumption of predicted day. The rule base of each layer contained two if–then rules; readers can refer to Tsekouras et al. (2007) for more details. Moreover, the calculation rules were different according to different clusters. This hybrid method outperformed regular BPNN in prediction accuracy.

2.2.1.2 Profile

The energy consumption profile in building is to quantify the total consumption contribution to sub-components, or further distinguish the usage characteristics. Regarding the positive influence for end-users, the capability of profiling the energy use as the feedback can educate the occupants on how to consume and change the consumption behaviours to certain extent. As for utility companies, DSM measures are implied after extracting load profiles in order to reach a proper load-shape objective, i.e. “peak clipping”, “valley filling”, “strategic conservation”, “flexible load shape”, “load building” and “load shifting” (Panapakidis et al. 2014). The commonly-used methods for energy and electricity profiling are clustering based method, which is detailed in Table 2.6.

Profile Application of Cluster Method

As one application of cluster method, analyzing electricity behaviour through pattern recognition and load curve classification has been investigated by massive researches. Tsekouras et al. (2007) developed a two-stage pattern recognition for customer’s classification. The first stage was

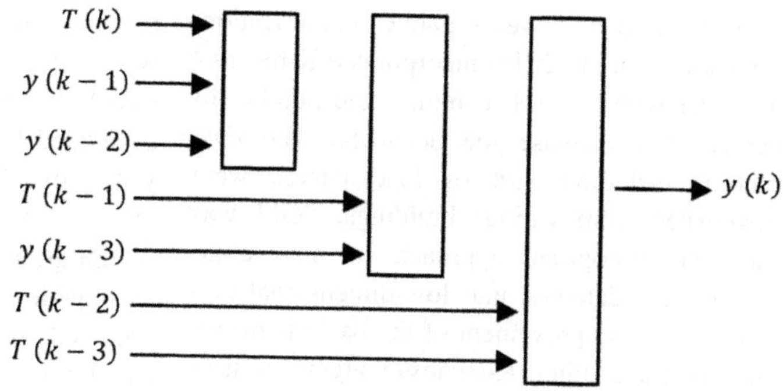

Fig. 2.10 Architecture of GA-HANFIS model with 3 layers (Li and Su 2010)

Table 2.6 Summary of data-driven approaches in building energy consumption profiling

Type of house	Scale	Inputs	Data source	Measure length	Algorithm
Multiple (Tsekouras et al. 2007)	94 buildings	Daily electricity consumption	Meters	10 months	K-means cluster, fuzzy K-means cluster, seven hierarchical cluster
University building (Panapakidis et al. 2014)	27 buildings	Daily electricity consumption	Meters	2 years	SOM combined with K means++
National office buildings (Xiao et al. 2012)	24 provinces	Annual electricity consumption	Survey	1 year	Hierarchical cluster
High performance buildings (Heidarinejad et al. 2014)	134 buildings	Energy end use	Simulated data	3 year	K-means cluster
Campus buildings (Arambula Lara et al. 2014)	85 buildings	Heating demand	School manage service	5 years	K-means cluster combined with MLR
Residential buildings (Tiedemann 2007)	791 customers	Weather data and energy consumption of residential appliances	Survey	2 years	Statistical regression
Residential buildings (Aydinalp-Koksal and Ugursal 2008)	8767 customers	Weather data and energy consumption of residential appliances	Survey	1 years	Statistical regression and ANN
Residential buildings (Aydinalp et al. 2002)	8767 customers	Appliance, lighting, cooling loads, space heating, domestic heat water	Survey	1 year	PBNN

to pattern load curves of each customer; the second stage was to cluster the customers according to pattern features. In their research, K-means cluster was proven by adequacy measures as the most appropriate approach compared to other methods. The function of adequacy measures is to evaluate the within-group similarity and between-group dissimilarity, in order to obtain a well-separated classification. Panapakidis et al. (2014) incorporated K-means++ cluster within SOM to reduce the number of centers and increase the accuracy. The data records including vast of load curves were aggregated from various buildings, SOM was thus an appropriate approach to map high-dimensional database into low-dimensional patterns. As the improvement of the basic K-means clustering algorithm, K-means++ algorithm tries to initialize the centroids that far from each other rather than random selection. The combination of SOM and K-means++ resulted in small errors in all cases.

Recently, cluster method becomes prevailing to profile EUI of buildings on large-scale. Xiao et al. (2012) conducted a study on EUI (excluding district heating) of business office buildings in China. Each data point was defined as (x_{1i}, x_{2i}) in which x_{1i} and x_{2i} refer to EUI and gross floor area of corresponding building. Eventually, two clusters were formed by using hierarchical cluster and the frequency distribution of EUI is illustrated in Fig. 2.11. The cluster results revealed the unique "dual section distribution" pattern which is different from developed countries. Heidarinejad et al. (2014) used K-means cluster algorithm to classify the EUI of 134 U.S. high-performance buildings (HPBs) by the squared Euclidean distance. These HPBs were well separated into three clusters, as high/medium/low EUI. Studies showed that unregulated loads which include various equipment and uncategorized loads, accounted for 30–40% total energy consumption that should be reduced specifically through effective programs and modification. It can be found out that studies mentioned above that analyze building energy issues on large scale, are greatly dependent on the clustering methodology.

Clustering technology can be also applied for heating/cooling demand classification. K-means cluster analysis combined with MLR were proposed by Arambula et al. (2014) to analysis the heating demand of 85 high schools. In their model, MLR analysis was firstly conducted to select 6 significant building thermal indicators according to R^2 value Eq. (2.12). Three clusters were developed by K-means cluster analysis based on these 6 indicators, while later R^2 was calculated for each cluster. The regression analysis showed that cluster 3 need to be further divided by clustering analysis since its low within-group similarity ($R^2 < 0.5$). Finally, more reasonable classification results could be obtained after such twice MLR analysis and twice clustering analysis when comparing to the sole clustering.

Profile Application of Regression

One regression method specialized for profiling energy consumption of residential buildings is conditional demand analysis (CDA). The basic idea of the CDA model is that total household consumption is the sum of various end-use consumptions.

CDA is frequently used to profile building energy consumption at national level (Tiedemann 2007). Aydinalp-Koksal and Ugursal (2008) used CDA to profile residential end-use energy consumption at national level, large-scale database including the surveys from occupants, weather conditions as well as historical energy bills were used. Their CDA model adopts 6 electricity end-uses including main and supplementary space heating, domestic heating water, space cooling, lighting, major and minor appliances. Meanwhile, they also developed neutral network model for comparison purpose (Aydinalp et al. 2002). In their research, BPNN outperformed CDA model in evaluating the effects of socio-economic factors, such as income, dwelling ownership and area sizes of residence. Because these socio-economic factors were considered as input variables in BPNN while CDA cannot not include comprehensive variables due to the limitation of statistical regression.

2.2.1.3 Energy Mapping

Energy mapping methods, usually based on the Geographic Information System (GIS) city building database, consider using data-driven technology for pre-and post-progressive

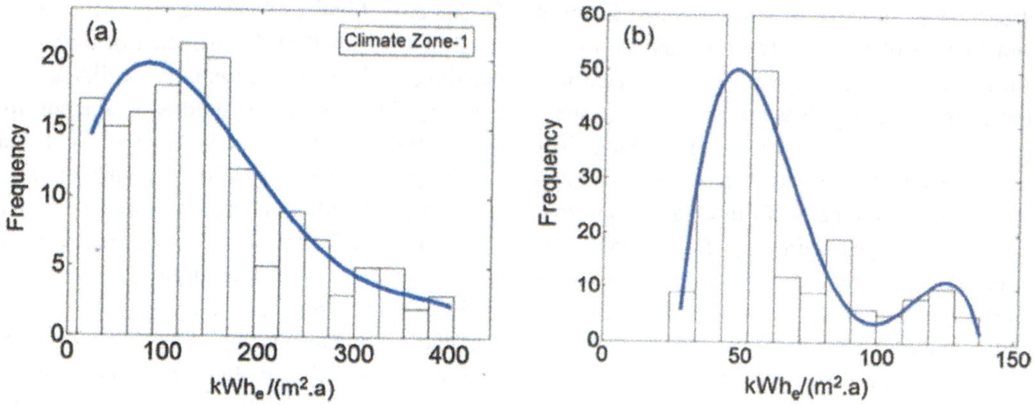

Fig. 2.11 Frequency distribution and polynomial fitting plot of EIU in office buildings. **a** US climate zone-1. **b** Certain city of China (excluding district heating) (Xiao et al. 2012)

operation (Caputo et al. 2013). Thanks to the capabilities of GIS, immediate updating of energy evaluation and visual representation via maps are both permitted in a user-friendly model, to provide energy consumption distribution within the city. Among the massive technologies for energy mapping, statistical regression (MLR) and clustering algorithm are the mostly utilized data-driven methods, as displayed in Table 2.7.

MLR is a traditional used approach in energy mapping of building section at zip-code level (Larivière and Lafrance 1999). Mastrucci et al. (2014) applied MLR model to map the energy consumption of dwellings in a city of Dutch. The contributing inputs included floor area, number

Table 2.7 Summary of data-driven approaches in energy mapping

Type of house	Scale	Energy consumption Inputs	Data source	Measure length	Algorithm
Multiple (Larivière and Lafrance 1999)	45 cities	Annual electricity, population density, age of inhabitants, weather condition, living expenditure	Organizations involved in city affairs	1 year	MLR
Multiple (Howard et al. 2011)	City	Annual electricity consumption, natural gas, steam, and fuel oil consumption	Utility company, RECS, CBECS, geo-rectified database	1 year	MLR
Residential buildings (Mastrucci et al. 2014)	City	Yearly consumption of natural gas and electricity	GIS database, Royal Netherlands Meteorological Institute, metering	Nearly 50 years	MLR
Residential buildings (Jones et al. 2007)	City	Energy consumption and carbon dioxide emission	GIS database, electricity map, surveys	18 months	Clustering algorithm
Multiple (Fonseca and Schlueter 2015)	City	Electricity load, thermal loads,	GIS database, simulated and measured data, building standards	1 year	K-means cluster

of occupants and type of house defined for each combination of type of dwelling and period of construction during 50 years. The predicted natural gas consumption was apportioned into space heating, domestic hot water and cooking. The results observed that space heating is the biggest contribution (average 50%) in energy consumption. After 50 years tracking from 1965, they assumed the percentage of energy reduction is nearly zero for dwellings after 2005. Besides single building function of residential family, energy mapping model has been expended to profile the building energy consumption of multiple functions. For example, Howard et al. (2011) calculated the annual EUI in New York City through MLR analysis, both the tax lot designations and building area categories were used to place the buildings into n building functions (e.g. residential family, office, warehouse, education and et al.). The MLR analysis is explained in Eq. (2.16), y_i is the energy consumption of ith zip-code, x_{in} is the total building area of each building function in ith zip-code. β_i is the coefficient need to be determined in MLR. On top of that, it was found out extra contributing inputs are needed in some regions to distinct the unique characteristics of energy consumption. However, the research excluded energy consumption for cooking, electrical heating and other end-uses, which inevitably causes errors in energy mapping.

$$y_i = \beta_1 x_{i1} + \beta_2 x_{i2} + \cdots + \beta_n x_{in} \tag{2.23}$$

Clustering algorithm is typically used as subsidiary approach for mapping the energy consumption at urban scale. In Jones et al. (2007) research, cluster analysis technique was adopted to classify 55,000 dwellings with similar energy consumption and carbon dioxide emission in a Local Authority of UK. The energy rating results and carbon dioxide emission results were profiled on the regional map for further retrofit purpose. Clustering algorithm is not limited to classification of energy consumption, also utilized to develop geographical clusters. Instance, Yamaguchi et al. (2007) proposed a district clustering model for commercial buildings in Osaka city. Firstly, clustering of district were presented by small grid cells, each of them was classified to certain representative building-type category. Then, EUI was used as evaluation for the district typology. Fonseca and Schlueter (2015) proposed a model for mapping the spatiotemporal building energy consumption in a city district of Switzerland. The model involved K-means cluster for spatial grouping in the band of 50–200 m, where spatial association of every variable of interesting buildings was strongly persistent (e.g. infrastructure types and temperature requirements). Two significant variables were used to measure the intensity of spatial clusters and similarity of groups. GIS framework gathered overall results and enabled 4D visualization that provides understandable display. The peak space heating demand of buildings in four zones at 10–11 am (April 1st, 2010) is presented as Fig. 2.12. The height and color code of buildings represent the demand level in relation to their associated zones.

2.2.1.4 Benchmarking of Buildings

Different from individual building energy analysis, benchmarking was used to address large-scale building energy related issues. Two fundamental issues in benchmarking are: (1) ascertaining the current energy performance of certain building (good, average or poor) compared to same types of building stock; (2) identifying the previous/current energy performance for energy saving potential and retrofit changes (Nikolaou et al. 2011). Regression based model, ANNs, cluster algorithms and DT are the typical data-driven techniques for building energy benchmarking. Table 2.8 provides the benchmarking pilots that usually adopt EUI as the single benchmarking index.

Regression technique is one popular method in building energy performance benchmarking. Chung et al. (2005) benchmarked the EUI of 30 supermarkets in Hong Kong. MLR model was established to calculate EUI based on nine significant variables. By using bootstrapping function (Efron and Tibshirani 1993) for the empirical sample $\{EUI_{(1)}, EUI_{(2)} \ldots EUI_{(30)}\}$, they

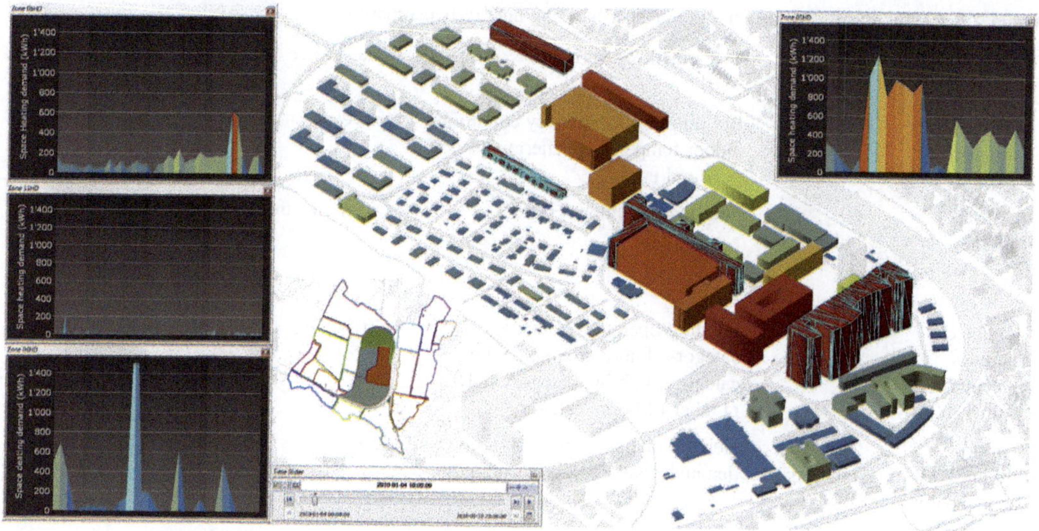

Fig. 2.12 Spatio-temporal energy map of space heating demand of a city district in Switzerland (Yamaguchi et al. 2007)

obtained the estimation of EUI cumulative distribution as percentiles $\{EUI_{10}, EUI_{20}...EUI_{90}\}$. Although conducting on small-scale samples, they formed a benchmarking table through the percentiles. The results showed that average value of energy consumption is greater than UK energy benchmarking. They also raised the suggestions that only unmanageable factors (e.g. building thermal characteristics) should be considered during benchmarking process while all manageable variables (e.g. occupancy behaviour) were set into average values, in order to present clearer improvement suggestions for government.

ANNs method in energy benchmarking was initially presented by Yalcintas (2006). He developed three sub-models to predict EUIs as output for the plug load, lighting and HVAC components over 60 mix-used buildings. The information from questionnaire includes lighting types, floor area, equipment types and hours were used as inputs. The elaborated ANNs model could identify the EUI if new data is entered. The most outstanding advantage of ANNs benchmarking method is to renew the algorithm itself rather than manual update. Yalcintas and Ozturk (2006) also developed a national energy benchmarking model for commercial buildings based on ANNs. Different from abovementioned ANN model which included continuous value of inputs/output, both input variables and output EUI were standardized into categorical forms for classification purpose in this model. In order to avoid inappropriate benchmarking results, database was firstly divided into 9 geographic regions. The results showed that ANN model provides more accurate EUI estimation and reasonable benchmarking result than MLR model in all cases except one.

Fuzzy cluster algorithm is a frequently-used methodology for energy benchmarking for buildings. Santamouris et al. (2007) proposed an energy rating system for 340 schools based on fuzzy clustering technology. Five classes of total and thermal energy consumption had been defined. Compared to frequency distribution rating system, fuzzy clustering rating system is more reasonable to avoid unbalanced classification, such as too small or too large range. Apart from building energy consumption benchmarking, thermal comfort rating system was also proposed by Nikolaou et al. (2012) based on FCM cluster. The predicted mean vote index, which represents mean response about thermal comfort from a larger group of people, was used as thermal comfort indicator. In their study, the thermal comfort of each climate zone was

Table 2.8 Summary of data-driven approach in building energy benchmarking

Type of house	Scale	Benchmarking variables	Data source	Measure length	Algorithm
Supermarkets (Chung et al. 2005)	30 buildings	Building age, occupancy condition, indoor temperature, energy system type and et al.	Survey	45 year	MLR
mix-used buildings (Yalcintas 2006)	60 buildings	plug load, lighting, HVAC	Questionnaire	1 year	BPNN
Commercial buildings (Yalcintas and Ozturk 2006)	National	building-operation hours, age category, building-area, cooling category, lighting category, CDD, number of floors category	CBECS	1 year	ANNs, MLR
School buildings (Santamouris et al. 2007)	340 buildings	Heating demand, electricity demand, total energy consumption	Energy bills	3 years	Fuzzy cluster
office buildings (Nikolaou et al. 2012)	30,000 buildings	Heating load, cooling load, thermal comfort	Simulated data: VBD	1 year	K-means cluster, SOM, FCM cluster
Residential buildings (Wang 2015)	324 buildings	EUI, CDD efficiency, HDD efficiency, bath room oriented, total room oriented efficiency and etc	Panel dataset	3 years	TOPSIS, PCA, K-means cluster
Commercial buildings (Park et al. 2016)	1072 buildings	EUI, gross floor area, building use ratio	Official building register, Korea Appraisal Board	3 years	DT

classified as three clusters, respectively. The majority commercial buildings in Greek were belong to class 2, while "best practice office buildings" were belong to class 1.

Another energy benchmarking method for improving energy efficiency of office building is DT. Park et al. (2016) developed DT model to benchmark the energy consumption of 1072 office buildings in South Korea. Gross floor area and building use ratio were identified as two significant predictor variables by correlation analysis, source EUI was defined as target variable. As the result, six rating groups of EUI were developed for each type of building use. After establishment of benchmarking model, analysis of variance was utilized to test the difference among groups. DT model was believed to improve the conventional baseline benchmarking system via a more reasonable and fair classification.

Although most benchmarking projects are developed based on single EUI indicator, there is much effort for the multi-criteria benchmarking indicators. Technique for Order Preference by Similarity to Ideal Solution (TOPSIS) based energy efficiency benchmarking approach using seven indicators was developed by Wang et al. (Wang 2015). The illustration of TOPSIS of two indicators (energy use per occupant and EUI) is showed as Fig. 2.13. When building A and building B have the same distances to the most energy efficient condition I_P, distances to the

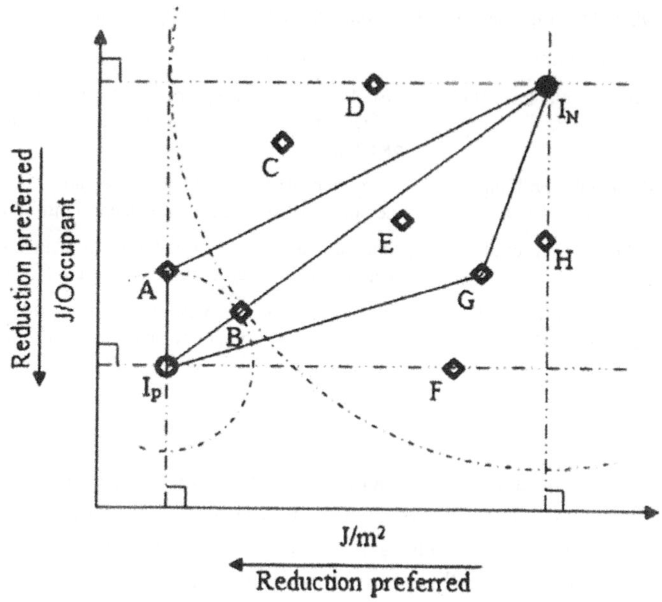

Fig. 2.13 Illustrative example of TOPSIS for building energy benchmarking (Wang 2015)

least energy efficient condition I_N were used to evaluate A and B. Since MLR cannot easily produce reliable weights among highly correlated indicator, principle component analysis (PCA) was adopted to weight the importance of seven energy indicators. PCA can transform a high-dimensional dataset consisting of possibly correlated variables into a less number of their linear combinations. Finally, K-means cluster was adopted to classify the TOPSIS space into six categories as benchmarking table. Without a doubt, the benefits were obvious compared to single-criteria benchmarking which is observed with collision during evaluation process.

2.2.1.5 Retrofit of Buildings

Retrofit is based on the knowledge of energy profiling and benchmarking on existing buildings, presenting the largest potential of incorporation of renewable energy technology and energy conservation after efficiency retrofit measures. ANNs and GA are the main data-driven approaches in building retrofit projects, within a brief introduction in Table 2.9.

ANNs are usually applied to predict energy-saving potential for single retrofit project. Yalcintas et al. (2008) developed BPNN model for two hotel equipment-retrofit projects. Energy usage data, weather data and occupancy data of post-retrofit period were used to train the neutral network model. It then estimated the energy consumption of pre-retrofit equipment as output. The difference between recorded and predicted energy consumption was regarded as the energy saving.

As a powerful optimization algorithm, GA has been frequently adopted as the evaluation tool in building retrofit project. Juan et al. (2009) presented a GA-based on-line decision support system to offer residents a series of optimal refurbishment actions considering two objectives, cost and quality. In GA, each chromosome represented a set of retrofit solutions, the distance between chromosome and trade-off curve of cost and quality was used as fitness function to select the parents for generation. With the process of evolution, the trade-off curve would gradually converge to the best retrofit solutions with higher quality and acceptable cost.

Developed based on two-objective optimization, multi-objective optimization model was conducted by Asadi et al. (2014). They adopted GA associated ANNs to study the interaction between three main conflicting target variables, including energy consumption (EC), retrofit cost (EC), thermal discomfort hours (TDH) and

Table 2.9 Summary of data-driven approach in building retrofit

Type of house	Retrofit measures	Algorithm
Hotels (Yalcintas 2008)	Install energy management systems and Variable Frequency Drives (VFDs) on the air-handling units. New cooling towers and VFDs on motor fans	BPNN
Residential buildings (Juan et al. 2009)	More than twenty retrofit measures under six main criterion, including safety, usage, convenient, comfortable, utility and health	GA
School (Asadi et al. 2014)	External wall insulation materials, roof insulation materials, the windows type, solar collector type, the HVAC systems	ANNs and GA
Residential buildings (Lannon et al. 2050)	Lower the carbon dioxide emission of grid, renewable resources, improve energy efficiency, and change occupancy behaviour	K-means cluster algorithm

assess their trade-offs in school retrofit project. First, the database was created in simulation tool for training and validating ANN model. BPNN model adopted in this study was composed of input layer representing different retrofit measures, one hidden layers and one output layer of energy consumption and thermal discomfort indicator. Then, the GA tool was used for minimize these three target variables as Eq. (2.17) and provide optimal combinations of retrofit measures.

$$\min y_1 = \mathrm{EC}(X) \quad (2.24a)$$

$$\min y_2 = \mathrm{RC}(X) \quad (2.24b)$$

$$\min y_3 = \mathrm{TDH}(X) \quad (2.24c)$$

$$X = \{\mathrm{xWALL}, \mathrm{xROOF}, \mathrm{xWINDOW}, \mathrm{xCOLLECTOR}, \mathrm{xHVAC}\}$$

where x represent different materials/types of alternative retrofit choices. The treat-off curves of multi-objective optimization could be available on 3D visualization. The proposed approach presented variety of recommendations with high computation efficiency. However, simultaneous optimizations of conflicting variables gave large diversity of retrofit choices, which are difficult to understand the impact of each retrofit action at whole level.

Cluster algorithm is usually adopted to make a distinction of retrofit measures among different buildings on large scale. Lannon et al. (2050) developed model of 55,000 houses over 50-year performance via cluster analysis, aiming to investigate the retrofit pathways to UK government's ambitious target of 80% reduction greenhouse gases emission by 2050. 100 clusters were developed to identify the dwelling with similar energy consumption and built age. Different combinations of retrofit measures were proposed and analyzed in the simulation tool. Overall, challenges and barriers in aggregate are still difficulties for individual family house.

2.2.2 Analyses of the Review Works

Data-driven approaches for predicting and classifying building energy consumption typically focus on total energy consumption, electricity demand, heating/cooling load and important energy parameters. The scopes of these researches are from sub-system level to single building level or even to national level.

Substantial up-to-date mythologies are proposed in order to enhance the accuracy and reliability of data-driven models, such as algorithm optimization and data pretreatment. As for algorithm optimization, micro-scale researches based on individual buildings are proposed with considerations to develop variants of basic algorithms and hybrids of several approaches (Zhao and Magoulès 2012; Kalogirou and Bojic 2000; Yang et al. 2005; Li et al. 2010; Li and Su 2010). The improvements of macro-scale analyses of building energy performance are invested to increase calculation efficiency when the raw data is large and chaos (Zhao and Magoulès

2010; An et al. 2013; Tsekouras et al. 2007; Wang 2015). In addition to algorithm optimization, data pretreatment is another focus for many researches (Yokoyama et al. 2009; Li et al. 2010; Amjady 2001; Arambula Lara et al. 2014). Appropriate pretreatment layered on the top of data-driven approach is the premise of accurate results and high computation efficiency. In short, high similarity between training and testing dataset is important for establishing a good model.

Meanwhile, substantial studies applied the simulated database to test model performance rather than the measured data. The analysis results of these models cannot be regarded persuasive enough since simulated data records are less fluctuant than real situation. In these scenarios, the question arises for reliability of simulated data again with no clear answer.

So far, the researches on residential buildings are not elaborated as researches of commercial buildings. The main reasons are including (1) lack of energy-use database from family-houses; (2) more freedom of occupancy behaviour in residential buildings. Hence, most researches on residential buildings are at low granularity, such as roughly profile energy consumption on regional level (Yu et al. 2010; Sadeghi et al. 2011; Mastrucci et al. 2014; Wang 2015; Lannon et al. 2050).

2.3 Opportunities for Further Works

As effective and useful techniques providing profound insights and possible strategic solutions in policy and management of building energy consumption, data-driven approaches have been deemed as favorable means for facilitate future in-depth studies on building energy performance. In this section, we tailor a few promising research directions of data-driven approaches applied in building energy.

The first direction suggested is to modify the framework of the current data-driven algorithms in the contexts of building performance to better fulfill the calculation of building energy consumption. This necessitates special optimization of the data-driven approaches responding to unique requirements of prediction and classification of building energy consumption. It also calls for reliable and sufficient data source to yield training in high quality. Well-designed pretreatments particularly compatible with data collection in buildings and high-granularity measurement from smart meters are strongly recommended.

Another direction is to enrich the applications of the current data-driven versions, enabling them to be viable for building energy characterization at different scales and various climate conditions. In the literature, a great number of efforts in this direction are limited to small-scale building stocks. Future focuses should be placed on energy-use evaluation of a large amount of building population. To be specific, the next generation of data-driven approaches in building industries should cover more building thermal characteristics and climate conditions, and are able to perform energy mapping and benchmarking at the macro-scale.

Furthermore, it has been widely noted that most building energy consumption models at the present stage only offer short-term analyses. They fail to discuss the long-term profiles and predictions, nonetheless. Therefore, strong drive exists to extend data-driven approaches with applications to building energy consumption with climate changes in a large temporal scale. This will play an important role in future building retrofitting, where long-term energy consumption is a crucial contributing factor for new HPB design when used to replace old buildings.

As pointed out above, building performance evaluation nowadays strives for high efficiency in building energy consumption. However, other factors, such as indoor air quality, occupant thermal comfort, occupancy behaviour interaction and equipment energy-performance coefficient, are also equally important. This indicates an ideal data-driven model should make use of multiple indexes to provide a comprehensive analysis of building performance, instead of the current single output of energy consumption or heating/cooling loads. Significantly, apart from the basic functions (i.e., prediction and classification) discussed in this review, future outlook of data-driven techniques targets decision-making machine, such as occupancy behaviour recommender and equipment operation instructor. These data-driven based developments in building industries would offer real-time on-site information for thermo-comfortable accommodation with minimum energy consumption.

2.4 Conclusion

Given that the growing concerns about natural resource shortages and eco-environment deterioration, today's building industry necessitates strongly innovative techniques to better optimize energy consumption in buildings. Data-driven approaches are one of the most effective advanced means to fulfill this goal in function of more accurate prediction and clearer classification.

This chapter reviewed the mainstream data-driven techniques currently applied in building energy studies. A comprehensive summary of each approach, including their basic thought, algorithm structure, strengths and weaknesses, was elaborated. In the category of data-driven approaches for prediction, ANNs gains a primary position in a large number of applications ranging from load forecasting, retrofit potential estimation. The challenges in this method are to choose an appropriate architecture and learning rate. As to SVM, the method has a simple training process requiring a few inputs whereas low calculation efficiency has become a roadblock in its application, especially for large-scale building energy analysis and high granularity. To shorten its calculation time, recent tremendous efforts have made to optimize SVM structure and develop hybrid models combing the method with clustering algorithms, DTW and et al. Statistical regression plays an important role in significant parameters evaluation, energy mapping and benchmarking at the urban scale. However, inaccuracy in short-term prediction and possible unforeseen correlations among the selected predictors greatly undermine the effectiveness of the regression models in practical applications. As to DT, it is regarded as a feasible technique for energy consumption prediction and building energy benchmarking provided that these cases do not involve sequence and nonlinear data. The major concern is DT results are primarily based on expectations, and thus validity of many DT models being questioned. GA excels in coping with sophisticated data and can provide optimal and multi-objective solutions, but GA suffers from low computational efficiency and non-unique deliveries. As far as data-driven approaches for classification are concerned, K-means clustering, self-organizing map and hierarchy clustering were exemplified with focuses on pattern recognition of load curves, energy mapping, benchmarking and retrofit at macro-scale. The K-means clustering algorithm achieves highly accurate data partition in many applications. A fundamental deficiency in this classification approach is its framework lacks a priori to specify the number of clusters and centroids yet when the classifying features are unavailable. Self-organizing map can effectively reduce the dimensions of input data to a feature map, but its accuracy relies on a careful parameter selection and an appropriate design of its algorithmic structure. As to hierarchical clustering, it is usually implemented by a number of diverse merging and terminating ways, which could lead to multiple clustering results even subject to the same similarity criterion. This greatly influences its accuracy and effectiveness when performed for many clustering problems.

Based on the above analysis, this review points out several future directions for data-driven approaches to better underpin building energy analysis. They include (1) modifying the frameworks of different data-driven approaches in the context building performance features to more accurately respond specific demands from building energy calculation; (2) enriching applications of data-driven approaches to cover building energy uses at different scales under a wide spectrum of weather conditions; (3) extending data-driven approaches serving building energy consumption investigation and in both the short- and long-term; (4) integrating multiple target indices in the data-driven framework to deliver a more balanced evaluation on building energy performance. It is believed that data-driven techniques with progress along these routes will offer more efficient and reliable support for energy management and optimization in future building industries.

References

Alhamazani K, Ranjan R, Mitra K et al (2015) An overview of the commercial cloud monitoring tools: research dimensions, design issues, and state-of-the-art. Computing 97(4):357–377

Al-Homoud MS (2001) Computer-aided building energy analysis techniques. Build Environ 36(4):421–433

Amjady N (2001) Short-term hourly load forecasting using time-series modeling with peak load estimation capability. IEEE Trans Power Syst 16(3):498–505

An N, Zhao W, Wang J et al (2013) Using multi-output feedforward neural network with empirical mode decomposition based signal filtering for electricity demand forecasting. Energy 49(1):279–288

Arambula Lara R, Cappelletti F, Romagnoni P et al (2014) Selection of representative buildings through preliminary cluster analysis

Asadi E, Silva MGD, Antunes CH et al (2014) Multi-objective optimization for building retrofit: A model using genetic algorithm and artificial neural network and an application. Energy Build 81(na):na

Aydinalp M, Ugursal VI, Fung AS (2002) Modeling of the appliance, lighting, and space-cooling energy consumptions in the residential sector using neural networks. Appl Energy 71(2):87–110

Aydinalp-Koksal M, Ugursal VI (2008) Comparison of neural network, conditional demand analysis, and engineering approaches for modeling end-use energy consumption in the residential sector. Appl Energy 85 (4):271–296

Azadeh A, Ghaderi SF, Tarverdian S et al (2007) Integration of artificial neural networks and genetic algorithm to predict electrical energy consumption. Appl Math Comput 186(2):1731–1741

Barnaby CS, Spitler JD (2005) Development of the residential load factor method for heating and cooling load calculations. ASHRAE Trans 111:291–307

Beyer HG (2000) Evolutionary algorithms in noisy environments: theoretical issues and guidelines for practice. Comput Methods Appl Mech Eng 186(2–4):239–267

Bojić M, Lukić N (2000) Numerical evaluation of solar-energy use through passive heating of weekend houses in Yugoslavia. Renew Energy 20(2):207–222

Canyurt OE, Ozturk HK, Hepbasli A et al (2005) Estimating the Turkish residential–commercial energy output based on genetic algorithm (GA) approaches. Energy Policy 33(8):1011–1019

Caputo P, Costa G, Ferrari S (2013) A supporting method for defining energy strategies in the building sector at urban scale. Energy Policy 55(249):261–270

Chung W, Hui YV, Lam YM (2005) Benchmarking the energy efficiency of commercial buildings. Appl Energy 83(1):1–14

Dong B, Cao C, Lee SE (2005) Applying support vector machines to predict building energy consumption in tropical region. Energy Build 37(5):545–553

Dunn JC (1973) A fuzzy relative of the ISODATA process and its use in detecting compact well-separated clusters. J Cybern 3(3):32–57

Efron B, Tibshirani R (1993) An introduction to bootstrap. Chapman & Hall, New York

Farahbakhsh H, Ugursal VI, Fung AS (1998) A residential end-use energy consumption model for Canada. Int J Energy Res 22(13):1133–1143

Fonseca JA, Schlueter A (2015) Integrated model for characterization of spatiotemporal building energy consumption patterns in neighborhoods and city districts. Appl Energy 142:247–265

Goldberg DE (1986) The genetic algorithm approach: why, how, and what next? In: Adaptive and learning systems. Springer US, pp 247–253.

Heidarinejad M, Dahlhausen M, Mcmahon S et al (2014) Cluster analysis of simulated energy use for LEED certified U.S. office buildings. Energy Build 85:86–97

Hong T, Le Y, Hill D et al (2014) Data and analytics to inform energy retrofit of high performance buildings. Appl Energy 126(C):90–106

Howard B, Parshall L, Thompson J et al (2011) Spatial distribution of urban building energy consumption by end use. Energy Build 45:141–151

Huang YJ (2000) A bottom-up engineering estimate of the aggregate heating and cooling loads of the entire US building stock. Escholarship University of California

ISO (2013) ISO Standard 12655: energy performance of buildings—presentation of real energy use of buildings

Jones P, Patterson J, Lannon S (2007) Modelling the built environment at an urban scale—energy and health impacts in relation to housing. Landsc Urban Plan 83 (1):39–49

Juan YK et al (2009) GA-based decision support system for housing condition assessment and refurbishment strategies. Autom Construct 18(4):394–401

Kalogirou SA (2001) Artificial neural networks in renewable energy systems applications: a review. Renew Sustain Energy Rev 5(4):373–401

Kalogirou SA, Bojic M (2000) Artificial neural networks for the prediction of the energy consumption of a passive solar building. Energy 25(5):479–491

Kang Z, Jin M, Spanos CJ (2014) Modeling of end-use energy profile: an appliance-data-driven stochastic approach. Statistics 5382–5388

Kuo WJ, Chang RF, Chen DR et al (2001) Data mining with decision trees for diagnosis of breast tumor in medical ultrasonic images. Breast Cancer Res Treat 66 (1):51

Lannon S, Georgakaki A, Macdonald S (2013) Modelling urban scale retrofit, pathways to 2050 low carbon residential building stock. Ibpsa

Larivière I, Lafrance G (1999) Modelling the electricity consumption of cities: effect of urban density. Energy Econ 21(1):53–66

Li K, Su H (2010) Forecasting building energy consumption with hybrid genetic algorithm-hierarchical adaptive network-based fuzzy inference system. Energy Build 42(11):2070–2076

Li Q, Meng Q, Cai J et al (2009a) Predicting hourly cooling load in the building: a comparison of support vector machine and different artificial neural networks. Energy Convers Manage 50(1):90–96

Li Q, Meng Q, Cai J et al (2009b) Applying support vector machine to predict hourly cooling load in the building. Appl Energy 86(10):2249–2256

Li Z, Han Y, Xu P (2014) Methods for benchmarking building energy consumption against its past or intended performance: an overview. Appl Energy 124(7):325–334

Li Z, Huang G (2013) Re-evaluation of building cooling load prediction models for use in humid subtropical area. Energy Build 62(3):442–449

Li X, Deng Y, Ding L et al (2010) Building cooling load forecasting using fuzzy support vector machine and fuzzy C-mean clustering[C]. In: International conference on computer and communication technologies in agriculture engineering (CCTAE 2010), vol 2010, pp 438–441

Magoules F, Zhao HX (2016) Data mining and machine learning in building energy analysis. Wiley

Mastrucci A, Baume O, Stazi F et al (2014) Estimating energy savings for the residential building stock of an entire city: a GIS-based statistical downscaling approach applied to Rotterdam. Energy Build 75 (2):358–367

Mathew PA, Dunn LN, Sohn MD et al (2015) Big-data for building energy performance: lessons from assembling a very large national database of building energy use. Appl Energy 140:85–93

Mathieu JL, Price PN, Kiliccote S, Piette MA (2011) Quantifying changes in building electricity use, with application to demand response. IEEE Trans Smart Grid 2:507–518

Mejri O, Barrio EPD, Ghrab-Morcos N (2011) Energy performance assessment of occupied buildings using model identification techniques. Energy Build 43 (2):285–299

Neto AH, Fiorelli FAS (2008) Comparison between detailed model simulation and artificial neural network for forecasting building energy consumption. Energy Build 40(12):2169–2176

Nikolaidis Y, Pilavachi PA, Chletsis A (2009) Economic evaluation of energy saving measures in a common type of Greek building. Appl Energy 86(12):2550–2559

Nikolaou T, Kolokotsa D, Stavrakakis G (2011) Review on methodologies for energy benchmarking, rating and classification of buildings. Adv Build Energy Res 5(1):53–70

Nikolaou TG, Kolokotsa DS, Stavrakakis GS et al (2012) On the application of clustering techniques for office buildings' energy and thermal comfort classification. IEEE Trans Smart Grid 3(4):2196–2210

Olofsson T, Andersson S (2002) Overall heat loss coefficient and domestic energy gain factor for single-family buildings. Build Environ 37(11):1019–1026

Panapakidis IP, Papadopoulos TA, Christoforidis GC et al (2014) Pattern recognition algorithms for electricity load curve analysis of buildings. Energy Build 73 (2):137–145

Park HS, Lee M, Kang H et al (2016) Development of a new energy benchmark for improving the operational rating system of office buildings using various data-mining techniques. Appl Energy 173:225–237

Paudel S, Nguyen PH, Kling WL et al (2015) Support vector machine in prediction of building energy demand using pseudo dynamic approach. Comput Sci

Pérez-Lombard L, Ortiz J, Pout C (2008) A review on buildings energy consumption information

Perino M, Tardioli G, Kerrigan R et al (2015) Data driven approaches for prediction of building energy consumption at urban level. Energy Proc 78:3378–3383

Quinlan JR (1986) Induction of decision trees machine learning. In: Data: goals and general description of the IN L.EN System, pp 257–264

Sadeghi H, Zolfaghari M, Heydarizade M (2011) Estimation of electricity demand in residential sector using genetic algorithm approach

Santamouris M, Mihalakakou G, Patargias P et al (2007) Using intelligent clustering techniques to classify the energy performance of school buildings. Energy Build 39(1):45–51

Setiawan A, Koprinska I, Agelidis VG (2009) Very short-term electricity load demand forecasting using support vector regression. In: International joint conference on neural networks, IJCNN 2009, Atlanta, Georgia, USA, 14–19 June. DBLP, pp 2888–2894

Shimoda Y, Fujii T, Morikawa T et al (2004) Residential end-use energy simulation at city scale. Build Environ 39(8):959–967

Sides J (2014) The victory lab: the secret science of winning campaigns. Public Opin Q 78(S1):363–364

Sözen A, Arcaklioglu E (2007) Prediction of net energy consumption based on economic indicators (GNP and GDP) in Turkey. Energy Policy 35(10):4981–4992

Swan LG, Ugursal VI (2009) Modeling of end-use energy consumption in the residential sector: a review of modeling techniques. Renew Sustain Energy Rev 13 (8):1819–1835

Tiedemann KH (2007) Using conditional demand analysis to estimate residential energy use and energy savings. In: Proceedings of the Cdeee

Tsekouras GJ, Hatziargyriou ND, Dialynas EN (2007) Two-stage pattern recognition of load curves for classification of electricity customers. IEEE Trans Power Syst 22(3):1120–1128

Tso GKF, Yau KKW (2007) Predicting electricity energy consumption: a comparison of regression analysis, decision tree and neural networks. Energy 32 (9):1761–1768

UNEP (2013) Energy efficiency for buildings. http://www.studiocollantin.eu/pdf/UNEP%20Info%20sheet%20-%20EE%20Buildings.pdf

Vapnik V, Golowich SE, Smola A (1996) Support Vector method for function approximation, regression estimation, and signal processing. Adv Neural Inf Process Syst 9:281–287

Vesanto J, Alhoniemi E (2000) Clustering of the self-organizing map. IEEE Trans Neural Networks 11 (3):586

Wang E (2015) Benchmarking whole-building energy performance with multi-criteria technique for order preference by similarity to ideal solution using a selective objective-weighting approach. Appl Energy 146:92–103

Wauman B, Breesch H, Saelens D (2013) Evaluation of the accuracy of the implementation of dynamic effects in the quasi steady-state calculation method for school buildings. Energy Build 65(10):173–184

Xiao H, Wei Q, Jiang Y (2012) The reality and statistical distribution of energy consumption in office buildings in China. Energy Build 50(50):259–265

Yalcintas M (2006) An energy benchmarking model based on artificial neural network method with a case example for tropical climates. Int J Energy Res 30 (14):1158–1174

Yalcintas M (2008) Energy-savings predictions for building-equipment retrofits. Energy Build 40 (12):2111–2120

Yalcintas M, Ozturk UA (2006) An energy benchmarking model based on artificial neural network method utilizing US Commercial Buildings Energy Consumption Survey (CBECS) database. Int J Energy Res 412–421

Yamaguchi Y, Shimoda Y, Mizuno M (2007) Proposal of a modeling approach considering urban form for evaluation of city level energy management. Energy Build 39(5):580–592

Yan CW, Yao J (2010) Application of ANN for the prediction of building energy consumption at different climate zones with HDD and CDD. In: International conference on future computer and communication. IEEE, pp V3-286–V3-289

Yang J, Rivard H, Zmeureanu R (2005) On-line building energy prediction using adaptive artificial neural networks. Energy Build 37(12):1250–1259

Yezioro A, Dong B, Leite F (2008) An applied artificial intelligence approach towards assessing building performance simulation tools. Energy Build 40(4):612–620

Yokoyama R, Wakui T, Satake R (2009) Prediction of energy demands using neural network with model identification by global optimization. Energy Convers Manage 50(2):319–327

Yu Z, Haghighat F, Fung BCM et al (2010) A decision tree method for building energy demand modeling. Energy Build 42(10):1637–1646

Zhao HX, Magoulès F (2010) Parallel support vector machines applied to the prediction of multiple buildings energy consumption. J Algorithms Comput Technol 4(2):231–250

Zhao HX, Magoulès F (2012) A review on the prediction of building energy consumption. Renew Sustain Energy Rev 16(6):3586–3592

3 Prediction of Occupancy Level and Energy Consumption in Office Building Using Blind System Identification and Neural Networks

Jinshun Wu, Yixuan Wei, and Xingxing Zhang

Abstract

Occupancy behaviour plays an important role in energy consumption in buildings. Currently, the shallow understanding of occupancy has led to a considerable performance gap between predicted and measured energy use. This chapter presents an approach to estimate the occupancy based on blind system identification (BSI), and a prediction model of electricity consumption by an air-conditioning system is developed and reported based on an artificial neural network with the BSI estimation of the number of occupants as an input. This starts from the identification of indoor CO_2 dynamics derived from the mass-conservation law and venting levels. The unknown parameters, including the occupancy and model parameters, are estimated by using a frequentist maximum-likelihood algorithm and Bayesian estimation. The second phase is to establish the prediction model of the electricity consumption of the air-conditioning system by using a feed-forward neural network (FFNN) and extreme learning machine (ELM), as well as ensemble models. To analyse some aspects of the benchmark test for identifying the effect of structure parameters and input-selection alternatives, three studies are conducted on (1) the effect of predictor selection based on principal component analysis, (2) the effect of the estimated occupancy as the supplementary input, and (3) the effect of the neural network ensemble. The result shows that the occupancy number, as the input, is able to improve the accuracy in predicting energy consumption using a neural-network model.

Keywords

Occupancy estimation · Blind system identification (BSI) · Prediction model for energy consumption · Feedforward neural network · Extreme learning machine

J. Wu (✉)
College of Architecture and Civil Engineering, North China Institute of Science & Technology, Hebei, China
e-mail: wujinshun2005@163.com

Y. Wei
School of Civil Engineering and Resources, University of Science & Technology Beijing, Beijing, China
e-mail: weiyixuan@ustb.edu.cn

X. Zhang
Department of Energy and Community Buildings, Dalarna University, 79188 Falun, Sweden
e-mail: xza@du.se

3.1 Introduction

The shortage of natural resources has compelled many countries to focus on reducing energy consumption as well as greenhouse gas emissions. Improving energy efficiency of buildings has become a priority for many nations.

X. Zhang (ed.), *Data-driven Analytics for Sustainable Buildings and Cities*, Sustainable Development Goals Series, https://doi.org/10.1007/978-981-16-2778-1_3

Improvement of the energy efficiency of a building can only be achieved with the accurate prediction of the building's energy consumption. Various physical-based and data-driven models have been proposed for predicting the cooling/heating load of heating, ventilating, and air-conditioning (HVAC) systems. Physical-based models, for example, EnergyPlus, IES, and TRYSYS, require comprehensive and detailed information about the complex thermal behaviour of the HVAC system and building envelope. However, the realisation of precise energy consumption analysis is still a formidable task, as it involves uncertain and complex interactions among weather conditions and stochastic occupant behaviours (Li et al. 2014; Cui et al. 2017).

Data-driven approaches are used to predict a building's energy consumption with limited physical information but with enormous operational data of the building's thermal performance. To further categorise the data-driven approach, there are black-box and grey-box models. The grey-box model is a hybrid model that combines a machine-learning algorithm and physical model. Grey-box models have been widely applied to estimate the heating/cooling load of buildings (Fux et al. 2014; Dong et al. 2016). The black-box model is a purely machine-learning approach that defines the relationships among different parameters. Historical data should be available for training the black-box model. Through the use of black-box models, the building energy behaviour can be analysed. The algorithm is excellent in illustrating the nonlinear relationship between input and output parameters; therefore, establishing explicit physical models to describe the complex interaction is unnecessary (Ahmad et al. 2014). In recent years, many practical researchers have identified that black-box models are cable of rapid and accurate calculation compared to physical models (Yezioro et al. 2008; Zhao and Magoulès 2012; Turhan et al. 2014).

3.1.1 State of the Art

Different statistical methods can be adopted for black-box approaches. The artificial neural network (ANN) is one popular data-driven approach for the purpose of electricity-consumption prediction for long-term (Günay 2016), short-term (Hooshmand et al. 2013) and mid-term applications (Hassan et al. 2015; Jovanović et al. 2015), as well as retrofitting (Asadi et al. 2014; Ascione et al. 2017; Gossard et al. 2013), fault detection and diagnoses (Du et al. 2014; Guo et al. 2017), control of air-conditioning (AC) systems (Afram et al. 2017; Magoulès et al. 2013), etc. In order to predict the energy consumption accurately using an ANN model, significant effort has been made to optimise the model parameters or select the suitable input parameters. For example, the modal trimming method (Yokoyama et al. 2009; Li et al. 2011), particle-swarm optimisation (Jiang et al. 2013), and genetic algorithm (Castelli et al. 2015) are usually used to find feasible solutions of ANN model parameters. Roldán-Blay et al. (2013) introduced a model to predict the outdoor temperature at each moment. It provides benefits when selecting the training days and improves the relationship between the outdoor temperature and energy consumption. In reference (Jetcheva et al. 2014), load and temperature clustering were carried out to develop the corresponding neural network model for each group; the suitable model was selected for each time point during the forecasting phase. Hence, the set of these forecasts (one for each time point during the day) forms the load forecast for the entire day. Signal filtering was introduced by Shao et al. (2015) to reduce the noise from the time-series data based on ensemble empirical mode decomposition. As expected, some potentially important and hidden characteristics of the original consumption data were extracted before input to the neural network. The forecasting accuracy was improved at the same time. In order to solve the problem of the vanishing/exploding gradient of recurrent neural networks, Rahman et al. (2018) adopted long short-term memory (LSTM) activation functions to remember/forget values at previous time steps, where long-term dependencies are considered. Comparison among regression analysis, neural networks, and least-squares support vector machine (SVM) were conducted by Kaytez et al. (2015) in predicting national energy consumption in Turkey. The analysed results suggested that both SVM and ANN were powerful forecasting tools for long-term electricity consumption. More dedicated predictive modelling for energy use in US commercial buildings was proposed by Deng et al. (2018). According to their analysis, SVM has superior performance to ANN, while the performance of ANN reflects the importance of regularisation for a generalised prediction model.

However, the downside of fundamental ANN and SVM is the long training time required for excessively large amounts of data. Another limitation of ANN and SVM is due to the difficulty in the determination of model parameters and kernel function. There is no uniform standard to select suitable model parameters and kernel functions. Researchers must make decisions based on the characteristics of the data and/or their experience. The ELM, as an emergent technology that overcomes the challenges faced by fundamental ANN and SVM, has been proposed recently. Unlike the fundamental ANN, the hidden layer of ELM does not need to be tuned. This approach is capable of solving the problem without a back-propagation training process. Naji et al. (2016) and Li et al. (2017) have developed a building energy estimating model based on ELM, the results revealed the robustness of the ELM method.

Many researchers have analysed the influence of input parameters on prediction accuracy and observed that the indoor occupancy plays a critical role in energy prediction model. A fixed occupancy schedule or simple day type (weekdays/weekends) was usually adopted as an input parameter by some models (Li et al. 2017; Neto and Fiorelli 2008; Cui et al. 2016). These results indicated that the fixed profile of the occupancy was able to increase the prediction accuracy compared to that of the cases without consideration of the occupancy schedule. Other approaches to simulate dynamic internal load variance caused by occupants were based on the time factor (Wong et al. 2010; Zuo et al. 2016) and historical load of electricity consumption (Osman et al. 2009; Mena et al. 2014). For example, a multi-layered feedforward network with Bayesian regularisation was adopted by Chae et al. (2016) to predict the sub-hourly energy consumption. In their work, both the day type and time of day were selected as input elements. As for the large-scale campus building, the fit of a load-forecasting model was dramatically improved when time of day and day of week were introduced as inputs (Powell et al. 2014). In another study (Leung et al. 2012; Kwok et al. 2011), three kinds of factors were introduced to represent the internal load features: the operating schedules of air units, hour of day, and occupancy space power demand. Although use of the occupancy space power demand would enhance the model accuracy, the prediction was not sufficiently accurate for winter. Concerning implementation of historical load as an influential factor of internal load, An et al. (2013) used previous electricity consumption as the input of a multi-output prediction model to forecast the energy demand for the next half hour. In Paudel et al. (2014), transitional characteristics and power-level characteristics of the heating system were adopted as the input variables of the ANN for load prediction. These variables were validated using a precise description of transitional delay for fluctuation of occupancy in buildings. In recent research (Deb et al. 2016), the authors concluded that the ANN was able to be used for forecasting the next day's energy use based on the five previous days' data with acceptable accuracy. In all the above studies, the fixed occupancy schedule, time factor, and historical internal load were considered as input parameters of the energy-prediction models.

However, variance of the internal load aroused by randomness due to occupancy variation cannot be represented comprehensively and accurately with a fixed occupancy schedule, time factor, or historical load. Most previous works show that there is a considerable performance gap between the predicted and measured energy use. One cause of poor performance is the interaction between occupancy and building systems. By introducing occupant information as a component of building performance, energy consumption can be very different according to the various occupant actions (Feng et al. 2016). Therefore, introducing a real occupant schedule into the building energy-prediction model has been brought to the forefront. Some energy-prediction models have been proposed based on the estimated occupancy profile (Virote and Neves-Silva 2012). However, in reality, few energy-prediction models use true occupant presence or interactions (adjusting thermostats, lighting control, etc.) as the inputs to the model design because the key parameters regarding occupancy is difficult to collect. Some basic models were proposed based on a few test buildings, such as an airport terminal (Huang et al. 2015) or single office room (Gruber et al. 2014).

The number of occupants is not easy to count and should be measured by intrusive sensors, such as cameras and pattern recognition (Labeodan et al. 2015; Liu et al. 2013). Personal privacy is one problem during implementation. The non-intrusive types of sensors, such as pyroelectric infrared (PIR), ultrasonic, and acoustic sensors, can only be used to determine whether the room is occupied, rather than determining the actual number of occupants (ul Haq et al. 2014; Sun et al. 2014). As an indirect estimation, the amount of the total energy consumed to supply fresh air was applied to simulate the occupancy area and occupancy rate in a commercial building (Kwok and Lee 2011). According to Hailemariam et al. (2011) and Amayri et al. (2016), occupancy-detection accuracy was improved when various sensors were associated with a motion sensor. However, overfitting is one problem for the occupancy-detection model. Probabilistic models have been usually adopted for the development of an accurate dynamic occupancy model, which use statistical data to predict the probability of certain behaviour, such as the stochastic presence of occupancy. Researchers proposed many probabilistic approaches to estimate the number of occupants based on the heterogeneous Markov chain model (Page et al. 2008; McKenna et al. 2015), inhomogeneous Markov chain model (Chen et al. 2015), and semi-Markov model (Dong and Lam 2014), which generate a time series of the state of occupancy. According to Li and Dong (2017), the inhomogeneous Markov chain model outperformed ANN and SVM for short-term prediction of occupancy in residential buildings. In addition, long-vacancy activities, e.g., vacations or absences greater than one week, were described by Stoppel and Leite (2014) using a probabilistic method. The simulation result was integrated with EnergyPlus for energy-consumption prediction. Liao and Barooah (2010 and Liao et al. (2012) conducted a study on occupancy modelling, and an agent-based model was developed to simulate the arbitrary behaviour of individual occupants. The survey and sensor data were collected as the initial agents' states (occupied/unoccupied). Apart from probabilistic models, a stochastic model was also utilised by researchers. Tahmasebi and Mahdavi (2015) generated daily occupancy data based on year-long observational data on occupancy. The results suggested that stochastic presence modelling could offer a better representation of occupants' presence rather than distribution and peak values (Sun et al. 2014; Mahdavi and Tahmasebi 2015). Unlike the probabilistic model and agent-based model, the random walking approach was proposed by Ahn and Park (2016). They regarded occupant presence as unpredictable in certain cases. Recently, a logistic regression model is proposed by Shi et al. (2017) to forecast the building's occupancy state. The proposed model has been validated with higher accuracy than the Markov chain algorithm. It should be noted that most models mentioned above require people-counting data as necessary known parameters, which is difficult collect due to the implementation cost and ease of constancy.

Many strategies have been proposed to estimate the indoor occupancy by utilising environmental sensors. As for a CO_2-based detection model, (Mumma et al. 2004; Ansanay-Alex 2013) developed an equation to calculate the number of occupants by using CO_2 concentration measurement, which is regarded the analytical method. However, the slow response with a delay and difficulty of physical parameter identification were the main drawbacks of this equation. Similarly, Yang et al. (2014) has pointed out that late responses of the CO_2 sensor is one limitation when applying CO_2 sensor for occupancy number calculation. The reliable correlating CO_2 level with actual occupancy is difficult when the office spaces are irregular and open (Meyn et al. 2009). When the CO_2 sensors are suitably installed and details about observed rooms (room volume and air flow rate) are known, this method performs well for the detection of indoor occupancy in closed-space offices. Latterly, more accurate methods were developed to compensate for these drawbacks. A CO_2-detection sensor network was installed to measure the gas concentration of 19 locations. SVMs, neural networks, and hidden Markov models were introduced as possible techniques for studying the occupancy detection (Dong et al. 2010). The autocorrelation function was applied by Szczurek et al. (2016) to examine the regularities in the CO_2 concentration and to identify the pattern related to each occupancy profile. They also found that the CO_2 variation corresponding to a fixed occupancy profile could be quite different in the same building. Hence, local receptive fields (LRF) with random weights were adopted as feature learning by Zhu et al. (2017), and later, a classifier was trained for division of the occupancy level. Díaz and Jiménez (2017) reported the trend of CO_2 concentration in the office building and compared different ways to estimate occupancy. The result indicated that the estimation errors for occupancy were higher using moving averages than with raw data. Jin et al. (2015) utilised sensing by proxy to develop an occupancy-estimation algorithm. The convection effect and dispersion rate of the CO_2 concentration and indoor air were considered in their algorithm. As for the probabilistic models, simulation was conducted by generating a grey-box model based on CO_2-concentration measurement to estimate the number of occupants in a multi-room case (Ebadat et al. 2015). In the model, the physical parameters and occupancy were estimated by using the maximum likelihood and regularised deconvolution approach. Similar work was proposed and compared to black-box methods in Ebadat et al. (2013). The Bayesian approach was used by Bottegal et al. (2015a) to estimate the occupancy signal; other unknown parameters are estimated by the expectation–maximisation algorithm.

The application of the above-mentioned model in occupancy estimation requires a careful training process wherein both people-counting data and environmental parameters are necessary. Collection of comprehensive data might be expensive or infeasible, especially for small time intervals.

3.1.2 Statements of Contribution

After reviewing peer research, a few key points are identified:

- Despite the extensive application of ANNs in prediction, ANN training is time-consuming and greatly affected by model parameters. It is necessary to improve the generality and prediction performance of the fundamental ANN models.
- As one of the leading influential factors for energy consumption in buildings, the occupancy factor is usually introduced as a fixed occupancy schedule, time factor, or historical load as an input parameter of the energy-prediction model. Few studies have focused on using a dynamic occupancy profile to predict the building load with ANN. The neglect of occupancy presence and behaviour is one cause of the performance gap between the predicted and measured energy use.
- As mentioned above, some occupancy-detection models, although providing precise prediction of the indoor occupancy level, depend on people-counting data or various sensors. Collection of comprehensive data

> might be expensive or infeasible, especially for small time intervals, because installation of new sensors is difficult due to privacy issue for most buildings in operation. Most frequently found sensors are temperature, humidity and CO2 sensors. Hence, investigating the occupancy pattern from environmental parameters is a challenging but significant task. However, the occupancy-estimation algorithm developed based on the analytical method is not as accurate as other models because the identification of physical parameters is not evaluated during calculation.

Hence, the objects of this study are manifold. First, the blind system identification (BSI) models are proposed to blindly compute the number of occupants based on CO_2 concentration. Unlike the aforementioned occupancy-estimation models, the BSI models require no prior knowledge of people-counting data or extra training steps. In addition, the estimation result is more accurate than that by using the analytical method because the measured error is eliminated and prior experience parameters are considered during calculation. Second, fundamental ANN models, ELM models, and ensemble models are constructed as the prediction performance of building energy consumption. Finally, we compare the performance of these three models with the different supplementary inputs, i.e., true occupancy, estimated occupancy from BSI models, and without inputs of occupancy. Overall, this chapter focuses on bridging the gap between energy-prediction models and the dynamic occupancy profile estimated from indoor CO_2 concentration.

The primary contribution of this chapter the presentation of an integrated approach to estimate the occupancy level based on BSI and develop a prediction model of the electricity consumption of an AC system based on an ANN with the occupancy input from the BSI estimation. As shown in Fig. 3.1, the estimation model of the current occupancy has been carried out based on measured CO_2 concentration and electricity consumption of the fresh-air system at tier 1. At tier 2, the prediction models are established based on the ANN model to predict the electricity consumption of the AC system at the next time step. The superiority of the ANN model with the supplementary input of estimated current occupancy is verified by comparing the ANN model results without the input occupancy.

The paper is structured as follows. In Sect. 3.2, we identify the occupancy-estimation model based on BSI. In Sect. 3.3, we provide an introduction to the energy demand-prediction model using ELM and FFNN. The reference room and datasets and the parameter analysis are introduced in Sect. 3.4 and Sect. 3.5, respectively. In Sect. 3.6, we discuss the results of the energy-consumption estimation with and without occupancy estimation as an input parameter. Finally, in Sect. 3.7, conclusions are made.

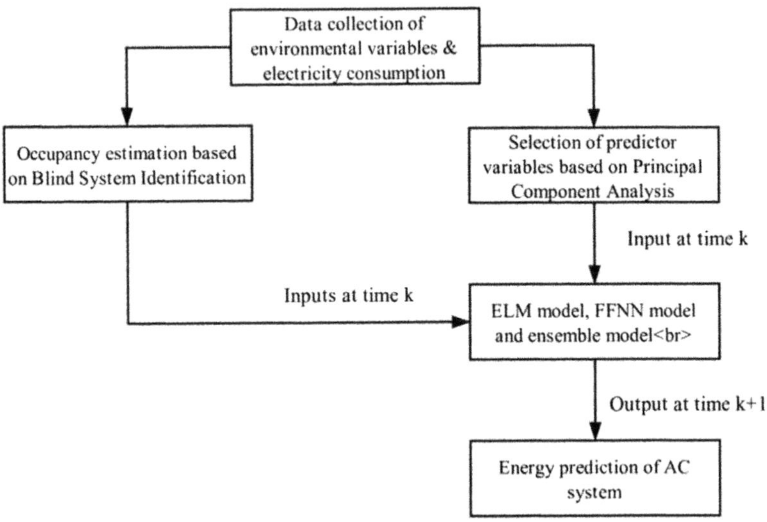

Fig. 3.1 Outline of the research methodology

3.2 Occupancy Estimation Methodologies

BSI is a technology aimed at retrieving a system's unknown information from its outputs only. The word blind means that the system's inputs are not available for analysis. The task of BSI is to identify the inputs or the system function from outputs as well as other available data obtained from the system (Abed-Meraim et al. 1997).

Usually, the occupant number can be roughly estimated from the mass-conservation equation of CO_2 concentration based on the analytical method. In this chapter, however, the measured error term and prior experience of parameters are considered in our model. The calculation process can be regarded as an optimisation process that is repeated iteratively until an accurate solution is found for optimisation. The superiority of BSI is that the calculations are completed automatically without parameter tuning or training process. In addition, the estimation result is more accurate than that of the analytical method because the measured error term and prior experience of parameters are considered during calculation (Abed-Meraim et al. 1997). Among different estimation approaches, ML and Bayesian estimation are adopted to the BSI problems.

3.2.1 Frequentist Maximum Likelihood (ML) Approach

The variation of indoor CO_2 concentration was, based on the mass-conservation law, mathematically derived as in Ebadat et al. (2013). We discretise the continuous-time model by standard backward Euler discretisation, and the indoor measured CO_2 concentration can be expressed as

$$\frac{y(k)-y(k-1)}{T}=\frac{\dot{Q}^{vent}(k)+\dot{Q}^{leak}(k)}{V}(C-y(k-1))+\frac{g}{V}O(k-1)+e(k-1), \tag{3.1}$$

where y is the measured CO_2 concentration, T is the sampling time, and k is the discrete time domain. $\dot{Q}^{vent}(k)$ is the supply fresh air rate, $\dot{Q}^{leak}(k)$ is the air leakage, C is the outdoor air CO_2 concentration and assumed equal to 450 ppm, and V is the volume of considered room.g is the CO_2-generation rate per person and assumed to be 0.005 L/s, O(t) is the number of occupants at time t, and $O(t) \in N_+$.e(t) is the measurement error following the Gaussian distribution $\mathcal{N}(0,\sigma^2)$.

Then, Eq. (3.1) can be rewritten as

$$(I-a\Delta)y=(1-a)C+b_o\Delta O+e, \tag{3.2}$$

$$\mathbf{y}:=\begin{bmatrix} y(1)\\ y(2)\\ \vdots\\ y(n)\end{bmatrix},\mathbf{O}:=\begin{bmatrix} O(1)\\ O(2)\\ \vdots\\ O(n)\end{bmatrix},\mathbf{e}:=\begin{bmatrix} e(1)\\ e(2)\\ \vdots\\ e(n)\end{bmatrix}, \tag{3.3}$$

where $a=1-\frac{\dot{Q}^{vent}(k)+\dot{Q}^{leak}(k)}{V}T$, $b_o=\frac{g}{V}T$, I is the n-dimensional identity matrix, and Δ is the n-1-dimentional identify matrix.

It should be noted that Eq. (3.1) is a simplified governing equation that does not account for the changes in parameters a and b_o due to human activities with time variations. In this study, the simplification is reasonable, as the human activity of all office workers does not vary greatly. However, in other types of buildings, these non-ideal issues would need to be considered. In this model, yielding the unknown parameters, which reflect the air leakage, CO_2-generation rate per person, measurement noise, and occupancy, correspond to θ as

$$\theta := \left[ab_o\sigma^2O(1)O(2)\ldots O(n)\right]^T \tag{3.4}$$

We assume that measured CO_2 concentration described in model (1) follows the distribution function as the normal distribution $N(\mu,\sigma^2)$. As Eq. (3.2) follows the form as $\mathbf{y}=\mathbf{A}+\mathbf{B}\mathbf{e}$ and **the** measurement error **e** is assumed as $\mathcal{N}(0,\sigma^2)$, the measured CO_2 concentration **y** must also follow the normal distribution function $\mathcal{N}(\mu, \sigma^2)$, which is expressed as

$$p(y; \theta) = N(m_y, cov_y), \tag{3.5}$$

where m_y is the expected value and cov_y is the variance value, which are expressed as

$$\begin{cases} m_y = (I - a\Delta)^{-1}(1 - a)C + (I - a\Delta)^{-1} b_o \Delta O \\ cov_y = \sigma^2 (I - a\Delta)^{-1}(I - a\Delta)^{-T} \end{cases} \tag{3.6}$$

When using a normal distribution to describe a given independent vector, the log-likelihood function is adopted to estimate the unknown parameters:

$$\log L(\theta) = log(cov_y) + \frac{1}{2\sigma^2} e^T e. \tag{3.7}$$

Thus, the realisation of ML estimation can be regarded as the optimisation problem of Eq. (3.7). The execution of the operation can be carried out by iterating of following two steps: (1) assign the initial values for parameter $[a, b_o, \sigma^2]$ and calculate the occupancy O by ML estimation; (2) fix the occupancy O and recalculate the parameter $[a, b_o, \sigma^2]$. Calculation is terminated when the required minimum cost function is reached.

3.2.2 Bayesian Estimation Approach

In the Bayesian estimation, we consider the occupancy-estimation problem as a multiple-input and single-output (MISO) linear time-invariant discrete-time dynamic system, as shown in Fig. 3.2 and Eq. (3.8).

$$y_{(t)} = \sum_{i=1}^{+\infty} g_{yi} y_{(t-i)} + \sum_{i=1}^{+\infty} g_{ui} u_{(t-i)} + \sum_{i=1}^{+\infty} g_{oi} o_{(t-i)} + e_{(t)}, \tag{3.8}$$

where $\sum_{i=1}^{+\infty} g_{yi}$, $\sum_{i=1}^{+\infty} g_{ui}$ and $\sum_{i=1}^{+\infty} g_{oi}$ are the transfer functions reflecting the dynamics of the model driven by the input signals. $y_{(t)}$ is the measured CO_2 concentration, $u_{(t)}$ is the actual level of the fresh-air system, which is presented as electricity consumption in kWh. $o_{(t)}$ represents the number of occupants in the room at each time instant and $e_{(t)}$ is the zero-mean Gaussian white noise with unknown variance σ^2.

Fig. 3.2 Block scheme of the MISO model

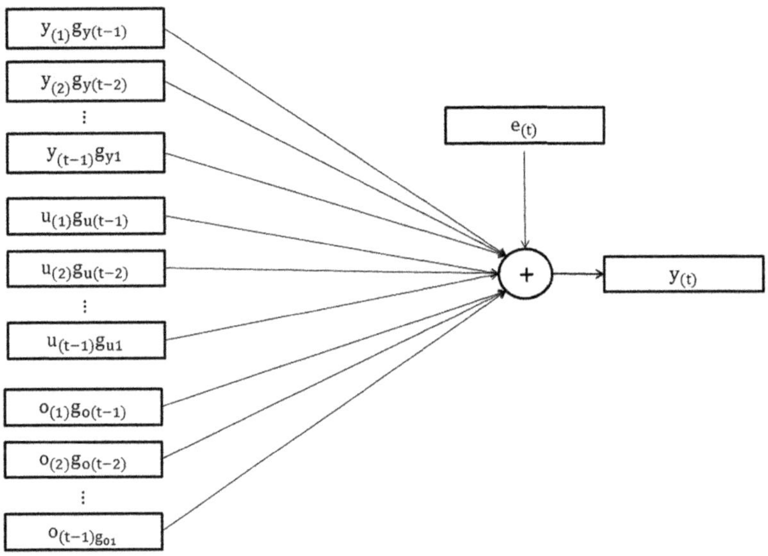

After revising all the input signals y,u, and o into the Toeplitz matrix as $T_n(y)$, $T_n(u)$, and $T_n(o)$, we reserve the symbols Y, U, and O for matrices $T_n(y)$, $T_n(u)$, and $T_n(o)$ respectively. Therefore, Eq. (3.8) can be rewritten as

$$y = Yg_y + Ug_u + Og_o + e \qquad (3.9)$$

The transfer functions g_y,g_u, and g_o follow the Gaussian distribution (Rasmussen 2004):

$$\begin{aligned} \mathbf{g}_y &\sim \mathcal{N}(0, \lambda_y K_{\beta y}), \\ \mathbf{g}_u &\sim \mathcal{N}(0, \lambda_u K_{\beta u}), \\ \mathbf{g}_o &\sim \mathcal{N}(0, \lambda_o K_{\beta o}) \end{aligned} \qquad (3.10)$$

where $K_{\beta y}$, $K_{\beta u}$, and $K_{\beta o}$ are $n \times n$ covariance matrices obeying the Weibull distribution, whose structure depends on the shaping parameters βy, βu, and βo, respectively, which regulate how fast the signal decays in the time series. βy, βu, and βo are scalars in the interval $[0, 1)$. The scaling factors λ_y, λ_u, and λ_o tune the amplitude of the responses from the corresponding input signals.

Then, the unknown parameters are defined as

$$\theta := \left[\mathbf{o}(t)\ \beta y\ \beta u\ \beta o\ \lambda_y\ \lambda_u\ \lambda_o\ \sigma^2\right]^T \qquad (3.11)$$

g_y follows the posterior distribution of a given y and θ as a Gaussian distribution, namely,

$$P(\mathbf{g}_y|y, \theta) = \mathcal{N}(C_y, P_y), \qquad (3.12)$$

where $P_y = (\frac{Y^TY}{\sigma^2} + K_{\beta y}{}^{-1})^{-1}$, $C_y = P_y \frac{Y^T}{\sigma^2}$. In Eq. (3.12), the transfer function $\mathbf{g}_y$ can be estimated as

$$\hat{\mathbf{g}}_y = E(\mathbf{g}_y|y, \theta) = C_y y \qquad (3.13)$$

Similarly, the other two transfer functions, $\mathbf{g}_u$ and $\mathbf{g}_o$, can also be derived as in Eq. (3.13). Clearly, such estimators are determined based on the function of θ.Thus, the initialisation of the unknown parameter θ is fundamental in the Bayesian estimation algorithm.

The method to estimate θ is to maximise the marginal likelihood (Labeodan et al. 2015) as follows.

$$\hat{\theta} = \arg{}_{\theta}^{max} \log p(y|\theta) \qquad (3.14)$$

However, solving the nonlinear Eq. (3.14) in that form is difficult, because **y**, **u**, and **o** and $\mathbf{g}_y$, $\mathbf{g}_u$, and $\mathbf{g}_o$ obey the Gaussian distribution. For this reason, an iterative solution is adopted by maximising the complete log-likelihood:

$$\begin{aligned} L(y, g|\theta) &= \log P(y, g|\theta) \\ &= \log P(y|g, \theta) + \log P(g|\theta). \end{aligned} \qquad (3.15)$$

The expansion of Eq. (3.13) is composed of eight components, leading to differentiable equations to update $\hat{\theta}^{k+1}$ from $\hat{\theta}^{k}$. For relevant calculation details, refer to Bottegal et al. (2015b).

3.2.3 Evaluation Criterion

The performance of BSI for occupancy estimation described in the previous section is evaluated. To assess the performance of the proposed models of the occupancy estimators, we consider the normalised root-mean-square error (NRMSE) (Yang et al. 2014) as the performance index, defined as

$$NRMSE = \frac{\sqrt{\frac{\|\hat{o} - o\|_2^2}{T}}}{o_{max} - o_{min}}, \qquad (3.16)$$

where $\hat{o} \in \mathbb{R}^n$ is the estimated number of occupants in n sampling time, o represents the true occupant count, $\|\cdot\|_2^2$ is the l_2-norm, and T is the number of samples per day. However, the estimation error would be underestimated when using abovementioned NRMSE in this case study because the maximum occupant number o_{max} is relatively large compared with other studies. Hence, NRMSE adopted in this chapter is revised as:

$$\mathrm{NRMSE}^* = \frac{\sqrt{\frac{\hat{\mathbf{o}}-\mathbf{o}_2^2}{\mathrm{T}}}}{\mathbf{o}_{\mathrm{mean}}}, \quad (3.17)$$

where o_{mean} is the mean value of occupants number.

Because the measured CO_2 concentration with spikes cannot correctly reflect the real environmental situation of the whole room, data smoothing for CO_2 concentration is conducted as important pre-treatment before calculating the number of occupants. Three methods have been adopted to smooth the CO_2 concentration, including two-hour moving average, two-hour bin, and globally smooth (Cho and Lee 2016).

3.3 Energy-Consumption Prediction Methodologies

Energy-consumption prediction is an important research topic in building energy management, which means to forecast future energy demand based on historical load data, weather data, social factors, etc. It is well recognised that the prediction of energy consumption in the long, medium, and short term are important for energy market planning and investments. In particular, a very-short-term (hours or minutes ahead) estimation of electricity demand is prerequisite to the anticipatory control model, which is able to improve the system performance of buildings. Therefore, a precise prediction in these scenarios would lead to more efficient energy management and result in considerable reduction in the operational cost for both energy suppliers and end users in buildings.

In this study, FFNN and ELM are the two neural networks used as very-short-term models to predict the electricity consumption of the AC system installed in the reference room. The structures of these neural networks are described in this section.

3.3.1 Architecture of FFNN Model

FFNNs were designed based on the biological nervous system, which deals with complex external things (Meyn et al. 2009). The network consists of a large number of neurons arrayed in layers, where the neurons in different layers are connected. Figure 3.3 (Yan and Yao 2010) shows a representative model that is used to map the relationship between the input and output. The whole problem can be described to find the function $F(F : x \rightarrow z)$, which estimates the output z based on input x.

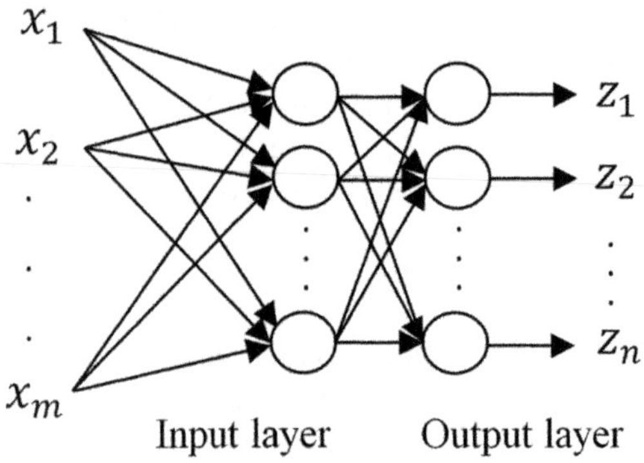

Fig. 3.3 The architecture of FFNN (Wei et al. 2018)

The training process of FFNN contains two stages. In the first stage, a set of inputs $x_{il}(i = 1, 2, \ldots\ldots, m)$ and the desired output $z_q(q = 1, 2, \ldots\ldots, n)$ are used. The summed output I_l of neuron l in the FFNN is determined by using Eq. (3.18).

$$I_l = f\left(\sum_{i=1}^{m} w_{il}x_i + b_l\right), \quad (3.18)$$

where b_l is a bias set specifically for each neuron, $w_{il}(i = 1, 2, \ldots\ldots, m)$ is the weight, and f denotes the activation function, commonly defined as the sigmoid function described by Eq. (3.19).

$$f(x) = \frac{1}{1 + e^{-x}} \quad (3.19)$$

The output I_l will be used as an input signal for the neurons in the next layer connecting to this neuron l. The input signals are transmitted from the input layer to the output layer through the network to determine the network outputs. In the second stage, the network output of network

$\widehat{z}_q$ is compared with the desired output, i.e., z_q. The comparison error is used to adjust the weights and biases during the back-propagation. Naturally, the sum of the error can be reduced after the iterative training process.

In this study, we use an FFNN with one hidden layers of neurons and a single linear output to predict the electricity consumption of an AC system. The number of hidden neurons is determined when the accuracy is satisfied as changing the structure of the FFNN. The activation function is a sigmoid function for all layers. During the back-propagation training process, the gradient-descent algorithm has been used to decrease the sum-of-the-squares of error to tune the weights and biases.

3.3.2 Architecture of ELM Model

As mentioned before, it is known that fundamental ANN faces some challenging issues, such as slow learning speed and necessity of human intervention (Huang et al. 2011). Compared with those traditional computational intelligence techniques, ELM provides better generalisation performance at a much faster learning speed and with less intervention. Unlike fundamental ANN, ELM not only tends to reach the smallest training error, but also the smallest norm of the output weights. According to the neural network theory (Bartlett 1997), the generalisation performance of the model would be improved when the smaller training error and smaller norm of weights are both considered.

ELM is a tool of learning algorithm for the single-layer FFNN architecture (Huang et al. 2006). The essence of ELM is that the hidden layer need not be tuned. Therefore, this algorithm requires less calculation time compared to the traditional FFNN because it determines the network weights and minimises the sum of the error simultaneously without iterative training. In the ELM, the least-squares method is adopted to optimise the output weighting matrix β.

A set of inputs $x_i (i = 1, 2, \ldots\ldots, m)$ corresponds to a known representation $\sigma_k(x_i)$. The desired output is represented as $z_q (q = 1, 2, \ldots\ldots, n)$. The estimated output $\widehat{z}_q$ is expected to be calculated to minimise the estimation error to zero. The process can be expressed as

$$\sum_{q=1}^{n} \hat{z}_q - z_q = 0. \tag{3.20}$$

The problem can be described as

$$\sigma_k(x_i)\beta = z_q \tag{3.21}$$

where $\sigma_k(x_i)$ is the input matrix of the ELM, which is represented as

$$\sigma_k(x_i) = [\sigma_k(x_1), \sigma_k(x_2)\ldots\ldots\sigma_k(x_m)] \begin{pmatrix} f(b_1^k + w_1^k f(\ldots w_{1j}^2 f(b_1^1 + w_1^1\, x_1))) \\ f(b_1^k + w_{1j}^k f(\ldots w_{1j}^2 f(b_1^1 + w_1^1\, x_2))) \\ \vdots \\ f(b_1^k + w_1^k f(\ldots w_{1j}^2 f(b_1^1 + w_1^1\, x_m))) \end{pmatrix}_{m\times n}^{T}, \tag{3.22}$$

$$\beta = [\beta_1, \beta_2 \ldots\ldots \beta_m]^T, \tag{3.23}$$

$$z_q = [z_1, z_2 \ldots\ldots z_n]^T. \tag{3.24}$$

According to matrix theory (Li et al. 2005), the optimal matrix β in Eq. (3.21) is derived as

$$\beta = \sigma_k(x_i)^+ z_q \tag{3.25}$$

where $\sigma_k(x_i)^+$ is the Moore–Penrose generalised inverse of $\sigma_k(x_i)$.

Hence, the extreme learning of the weighting matrix takes place of the iterative training process of traditional back-propagation neural networks, such like FFNN. ELM tends to obtain the least training error at once. Ease of use, compatibility with various types of activation functions, fast calculating speed, and superior performance are the advantages of this algorithm.

3.3.3 Architecture of Ensemble Model

Engineering problems such as energy-use prediction is perhaps difficult for a single neural network. An ensemble of results from two neural networks is conducted for the possible improvement of prediction accuracy. In this research, the neural-network ensemble is developed based on the average value of model outputs from FFNN and ELM. In Fig. 3.4, the neural-network ensemble in this study is shown.

3.3.4 Evaluation Criteria

To evaluate the obtained results from prediction model and compare the performance of different FFNN models, the coefficient of determination (R^2), root-mean-square error (RMSE), and mean absolute percentage error (MAPE) are usually adopted as evaluating criteria. These coefficients are defined as

$$R^2 = \frac{\sum_{q=i}^{n}(\hat{y}_q - y_q)^2}{\sum_{q=i}^{n}(y_q)^2}, \quad (3.26)$$

$$\mathrm{RMSE} = \sqrt{\frac{\sum_{q=i}^{n}(\hat{y}_q - y_q)^2}{n}} \times 100, \quad (3.27)$$

$$\mathrm{MAPE} = \frac{1}{n}\sum_{q=1}^{n}\left|\frac{\hat{y}_q - y_q}{y_q}\right| \times 100\%. \quad (3.28)$$

In addition to MAPE, defined by Eq. (3.28), two other indices are adopted for evaluate the model performance: $\mathrm{MAPE}_{\mathrm{peak}}$ and $\mathrm{MAPE}_{\mathrm{simple-peak}}$.

$$\mathrm{MAPE}_{\mathrm{peak}} = \left|\frac{L_{mpl} - \hat{L}_{mpl}}{L_{mpl}}\right| \times 100, \quad (3.29)$$

where L_{mpl} and $\hat{L}_{mpl}$ are the actual and predicted electricity consumption of the AC system. $\mathrm{MAPE}_{\mathrm{peak}}$ is introduced to locate the actual daily consumption peak and its occurrence time, and to compare the magnitude to the predicted value at the same time. The relative error is calculated as the performance index.

Then, the $\mathrm{MAPE}_{\mathrm{simple-peak}}$ is adopted to compare the daily peak value of the actual and predicted electricity consumption, without considering the occurrence time, defined as

$$\mathrm{MAPE}_{\mathrm{peak}} = \left|\frac{L_{opl} - \hat{L}_{ppl}}{L_{opl}}\right| \times 100, \quad (3.30)$$

where L_{opl} and $\hat{L}_{ppl}$ are the peak values of the actual and predicted electricity consumption of the AC system.

3.4 Reference Room and Dataset

The reference room, in this investigation, is located in a commercial building in Beijing. The performance of the occupancy-estimation model and that of the energy-prediction model are evaluated using a 36-seat office room with a floor area of 152 m^2 and height of 2.9 m, as depicted in Fig. 3.5. Ventilation of the office room is provided by an independent fresh-air system combined with a variable-refrigerant-volume (VRV) AC system. According to the design standard, the indoor temperature and relative

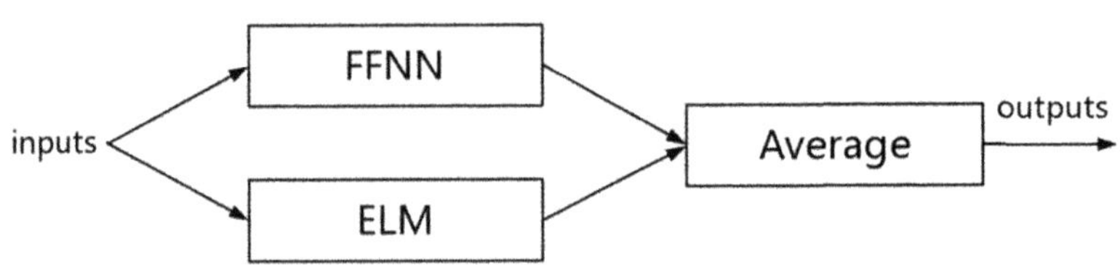

Fig. 3.4 The neural-network ensemble structure

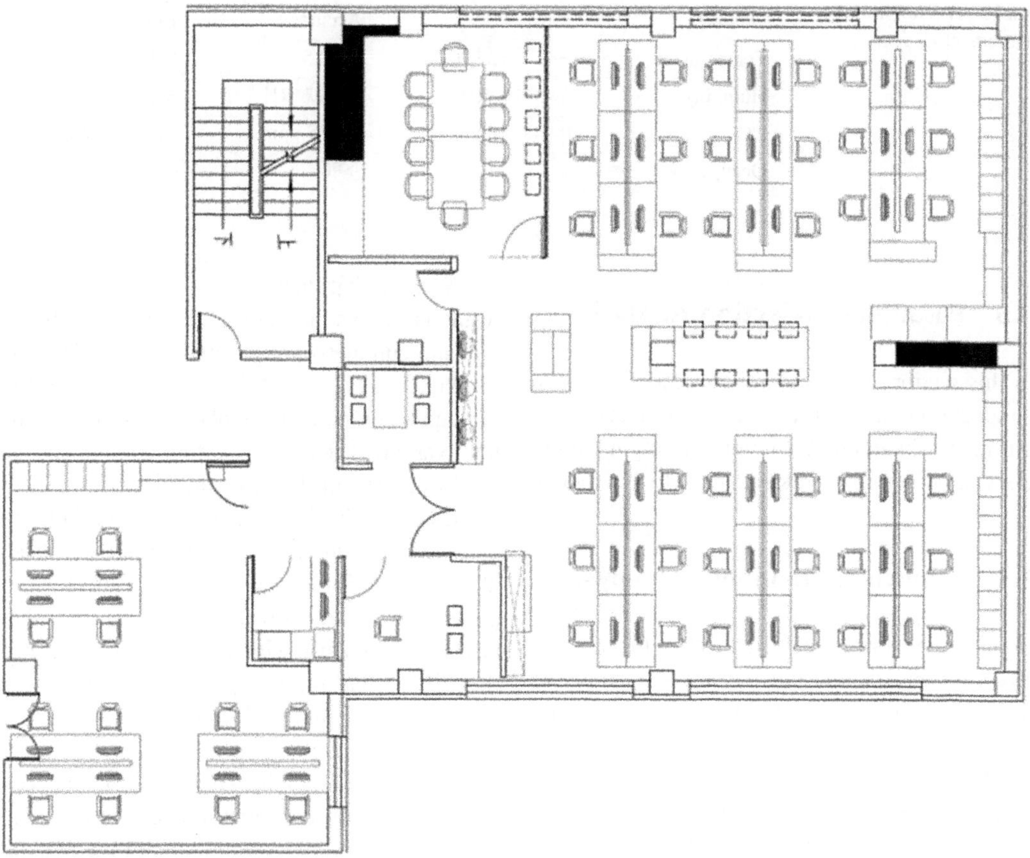

Fig. 3.5 Reference office room for model test

humidity are required to be maintained within the range of 26–28 °C and 50–65%, respectively. The occupancy schedule on weekdays is from 9:00 a.m. to 17:00 p.m., Monday–Friday. The door is kept closed most of time due to the entrance guard system.

Real-time measurements of the CO_2 concentration, indoor temperature, and indoor relative humidity are continuously recorded and fed into a database. In addition to thermal-comfort-related parameters, additional measurements, including the energy consumption of the appliances, lighting, and AC system, are also recorded. In this work, we did not measure the flow rate of the fresh air, but the corresponding value is obtained indirectly from the electricity consumption of the fresh-air system. For example, when the fresh-air system is set to the medium, high, or off state, the corresponding flow rate of each state is estimated to be 400 , 700, and $0\text{m}^3/\text{h}$, respectively. The climatic parameters are obtained from meteorological data acquisition system. The data set covers the period of 06/2017–09/2017 with a resolution time of 10 min. Table 3.1 presents the main features of the monitoring sensors.

The output value of sigmoid function is 1.0 when the input is a large negative number and 0.0 for a large positive input (Krarti et al. 1998). All inputs and outputs are therefore normalised to the interval [0,1] and fed into the FFNN. This is done to give all inputs an equal chance for contribution in model development.

Table 3.1 Monitoring equipment

Parameter	Range	Accuracy	Brand
Temperature	−40 – 125 °C	± 0.5 °C	SHT20
Humidity	0–100% RH	± 3% RH	SHT20
CO_2	0–5000 ppm	± 75 ppm	Telaire T6703
Occupancy	–	>90%	–

3.5 Parameter Selection Analysis

In this section, we discuss the effect of different parameter choices on the prediction accuracy of our model, including the number of input parameters and structure parameters of the FFNN model.

3.5.1 Principal Component Analysis

In this study, a full-scale site measurement on the environment parameters, energy-consumption data, and estimated number of occupants, including 12 variables, for the reference office room is conducted. In reality, it is not always possible to obtain all the variables that are collected in this study. Hence, the importance is that prediction of the energy consumption with acceptable accuracy is still achievable even if using only a few of the most influencing parameters.

In order to identify the most important factors which have large influences on the power consumption of AC system, principal component analysis (PCA) has been used to assess these factors. PCA is a popular multivariate statistical analysis (Jackson 2005) method and has been successfully adopted in various applications (Qin and Wang 2005; Platon et al. 2015). The analysis result is shown in Table 3.2. The contribution rate and cumulative contribution rate of PCA are displayed in Fig. 3.6. It is indicated that the four principal components explain 81.479% of the total variance. Hence, these four components are used to represent the raw variables.

Based on the PCA result, these 12 independent variables are listed with the most significant at the top in Table 3.3. The values of the component matrix are shown in Table 3.4. Variables having coefficients with absolute values greater than 0.7 from these top 5 components were considered significant. In this study, the former six independent variables, including electricity consumption of appliances, number of occupants, electricity consumption of lighting, solar radiation, electricity consumption of the fresh-air system, and outdoor temperature are selected as the inputs to the predication model. The reason for omitting the other six variables is that certain variables have less effect on building energy consumption and too many predictors that are highly correlated may cause a decrease in the accuracy.

Furthermore, the electricity consumption of the AC system with three time-step delays are also parameterised as inputs to predict the energy

Table 3.2 Principal component analysis results

Component	Eigenvalues	Contribution rate (%)	Cumulative contribution rate (%)
1	5.474	45.615	45.615
2	2.369	19.742	65.357
3	1.059	8.822	74.179
4	0.876	7.300	81.479
5	0.656	5.470	87.060

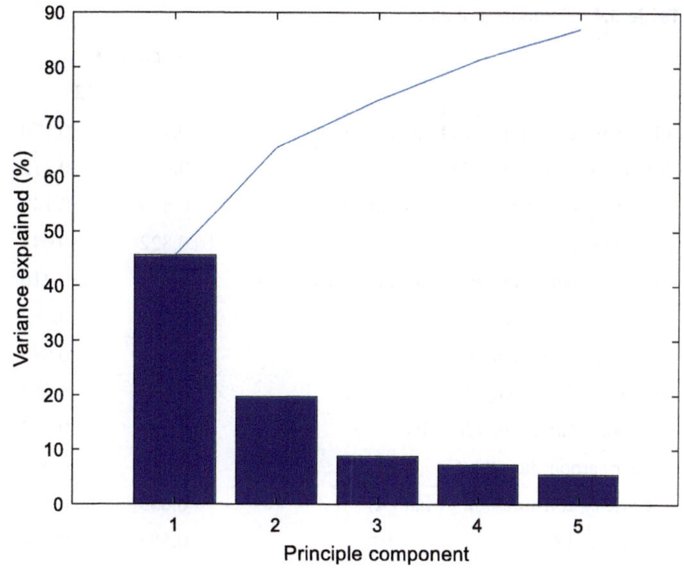

Fig. 3.6 The contribution rate and cumulative contribution rate of PCA

Table 3.3 Variable importance ranked by PCA

Variables	Unit/index
Electricity consumption of appliances (P)	kWh
Number of occupants (O)	0 – 36
Electricity consumption of lighting (L)	kWh
Solar radiation (S)	W/m^2
Electricity consumption of fresh air system (F)	%
Outdoor dry-bulb temp. (T)	°C
Indoor relative humidity (r)	%
Indoor dry-bulb temp. (b)	°C
Outdoor relative humidity (R)	%
Indoor carbon dioxide (C)	ppm
Electricity consumption of fan (A)	kWh
Wind speed (W)	m/s

consumption at the current time. Additionally, the hour of the day is coded by a sine value, as below, and fed into the energy-prediction model.

$$\text{sh} = \sin\left(\frac{\pi t}{48}\right) \quad (3.31)$$

3.5.2 Effect of Structure Parameters of FFNN Model

To save computational time, an FFNN model with fewer neurons in the hidden layer is more favourable if it can meet the requirement of accuracy. The values of the structure parameters

Table 3.4 Component matrix

	Components				
	1	2	3	4	5
Electricity consumption of appliances (P)	0.928	0.170	−0.031	0.002	0.095
Number of occupants (O)	0.925	0.212	0.032	−0.043	0.105
Electricity consumption of lighting (L)	0.859	−0.142	0.144	0.052	−0.164
Solar radiation (S)	0.822	−0.041	0.248	0.019	0.179
Electricity consumption of fresh air system (F)	0.817	−0.416	−0.086	0.039	0.090
Outdoor dry-bulb temp. (T)	0.754	0.382	−0.221	−0.277	0.109
Indoor relative humidity (r)	−0.156	0.643	0.179	0.342	0.145
Indoor dry-bulb temp. (b)	−0.556	0.552	0.147	0.219	0.150
Outdoor relative humidity (R)	−0.138	−0.543	−0.032	0.305	0.549
Indoor carbon dioxide (C)	0.458	0.626	−0.460	−0.078	0.159
Electricity consumption of fan (A)	0.653	0.082	0.669	0.061	−0.089
Wind speed (W)	0.583	−0.069	−0.363	0.632	−0.310

are identified by changing the neurons in the hidden layer.

In order to find the effect of the FFNN structure on its accuracy in prediction, we investigate the performance of different neuron sizes. We start from three-layer FFNNs, for which the number of neurons in the hidden layer increases from 10 to 160. Each network configuration has been tested by repeating 10 times to evaluate the robustness of the performance, which should eliminate the effect of the randomness of the initial setting of weights and bias on the prediction accuracy. The performance of these FFNNs is identified by using the value of MAPE after 1500 iterations. Figure 3.7 shows the value of MAPE in training (80% datasets)

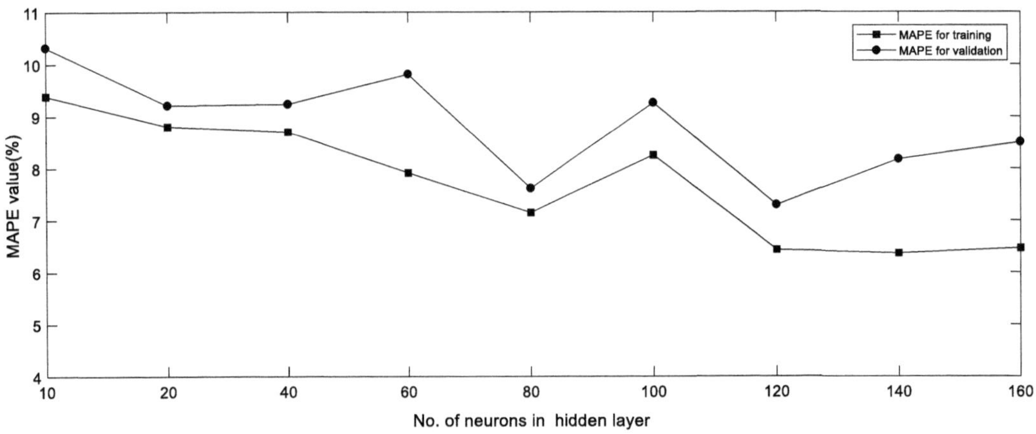

Fig. 3.7 MAPE performance with different number of neurons in the hidden layer

and validating (20% datasets) for all these FFNNs. It is shown that MAPE is the lowest (6.43% for training, 7.29% for validation) when the hidden-neuron size is 120, which is regarded as the best structure of the model.

3.6 Prediction Results

In this section, the aforementioned FFNN and ELM are applied to predict the electricity consumption of the AC system. To highlight some aspects of the benchmark test and introduce occupancy as the input of the energy-prediction model, we conduct following three studies:

(1) Occupancy-estimation results.
(2) Energy-prediction result with the input of true occupant counts.
(3) Energy-prediction result with the input of estimated occupant counts.

3.6.1 Occupancy-Estimation Results

Table 3.5 shows the NRMSE of two BSI models, where the average daily results are reported. The Bayesian estimation outperforms frequentist ML in all cases except Thursday, yielding a low error ranging from 0.177 to 0.2782 in NRMSE*, whereas the NRMSE* of the frequentist ML model ranges from 0.1874 to 0.3228. Smoothing the CO_2 concentration data by the moving-average approach and globally smoothing approach can reduce the estimation errors of both BSI models. However, the binned CO_2 data fails to provide an improvement in terms of NRMSE* with Bayesian estimation.

The visualisation comparisons of the occupancy estimated by two BSI models are presented in Fig. 3.8. Figure 3.8 indicates that the two models have satisfactory accuracy. It also needs to be mentioned that there are several spikes in the true occupancy that are not identified by either BSI model. These spikes represent a sudden change in the number of occupants in short time, such as stopping by or temporary departures. In addition, the indoor CO_2 concentration is a cumulative parameter with time lags; it is difficult to track every spike in the occupancy profile. As for Monday and Thursday, the estimated profiles present the smoothed version of true occupancy. Furthermore, the two BSI models fail to track the time of departure on Monday evening, because occupants may work overtime occasionally.

Because the experimental data reported in the literature is not available and the metering

Table 3.5 NRMSE of frequentist ML and Bayesian estimation

Methods	CO_2 concentration data	NRMSE					Average
		Mo	Tu	We	Th	Fr	
Frequentist ML	Raw data	0.3026	0.2210	0.3028	0.2282	0.2368	0.2582
	Moving average data	0.2798	0.1978	0.2912	0.1952	0.2324	0.2392
	Binned data	0.3228	0.2216	0.2966	0.1926	0.2156	0.2498
	Globally smoothed data	0.3124	0.2138	0.2808	0.1874	0.2086	0.2406
	Average	0.3044	0.2134	0.2928	0.2008	0.2232	0.2470
Bayesian estimation	Raw data	0.2782	0.1834	0.2046	0.2596	0.1856	0.2228
	Moving average data	0.2324	0.2060	0.2062	0.2624	0.1738	0.2160
	Binned data	0.2818	0.2202	0.2290	0.2504	0.1628	0.2288
	Globally smoothed data	0.2380	0.1950	0.2130	0.2422	0.1864	0.2250
	Average	0.2576	0.2010	0.2258	0.2536	0.1770	0.2230

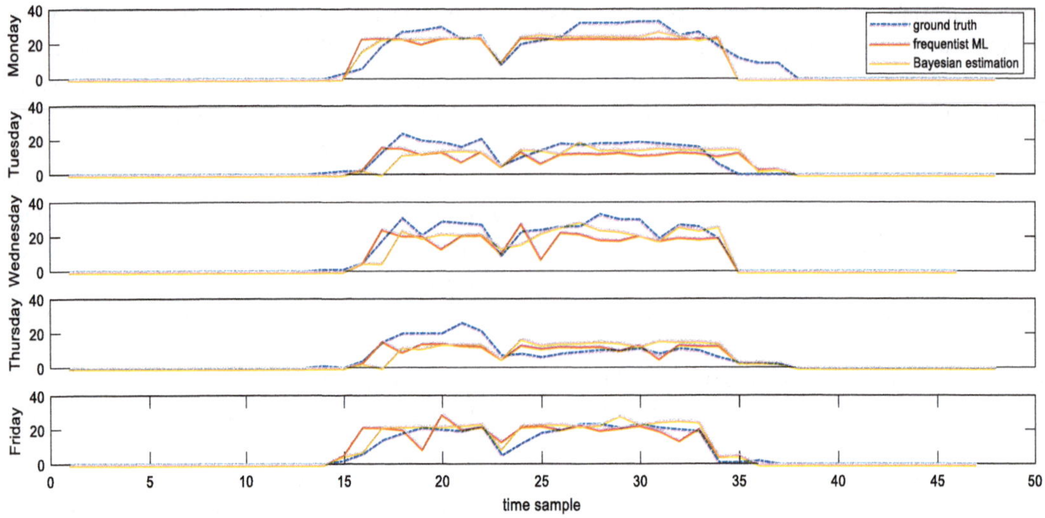

Fig. 3.8 The occupancy estimation results of the five working days by using moving-average data of CO_2 concentration

condition and equipment are not the same, it is not possible to fully compare all the models and their performance indices. Several proposed models [74, 76 and 83] report an NRMSE between 0.1912 and 0.2509, which is similar with those of the frequentist ML model (0.2470) and the Bayesian estimation (0.2230). In summary, BSI produces the accurate estimation for real-time occupancy. The dynamic occupancy level is regarded as the significant indicator that can affect the indoor load variance. Thus, the simulation result obtained from the Bayesian estimation model is adopted as the input variable to predict the electricity consumption of the AC in the room.

3.6.2 Energy-Prediction Result with True Occupant Counts

In order to investigate the prediction performance of the models with the input of true occupant counts, other models without the input of occupant number are also constructed for comparison. Additionally, the effects of PCA pre-treatment on prediction accuracy are analysed, and a total of twelve models are proposed to predict the energy consumption of the AC system in the next time-step $z(k+1)$ shown in Table 3.6. As shown in Table 3.6, the model named FFNN10 refers to the energy-prediction model with 10 inputs that adopts FFNN as the neural-network structure.

In Table 3.6, $z(k), z(k-1)$, and $z(k-2)$ are the electricity consumption of the AC system at times k, $k-1$, and k-2. $O(k)$ is the true occupant count and sh is the sine value representing the hour of the day.

Electricity consumption of the AC system and true occupancy profile during the workday is depicted in Fig. 3.9. From Fig. 3.9, the occupancy profile provides major information about electricity-consumption variances. However, the electricity consumption is not in accordance with the occupancy profile at 18:00. Thus, it further shows that occupancy cannot be regarded as the only indicator to predict energy demand; other parameters such as time indicator and electricity consumption of the appliance might be considered as indicators as well.

Tables 3.7 and 3.8 show the comparison of the performance of the proposed models, including FFNN, ELM, and the ensemble models in training datasets and validation datasets, respectively. Average results from cross validation are reported to avoid the biased condition. It

Table 3.6 Input parameters of prediction models (with true occupant counts)

Model	Input parameters	Output parameter
FFNN10, ELM10, Ensemble10	P(k), O(k), L(k), S(k), F(k), T(k), z(k), z(k − 1), z(k − 2), sh	$z(k-1)$
FFNN9, ELM9, Ensemble9	P(k), L(k), S(k), F(k), T(k), z(k), z(k − 1), z(k − 2), sh)	
FFNN15, ELM15, Ensemble15	P(k), L(k), S(k), F(k), T(k), W(k), r(k), b(k), R(k), C(k), A(k), z(k), z(k − 1), z(k − 2), sh	
FFNN16, ELM16, Ensemble16	P(k), O(k), L(k), S(k), F(k), T(k), W(k), r(k), b(k), R(k), C(k), A(k), z(k), z(k − 1), z(k − 2), sh	

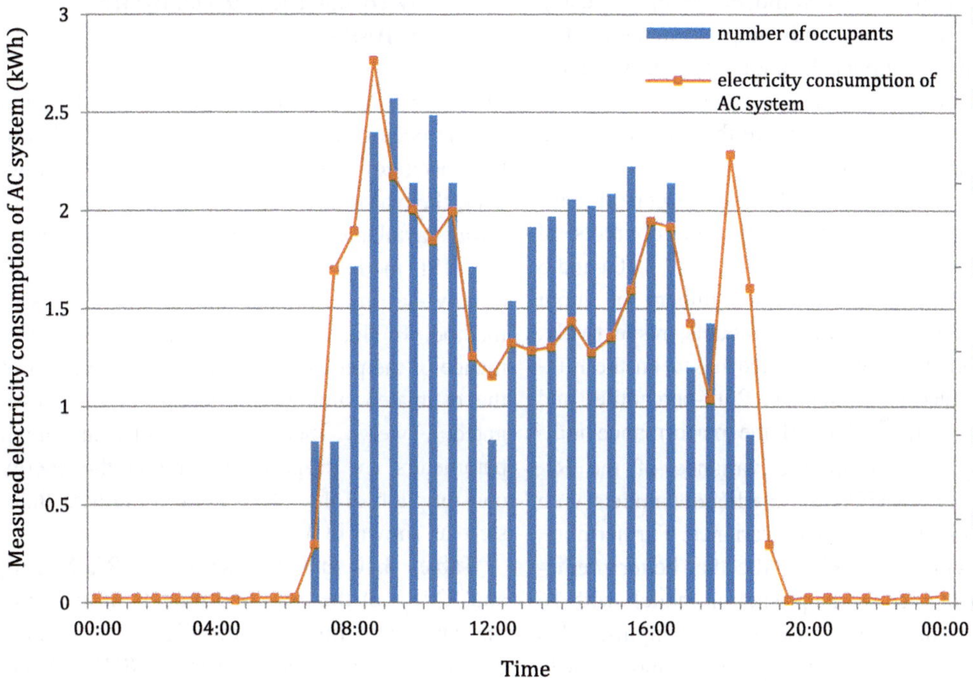

Fig. 3.9 Measured electricity consumption of AC system and true occupancy profile during workday

Table 3.7 R^2, MAPE, and RMSE of the prediction models in the training datasets

	R^2				MAPE (%)				RMSE			
Number of inputs	9	10	16	15	9	10	16	15	9	10	16	15
FFNN	0.9035	0.9463	0.9329	0.9268	8.9297	5.3017	6.4489	7.4219	4.2430	2.3001	3.3053	3.3711
ELM	0.9222	0.9773	0.9542	0.9466	7.8592	4.1950	4.4273	7.7359	3.7112	1.5749	2.6393	4.0630
Ensemble	0.9363	0.9831	0.9624	0.9048	5.7525	3.0384	3.8203	7.1314	3.2596	1.5230	2.3353	3.3028

Table 3.8 R^2, MAPE, and RMSE of the prediction models in the validation datasets

	R^2				MAPE (%)				RMSE			
Number of inputs	9	10	16	15	9	10	16	15	9	10	16	15
FFNN	0.9174	0.9398	0.9072	0.8758	7.7964	6.9675	8.6625	10.3566	3.9548	2.1590	2.7754	4.9012
ELM	0.9170	0.9501	0.9355	0.9330	7.2806	5.8652	7.7034	8.1813	3.8555	2.2792	2.9170	3.4351
Ensemble	0.9315	0.9639	0.9384	0.9231	6.6259	4.7923	5.6668	8.4668	3.4996	1.9179	2.6227	3.7502

is clear that the ensemble models generate the best prediction results and validation results with 10 inputs. The reason may be that the data of true occupant counts is necessarily close to the electricity consumption of AC system and hence improves the prediction performance. It could be seen from the result that models developed using the PCA selection slightly outperform the models using all the inputs. In addition, it is validated that the ensemble model in general can improve the prediction accuracy more than the individual neural-network model.

If we look more deeply into the performance indices of prediction models, the boxplots of Tables 3.7 and 3.8 are plotted in Fig. 3.10a and b according to the different numbers of input parameters and prediction models, respectively. The predictions are evaluated based on the minimum, maximum, median, 25th-percentile, and 75th-percentile values of the performance indices. By considering the variance of the R^2, MAPE, and RMSE, the prediction models with 10 inputs yields the least variance. Furthermore, it is obvious from Fig. 3.10a that better prediction is obtained by the prediction models with 16 inputs rather than those with 9 inputs. This means that models with 10 inputs > models with 16 inputs > models with 9 inputs > models with 15 inputs, where > indicates 'performs better than'. The results illustrate that the model performance is more sensitive to the number of occupants rather than the PCA selection. As for the performance of different prediction models shown in Fig. 3.10b, it can be easily seen that FFNN models provide the lowest R^2 value and highest MAPE and RMSE. However, it is difficult to draw the conclusion for the best-performing method between the ELM model and ensemble model from Fig. 3.10b, as ELM models provide more stable performance with smaller variances of R^2 and MAPE.

3.6.3 Energy Prediction Result with Estimated Occupant Counts

In Sect. 3.6.2, the ensemble model with the input of true occupant counts has been validated with high accuracy compared to other models. However, the real-time data of occupant counts is not always available for an office building. By using the BSI model, the number of occupants is blindly estimated through a non-intrusive method. In this section, the prediction performance of the prediction models with the input of the estimated number of occupants is investigated. Twelve models with different input parameters are proposed to predict the energy consumption of the AC system in the next time-step, as shown in Table 3.9.

Tables 3.10 and 3.11 show the R^2, MAPE, and RMSE of the proposed models in the training and validation datasets. The best training performance with the highest R^2 (0.9594) and lowest MAPE (4.7972%) and RMSE (2.3977) is obtained with the ensemble model, which outperforms FFNN and ELM in all configurations. As illustrated in Table 3.10, the overall training performance of ELM is superior to all FFNNs with different numbers of input variables. Furthermore, when using 10 variables, including the estimated number of occupants, as inputs, the FFNN, ELM, and ensemble models yield better training qualities than when using 9, 15, or 16 inputs. It is suggested in this research that using

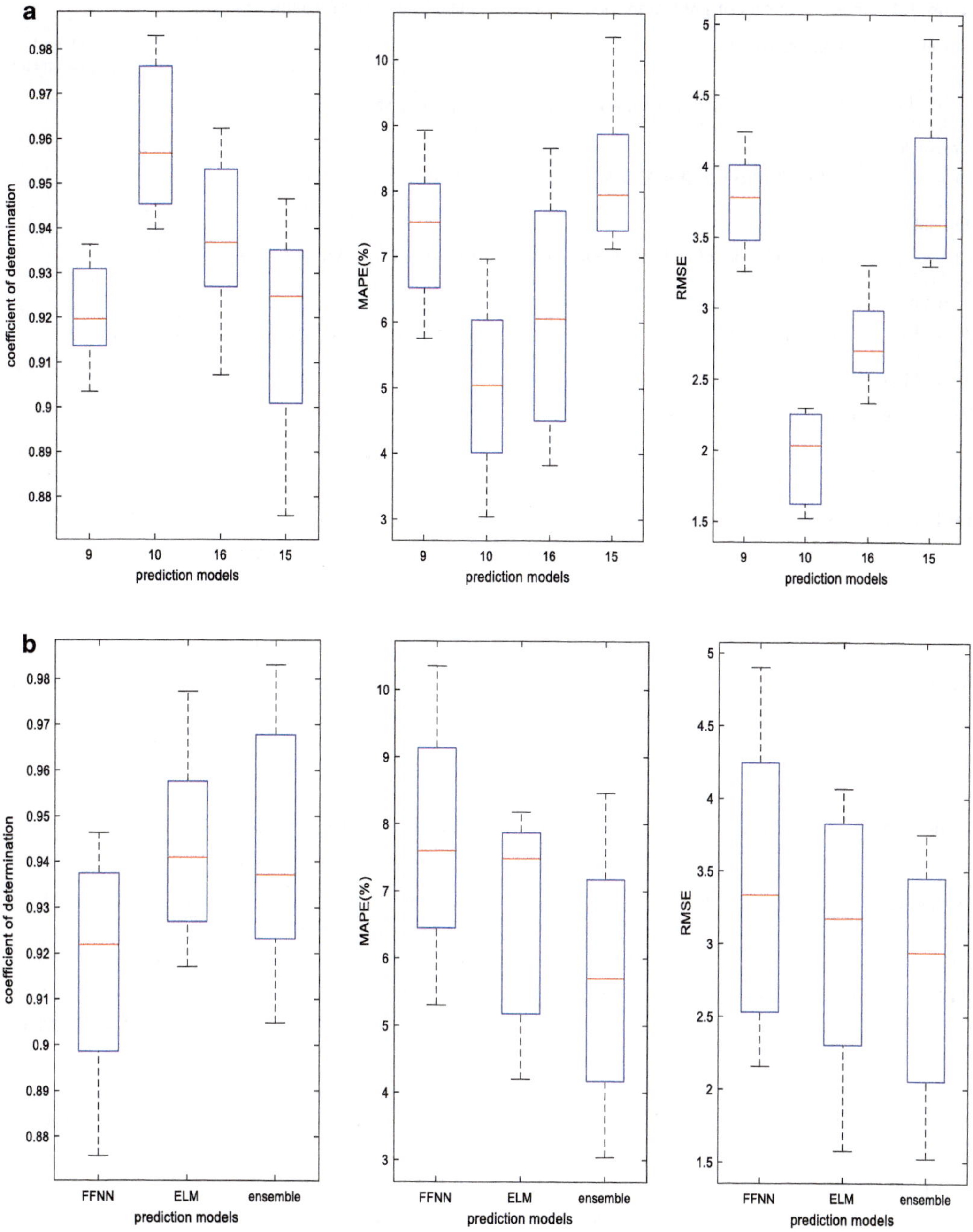

Fig. 3.10 **a** Boxplot R^2 (left), MAPE (middle), and RMSE (right) according to different number of input parameters. **b** Boxplot R^2 (left), MAPE (middle), and RMSE (right) according to different prediction models

Table 3.9 Input parameters of prediction models (with estimated number of occupants)

Model	Input parameters	Output parameter
FFNN10, ELM10, Ensemble10	$\mathrm{P(k)}, \hat{\mathrm{O}}\mathrm{(k)}, \mathrm{L(k)}, \mathrm{S(k)}, \mathrm{F(k)}, \mathrm{T(k)}, \mathrm{z(k)}, \mathrm{z(k-1)}, \mathrm{z(k-2)}, \mathrm{sh}$	$\mathrm{z(k-1)}$
FFNN9, ELM9, Ensemble9	$\mathrm{P(k)}, \mathrm{L(k)}, \mathrm{S(k)}, \mathrm{F(k)}, \mathrm{T(k)}, \mathrm{z(k)}, \mathrm{z(k-1)}, \mathrm{z(k-2)}, \mathrm{sh)}$	
FFNN15, ELM15, Ensemble15	$\mathrm{P(k)}, \mathrm{L(k)}, \mathrm{S(k)}, \mathrm{F(k)}, \mathrm{T(k)}, \mathrm{W(k)}, \mathrm{r(k)}, \mathrm{b(k)}, \mathrm{R(k)}, \mathrm{C(k)}, \mathrm{A(k)}, \mathrm{z(k)}, \mathrm{z(k-1)}, \mathrm{z(k-2)}, \mathrm{sh}$	
FFNN16, ELM16, Ensemble16	$\mathrm{P(k)}, \hat{\mathrm{O}}\mathrm{(k)}, \mathrm{L(k)}, \mathrm{S(k)}, \mathrm{F(k)}, \mathrm{T(k)}, \mathrm{W(k)}, \mathrm{r(k)}, \mathrm{b(k)}, \mathrm{R(k)}, \mathrm{C(k)}, \mathrm{A(k)}, \mathrm{z(k)}, \mathrm{z(k-1)}, \mathrm{z(k-2)}, \mathrm{sh}$	

$^*\hat{O}(k)$ is the estimated number of occupants

Table 3.10 R^2 MAPE, and RMSE of the prediction models in the training datasets

	R^2				MAPE (%)				RMSE			
Number of inputs	9	10	16	15	9	10	16	15	9	10	16	15
FFNN	0.9035	0.9350	0.9326	0.9273	8.9297	6.4329	8.0790	7.4219	4.2430	3.3526	3.2626	3.3711
ELM	0.9222	0.9682	0.9400	0.9466	7.8592	5.2422	6.3590	7.7359	3.8555	3.1633	3.2824	3.4351
Ensemble	0.9363	0.9594	0.9607	0.9048	5.7525	4.7972	6.0657	7.1314	3.2596	2.3977	2.8677	3.3028

Table 3.11 R^2 MAPE, and RMSE of the prediction models in the validating datasets

	R^2				MAPE (%)				RMSE			
Number of inputs	9	10	16	15	9	10	16	15	9	10	16	15
FFNN	0.9174	0.9224	0.8884	0.8758	7.7964	7.2981	8.7161	10.3566	3.9548	3.7841	4.5786	4.9012
ELM	0.9170	0.9412	0.9412	0.9330	7.2806	7.0690	7.8973	8.1813	3.7112	2.4239	4.4015	4.0630
Ensemble	0.9315	0.9435	0.9272	0.9231	6.6259	6.1852	7.6433	8.4668	3.4996	3.1280	3.6373	3.7502

too many predictors that are highly correlated is likely to cause a decrease in the accuracy.

In the validation dataset shown in Table 3.11, adding the estimated number of occupants as an input variable leads to higher prediction accuracy as well. For example, R^2, MAPE, and RMSE corresponding to the ensemble model with 10 input variables are 0.9435, 6.1852%, and 3.1280, respectively, whereas using only 9 input variables gives performance indices as 0.9315, 6.6259%, and 3.4996. Additionally, the prediction accuracy decreases with increasing numbers of input variables. It is shown that the predictor selection conducted by PCA plays a crucial role in the energy-prediction model. Therefore, using fewer input variables has several advantages, such as high accuracy and simple measurement. In addition, it is validated that the ensemble model can compensate for the predicting error of the single-FFNN or ELM model.

Tables 3.8 and 3.11 show the comparison of the performance of prediction models with the input of true occupant counts and the prediction models with the input of estimated number of occupants. Although the neural-network models using the estimated number of occupants as input have improved the prediction accuracy to some extent, the R^2 of the ensemble model with true occupant counts is higher (0.9639), which is strong evidence supporting that the indoor occupancy is an important factor for the electricity-consumption prediction of the AC system.

Figure 3.11a and b show an overview of R^2, MAPE, and RMSE of the prediction models in the training and testing datasets. It can be seen that the variance of the RMSE value is the largest among these three indices. When the number of inputs is relatively large, such as 15 or 16, the RMSE value of FFNN models is extremely large in testing datasets. The results show that FFNN models probably suffer an overfitting problem when the number of inputs is large.

Figure 3.12a shows the comparison results of the measured and predicted electricity consumption of AC system for training process (27 Aug. 2017 to 31 Aug. 2017). The measurement and prediction are in good agreement for the three comparative methods during the training process. However, the measured data fluctuates uncertainly during the day, especially on the mornings of the second, third, and fourth days. Hence, the absolute error between the measured and predicted values sometimes becomes large. This is because some unexpected sudden changes in input variables deteriorate the prediction performance of the models, especially when the electricity consumption becomes zero or suffers unpredictable spikes. For example, the electricity consumption of the VRV system became zero at 18:20 on the second training day, when the indoor lighting was not switched off until 18:50. Therefore, the prediction value of the electricity consumption of the VRV system fluctuated slightly and was non-zero during this half hour, because the electricity consumption of lighting was adopted as one input parameter.

Figure 3.12b shows the comparison of the measured and predicted electricity consumption of the AC system for the validation process (1 Sep. 2017). It is shown that the predicted result of the ensemble model is in the best accordance with the actual target, even though there are some fluctuations (e.g., from 11:00 to 14:00). Table 3.12 lists the actual electricity consumption, predicted electricity consumption, and predicted error in detail. The improvement is expected because the ensemble model combines outputs from different networks, which may be offset; hence, the average error, which is the error for ensemble network, is the lowest, as shown in Fig. 3.13. Hence, a better result with higher accuracy is provided by the ensemble model.

In order to demonstrate the merits of the proposed ensemble model on a more definite basis, Fig. 3.14 depicts the overall validation performance of ensemble models with various numbers of input variables. Compared to ensemble models with 9 and 10 inputs, the ensemble models with 15 and 16 inputs fail to offer highly accurate prediction of electricity consumption. Specifically, the estimated energy uses of these two models are lower than the actual measurement during the day. Although high correlation of the measured and predicted results is verified by using an ensemble model with 10 inputs, the peaks observed at samples 17, 21, and 35 are not captured effectively by this model. This suggests that unusual data spikes in the testing day that were not observed during the training days are more difficult for prediction models to manage.

As for the peak forecasting shown in Table 3.13 and Fig. 3.12b, it is evident that the peak value (0.61 kWh) of measured electricity consumption is captured effectively by the ELM model. Additionally, the improvements are more dramatic with ELM models, such that the $MAPE_{peak}/MAPE_{simple-peak}$ is the lowest, i.e., 1.9770%.

As in the case of overall performance, the ELM models are more able to learn peak behaviour than the FFNN and the ensemble models, whereas ensemble models provide best prediction performance in evaluating criteria, such as R^2, MAPE, and RMSE. In summary, the advantages of the various network models are different; the model selection should be determined case by

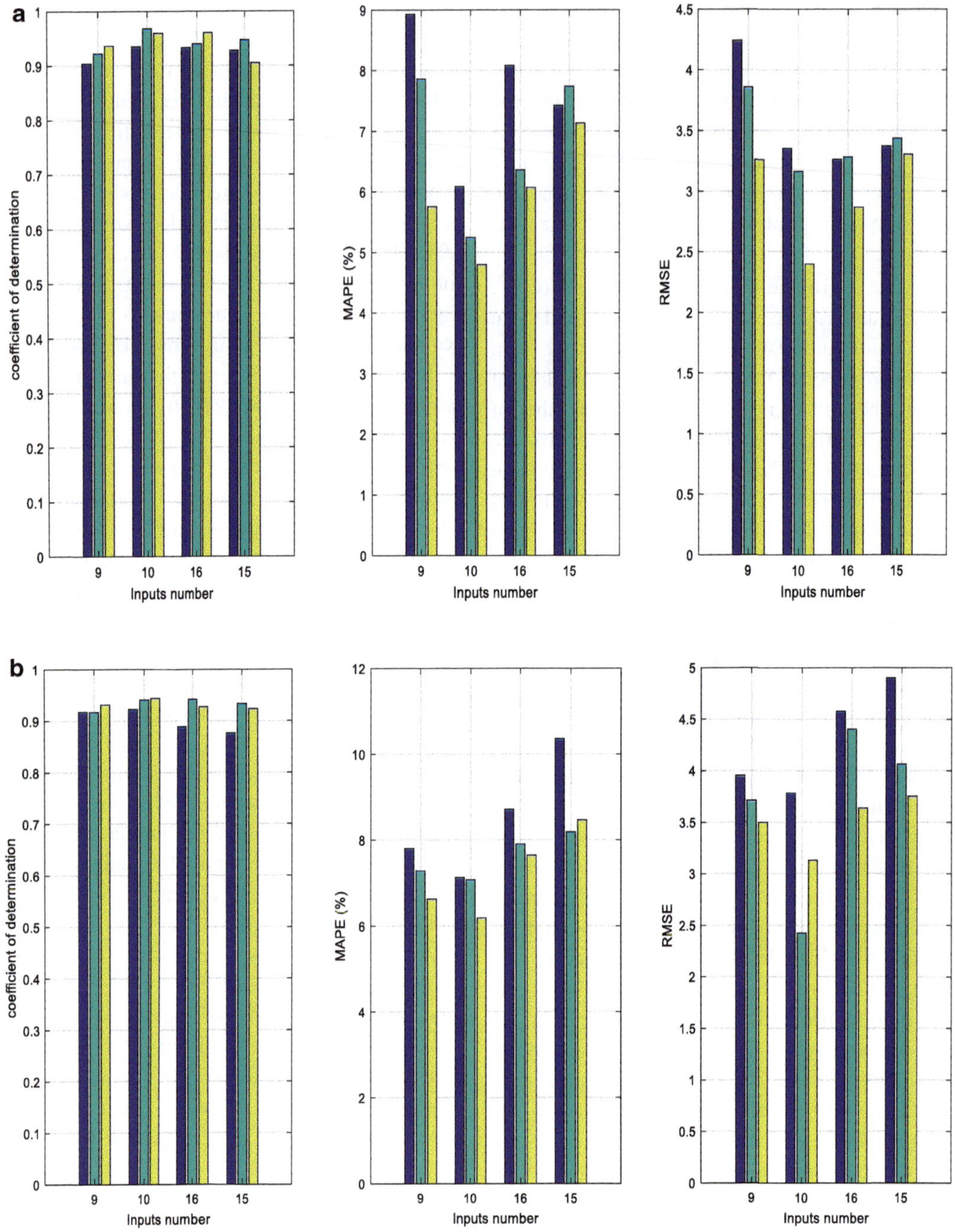

Fig. 3.11 **a** Graphic representation of R^2 (left), MAPE (middle), and RMSE (right) of the prediction models in the training datasets. **b** Graphic representation of R^2 (left), MAPE (middle), and RMSE (right) of the prediction models in the testing datasets

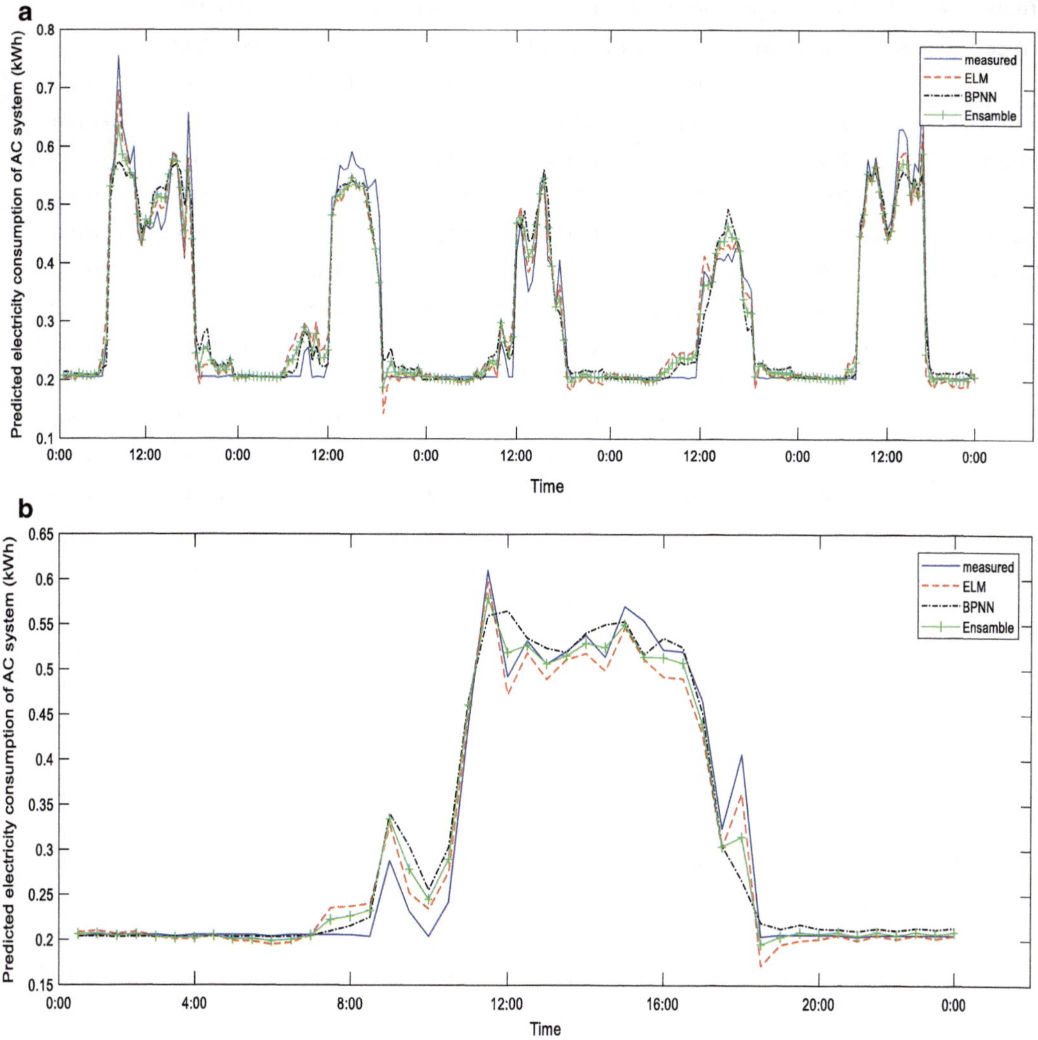

Fig. 3.12 **a** Comparison of measured and predicted electricity consumption for training process (10 input variables). **b** Comparison of measured and predicted electricity consumption for validation process (10 input variables)

case. For example, the ELM models could be used to predict the peak electricity demand. Based on the load forecasting of regional buildings, the energy market operator could determine the dispatch strategy of electricity to match the maximum supply capacity of generators and regional peak demand. In addition, the estimation of electricity usage at off-peak hours could be conducted based on ensemble models. This information is also important to guide consumers to develop their own energy-saving plans and improve the reliability of the power network at the same time.

Figure 3.15 shows the comparison of the measured and predicted electricity consumption of the twelve proposed models. It can be noted that, for ensemble models, 97.8% of the datasets have been included in the error range of ± 10%, whereas 93.5% and 86.9% of the datasets are within the range for ELM models and FFNN models, respectively. Specifically, the validation datasets of the ensemble model with 10 inputs all

Table 3.12 The actual value, predicted value, and predictor error on 1 Sep. 2017 (10 input variables)

Time	Actual value (kWh)	Predicted value (kWh)			Predicted error (kWh)		
		ELM	FFNN	Ensemble	ELM	FFNN	Ensemble
00:30	0.206	0.210	0.204	0.207	−0.004	0.002	−0.001
01:00	0.204	0.209	0.204	0.207	−0.005	0.000	−0.003
01:30	0.206	0.211	0.204	0.207	−0.005	0.002	−0.001
02:00	0.206	0.207	0.203	0.205	−0.001	0.003	0.001
02:30	0.206	0.210	0.204	0.207	−0.004	0.002	−0.001
03:00	0.206	0.204	0.203	0.204	0.002	0.003	0.002
03:30	0.204	0.201	0.204	0.203	0.003	0.000	0.001
04:00	0.206	0.202	0.203	0.203	0.004	0.003	0.003
04:30	0.206	0.206	0.204	0.205	0.000	0.002	0.001
05:00	0.206	0.200	0.203	0.201	0.006	0.003	0.005
05:30	0.206	0.199	0.204	0.201	0.007	0.002	0.005
06:00	0.204	0.195	0.203	0.199	0.009	0.001	0.005
06:30	0.206	0.198	0.204	0.201	0.008	0.002	0.005
07:00	0.206	0.206	0.205	0.205	0.000	0.001	0.001
07:30	0.206	0.236	0.210	0.223	−0.030	−0.004	−0.017
08:00	0.206	0.238	0.216	0.227	−0.032	−0.010	−0.021
08:30	0.204	0.241	0.225	0.233	−0.037	−0.021	−0.029
09:00	0.288	0.329	0.340	0.334	−0.041	−0.052	−0.046
09:30	0.232	0.252	0.304	0.278	−0.02	−0.072	−0.046
10:00	0.204	0.235	0.255	0.245	−0.031	−0.051	−0.041
11:30	0.242	0.276	0.303	0.290	−0.034	−0.061	−0.048
11:00	0.438	0.455	0.466	0.461	−0.017	−0.028	−0.023
11:30	0.610	0.598	0.560	0.579	0.012	0.050	0.031
12:00	0.492	0.473	0.565	0.519	0.019	−0.073	−0.027
12:30	0.532	0.519	0.535	0.527	0.013	−0.003	0.005
13:00	0.506	0.490	0.523	0.506	0.016	−0.017	0.000
13:30	0.520	0.511	0.519	0.515	0.009	0.001	0.005
14:00	0.538	0.518	0.540	0.529	0.020	−0.002	0.009
14:30	0.514	0.499	0.550	0.525	0.015	−0.036	−0.011
15:00	0.570	0.547	0.553	0.550	0.023	0.017	0.020
15:30	0.554	0.512	0.516	0.514	0.042	0.038	0.040
16:00	0.522	0.492	0.535	0.514	0.030	−0.013	0.008
16:30	0.520	0.490	0.525	0.507	0.030	−0.005	0.013
17:00	0.466	0.428	0.452	0.440	0.038	0.014	0.026
17:30	0.324	0.303	0.305	0.304	0.021	0.019	0.02
18:00	0.406	0.363	0.266	0.315	0.043	0.140	0.091
18:30	0.204	0.172	0.219	0.195	0.032	−0.015	0.009
19:00	0.206	0.195	0.212	0.204	0.011	−0.006	0.002
19:30	0.206	0.200	0.217	0.209	0.006	−0.011	−0.003

(continued)

Table 3.12 (continued)

Time	Actual value (kWh)	Predicted value (kWh)			Predicted error (kWh)		
		ELM	FFNN	Ensemble	ELM	FFNN	Ensemble
20:00	0.206	0.201	0.213	0.207	0.005	−0.007	−0.001
20:30	0.206	0.206	0.212	0.209	0.000	−0.006	−0.003
21:00	0.204	0.200	0.210	0.205	0.004	−0.006	−0.001
21:30	0.206	0.206	0.213	0.209	0.000	−0.007	−0.003
22:00	0.206	0.202	0.210	0.206	0.004	−0.004	0.000
22:30	0.206	0.206	0.213	0.210	0.000	−0.007	−0.004
23:00	0.206	0.202	0.212	0.207	0.004	−0.006	−0.001
23:30	0.206	0.206	0.214	0.210	0.000	−0.008	−0.004

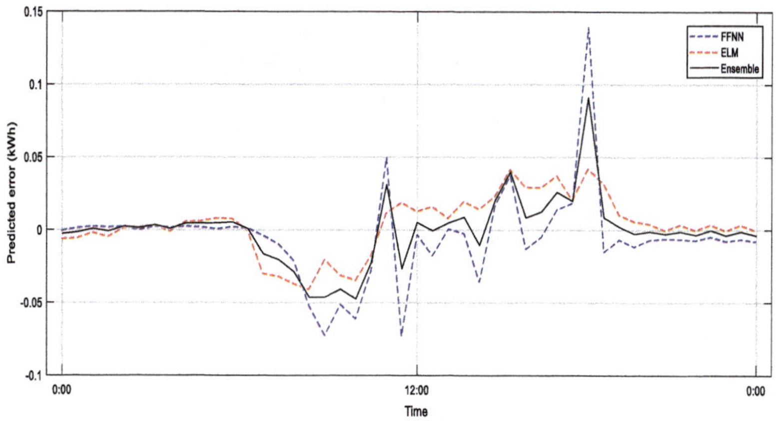

Fig. 3.13 Predicted errors of different models for validation process (10 input variables)

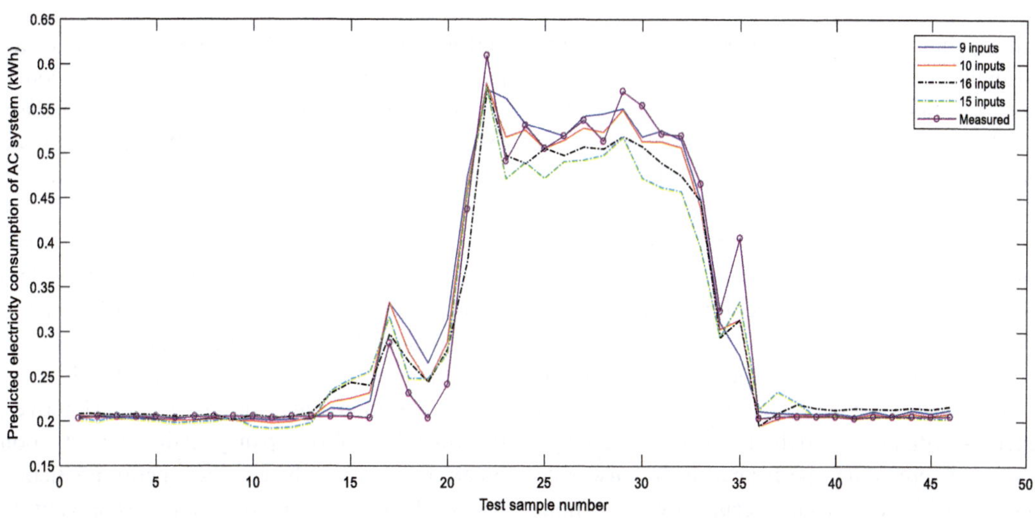

Fig. 3.14 Comparison of measured and predicted electricity consumption for ensemble models

Table 3.13 $MAPE_{peak}$ and $MAPE_{simple-peak}v$ of the prediction models in the validating datasets

	$MAPE_{peak}/MAPE_{simple-peak}$			
Number of inputs	9	10	16	15
FFNN	14.7253/10.0687	8.2606/7.4109	15.8267/15.8267	13.6686/12.2969
ELM	2.0219/2.0219	1.9770/1.9770	3.8770/3.8770	2.3503/2.3503
Ensemble	6.2041/6.2041	2.6054/2.6054	5.9748/5.9748	5.6591/5.6591

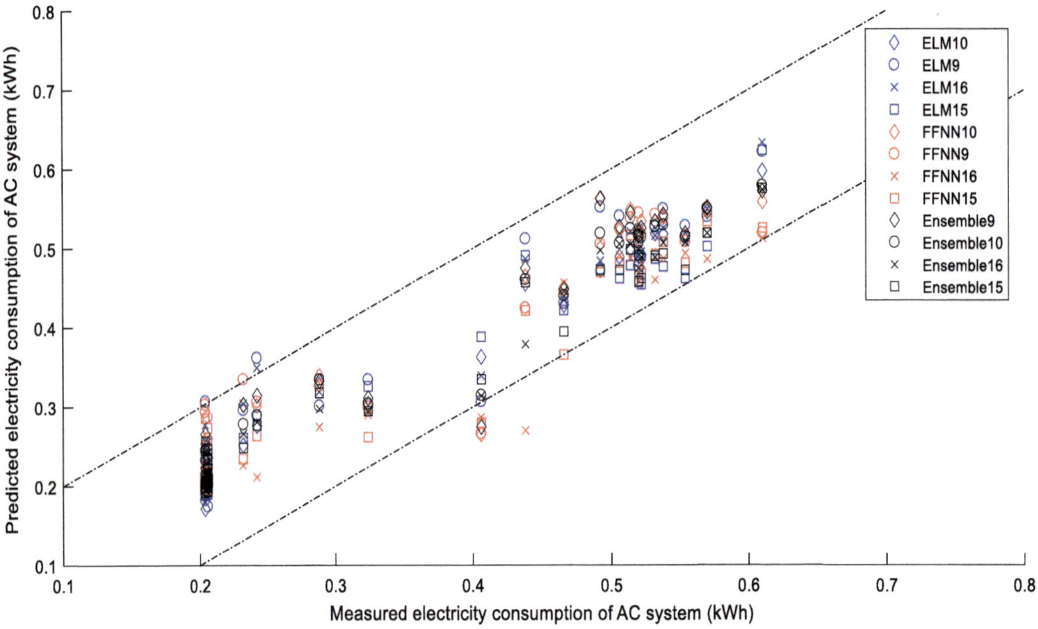

Fig. 3.15 Comparison between the measured and predicted electricity consumption of twelve models

fall within the error range. However, the ensemble models with 9, 15, and 16 inputs are not good compared to that with 10 inputs. The results show that the electricity-consumption prediction model requires the predictor selection conducted by PCA and the input of occupant numbers so as to capture the characteristic of the indoor load variance, which has significant influence on the energy consumption of the building.

In order to demonstrate the effects of estimation error of occupant counts on the accuracy of energy-prediction models in a more tangible way, the predicting performance of ensemble model with estimated occupancy as input parameter was compared to the predicting performance of ensemble model with true occupancy, which was used as benchmark model. Figure 3.16 graphically shows the comparison results for the testing day. Differences can be noticed between the ensemble models and the benchmark model at morning peak, noon break and afternoon peak (point 1, 2 and 3). Concretely, the estimated occupant number is less than true value at 9:30 (point 1) and 18:30 (point 3), corresponding energy consumption predicted by ensemble model is slightly smaller than the benchmark model and the true value. Moreover, the time lag of occupancy estimation at noon break is obvious, hence the energy-prediction result of ensemble model is not satisfying as benchmark model during lunch time.

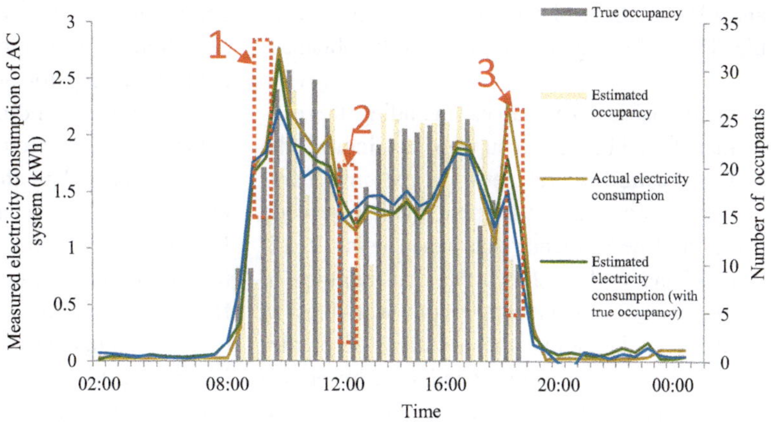

Fig. 3.16 Comparison of ensemble models (10 inputs) with true/estimated occupant counts as input parameter

3.6.4 Limitations

One limitation of this study is that the there is only one indoor sensor measuring the indoor CO_2 concentration. However, it is difficult for one sensor to capture the dynamics of the entire indoor environment. A comprehensive sensor network covering centre and corners of the room should be installed to supply sufficient information about the real-time variance of the indoor environment. The estimation result in this study is acceptable because the test-bed office is a closed and regular space, which is easily measured by a single sensor. A similar study conducted with a comprehensive sensor network may produce better results. Because the placement of sensors is vital, we shall investigate the optimal placement of sensors in our future works.

The second limitation of this study is that the application of the energy-prediction model combining the dynamic occupant profile is only validated during the summer. The model performance should be deeply investigated for a one-year period, as the occupancy interactions with the building system and energy profiles of subcomponents would vary according to the season. A comprehensive study and more elaborate approaches should be carried out to investigate the model's prediction performance.

3.7 Conclusion

In this chapter, we present an integrated approach to estimate the occupancy level based on BSI and develop a prediction model of electricity consumption of an AC system based on a neural network with an input of the occupancy determined by BSI estimation. At tier 1, the model starts from the identification of indoor CO_2 dynamics, which is derived from the mass-conservation law and venting level. The unknown parameters, including the model parameters and occupancy level, are estimated by BSI models. The calculation results show that both frequentist maximum likelihood and Bayesian estimation can provide reliable and accurate estimation in a real-time and non-iterative way.

After estimation of the occupancy, tier 2 is to establish the prediction model of the electricity consumption of the AC system using feed-forward neural network, extreme learning machine, and ensemble models. To analyse some aspects of the benchmark test for the effect of structural parameters and input-selection alternatives, three studies are conducted: (1) the effect of the predictor selection conducted by principal component analysis, (2) the effect of the estimated occupancy as the supplementary input, and (3) the effect of the neural-network ensemble. The following are the main results obtained:

- When using 10 variables, including the estimated number of occupants, as inputs, the feed-forward neural network, extreme learning machine, and ensemble models provide the best performance than when using 9, 15, or 16 inputs. It is shown that consideration of too many supplemental input variables is likely to deteriorate the accuracy of prediction. Predictor selection conducted by principal component analysis plays crucial role in the energy-prediction model.
- The consideration of extra inputs of occupancy can improve the accuracy of training and validation for the proposed models. Hence, the electricity-consumption prediction model requires the occupant number to be input so as to capture the characteristic of the indoor load variance, which has a significant influence on the energy consumption of the building.
- The best performance with the highest R^2 and lowest MAPE and RMSE can be obtained with the ensemble model, which is better than the feed-forward neural network and extreme learning machine in all neural-network structures. As for the peak forecasting, the extreme learning machine is better able to learn peak behaviour than the feed-forward neural network and ensemble models.

From this work, it is generally concluded that occupancy greatly impacts the electricity consumption of the AC system. Further optimisation of building system operation should be guided by occupant interaction. The present work is the first step in addressing a challenging problem in short-term prediction models for electricity consumption of HVAC system with occupant profile. Prediction of energy consumption of air conditioner is an important part of prediction of the total building energy consumption. Prediction of energy consumption of air conditioner offers us an index to evaluate the performance of air condition and check the operation status of air conditioner. The prediction of the energy consumption is also applied in designing optimal control strategy for air conditioner, contributing to reduction of building energy consumption and environment protection. In future work, the results obtained in this study can be beneficial toward developing a predictive controller of HVAC systems for energy conservation and thermal comfort. Compared with the traditional feedback controller, the predictive controller is designed to maintain the indoor temperature with lower thermal violations and minimizing the energy consumption. Although the time step of current prediction model is 10 min, the selection of control period for improving the operating performance of HVAC system would be investigated in the future work. As for energy usage in large-scale buildings, utility companies could apply demand-side management measures after extracting load forecasting to reach a proper load-shape objective, such as load shifting and strategic conservation. Hence, dedicated energy-prediction models with consideration of occupancy provide an opportunity to couple the electric grid and the building's control actions, and to be utilised by buildings and utility companies to simultaneously optimise their performance.

References

Abed-Meraim K, Qiu W, Hua Y (1997) Blind system identification. Proc IEEE 85(8):1310–1322

Afram A, Janabi-Sharifi F, Fung AS, Raahemifar K (2017) Artificial neural network (ANN) based model predictive control (MPC) and optimization of HVAC systems: a state of the art review and case study of a residential HVAC system. Energy Build 141:96–113

Ahmad AS, Hassan MY, Abdullah MP, Rahman HA, Hussin F, Abdullah H, Saidur R (2014) A review on applications of ANN and SVM for building electrical energy consumption forecasting. Renew Sustain Energy Rev 33(1):102–109

Ahn KU, Park CS (2016) Different occupant modeling approaches for building energy prediction. Energy Procedia 88:721–724

Amayri M, Arora A, Ploix S, Bandhyopadyay S, Ngo QD, Badarla VR (2016) Estimating occupancy in heterogeneous sensor environment. Energy Build 129:46–58

An N, Zhao W, Wang J, Shang D, Zhao E (2013) Using multi-output feedforward neural network with empirical mode decomposition based signal filtering for electricity demand forecasting. Energy 49:279–288

Ansanay-Alex, G (2013) Estimating occupancy using indoor carbon dioxide concentrations only in an office building: a method and qualitative assessment. In: REHVA world congress energy efficient, smart and healthy buildings (CLIMA)

Asadi E, da Silva MG, Antunes CH, Dias L, Glicksman L (2014) Multi-objective optimization for building retrofit: a model using genetic algorithm and artificial neural network and an application. Energy Build 81:444–456

Ascione F, Bianco N, De Stasio C, Mauro GM, Vanoli GP (2017) Artificial neural networks to predict energy performance and retrofit scenarios for any member of a building category: a novel approach. Energy 118:999–1017

Bartlett PL (1997) For valid generalization the size of the weights is more important than the size of the network. In: Advances in neural information processing systems, pp 134–140

Bottegal G, Risuleo RS, Hjalmarsson H (2015a) Blind system identification using kernel-based methods. IFAC PapersOnLine 48(28):466–471

Bottegal G, Risuleo RS, Hjalmarsson H (2015b) Blind system identification using kernel-based methods . Ifac Pap 48(28):466–471

Castelli M, Trujillo L, Vanneschi L, Popovič A (2015) Prediction of energy performance of residential buildings: a genetic programming approach. Energy Build 102:67–74

Chae YT, Horesh R, Hwang Y, Lee YM (2016) Artificial neural network model for forecasting sub-hourly electricity usage in commercial buildings. Energy Build 111:184–194

Chen Z, Xu J, Soh YC (2015) Modeling regular occupancy in commercial buildings using stochastic models. Energy Build 103:216–223

Cho CS, Lee S (2016) Effective five directional partial derivatives-based images moothing and a parallel structure design. IEEE Trans Image Process 25 (4):1617–1625

Cui C, Wu T, Hu M, Weir JD, Li X (2016) Short-term building energy model recommendation system: a meta-learning approach. Appl Energy 172:251–263

Cui Y, Yan Da, Hong T, Xiao C, Luo X, Zhang Qi (2017) Comparison of typical year and multiyear building simulations using a 55-year actual weather data set from China. Appl Energy 195:890–904

Deb C, Eang LS, Yang J, Santamouris M (2016) Forecasting diurnal cooling energy load for institutional buildings using artificial neural networks. Energy Build 121:284–297

Deng H, Fannon D, Eckelman MJ (2018) Predictive modeling for US commercial building energy use: a comparison of existing statistical and machine learning algorithms using CBECS microdata. Energy Build 163:34–43

Dong B, Andrews B, Lam KP, Höynck M, Zhang R, Chiou YS, Benitez D (2010) An information technology enabled sustainability test-bed (ITEST) for occupancy detection through an environmental sensing network. Energy Build 42(7):1038–1046

Dong B, Li Z, Rahman SM, Vega R (2016) A hybrid model approach for forecasting future residential electricity consumption. Energy Build 117:341–351

Dong B, Lam KP (2014) A real-time model predictive control for building heating and cooling systems based on the occupancy behaviour pattern detection and local weather forecasting. Build Simul 7(1):89–106

Du Z, Fan B, Jin X, Chi J (2014) Fault detection and diagnosis for buildings and HVAC systems using combined neural networks and subtractive clustering analysis. Build Environ 73:1–11

Díaz JA, Jiménez MJ (2017) Experimental assessment of room occupancy patterns in an office building. Comparison of different approaches based on CO_2,concentrations and computer power consumption. Appl Energy 199:121–141

Ebadat A, Bottegal G, Molinari M, Varagnolo D, Wahlberg B, Hjalmarsson H, Johansson KH (2015) Multi-room occupancy estimation through adaptive gray-box models. In: 2015 IEEE 54th annual conference on decision and control (CDC). IEEE, pp 3705–3711

Ebadat A, Bottegal G, Varagnolo D, Wahlberg B, Johansson KH (2013) Estimation of building occupancy levels through environmental signals deconvolution. In: Proceedings of the 5th ACM workshop on embedded Systems for energy-efficient buildings. ACM, pp 1–8

Feng X, Yan Da, Wang C, Sun H (2016) A preliminary research on the derivation of typical occupant behaviour based on large-scale questionnaire surveys. Energy Build 117:332–340

Fux SF, Ashouri A, Benz MJ, Guzzella L (2014) EKF based self-adaptive thermal model for a passive house. Energy Build 68:811–817

Gossard D, Lartigue B, Thellier F (2013) Multi-objective optimization of a building envelope for thermal performance using genetic algorithms and artificial neural network. Energy Build 67:253–260

Gruber M, Trüschel A, Dalenbäck JO (2014) Model-based controllers for indoor climate control in office buildings–complexity and performance evaluation. Energy Build 68:213–222

Guo Y, Li G, Chen H, Wang J, Guo M, Sun S, Hu W (2017) Optimized neural network-based fault diagnosis strategy for VRF system in heating mode using data mining. Appl Therm Eng 125:1402–1413

Günay ME (2016) Forecasting annual gross electricity demand by artificial neural networks using predicted values of socio-economic indicators and climatic conditions: case of Turkey. Energy Policy 90:92–101

Hailemariam E, Goldstein R, Attar R, Khan A (2011) Real-time occupancy detection using decision trees with multiple sensor types. In: Proceedings of the 2011 symposium on simulation for architecture and

urban design. Society for Computer Simulation International, pp 141–148
Hassan S, Khosravi A, Jaafar J(2015) Examining performance of aggregation algorithms for neural network-based electricity demand forecasting. Int J Electr Power Energy Syst 64:1098–1105
Hooshmand RA, Amooshahi H, Parastegari M (2013) A hybrid intelligent algorithm based short-term load forecasting approach. Int J Electr Power Energy Syst 45(1):313–324
Huang H, Chen L, Hu E (2015) A neural network-based multi-zone modelling approach for predictive control system design in commercial buildings. Energy Build 97:86–97
Huang GB, Wang DH, Lan Y (2011) Extreme learning machines: a survey. Int J Mach Learn Cybern 2(2): 107–122
Huang GB, Zhu QY, Siew CK (2006) Extreme learning machine: theory and applications. Neuro Comput 70:489e501
Jackson JE (2005) A user's guide to principal components. J Oper Res Soc 43(6):641–641
Jetcheva JG, Majidpour M, Chen WP (2014) Neural network model ensembles for building-level electricity load forecasts. Energy Build 84:214–223
Jiang X, Ling H, Yan J, Li B, Li Z (2013) Forecasting electrical energy consumption of equipment maintenance using neural network and particle swarm optimization. Math Probl Eng
Jin M, Bekiaris-Liberis N, Weekly K, Spanos C, Bayen A (2015) Sensing by proxy: occupancy detection based on indoor CO_2 concentration. UBICOMM 2015:14
Jovanović RŽ, Sretenović AA, Živković BD (2015) Ensemble of various neural networks for prediction of heating energy consumption. Energy Build 94:189–199
Kaytez F, Taplamacioglu MC, Cam E, Hardalac F (2015) Forecasting electricity consumption: a comparison of regression analysis, neural networks and least squares support vector machines. Int J Electr Power Energy Syst 67:431–438
Krarti M, Kreider JF, Cohen D, Curtiss P (1998) Prediction of energy saving for building retrofits using neural networks. ASME J Sol Energy Eng 120(3)
Kwok SS, Yuen RK, Lee EW (2011) An intelligent approach to assessing the effect of building occupancy on building cooling load prediction. Build Environ 46 (8):1681–1690
Kwok SS, Lee EW (2011) A study of the importance of occupancy to building cooling load in prediction by intelligent approach. Energy Convers Manag 52 (7):2555–2564
Labeodan T, Zeiler W, Boxem G, Zhao Y (2015) Occupancy measurement in commercial office buildings for demand-driven control applications—a survey and detection system evaluation. Energy Build 93:303–314
Leung MC, Norman CF, Lai LL, Chow TT (2012) The use of occupancy space electrical power demand in building cooling load prediction. Energy Build 55:151–163
Li C, Ding Z, Zhao D, Yi J, Zhang G (2017) Building energy consumption prediction: an extreme deep learning approach. Energies 10(10):1525
Li Z, Dong B (2017) A new modeling approach for short-term prediction of occupancy in residential buildings. Build Environ 121:277–290
Li Z, Han Y, Xu P (2014) Methods for benchmarking building energy consumption against its past or intended performance: an overview. Appl Energy 124(7):325–334
Li MB, Huang GB, Saratchandran P, Sundararajan N (2005) Fully complex extreme learning machine. Neurocomputing 68:306–314
Li K, Su H, Chu J (2011) Forecasting building energy consumption using neural networks and hybrid neuro-fuzzy system: A comparative study. Energy Build 43 (10):2893–2899
Liao C, Barooah P (2010) An integrated approach to occupancy modeling and estimation in commercial buildings. In: American control conference. IEEE, pp 3130–3135
Liao C, Lin Y, Barooah P (2012) Agent-based and graphical modelling of building occupancy. J Build Perform Simul 5(1):5–25
Liu D, Guan X, Du Y, Zhao Q (2013) Measuring indoor occupancy in intelligent buildings using the fusion of vision sensors. Meas Sci Technol 24(7):074023
Magoulès F, Zhao HX, Elizondo D (2013) Development of an RDP neural network for building energy consumption fault detection and diagnosis. Energy Build 62(8):133–138
Mahdavi A, Tahmasebi F (2015) Predicting people's presence in buildings: An empirically based model performance analysis. Energy Build 86:349–355
McKenna E, Krawczynski M, Thomson M (2015) Four-state domestic building occupancy model for energy demand simulations. Energy Build 96(8):30–39
Mena R, Rodríguez F, Castilla M, Arahal MR (2014) A prediction model based on neural networks for the energy consumption of a bioclimatic building. Energy Build 82:142–155
Meyn S, Surana A, Lin Y, Oggianu SM, Narayanan S, Frewen TA (2009) A sensor-utility-network method for estimation of occupancy in buildings. In: Proceedings of the 48th IEEE conference on decision and control, 2009 held jointly with the 2009 28th Chinese control conference (CDC/CCC 2009). IEEE, pp 1494–1500
Mumma S (2004) Transient occupancy ventilation by monitoring CO_2. ASHRAE IAQ Appl 5(1):21–23
Naji S, Keivani A, Shamshirband S, Alengaram UJ, Jumaat MZ, Mansor Z, Lee M (2016) Estimating building energy consumption using extreme learning machine method. Energy 97:506–516
Neto AH, Fiorelli FA (2008) Comparison between detailed model simulation and artificial neural network for forecasting building energy consumption. Energy Build 40(12):2169–2176

Osman ZH, Awad ML, Mahmoud TK (2009) Neural network based approach for short-term load forecasting. In: 2009 IEEE/PES power systems conference and exposition (PSCE '09). IEEE, pp 1-8

Page J, Robinson D, Morel N, Scartezzini JL (2008) A generalised stochastic model for the simulation of occupant presence. Energy Build 40(2):83–98

Paudel S, Elmtiri M, Kling WL, Le Corre O, Lacarrière B (2014) Pseudo dynamic transitional modeling of building heating energy demand using artificial neural network. Energy Build 70:81–93

Platon R, Dehkordi VR, Martel J (2015) Hourly prediction of a building's electricity consumption using case-based reasoning, artificial neural networks and principal component analysis. Energy Build 92(1):10–18

Powell KM, Sriprasad A, Cole WJ, Edgar TF (2014) Heating, cooling, and electrical load forecasting for a large-scale district energy system. Energy 74:877–885

Qin J, Wang S (2005) Wang. A fault detection and diagnosis strategy of VAV air-conditioning systems for improved energy and control performances. Energy Build 37(10):1035–1048

Rahman A, Srikumar V, Smith AD (2018) Predicting electricity consumption for commercial and residential buildings using deep recurrent neural networks. Appl Energy 212:372–385

Rasmussen CE (2004) Gaussian processes in machine learning

Roldán-Blay C, Escrivá-Escrivá G, Álvarez-Bel C, Roldán-Porta C, Rodríguez-García J (2013) Upgrade of an artificial neural network prediction method for electrical consumption forecasting using an hourly temperature curve model. Energy Build 60:38–46

Shao Z, Gao F, Yang SL, Yu BG (2015) A new semiparametric and EEMD based framework for mid-term electricity demand forecasting in China: hidden characteristic extraction and probability density prediction. Renew Sustain Energy Rev 52:876–889

Shi J, Yu N, Yao W (2017) Energy efficient building HVAC control algorithm with real-time occupancy prediction ☆. Energy Procedia 111:267–276

Stoppel CM, Leite F (2014) Integrating probabilistic methods for describing occupant presence with building energy simulation models. Energy Build 68 (68):99–107

Sun K, Yan D, Hong T, Guo S (2014) Stochastic modeling of overtime occupancy and its application in building energy simulation and calibration. Build Environ 79:1–12

Szczurek A, Maciejewska M, Wyłomańska A, Zimroz R, Żak G, Dolega A (2016) Detection of occupancy profile based on carbon dioxide concentration pattern matching. Measurement 93:265–271

Tahmasebi F, Mahdavi A (2015) The sensitivity of building performance simulation results to the choice of occupants' presence models: a case study. J Build Perform Simul

Turhan C, Kazanasmaz T, Uygun IE, Ekmen KE, Akkurt GG (2014) Comparative study of a building energy performance software (KEP-IYTE-ESS) and ANN-based building heat load estimation. Energy Build 85:115–125

ul Haq MA, Hassan MY, Abdullah H, Rahman HA, Abdullah MP, Hussin F, Said DM (2014) A review on lighting control technologies in commercial buildings, their performance and affecting factors. Renew Sustain Energy Rev 33:268–279

Virote J, Neves-Silva R (2012) Stochastic models for building energy prediction based on occupant behaviour assessment. Energy Build 53:183–193

Wei Y, Zhang X, Shi Y, Xia L, Pan S, Wu J, Han M, Zhao X (2018) A review of data-driven approaches for prediction and classification of building energy consumption. Renew Sustain Energy Rev 82:1027–1047

Wong SL, Wan KKW, Lam TNT (2010) Artificial neural networks for energy analysis of office buildings with daylighting. Appl Energy 87(2):551–557

Yang Z, Li N, Becerik-Gerber B, Orosz M (2014) A systematic approach to occupancy modeling in ambient sensor-rich buildings. Simulation 90(8):960–977

Yezioro A, Dong B, Leite F (2008) An applied artificial intelligence approach towards assessing building performance simulation tools. Energy Build 40(4):612–620

Yokoyama R, Wakui T, Satake R. Prediction of energy demands using neural network with model identification by global optimization. Energy Convers Manag 50(2):319–327

Zhao HX, Magoulès F (2012) A review on the prediction of building energy consumption. Renew Sustain Energy Rev 16(6):3586–3592

Zhu Q, Chen Z, Masood MK, Soh YC (2017) Occupancy estimation with environmental sensing via non-iterative LRF feature learning in time and frequency domains. Energy Build 141:125–133

Zuo W, Huang S, Sohn MD (2016). A bayesian network model for predicting the cooling load of educational facilities. IBPSA-USA J 6(1)

4 Cluster Analysis for Occupant-Behaviour Based Electricity Load Patterns in Buildings: A Case Study in Shanghai Residences

Song Pan, Da Yan, Xingxing Zhang, and Yixuan Wei

Abstract

In building performance simulation, occupant behaviour contributes to large uncertainties, which often lead to considerable discrepancies between actual energy consumption and simulation results. This chapter aims to extract occupant-behaviour related electricity load patterns using classical K-means clustering approach at the initial investigation stage. Smart-metering data from a case study in Shanghai, China, was used for the load pattern analysis. The electricity load patterns of occupants were examined on a daily/weekly/seasonal basis. According to their load patterns, occupants were categorized as (a) white-collar workers, (b) poor or older families and (c) rich or young families. The daily patterns indicated that electricity use was much more random and fluctuated over a wide range. Most households of the monitored communities consumed relatively-low electricity; the characteristic double peak with higher level of consumption in the morning and evening were only apparent in a relatively small subset of residents (mostly white-collar workers). The weekly analysis found that significant load shifting towards weekend days occurred in the poor or old family group. The electricity saving potential was greatest in the white-collar workers and the rich or young family groups. This study concludes with recommendations to stakeholders utilizing our load profiling results. The research provides a rare insight into the electricity-use-related occupant behaviours of Shanghai residents through the case study of two communities. The findings of the study are also presented in a meaningful way so that they can directly aid the decision-making of governments and other stakeholders interested in energy efficiency. The research results are also relevant to the building energy simulation community as they are derived from observations, and thus can have the potential to improve the efficiency and accuracy of numerical simulation results.

S. Pan (✉)
Beijing Key Laboratory of Green Built Environment and Energy Efficient Technology, Beijing University of Technology, Beijing, China
e-mail: pansong@bjut.edu.cn

D. Yan
Building Energy Conservation Research Center, Tsinghua University, Beijing, China
e-mail: yanda@tsinghua.edu.cn

X. Zhang
Department of Energy and Community Buildings, Dalarna University, 79188 Falun, Sweden
e-mail: xza@du.se

Y. Wei
School of Civil Engineering and Resources, University of Science & Technology Beijing, Beijing, China
e-mail: weiyixuan@ustb.edu.cn

X. Zhang (ed.), *Data-driven Analytics for Sustainable Buildings and Cities*, Sustainable Development Goals Series, https://doi.org/10.1007/978-981-16-2778-1_4

Keywords

Occupancy behavior · *K*-means cluster · Electricity · Load profile · Residential building

Nomenclature

C	Cluster
CV	Coefficient of variation
d	Euclidean distance
K	Number of clusters
N	Number of input vector
MLR	Multiple linear regression
RMSD	Root-mean-square deviation
t	Time
x	Object in a Euclidean n-space
y	Object in a Euclidean n-space

4.1 Introduction

The building sector is responsible for the largest share of energy consumption and greenhouse emissions in the world. Achieving high energy-efficiency in buildings has become a primary objective that many stakeholders strive for (Mathew et al. 2015). Apart from the advancement of energy technologies, another challenge in achieving higher energy-efficiency in buildings is influencing occupant behaviour, which is a key driver of building energy consumption. Past studies demonstrated that changes in occupant-behaviour have a large energy-saving potential during the life-cycle of buildings, i.e. about 7.4% in US (Dietz et al. 2009), 9.1% in China (Zhang et al. 2016) and 11.3% in UK (Philip et al. 2014).

Over the last decades, significant research effort has been devoted to energy load prediction and pattern profiling/classification. In this regard, IEA Annex 66 participants achieved a breakthrough in defining and predicting the occupant-behaviour related energy loads in buildings (they defined and standardized the energy impact of actions such as the adjustment of thermostats for comfort, switching lights, opening/closing windows, pulling up/down window blinds, and moving between spaces (Yan et al. 2017; Zhao et al. 2014; Wang et al. 2016; Reinhart 2004; D'Oca and Hong 2014). However, it is still a long-way to fully understand how energy is used by the end user on the demand side, since the phenomenon is complex, stochastic and requires a multi-disciplinary approach. Profiling the energy load of occupants qualitatively and quantitatively is another way to improve load prediction (Tsekouras et al. 2007; Chicco et al. 2006; Nikolaou et al. 2012; Ayodele et al. 2017; López et al. 2011). The identification of characteristic energy load patterns could make occupants aware of their energy-intensive behaviours, and provide the social-technological basis for leveraging the economic benefits and enhance the competitiveness of utility companies.

For this purpose, the application of clustering algorithms to analyze and classify the energy consumption behaviour of a building was proposed (Tsekouras et al. 2007). Owing to their effectiveness, the most common clustering methods in load profiling are the *K*-means, the self-organizing map (SOM), the minimum variance criterion (MVM), and the fuzzy C-means (FCM) (Tsekouras et al. 2007). Other less used approaches include the Hopfield neural network (López et al. 2011), the ISODATA algorithm and the Support Vector Clustering (SVC). In addition, different combinations of these algorithms are also found in past studies (Tsekouras et al. 2007).

Since load profiling in residential buildings are generally hindered by privacy concerns, most research has only focused on office and public buildings—despite the fact that residential occupant behaviour is more complex and characterized by randomness (Chen et al. 2012). Due to the recent surge of interest in residential energy consumption, the amount of energy data harvested through the growing installation of smart meters has also increased. Consequently, it is now necessary to develop valid and beneficial methods for presenting such data in meaningful ways to both the occupants and other stakeholders. This information is of great importance to developing countries, such as China, where residential occupant behaviour is more diverse and also contributes to considerable carbon emissions.

As a result, this chapter aims to fill the research gap of load profiling in residential buildings by exploring the situation in a developing country utilizing long-term electricity. At the initial investigation stage, this study characterized residential electricity load pattern utilizing the standard *K*-means clustering approach through a case study in Shanghai, China. This research presents the complex electricity behaviour of two residential communities in Shanghai in a meaningful way and provides recommendations for different stakeholders based on the findings.

Procedures for generating profiles of other energy loads—i.e. hot water, space heating, etc. —from empirical data could be developed in a similar fashion and applied to synthetic activity patterns without much modification. The activity patterns can also be applied to occupant behaviours in various types of building simulation by considering the unique energy characteristics of occupants. The research results could also be utilized to gauge the energy load pattern of a district or the entire city by appropriately weighing the different occupancy groups and superimposing the results.

4.2 Cluster Concept

Cluster analysis is defined by Han, et al. (2006) as a process of partitioning a set of observations into subsets (clusters) in a way that objects belonging to the same cluster have high similarity, while objects belonging to different clusters have low similarity. The clusters are established according to a "dissimilarity function" based on distances, so as to achieve both maximum internal cohesion within each cluster and maximum external separations among different clusters. The clustering analysis has the main objective to obtain sets of occupants and load patterns. Cluster algorithms are generally divided into six families: *K*-means, fuzzy clustering, ISODATA, hierarchical clustering, follow-the-leader clustering and self-organizing map clustering (Aldenderfer and Blashfield 1985), among which *K*-means is recognized as one of the most solid cluster analysis methods. With a large number of variables, *K*-means can be computationally faster as it doesn't require the calculation all of distances between each observation and every other observation. Since it can be written to efficiently deal with large data sets, it is efficient for initial-stage investigations.

The classical *K*-means clustering method groups a dataset of *N* input vectors to *C* clusters using an iterative procedure. Initially the weights of the *C* clusters are determined and a random selection among the *N* input vectors is made for the cluster centroids (Tsekouras et al. 2007). The estimated centroids are then used to classify occupants into clusters through Euclidean distance, expressed by

$$d(x,y) = \sqrt{(x_1 - y_1)^2 + (x_2 - y_2)^2 + \ldots + (x_n - y_n)^2} \tag{4.1}$$

where $x = (x_1, x_2, \ldots, x_n), y = (y_1, y_2, \ldots, y_n)$ are two objects in a Euclidean *n*-space.

In the next step, the Euclidean distances of each object of the centroid are recalculated in such a way that each object of the centroid is the average of the object of the load patterns within the cluster. The procedure is repeated until the stabilization of the cluster centroids. The optimal number of clusters is not known a priori, and the clustering quality depends on the value of *K* (Tsekouras et al. 2007; Anderberg 1973).

The initial selection of cluster centroids is of high importance as it influences whether convergence of the algorithm will occur to a (local) optima. Similarly, since a different number of clusters leads to different kinds of occupancy patterns, choosing an appropriate *K* value with some knowledge of the application domain is critical to obtaining meaningful results. Consequently, different centroid and *K* values should be tested to achieve the highest intra-cluster similarity and the most suitable cluster result.

4.3 Case Study of Occupant Behaviour Patterns in Residential Buildings

4.3.1 Data Source

The meteorological data for the period May 2013 to December 2015 was obtained from two public housing communities in Shanghai:

- Community A was built in 2012 and is located in Yangpu District. The 2 air-conditioning units (671 W), the washing machine, and the refrigerator (0.49kWh per day) were pre-installed. There are 40 metered households in this community with 2–3 residents and 70 m^2 floor area on average.
- Community B was built in 2013 and is located in Putuo District. Here too, the basic domestic appliances were pre-installed and the households were equipped the same appliances as Community A. There are 132 metered households in this community with two apartment configurations of 45 m^2 and 60 m^2 size and with an average of 2–3 residents per household.

According to the Köppene-Geiger climate classification, Shanghai has a humid subtropical climate with four distinct seasons. Hence, Shanghai's climate corresponds well to the rules specified in the Energy Efficiency Design Standards for Residential Buildings in the Hot Summer and Cold Winter Zone. The set-point heating temperature is about 18 °C and the heating season is from December to February. The set-point cooling temperature is around 26 °C and the cooling season is from June to August. The average outdoor temperature in July and January is 27.8 °C and 3.7 °C, respectively. Neither of the observed districts are supplied with centralized heating systems.

Considering the complexity of the energy performance of an actual building, it is difficult to separate the influence of the occupant behaviour from other factors, such as the climate, the physical characteristics of the building, and the type of the installed appliances. However, since our data is obtained from the same city, the effect of climate is expected to be weak. Likewise, since both communities were built about the same time with similar technologies, their building characteristics (such as house type, floor area, equivalent leakage areas and heat loss coefficient) can be assumed to be similar as well. Additionally, as appliances were uniformly pre-installed, their impact on the electricity loads due to their technological differences is expected to be also marginal. Consequently, our collected dataset enables us to examine the influence of end-use behaviours on energy use patterns.

4.3.2 Methodology

Our analytical approach consists of three steps (pre-processing, cluster analysis and correlation analysis), which are described in detail below.

Pre-processing, the first step, is aimed at restructuring the dataset and eliminating invalid data. First, the original data collected at 15-min intervals was processed in MATLAB and structured into hourly electricity consumption from 0:00 till 23:00 for each household. Then, the datasets were analyzed with MATLAB and 34 curves were discarded as corrupted data (due to noise and network failures). At the end, 36,318 days of data from 138 households were deemed suitable for further analysis.

The second step encompassed the *K*-means cluster analysis. Deciding the number of clusters (the K value) is of great importance, as too few clusters would affect the accuracy and too many clusters would reduce the calculation efficiency. Consequently, we tested several *K* values and selected 10 as the most reasonable for our purpose. Root-mean-square deviation (*RMSD*) was used to test the difference between the measured data and the centroid data, whereas the coefficient of variation (*CV*) was used to compare several datasets with different sample sizes.

$$RMSD = \sqrt{\sum_{t=1}^{n} \left(x_{1,t} - x_{2,t}\right)^2 / n} \tag{4.2}$$

$$CV(RMSD) = RMSD/\bar{x} \quad (4.3)$$

The initial centroids of K clusters were randomly chosen by MATLAB from the electricity consumption dataset. In Eq. (4.2), $x_{1,t}$, is the measured data and $x_{2,t}$ is the cluster centroid data for epoch t, respectively, and $\bar{x}$ is the mean value of within-cluster data. K-means clustering used *RMSD* to minimize the distance between measured data with their cluster centroids. Figure 4.1 shows *CV (RMSD)* values against different cluster number. It is worth noting that the rate of decrease in the case of *CV (RMSD)* is gradual, while the calculation time doubles as the cluster number exceeded 15. As a result, we decided to adopt a 10-cluster approach, as this configuration performed well both in terms of efficiency and accuracy (Tsekouras et al. 2007).

In step three, correlation analysis was used to find relationship between the clusters and other influencing factors (such as time of the day, week or season) and to test the influence of different factors on occupant electricity consumption patterns. When two variables did not have normal distributions, Spearman's rank correlation coefficient was applied as a nonparametric measure of rank correlation. The correlation analysis was performed with SPSS software.

4.3.3 Results and Analysis

4.3.3.1 Analysis of Daily Consumption

Figure 4.2 presents the centroids of hourly electricity consumption patterns over a day for each cluster. The 10 clusters, derived from the 138-household data, represent 10 different electricity consumption behaviours. The breakdown distribution of the ten clusters is illustrated in Fig. 4.3. C2 represents 51% of residents, followed by C9 and C8 with 23% and 6%, respectively.

Daily electricity load profiles are shown in Fig. 4.4. A double peak, with low morning and pronounce evening consumption levels, characterizes the electricity use profiles of the C1, C3 and C8 clusters—categorized as the clusters of mostly white-collar workers. They regularly use lighting and other appliances in kitchens, bathrooms and living rooms in the morning and evening, before and after office hours. In contrast, the households of C2, C5, C7 and C9—believed to be predominantly poor or older families—exhibit pronounced mid-day energy demand that extends well into the evening. These occupants demonstrate energy-conscious behaviour owing to their culture. Particularly, the load profiles of C2 and C9 (poor and/or elderly occupants) are distinguished by their relatively evenly distributed low electricity consumption levels. In addition, load curves from C4, C6 and C10 clusters—exemplifying rich and/or young families—have extremely high nighttime electricity consumption levels due to air conditioning. Differences between the 10 identified clusters are large owing to the diversity of occupant behaviours. Consequently, considerable energy-saving potential can be identified among these groups. Unlike to residential communities in developed countries—with comparable social status, climate and time of construction, which are characterized by fully

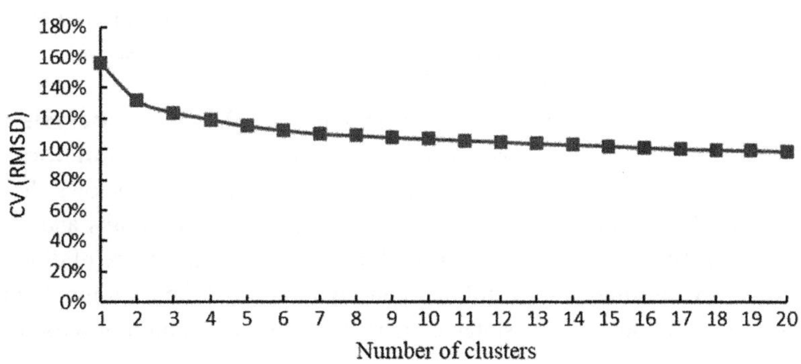

Fig. 4.1 CV(RMSD) value at different number of clusters

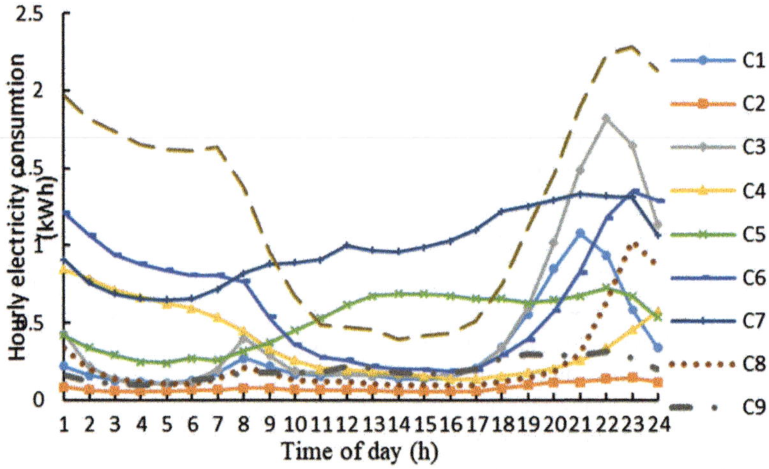

Fig. 4.2 The diurnal pattern of hourly electricity consumption as described by the centroids of the 10 identified clusters

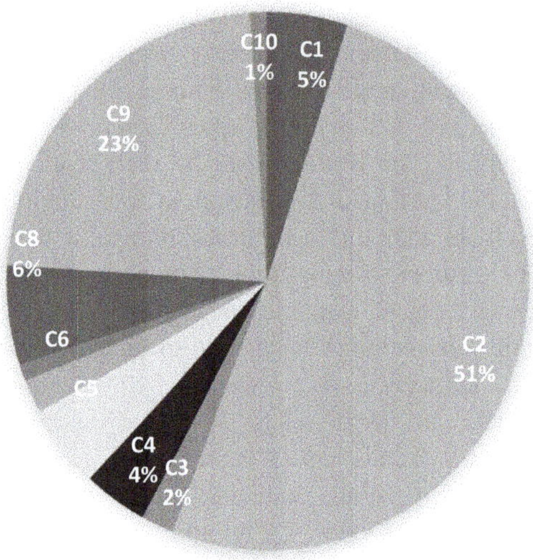

Fig. 4.3 The percentage shares of the 10 clusters

automatized control systems and full-time, full-space mode of operation—occupant behaviour is extremely random and fluctuates over a wide range in emerging economies.

As part of demand-side management (DSM) measures, a few load shaping strategies could be recommended on the basis of each clusters' diurnal load profile. Some profiles have low load values (according to Fig. 4.2), in which case load shifting, valley filling, conservation and peak clipping are suitable measures to smooth the profile shape, and hence to improve the building energy efficiency. The prolonged high daytime consumption pattern charactering C5 and C7 could be addressed both by peak clipping or conservation methods. Compared to compulsive peak clipping measures that generally impair the quality of life by reducing the thermal comfort of residents, visible smart metering could encourage occupants to improve their energy efficiency. Besides behaviour change, replacing old plug-in equipment with high efficiency appliances is also an effective DSM approach. However, when it comes to necessary loads during morning and evening periods that cannot be reduced further—as in the case of in C1, C3, C6 and C10—load shifting and valley filling might be appropriate measures. In these cases, dynamic Time-Of-Use (TOU) pricing plays a critical role in DSM, as it enables residents to optimize their energy use by selecting appropriate periods/tariffs for operating their appliances.

4.3.3.2 Analysis of Seasonal Consumption

Table 4.1 shows the correlation analysis of different months and clusters. In order to minimize the influence of different amount of days and thus data in each month and to perform a more direct evaluation of seasonal impacts, the collected data

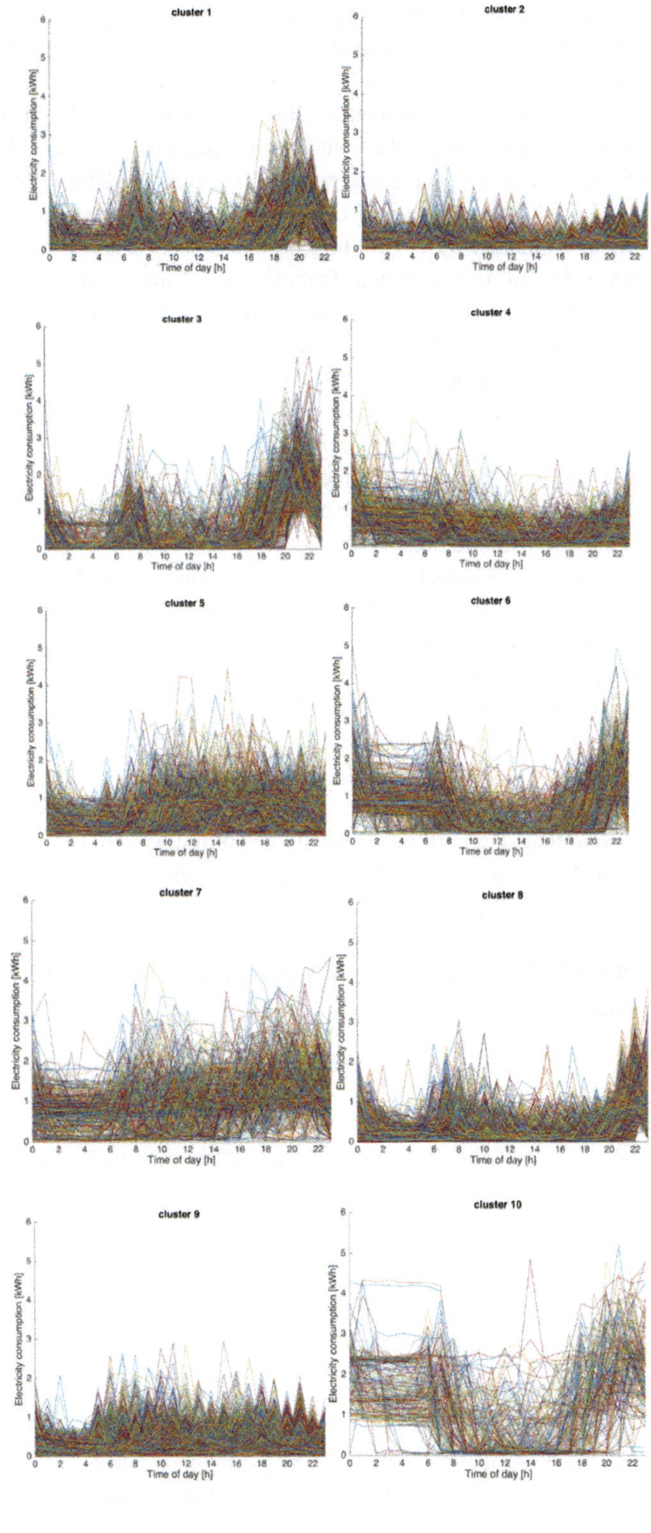

Fig. 4.4 Results of the 10 cluster analysis

was normalized to percentage values. As presented in Table 4.1, seasonal changes in electricity consumption are indicated by increased heating and cooling demands. Based on their seasonal energy usage, the consumption patterns of the 10 clusters can be categorized as: dominated by the heating period (C1, C3, C6, C7 and C10), dominated by the cooling period (C4, C5 and C8), no distinguished features (C2) and dominated by the transitional-seasons (C9). According to the clustering results, the overall electricity consumption in the two communities was slightly higher in winter than in summer.

The lifestyle and behaviour of southern Chinese residents are unique. According to a relevant survey (Hartkopf et al. 1986), it is common to use air conditioning combined with partially heating. In residential buildings, the percentage of effective heating (when the air-conditioning system is turned on) to occupancy period is around 20% in winter. In other words, residents rarely keep their AC running all the time for heating. Furthermore, nearly 85% of occupants prefer to open their windows for fresh air (Hartkopf et al. 1986). Jian's (2011) research showed that summertime electricity consumption from AC is strongly influenced by occupancy patterns. The seasonal electricity consumption within the 10 clusters ranges from less than 0.1 kWh /m^2 to 7.4 kWh /m^2. Figure 4.5 depicts each cluster's electricity load in the form of boxplots based on the seasonal energy use levels. The centroid values of the 10 clusters are regarded as reference values, while minimum and maximum consumptions are given in average values. The average seasonal electricity load in the identified clusters varies significantly, from nearly 0.01kWh to 3.5 kWh. These diverse energy consumption levels indicate great potentials for electricity saving by improved occupant behaviour, especially for clusters belonging to the white-collar workers and rich and/or young family category.

4.3.3.3 Analysis of Weekly Consumption

Table 4.2 shows the correlation analysis between the energy consumption of each cluster and different periods of the week. Because weekend days comprise 28.57% (2/7) of a week and weekdays comprise 71.4% (5/7), C5, C7 and C9 indicate significant load shifting to weekend days. A comprehensive summary of the electricity consumption characteristics is shown in

Table 4.1 Adjusted percentage of correlation analysis between cluster and month

Month\Cluster	1	2	3	4	5	6	7	8	9	10
1	19%	5%	27%	17%	10%	31%	24%	16%	6%	35%
2	15%	7%	24%	11%	12%	21%	20%	13%	5%	33%
3	9%	8%	8%	7%	5%	8%	8%	11%	9%	6%
4	3%	10%	3%	5%	1%	0	1%	2%	11%	0
5	1%	11%	0%	1%	1%	0	0	1%	9%	0
6	2%	11%	0	1%	2%	0	0	1%	9%	0
7	13%	4%	2%	24%	30%	9%	15%	19%	7%	0
8	9%	7%	1%	15%	20%	3%	4%	12%	9%	0
9	4%	10%	1%	3%	3%	0	0	3%	10%	0
10	1%	12%	0%	0%	0%	0	0	0%	8%	0
11	4%	10%	1%	2%	3%	0%	2%	4%	11%	0
12	21%	4%	32%	13%	13%	27%	27%	17%	6%	27%
	100%	100%	100%	100%	100%	100%	100%	100%	100%	100%
Characteristics	W&S	\	W	S&W	S&W	W&S	W&S	W = S	\	W

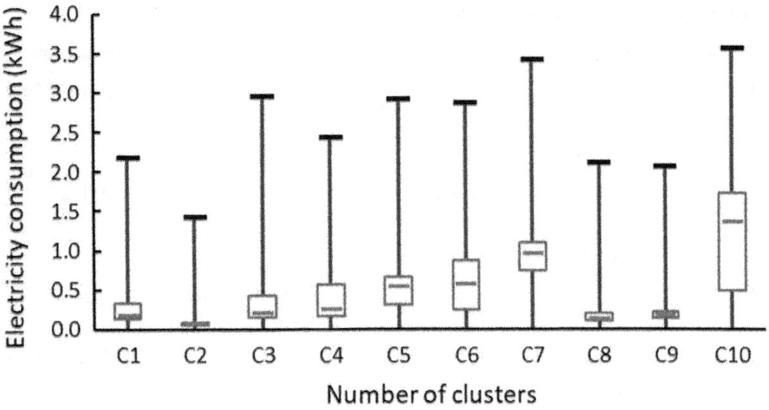

Fig. 4.5 Boxplot of seasonal electricity load of each cluster

Table 4.3. The combination of *K*-means clustering and statistical analysis presents a viable approach to identifying and characterizing different electricity load patterns. The cooling season in Shanghai lasts from July to August. However, according to our results, typical cooling season characteristics are not reflected in C2. This can be partly explained by the fact that community B is located near a campus and most of its residents are lecturers who are generally away for holiday in August. Another possible explanation is the occupants' strong energy conservation awareness. These observations indicate that even a limited amount of information can affect the assumed electricity load pattern and hence the accuracy of the predicted energy demand. Therefore, future studies should place more emphasis on gathering additional information about the occupants.

In order to balance the effects of the seasonal variability of electricity intensity distribution owing to varying heating and cooling loads (see Tables 4.1 and 4.3), electricity suppliers should improve the reliability of the national grid and avoid compulsory peak clipping measures. In general, energy use associated with heating and cooling in residential buildings has the greatest potential for conservation, since different set-point temperatures resulting from distinct occupant behaviours and preferences can cause markedly different energy consumption patterns. Hence, energy awareness education and outreach programs aiming to increasing building energy efficiency without negatively affecting the thermal comfort of residents should be widely implemented to reduce peak demands.

4.4 Further Work

The results of the daily/weekly/seasonal electricity consumption patterns derived by *K*-mean clustering and statistical analysis provides avenues for future research in occupant-behaviour.

Table 4.2 Correlation analysis between cluster and weekday/weekend

Day/Cluster	1	2	3	4	5	6	7	8	9	10	Total
Weekday	1449	13,394	573	1006	947	579	306	1788	5256	128	25,426
Percentage	75%	73%	75%	71%	57%	73%	58%	76%	63%	70%	
Weekend	488	4949	187	404	715	215	226	563	3090	55	10,892
Percentage	25%	27%	25%	29%	43%	27%	43%	24%	37%	30%	
Total	1937	18,343	760	1410	1662	794	532	2351	8346	183	36,318
Percentage	100%	100%	100%	100%	100%	100%	100%	100%	100%	100%	
Characteristics					Weekend		Weekend		Weekend		

Table 4.3 Electricity consumption characteristics of each cluster

Cluster	Night period	Morning peak	During daytime	Night peak	Characteristics
1		3kWh@7:00		4kWh@22:00	Normal weekday
2					No significant consumption
3		4kWh@7:00		5kWh@21:00	Normal weekday, higher than c1
4	Constant				Winter/summer, weekday
5			Peak at different time		Weekend
6	Constant	3kWh@7:00		5kWh@23:00	Winter/summer, weekday
7	Constant		Peak at different time		Winter/summer, weekend
8		2.5kWh@7:00		3kWh@23:00	Normal weekday, later peak
9			Peak at different time		Weekend, lower than c5
10	Constant, and high			5kWh@23:00	Winter/summer, extreme weather

On one hand, this study was subject to some limitations, which should be addressed by additional benchmarking research. Possibilities for improvements include: (1) optimizing the placement of the first centroids—which in our case were automatically positioned and then moved in MATLAB and might have affected the cluster results; (2) the isolation of occupant behaviours from other influencing factors—in this regard, further information needs to be collected about the physical characteristics of buildings; and (3) increasing the size of the dataset to enhance the accuracy and reliability of model predictions. On the other hand, future research should focus on the quantification of each end-users' contribution and on the identification of other parameters that significantly influence occupant behaviour (e.g. thermal comfort, social and economic factors, etc.).

Building energy simulation plays a significant role in predicting the energy performance for domestic buildings and in assessing the efficacy of specific policy targets. Occupant behaviour influences energy consumption greatly. However, there is often a mismatch between calculated and real energy savings due to a performance gap that is frequently caused by the "Prebound Effect" (Sunikka-blank and Galvin 2012)—which refers to the difference between the standardized/projected and the actual energy use. In building performance simulation, inaccurate results and mismatched policy targets are the outcomes of too many assumptions about the occupant behaviour. One possible solution is to introduce energy use patterns derived from field data instead of relying on generic assumptions about the behaviour of occupants, which is usually the case in numerical simulations. Thus, characterizing the clustering-analysis-derived occupant behaviour patterns of various energy sources (such as electricity, hot water, space heating etc.) and integrating them into building simulation software should be a future goal. In this regard, different levels of simplifications should be tested to identify the appropriate level of smart metering required to balance accuracy and efficiency.

Clustering analysis could also be used with different levels of design and planning targets, especially during the initial stage. Firstly, in the thermal retrofit of single-family homes, the analyst could use the clustering benchmarking results of the household in question to test the cost-effectiveness of different retrofit plans under different scenarios and thus customize their design and recommendations. Secondly, in the design of low-carbon communities, the analyst could use clustering results from similar groups of residents to test different design alternatives and control strategies.

4.5 Conclusion

This chapter presented a systematic approach to characterizing the electricity load patterns of two residential communities in Shanghai on the basis of occupant behaviours using the standard *K*-means clustering method. Daily/weekly/seasonal electricity consumption patterns have been profiled and analyzed.

Occupants were categorized as white-collar workers, poor or older families and rich or young families owing to their load patterns. In our study, the group of poor or older families constituted the largest group, accounting for nearly 80% of the total sample. In our case, the observed occupant behaviours were much more random and fluctuated over a wide range. The majority of metered households are characterized by continuous low consumption levels. Only a small proportion of households displayed the dual peak pattern with increased morning and evening consumption levels. The weekly analysis found significant load shifting towards weekend days in the case of the poor or old family group. Based on the seasonal electricity loads patterns, the clusters could be classified as dominated by heating period, dominated by cooling period, no distinguished features and dominated by energy use during the transitional seasons. The seasonal electricity consumption ranged from less than 0.1 kWh /m^2 to 7.4 kWh / m^2. The overall electricity consumption of the observed communities was slightly higher in winter than in summer. Great electricity-saving potential was observed within the group of white-collar workers and among the rich or young families as the individual loads varied a lot.

Based on the load profiling results, our recommendations to stakeholders for smoothing the load shape and improving building energy efficiency include approaches such as load shifting, valley filling, conservation, peak clipping and TOU.

Future works should characterize load curves of various energy sources (such as electricity, hot water, space heating) and integrate them with building energy simulation engines to aid the effective formulation of design and planning targets at different levels. Out research is expected to provide a valid and systematic approach to profiling occupant behaviour in selected residential settings and therefore to improve the efficiency and accuracy of building energy simulation in the contemporary energy-conscious environment.

References

Aldenderfer MS, R. K., Blashfield RK, (1985) Cluster Analysis. Sage Publications, Los Angeles

Anderberg MR (1973) Cluster Analysis for Applications. Academic Press, New York

Ayodele TR, Ogunjuyigbe ASO, Atiba IA (2017) Assessment of the impact of information feedback of prepaid meter on energy consumption of city residential buildings using bottom-up load modeling approach. Sustain Cities Soc 30:171–183

Chen P, Yan D, Wu R, Wang C, Zhou X, Jiang Y (2012) Quantitative description and simulation of human behaviour in residential buildings. Build Simul 5:85–94

Chicco G, Ilie IS (2009) Support vector clustering of electrical load pattern data. IEEE Trans Power Syst 24:1619–1628

Chicco G, R. Napoli R, and Piglione F, (2006) Comparisons among clustering techniques for electricity customer classification. IEEE Trans Power Syst 21 (2):933–940

D'Oca S, Hong T (2014) A data mining approach to discover patterns of window opening and closing behaviour in offices. Built Building and Environment 82(11):726–739

D'Oca S, Hong T (2015) Occupancy schedules learning process through a data mining framework, Energy and Buildings. Energy and Buildings 88:395–408

Dietz T, Gardner GT, Gilligan J, Stem PC, Vandenbergh MP (2009) Household actions can provide a behavioural wedge to rapidly reduce US carbon emissions. Proc Natl Acad Sci USA 106(44):18452–18456

Han J, Kamber M (2006) Data Mining Concept and Techniques. Elsevier Inc., San Francisco

Hartkopf V H, Loftness V E, Mill P A D (1986). Concept of total building performance and building diagnostics. Building Performance: Function, Preservation, and Rehabilitation: STP901, ISBN-EB: 978–0–8031–4968–7.

Jian Y, Li Q, Bai Z, Kong X (2011) Study on Influences of Usage Behaviour of Residential Air Handling Unit on Energy Consumption in summer. Build Sci 27 (12):16–19

Kazas G, Fabrizio E, Perino M (2017) Energy demand profile generation with detailed time resolution at an urban district scale: A reference building approach and case study. Appl Energy 193:243–262

López JJ, Aguado JA, Martín F, Munoz F, Rodríguez A, Ruiz JE (2011) Hopfield–K Means clustering algorithm: a proposal for the segmentation of electricity customers. Electric Power System Research 81:716–722

Marszal-Pomianowska AM, Heiselberg P, Larsen OK (2016) Household electricity demand profiles a high-resolution load model to facilitate modelling of energy flexible buildings. Energy 103:487–501

Mathew PA, Dunn LN, Sohn MD, Mercado A, Custudio C, Walter T (2015) Big-data for building energy performance: Lessons from assembling a very large national database of building energy use. Appl Energy 140:85–93

Nikolaou TG, Kolokotsa DS, Stavrakakis GS, Skias IO (2012,). On the application of clustering techniques for office buildings. *IEEE Transactions on Smart Grid*; 3: 2196–2210.

Panapakidis IP, Papadopoulos TA, Christoforidis GC, Papagiannis GK (2014) Pattern recognition algorithms for electricity load curve analysis of buildings. Energy and Buildings 73:137–145

Philip L, Bogacka A, Grigoriou R, Xu S (2014). Assessing the Use and Value of Energy Monitors in Great Britain. VaasaETT

Ruska MAM, M S. Repo S, P. Järventausta P, (2011) Customer classification and load profiling method for distribution systems. IEEE Trans Power Delivery 26:1755–1763

Reinhart CF (2004) Lightswitch-2002: a model for manual and automated control of electric lighting and blinds. Sol Energy 77(1):15–28

Sunikka-blank M, Galvin R (2012) Introducing the Prebound Effect: the gap between performance and actual energy consumption. Building Research and Information 40(3):260–327

Tsekouras GJ, Hatziargyriou ND, Dialynas EN (2007) Two-stage pattern recognition of load curves for classification of electricity customers. IEEE Trans Power Syst 22:1120–1128

Wang C, Yan D, Sun H, Jiang Y (2016) A generalized probabilistic formula relating occupant behaviour to environmental conditions. Build Environ 95:53–62

Yan D, Hong T (2017). IEA EBC Annex 66: Definition and Simulation of Occupant Behaviour in Buildings. (2013). Accessed February 21, 2017. http://annex66.org

Yu Z, Haghighat F, Fung BCM, Zhou L et al (2012) A novel methodology for knowledge discovery through mining associations between building operational data. Energy & & Buildings 47:430–440

Zhang X, Shen J, Yang T, Tang L, Wang L, Liu Y, Xu P, (2016). Smart meter and in-home display for energy savings in residential buildings: a pilot investigation in Shanghai, China. Intelligent Buildings International, 1–23. https://doi.org/10.1080/17508975.2016.1213694.

Zhao J, Lasternas B, Lam KP, Yun R, Loftness V (2014) Occupant behaviour and schedule modeling for building energy simulation through office appliance power consumption data mining. Energy & and Buildings 82:341–355

A Data-Driven Model Predictive Control for Lighting System Based on Historical Occupancy in an Office Building: Methodology Development

5

Yuan Jin, Da Yan, Xingxing Zhang,
Jingjing An, and Mengjie Han

Abstract

The lighting system accounts for 8% of the total electricity consumption in commercial buildings in the United States and 12% of the total electricity consumption in public buildings globally. This consumption share can be effectively reduced using the demand-response control. The traditional lighting system control method commonly depends on the real-time occupancy data collected using the passive infrared (PIR) sensor. However, the detection inaccuracy of the PIR sensor usually results in false-offs. To diminish the false-error frequency, the existing lighting system control simply deploys a delayed reaction period (e.g., 5–20 min), which is not sufficiently accurate for the demand-response operation. Therefore, in this research, a novel data-driven model predictive control (MPC) method that is based on the temporal sequential-based artificial neural network (TS-ANN) is proposed to overcome this challenge using an updated historical occupancy status. Using an office as case study, the proposed model is also compared with the traditional lighting system control method. In the proposed model, the occupancy data was trained to predict the occupancy pattern to improve the control. It was found that the occupancy prediction mainly correlates with the historical occupancy ratio and the time sequential feature. The simulation results indicated that the proposed method achieved higher accuracy (97.4%) and fewer false-offs (from 79.5 with traditional time delay method to 0.6 times per day) are achieved by the MPC model. The proposed TS-ANN-MPC method integrates the analysis of the occupant behaviour routine into on-site control and has the potential to further enhance the control performance practice for maximum energy conservation.

Y. Jin · D. Yan (✉)
School of Architecture, Tsinghua University, Beijing 100084, China
e-mail: yanda@tsinghua.edu.cn

Y. Jin
e-mail: jin-y17@mails.tsinghua.edu.cn

X. Zhang
Department of Energy and Community Buildings, Dalarna University, 79188 Falun, Sweden
e-mail: xza@du.se

J. An
Beijing Key Lab of HVAC, School of Environment and Energy Engineering, Beijing University of Civil Engineering and Architecture, Beijing 100044, China
e-mail: anjingjing@bucea.edu.cn

M. Han
Micro Data Analysis, Dalarna University, 79188 Falun, Sweden
e-mail: mea@du.se

X. Zhang (ed.), *Data-driven Analytics for Sustainable Buildings and Cities*,
Sustainable Development Goals Series, https://doi.org/10.1007/978-981-16-2778-1_5

Keywords

Lighting system · Occupant behaviour · Data-driven · Model predictive control

5.1 Introduction

Lighting is one of the prominent power-demand end-use (Mandil 2006). Grid-based electric lighting consumption accounts for 8% of the total electricity consumption in commercial buildings in the United States (EIA, Annual Energy Outlook 2019) and 12% of the total energy use in public buildings globally (IEA, ENERGY TECHNOLOGY PERSPECTIVES 2017). Related research shows that lighting accounts for 20%–45% of the energy consumption in office buildings (Dubois and Blomsterberg 2011). In the US, lighting accounts for approximately 14% of the total electricity consumption in residential and commercial buildings (Yin et al. 2017). However, significant energy conservation can be achieved by improving the energy efficiency of the lighting system (IEA, Energy Efficiency 2017). Besides the popularization of LED lighting (Magno et al. 2015) and dimming illuminance levels (Beccali et al. 2019), the demand-response lighting system control can contribute to energy conservation significantly. One of the effective ways of achieving demand-response control is by reducing the illumination time.

For this purpose, information on occupant behaviour pattern is essential. Li (2006) point out that even in the same building, the occupant behaviour with different household induces large discrepancy among energy consumption. The IEA (International Energy Agency) EBC (Energy in Buildings and Communities Programme) ANNEX 66 (Yan et al. 2017) gives standard definition and simulation methods for occupant behaviour. Yan (2015) and Hong (2017) have summarized the current state and future challenge of occupant behaviour research. Hong (2018) also conduct a critical review to analyze and emphasize the implementation and representation approaches of occupant behaviour models in building performance simulation programs.

As for the occupancy routine in office buildings, besides the arriving and leaving time, the intermediate activities also affect the control strategy of lighting system. According to the research of Tabak (2010), there are several intermediate activities during office time, such as having lunch, sport, get a drink/break or smoke, the duration of which lasts from 3 to 40 min. The intermediate activities bring about stochastic vacancy during office time, which needs careful analysis and helps to guide control. Chen (2018) also proposed an agent-based stochastic occupancy simulator to simulate the status transition events, random moving events and meeting events in office buildings. If combined with occupant behaviour based control, the lighting system may have great potentials in energy conservation together with greater precision. The data analysis of measured lighting energy consumption data from 15 large office buildings in Beijing and Hong Kong shows that that the 24-hourly variation in lighting energy use was mainly driven by the schedules of the building occupants (Zhou et al. 2015). Tetlow (2014) conduct experiments by setting visual prompt to avoid inadvertent lighting consumption. Manzoor (2012) use both PIR sensors as RFID tags to control lighting, with which they achieved an accuracy of 91.43%. Chenaru (2019) develop a virtualized sensor network architectural model for behaviour identification regarding lighting control. With the development of digital analysis, camera-based occupant behaviour recognition is also used for lighting control (Pham et al. 2019). Furthermore, each lighting circuit can be operated separately in office buildings, according to the occupant behaviour (Galasiu et al. 2007). The lighting control system based on the occupant behaviour is expected to save energy by approximately 60% (Bakker et al. 2017).

Many studies demonstrate the potential of energy conservation for occupancy-based lighting control. The future smart lighting scheme shall integrate smart algorithm and grasp the opportunity of Internet-of-Things (Chew et al. 2017). Bourgeois proposes a control model based on real-time occupancy data to compare both manual and automatic lighting system controls (Bourgeois et al. 2006); they note that their method achieves a lower energy consumption compared to the lighting system control with the fixed schedule. Following these findings, many researchers started to consider the lighting system control based on occupancy data (Guo et al. 2010). Nagy (2015) present an adaptive control strategy for lighting control in office buildings and realize 37.9% energy saving compared with the standard settings. Araji (2012) conduct a survey and discover that the lighting energy saving by occupant sensors in office buildings varies from 7 to 45%. A case study of split three lamps control presents 40% energy saving than the baseline (Rubinstein and Enscoe 2010). Aghemo (2014) conduct a case study of ten offices by both occupancy-based and dimming-based control, to proof the energy conservation potential for lighting system.

Occupancy data is obtained using different sensors, such as passive infrared sensor (PIR), ultrasonic sensor, microwave sensor, acoustic control sensor, and pressure sensor. The PIR is the most commonly used for occupancy detection in commercial buildings, because of sensor precision, robustness, stability, cost, and privacy. In recent years, there have been a proliferation of occupancy motion sensors based on PIR principles (Kumar 2017). Most of these methods utilize the algorithm of time-delay judgment. For instance, when it is determined that the room is occupied, the light will be turned on directly; however, if the room is determined to be unoccupied, the light is turned off until a certain delay period has passed. In view of this, Richman studied the impact of different values of delay period on energy conservation in relation to the lighting system (Richman et al. 1996). Maniccia and Von Neida conduct a similar research (Maniccia et al. 2001), and conclude that when the delay period varies from 20 to 5 min, the energy conservation changes from 28 to 38% for the private office room, while it ranges from 17 to 60% for the open office area.

However, the current lighting control, which is based on the occupancy state from PIR sensor, may not always preserve energy consumption. Tetlow (2014) point out that this kind of automation may make people consider PIR sensor-based control as an alternative of turning off the light, thus the delay period of lighting will lead to extra consumption. Furthermore, many on-site measurements and evaluation of lighting control systems expose an inevitable bottle-neck, the frequent false-offs. According to the motion-sensor performance study, when single PIR sensors are used, there is the tendency for occupancy to go unrecognized, almost as much as 20% of the time, leading to frequent false-offs (Guo 2007). This is attributable to the limited motion-sensing precision. The sensing range of the PIR sensor is shown in Fig. 5.1 (Benya 2001), and as may be observed, as the range of movement decreases, the detectable range of motion sensor also shrinks. The detectable range of the PIR sensor for whole body movements is 12 m; this range shrinks to 6 m for upper body or arm movements. Furthermore, in the practical application, the detection range tends to be even less. This will result in a high percentage of false-offs.

Thus, the research gap has been identified. There are stochastic intermediate activities

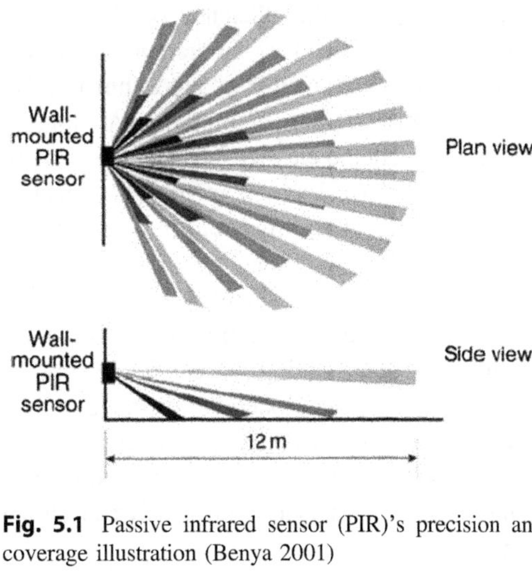

Fig. 5.1 Passive infrared sensor (PIR)'s precision and coverage illustration (Benya 2001)

during office time, which will cause vacancy of different durations. The occupancy routine may change with different periods or occupants, and with the prevailing decentralized personal lighting system development, it is crucial to update and learn occupancy routine for lighting system control. What's more, previous studies merely apply the control algorithm of the delay period. This strategy still cannot avoid several false-offs, because of the current lack of sensor precision, thereby resulting in the discomfort of the occupants. The mentioned research gap can be solved by harnessing the insight of occupancy routine to incorporate occupancy prediction into the lighting system control. Extensive research has been conducted on occupancy prediction, mostly for the heating, ventilation, and air controlling (HVAC) system control. Model predictive control is a comprehensive optimal control strategy to use prediction method taking occupancy and other factors into consideration. It compute optimal control actions by reducing the energy consumption and ensuring the thermal comfort as well as IAQ (indoor air quality) (Goyal et al. 2013). Kwok (2011) emphasize that occupancy routine play a critical role in cooling-load prediction, and they propose a probabilistic entropy-based neural (PENN) model for predicting the occupancy and cooling load of a building. Peng (2018) propose an occupant-behaviour-based model to increase the efficiency of HVAC systems, whereby occupancy prediction is achieved using the k-nearest neighbor algorithm (KNN). Kim (2018) propose a multinomial logistic regression model to classify the occupancy status in the room as "away, active, and inactive (sleep)". Kim (2019) propose a recurrent neural network (RNN) with long short-term memory (LSTM) units for predicting the number of occupants in a large exhibition hall. Qiu (2019) establish a data mining based framework to identify rule based operation strategies for buildings with power metering system, including lighting system on/off control, chiller sequencing control, and coordinated control between chiller and pump.

Therefore, this study investigates the historical occupancy state, and proposes a data-driven model predictive control (MPC) for lighting system based on occupancy prediction (Mirakhorli and Dong 2016; Oldewurtel et al. 2013; Wanjiru et al. 2016). The research utilized only the PIR sensor for cost effectiveness. A data-driven approach using the temporal sequential-based artificial neural network (TS-ANN) was proposed to discover the temporal sequential patterns of the occupancy state. By incorporating the occupancy prediction into the PIR sensor detection, it is anticipated that control will be improved, and result in fewer false-offs. It should be noted that this study will concentrate on the development and analysis of the TS-ANN MPC method for lighting system control.

The rest of this chapter is structured as follows. First, the methodology of the research, concerning the prediction, control algorithm, and evaluation index will be introduced. The details of an experiment conducted to present the control performance of the proposed method is presented next. Then, the results of the parameter optimization for the different methods are discussed. Finally, the conclusion and recommendation for future research directions are presented.

5.2 System Description and Research Methodology

5.2.1 Description of the Proposed Control System

To implement the PIR-MPC lighting system, the occupants, sensors, controllers, and lighting system interact, as shown in Fig. 5.2. The general working principle can be elaborated as follows. The PIR sensor detects occupancy through the occupant's movement and transmits the data to the controller, where the occupancy data is stored as the historical data. The occupancy data is subsequently processed and used for occupancy prediction for the lighting system control. Then the lighting system is controlled according to the ON/OFF signal transmitted from the controller. Meanwhile, the operation and status of the lighting system also have an impact on the occupant. If a false-off occurs, the occupant may

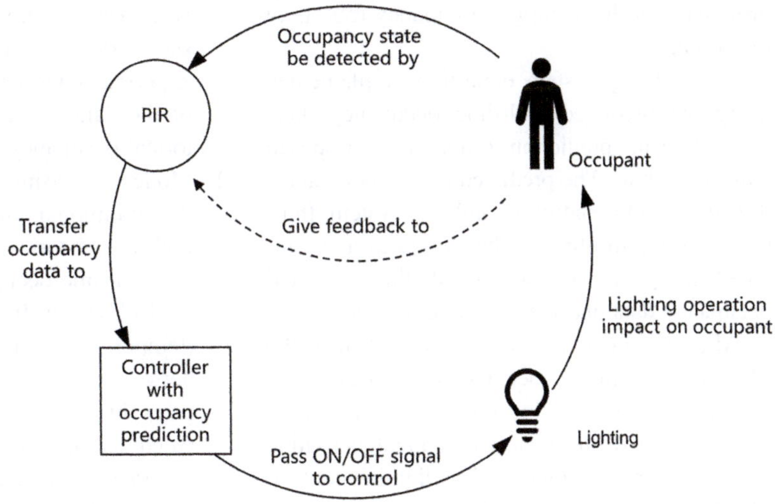

Fig. 5.2 Illustration of lighting system control working principle

act by waving his hand to turn on the lighting system once again. This is defined as the occupant's feedback to the lighting control system.

The false-off occurs because the occupancy data from the PIR sensor may deviate from the actual occupancy data. Hence, in this research, the work is divided into two aspects. Firstly, the occupancy data from the PIR sensor will be revised to represent the real occupancy state of the room as the basis of occupancy prediction. Secondly, the occupancy will be predicted to enhance the control accuracy and reduce false-offs.

5.2.2 Overall Methodology

5.2.2.1 Data Composition

For better understanding, the data composition is firstly explained (Fig. 5.3). The data has mainly three compositions, (1) real occupancy data, (2) occupancy data from the PIR sensor, and (3) predicted occupancy data.

The real occupancy data refers to the true occupancy condition of the room, and it can be used to evaluate and validate the control algorithm. In this research, an on-site survey was conducted to obtain the real occupancy data. A camera was used to obtain image data, and the occupants were requested to take notes that included the time point and the events (arrival or departure).

The occupancy data from the PIR sensor refers to the occupancy state as recorded by the PIR sensor. The raw occupancy data from the PIR sensor are not temporally equidistant. Through preprocessing, the occupancy data were in a temporal resolution of 30 s. The data quality of the sensor can be analyzed by comparing the occupancy data from the PIR sensor and the real occupancy data. This will contribute to the data

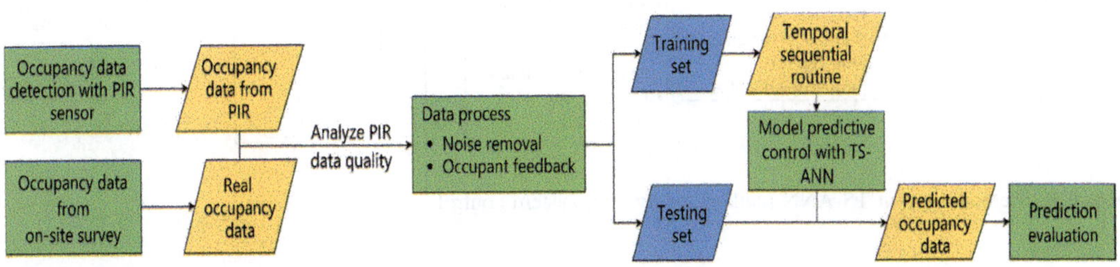

Fig. 5.3 Illustration of data composition of this study

processing of the historical occupancy routine of the room.

The lighting system control is implemented using the predicted real-time occupancy data. The data is the prediction results of the proposed control method. The predicted occupancy data is directly used to control the lighting system; thus, the control algorithm can be evaluated and validated using a combination of the predicted occupancy data and real occupancy data.

After data process, the occupancy data from the PIR sensor and the real occupancy data are divided as training set and testing set. The training set is used to analyze the temporal sequential routine and train the prediction model for lighting control. The prediction model uses the testing set to generate the predicted occupancy data, which is further used for prediction evaluation.

5.2.2.2 Overall Methodology Elaboration

The methodology of this research is divided into four parts (Fig. 5.4):

(1) Data acquisition and preprocessing: In implementing the lighting system control, the occupancy data is drawn from one PIR sensor. The ground truth data is collected from the on-site survey. The camera is installed, and the occupancy are recorded manually from the video, and the occupants were also requested to take notes for the presence as assistance data. The privacy issue has been noticed and avoided. The on-site survey is only conducted during the experiment and validation test. The objective of the data processing is to obtain the historical occupancy routine of the room.

(2) Model establishment: These methods, unlike the traditional time-delay method, utilize both the historical occupancy condition and the real-time occupancy data obtained using the PIR sensor. In this research, the TS-ANN (temporal sequential-based artificial neural network) model was proposed. The model takes historical occupied routine as inputs (such as time point, occupied state ratio, historical data, and number of continuous timestep for (un)occupied states) and takes MLP (multi-layer perceptron) network as the model format.

(3) Prediction-based control: turn on the lighting system when the room is detected as occupied by the PIR sensor; once the room is detected to be unoccupied, the occupancy prediction was triggered, if the room is predicted to be unoccupied, the lighting system will be turned off.

(4) Model evaluation: The main target of this research is to reduce or avoid false-offs. Therefore, the control evaluation focuses on the frequency of false-offs under the different control methods. The accuracy from confusion matrix is also calculated for each method.

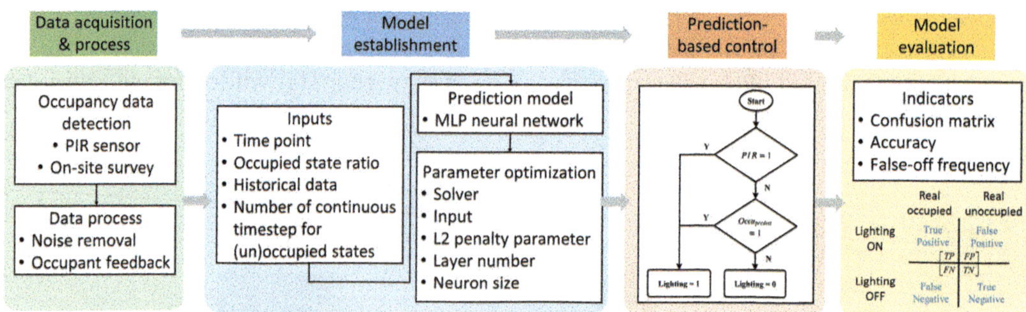

Fig. 5.4 Methodology of TS-ANN method for lighting system control

5.2.3 Data Acquisition and Processing

5.2.3.1 PIR Sensor Calibration

The quality of the occupancy data was analyzed using two types of PIR sensors to avoid distinct difference of accuracy among PIR sensors. A preliminary experiment was conducted with both sensors. The occupancy data in one office room was recorded using both types of PIR sensor simultaneously. The real occupancy data from the sensors was saved as well. The ground truth occupancy data was obtained through the on-site survey conducted with the aid of the camera and notes. The confusion matrix (Powers 2011) can be used to describe the data quality of the PIR sensors. 'Condition occupied' means the room is actually occupied. Thus, if the room is (un)occupied, the PIR sensor detects it accordingly. Consequently, the sensor is considered to be accurate. In other circumstances, the sensor is considered inaccurate.

From Tables 5.1 and 5.2, the two types of PIR sensors perform similarly in terms of occupancy detection. According to the measurement data, the accuracy of Type 1 is 81.9%, while that of Type 2 is 80.2%. Hence, the following case studies are all made using Type 1 because it is slightly more accurate.

5.2.3.2 Noise Removal

The comparison of the real occupancy condition and the occupancy condition as detected by the PIR sensor is illustrated in Fig. 5.5. The measurement duration was one hour, during which the room was continuously occupied. However, the PIR sensor continually reported that the room was unoccupied.

Based on this data feature from the PIR sensor, the duration of both the occupied condition and the unoccupied conditions, as detected, are defined. As is shown in Fig. 5.6, Duration (a) represents the former, and Duration (b) represents the latter.

Furthermore, from the statistical analysis, the frequent occupied–unoccupied switch was demonstrated. The cumulative distribution of Durations (a) and (b) from the PIR sensor, for the original data, is shown in Fig. 5.7. Here the cumulative distribution means the frequency of different duration values for duration (a) and (b) in the original PIR occupancy data. There was a tendency for the 'occupied' and 'unoccupied' durations in the measured office to condense to less than 3 min, which is unusually small for a regular office. The occupied-state data from the PIR sensor is considered to include white noise. Thus, noise removal during data processing was proposed.

According to the routine of official business as a processing parameter, it is assumed that the lighting system is unnecessarily turned off if the room is unoccupied for merely less than 5 min. Thus, the noise removal method involved revising the occupancy data to read occupied: if the room was unoccupied for less than 5 min, the occupancy data will be revised to be occupied.

5.2.3.3 Occupant Feedback

During the on-site lighting system control, in the event of a false-off, the occupant reacted and turned on the lights immediately. This feedback was also taken into consideration to revise the real-time occupancy data during the on-site control.

According to the judgment algorithm, false-off was considered to have occurred if the

Table 5.1 Confusion matrix results for PIR sensor Type 1

	Condition occupied (%)	Condition unoccupied (%)
Occupied, according to PIR sensor	48.4	2.4
Unoccupied, according to PIR sensor	51.6	97.6
Ratio	32.0	68.0

Table 5.2 Confusion matrix results for PIR sensor Type 2

	Condition occupied (%)	Condition unoccupied (%)
Occupied, according to PIR sensor	41.7	1.8
Unoccupied, according to PIR sensor	58.3	98.2
Ratio	31.9	68.1

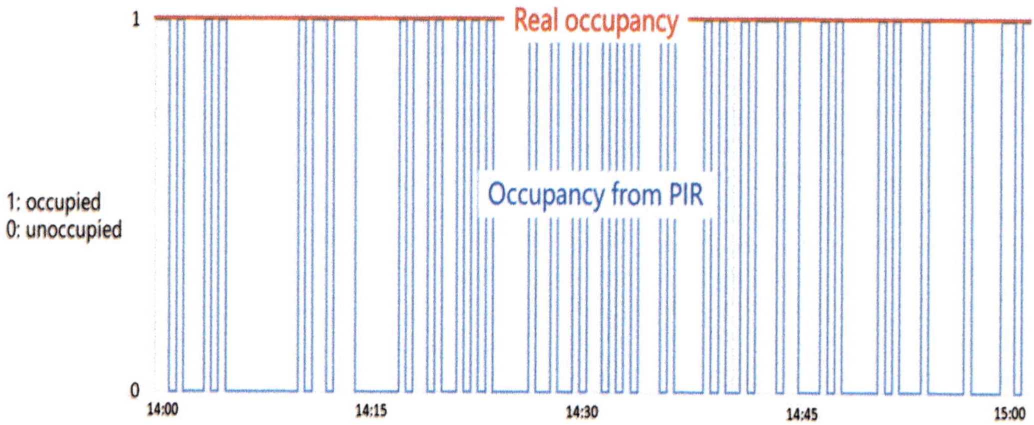

Fig. 5.5 Real occupancy condition and PIR sensor occupancy data comparison

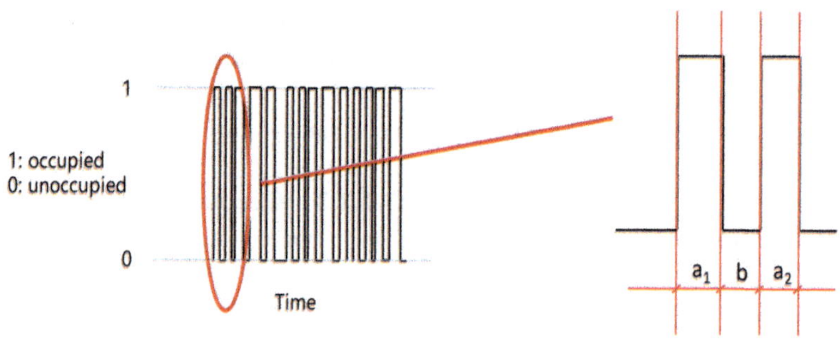

Fig. 5.6 Duration definitions during PIR occupancy data processing: a, duration of occupied time according to PIR sensor, b, duration of unoccupied time according to PIR sensor

lighting system goes from off to on between two consecutive time steps, and the historical occupancy data was revised to read occupied.

5.2.4 Model Predictive Control Method

In this study, the TS-ANN method, which takes MLP as the basic algorithm, was applied (Gardner and Dorling 1998). Based on the historical occupancy routine, this method can effectively predict the real-time occupancy condition of the room. Using the updated prediction of the occupancy state, the lighting system can then be controlled accurately. After data processing, the occupancy data for several days before was obtained, and the historical occupancy routine was analyzed. The historical occupied rate was used to determine the historical occupancy routine. The control method retrospected the historical occupancy state or the

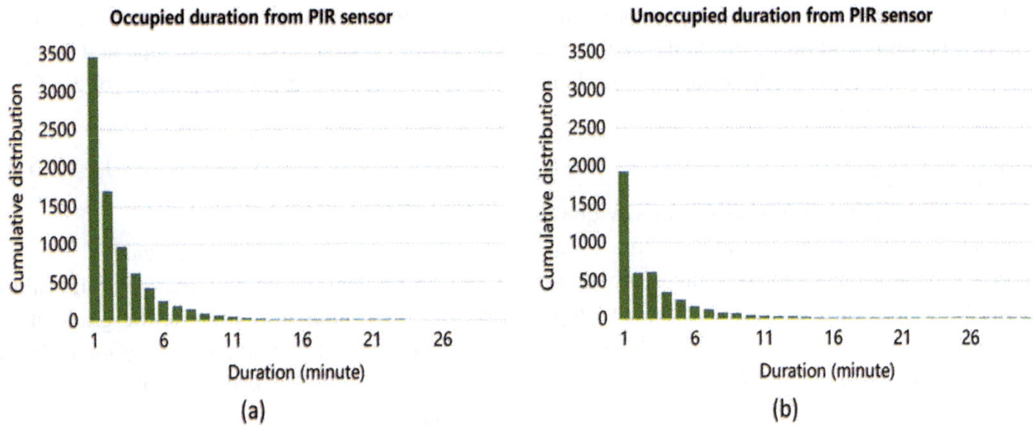

Fig. 5.7 Cumulative distribution of durations (**a**) and (**b**) according to the PIR sensors

momentary occupied rate, thereby making it possible to instruct the lighting system control using a prediction about the real-time occupancy.

It is possible for the PIR sensor to constantly erroneously detect the room to be unoccupied; this is a critical issue of detection. Therefore, the control algorithm commonly contains the logic of the lighting system on-and-off represented as follows: (1) turn on the lighting system when the room is detected as occupied by the PIR sensor; (2) once the room is detected to be unoccupied, the occupancy prediction was triggered to avoid false-off, if the room is predicted to be unoccupied, the lighting system will be turned off.

In the subsequent subsections, the basic multi-layer algorithm and the TS-ANN method for the lighting system control are introduced. In this study, the dataset is divided into training and testing sets. The indicator of the accuracy rate is selected according to the criteria of parameter optimization for better occupancy prediction performance and lighting system control.

5.2.4.1 Multi-layer Perceptron Algorithm

Neutral networks are a series of artificial intelligence; the MLP, defined as a feed-forward neural network is one of the most useful types of network (Schalkoff 1992; Jain et al. 1996). The model of the MLP neutral network also contains inputs and outputs, and it is necessary to train the relationship of the variables using different algorithms. The model can be adapted to nonlinear functions and accurately trained. It consists of a series of inner system with connected neurons. With multiple input and output variables, the neurons in-between are called the hidden layers. The connection refers to different simple nonlinear transfer functions, with different connection weight coefficients. The function can be depicted as follows:

$$[o_1, \ldots] = f([i_1, i_2, i_3, \ldots]) \tag{5.1}$$

here, the input and output variables can be represented as vectors. During the model training, the data of transfer function and connection weight coefficients are repeatedly determined and modified to optimize the input–output mapping.

The process of optimization involves minimizing the error values. The most basic training algorithm is back-propagation based on the gradient descent (GD) procedure, the most computationally straightforward algorithm for training the MLP, to locate the global minimum of the error surface (Rummelhart 1986).

In this study, three solvers are chosen for error optimization, the limited-memory Broyden–Fletcher–Goldfarb–Shanno (L-BFGS) method (Le et al. 2011; Liu and Nocedal 1989), stochastic gradient descent (SGD) method (Bottou 2010), and optimized SGD (adaptive moment estimation (ADAM)) method (Kingma and Ba 2015). Although the L-BFGS method is highly preferable for low-dimensional problems, ADAM, an optimization of the SGD method, is more applicable to high-dimensional problems where it outperforms the L-BFGS solver.

L-BFGS Method

The L-BFGS method is a limited-memory quasi-Newton method (Buckley and LeNir 1985); however, in addition to utilizing the combination of the BFGS cycle, it conjugates gradient steps. The convergence process is considered a smooth nonlinear function $f : R^n \rightarrow R$,

$$\min f(x) \tag{5.2}$$

The gradient, denoted by g, is available. The iterates are denoted by x_k; it is defined as follows: $s_k = x_{k+1} - x_k$, and $y_k = g_{k+1} - g_k$. The inverse BFGS formula is approximated using the sparse symmetric and positive definite matrix, H. The method uses the inverse BFGS formula in the following form:

$$H_{k+1} = V_k^T H_k V_k + \rho_k s_k s_k^T \tag{5.3}$$

where $\rho_k = 1/y_k^T s_k$, and $V_k = I - \rho_k y_k s_k^T$.

The difference between the BFGS and L-BFGS methods lies in the matrix update. The L-BFGS method stores the corrections to the initial matrix separately to avoid using the O(n2) storage. When the available storage is used up, the oldest correction is deleted to create space for the new one.

SGD Method

In the process of model training and parameter fitting, a loss function $l(\hat{y}, y)$ is used to depict the cost of predicting the output y. A set of functions $f_\omega(x)$ are parametrized using the weight vector, ω. Hence, the objective of model training is to minimize the loss function, $Q(z, \omega) = l(f_\omega(x), y)$. It is necessary that the model training minimizes the empirical risk, $E_n(f)$, which measures the performance of the training set:

$$E_n(f) = \frac{1}{n}\sum_{i=1}^{n} l(f(x_i), y_i) \tag{5.4}$$

Rummelhart (1986) used the GD to minimize the empirical risk, $E_n(f)$; the updated weight factor is generated from the gradient of $E_n(f_\omega)$:

$$\omega_{t+1} = \omega_t - \gamma \frac{1}{n}\sum_{i=1}^{n} \nabla_\omega Q(z_i, \omega_t) \tag{5.5}$$

where γ is an adequately chosen gain.

Based on the GD method, the SGD algorithm (Bottou 2010) does not compute the gradient of $E_n(f_\omega)$ precisely, but estimates the gradient from a single randomly picked example z_t at each iteration:

$$\omega_{t+1} = \omega_t - \gamma_t \nabla_\omega Q(z_t, \omega_t) \tag{5.6}$$

The stochastic algorithm does not remember the examples from the previous iterations, which improves the ability of processing data on a larger scale.

ADAM Method

ADAM method, a gradient-based optimization, is an algorithm for first-order gradient-based optimization of stochastic objective functions. This method is straightforward to implement, and computationally efficient. This method has little memory requirements; thus, it is suitable for large data size problems (Kingma and Ba 2015).

The update of the exponential moving average at time step, t (v_t), can be represented as the gradient function of all previous time steps:

$$v_t = (1 - \beta_2)\sum_{i=1}^{t} \beta_2^{t-i} \cdot g_i^2 \tag{5.7}$$

where β_2 is the decay rate, g indicates gradient.

In case of sparse gradients, for a reliable estimate of the second moment, it is necessary to find the average of many gradients by choosing a small value of decay rate, β_2.

5.2.4.2 Temporal Sequential-Based ANN Method

In this study, the ANN method was combined with the temporal sequential routine. Based on the unaltered sensor technique, the study focused on the temporal sequential routine of the occupancy state. The proposed TS-ANN method achieved real-time room occupancy prediction, and the prediction result further contributed to the lighting system control. One program package from Python (Pedregosa et al. 2011) was introduced to perform the neutral network learning.

In this study, five inputs of the prediction model were proposed. To determine the suitable ones, the prediction and control performance of the input combinations were compared and analyzed. The inputs are:

- Historical occupancy state for several time steps
- Time point of each time step
- Historical occupied ratio of each time step
- The continuous time steps for occupied state
- The continuous time steps for unoccupied state.

The proposed TS-ANN method is as shown in Fig. 5.8.

As for the model and parameter optimization, five dimensions were considered, and the most optimal for the model were determined using the grid search method. The prediction model was optimized from solver type, model inputs, L2 penalty parameter (α), layer number of the network, and neuron size of each layer (Fig. 5.9). The dataset was divided into two equal parts, the training and testing sets. The model was trained using the data from the training set, following which the model was used to predict the occupancy state, and instruct the lighting system control using the data the from testing set. The accuracy rate was selected as the criteria for determining the best model.

5.2.5 Evaluation of Predictive Control

The PIR-MPC lighting system was evaluated according to the control accuracy. The control accuracy was evaluated in the formation of the confusion matrix (Fawcett 2006):

$$\begin{bmatrix} \text{True positive} & \text{False positive} \\ \text{False negative} & \text{True negative} \end{bmatrix} \tag{5.8}$$

where the occupancy state is the judgment condition. “True positive” refers to when the room is rightly determined to be occupied, and lighting system is properly turned on. “False negative” refers to when the room is erroneously adjudged to be unoccupied, resulting in lighting system false-offs. “True negative” refers to when the

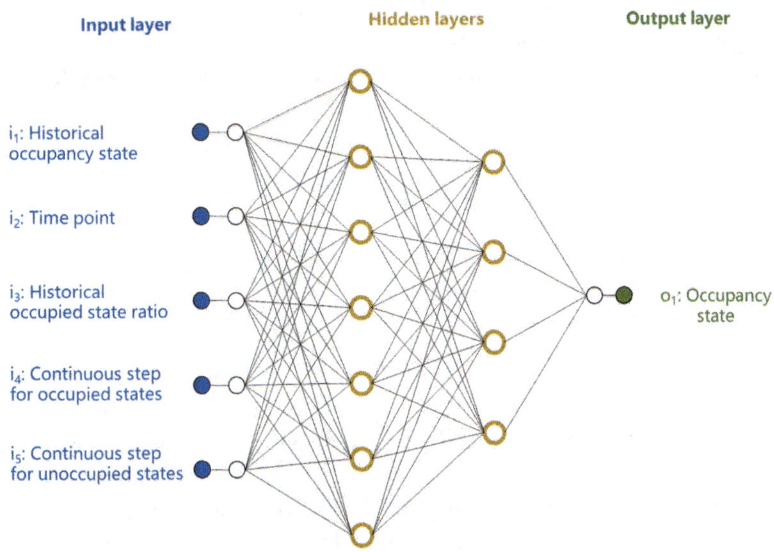

Fig. 5.8 Structure illustration of TS-ANN method

Solver	Train X	L2 penalty α	Layer number	Neuron size
"lbfgs" an optimizer in the family of quasi-Newton methods "sgd" a stochastic gradient descent algorithm "adam" a stochastic gradient-based optimizer	1. Historical occupancy state 2. Time point 3. Occupied state ratio 4. Continuous timesteps for occupied states 5. Continuous timesteps for unoccupied states	0.001 0.1 10	1 layer 2 layers	For different sizes

Fig. 5.9 Five dimensions of model optimization

room is rightly determined to be unoccupied, and lighting system is properly turned off, and "false positive" refers to when the room is wrongly adjudged to be occupied, and the lighting system remains on while the room is unoccupied.

The ratio of the confusion matrix is written as follows:

$$\begin{bmatrix} \text{TPR} & \text{FPR} \\ \text{FNR} & \text{TNR} \end{bmatrix} \tag{5.9}$$

where TPR refers to the true positive ratio, FNR refers to the false negative ratio, FPR refers to the false positive ratio, and TNR refers to the true negative ratio. From the confusion matrix, the accuracy rate can be computed to evaluate the control performance:

$$accuracy = \frac{TP + TN}{TP + FN + FP + TN} \tag{5.10}$$

Furthermore, the frequency of false-offs is evaluated to represent the comfort of the occupants depending on choice of lighting system control.

5.3 Case Study of an Office Building

An office in the campus of Tsinghua University, China, was selected as the case study for the purpose of comparing the performance of the traditional time-delay method and PIR-MPC methods based on historical occupancy state. This office room supplies the working conditions for three colleagues with regular working schedule. The room is approximately 20 m^2, and has no work partition, with one PIR sensor installed.

The PIR sensor (offset by the red box) was installed on the ceiling of the office (Fig. 5.10). The sensor was connected to the controller, which can store the occupancy data and control algorithms. The processed data is in 30-s temporal resolution.

5.3.1 Historical Occupancy Data

The occupancy detection work began from August 2017. In the beginning, the PIR sensor was calibrated. The occupancy database was established using occupancy data from November 1, 2017 to December 31, 2017. The occupancy dataset was divided evenly into training and testing sets. The training set was used for analyzing the historical occupied characteristic and model training. The original occupied ratio was computed from the historical occupancy data of the training set, and only the weekdays were considered.

After preprocessing, we can obtain the original occupied ratio for the data in the training set

Fig. 5.10 Illustration of installed PIR sensor

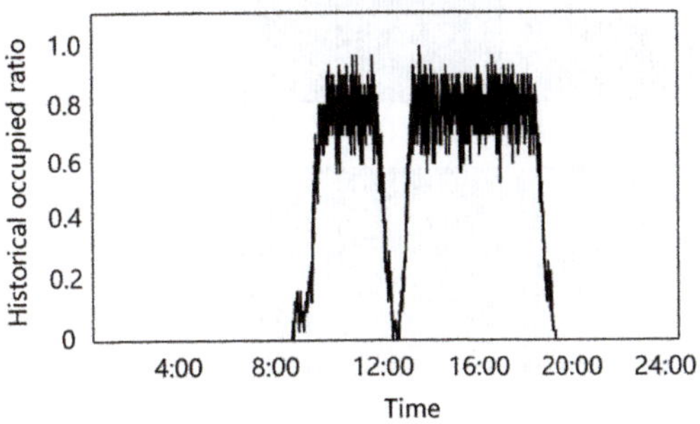

Fig. 5.11 Historical occupied ratio from occupancy data processing

(Fig. 5.11). The routine of this office can be inferred from this figure. The working hours are between 8.00 am and some minutes short of 7.00 pm. The occupied ratio increases and decreases sharply, indicating that the working routine is quite regular. Another remarkable feature may be observed at the noon time when the occupants go out for lunch. In the working period, the occupied ratio fluctuates, indicating that the occupants were not continuously in the room for reasons such as going to restroom or having a meeting. The historical occupied ratio depicts the working characteristic of the office room and contributes to the occupancy prediction for lighting system control.

5.3.2 Control Method

In this experiment, the occupancy data from the training set was used for model training. The study deploys the testing set as the ground-truth data to compare the control performance of different control methods and different model settings.

Two control methods were compared:

- Traditional time-delay method
- Machine learning method using different model input combinations.

For the traditional time-delay method, the delay period was 5 min. Each potential combination of the model inputs was evaluated. In this study, the input combination of "time point + occupied ratio + continuous time steps for occupied and unoccupied states was determined to be the optimized method. It was named the "time point-ratio-continuous method", abbreviated as PRC-TS-ANN. The detailed optimized parameters are summarized and listed in Table 5.3. In terms of the input of the time point, the time sequence number in each day was used to depict the time point and the temporal sequence characteristic of the occupancy state.

5.3.3 Control Performance Comparison

The following control performance results were obtained from the control of the occupancy data from the testing set. The occupant's feedback during the implementation of the lighting system control was taken into consideration.

Figure 5.12 illustrates the occupancy data collected over the course of one day and the lighting control results with two methods in a 30-s temporal resolution. From the real occupancy data, the main office hours were divided into

Table 5.3 Input and parameter setting selected for optimized TS-ANN method

Input	Solver	α	Neuron size
Time point + Occupied ratio + Continuous time steps for occupied and unoccupied states	LBFGS	0.001	(100, 70)

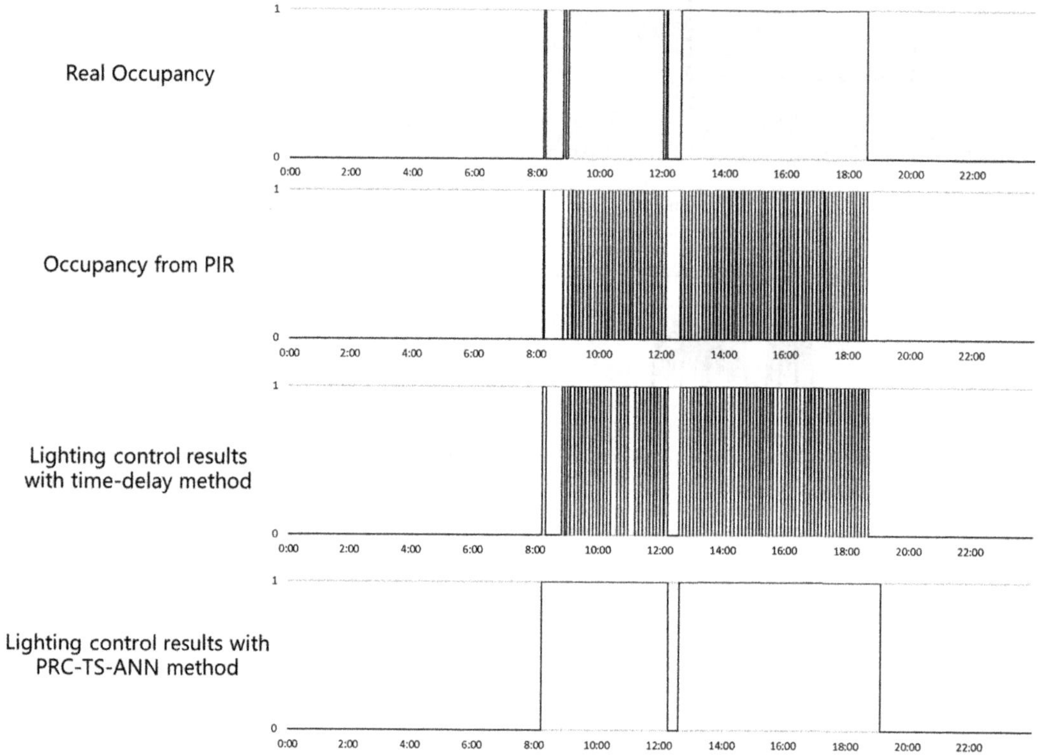

Fig. 5.12 Occupancy data of (1) real occupancy data, and (2) occupancy data from PIR sensor; and (3) lighting control results with time-delay method and proposed PRC-TS-ANN method

morning and afternoon. The occupancy data obtained from the PIR sensor frequently changed from the occupied (1) to the unoccupied state (0). The control results for one day of implementing the traditional and newly proposed PRC-TS-ANN methods were compared. From the comparison, the proposed machine learning method achieved better control performance, with less false-offs during the office hours.

Table 5.4 illustrates the confusion matrix results of the traditional time-delay control method and the newly proposed method. False-offs occurred less frequently while proposed method was being utilized. The false-off rate, at 0.05%, was much lower in the case of the newly proposed method, compared to that of the traditional time-delay method at approximately 6.92%. The higher FPR means the longer duration of the lighting system turning on under vacancy conditions. Compared with traditional time delay method, the FPR of the proposed method is 3% higher. Actually, the FPR and FNR are two contradictory reacting variables, the proposed method tries to avoid false-off of lighting system by decreasing FNR and to balance the rising duration and false-off. The performance of the two methods shall be comprehensively compared according to the evaluation metrics. The accuracy of the time-delay method was 96.4%, while the proposed

Table 5.4 Confusion matrix of lighting system control performance for time-delay method and TS-ANN method

Method	TPR (%)	FNR (%)	FPR (%)	TNR (%)
Time-delay method	93.08	6.92	1.32	98.68
PRC-TS-ANN method	99.95	0.05	4.34	95.66

(Remarks: TPR—true positive rate, referring to light on under occupied condition; FNR—false negative rate, referring to light off under unoccupied condition; FPR—false positive rate, referring to light on while the room is unoccupied; TNR—true negative rate, referring to light off when the room is unoccupied)

Table 5.5 False-off frequency and illumination duration for both methods

Method	Frequency of false-offs (number of times per day)	Duration (hour per day)
Time-delay method	79.5	9.1
PRC-TS-ANN method	0.6	10.2

method achieved up to 97.4% accuracy. The comparison of the false-off frequency and illumination duration under both methods are shown in Table 5.5. False-offs occurred 79.5 times per day under the time-delay method; the proposed method effectively enhances the operation accuracy by decreasing the occurrence of false-offs to 0.6 times per day. The illumination duration under the time-delay method was 9.1 h per day, the proposed duration extends it to 10.2 h per day. The lower illumination duration with the time-delay method is attributable to the frequent false-offs. Thus, the proposed method outperformed the traditional time-delay method by enhancing the accuracy of the lighting system control, and decreasing the frequency of the false-offs.

5.4 Discussion and Further Work

The proposed TS-ANN method considers the temporal sequential characteristic of the occupancy state, and attempts to incorporate this into occupancy prediction. This consequently constitutes the basis of the model predictive control of lighting system. From the former experiment, the TS-ANN method achieves better control performance than the traditional time-delay method. During the model optimization, it was discovered that the occupancy state prediction has different outcomes according to the different temporal sequential elements inputted into the predictive model. In this section, three different methods of TS-ANN are discussed, and their prediction and control performance are compared.

Besides the PRC method used in the experiment, there are two other TS-ANN methods according to the different temporal sequential elements through which they achieve the occupancy prediction. The detailed inputs are shown in Table 5.6.

In terms of the historical occupancy state, the last time step of the historical occupied state is selected. The control performance of different input combinations and parameter settings were simulated and compared. The control performance was also demonstrated from two

Table 5.6 Input detail of three TS-ANN methods

Methods	Inputs
PR-TS-ANN	Time point + occupied ratio
PRC-TS-ANN	Time point + occupied ratio + continuous time steps for occupied and unoccupied states
PRH-TS-ANN	Time point + occupied ratio + historical occupied state data

Fig. 5.13 Control performance results of PR-TS-ANN method

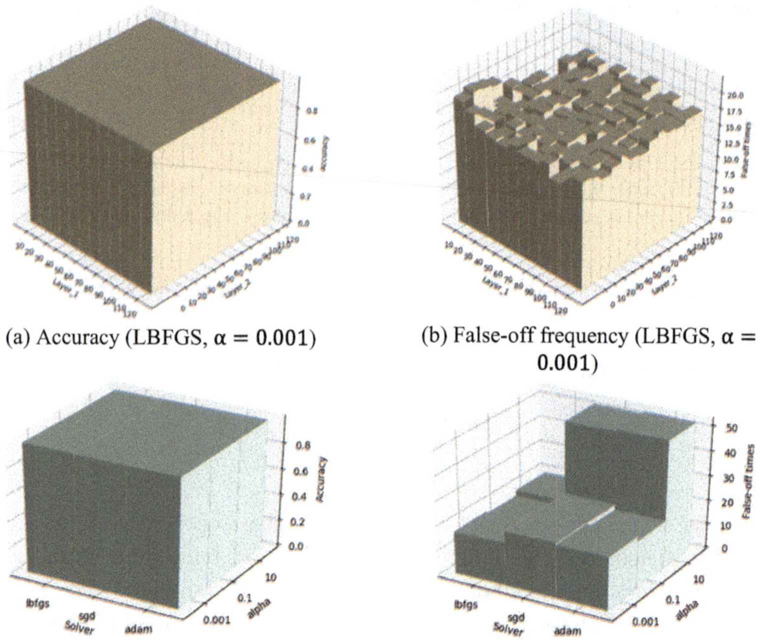

(a) Accuracy (LBFGS, α = 0.001)

(b) False-off frequency (LBFGS, α = 0.001)

(c) Accuracy (neuron size: 120,120)

(d) False-off frequency (neuron size: 120,120)

perspectives: occupancy prediction accuracy and false-off frequency of lighting system control. During the grid search for model optimization, there are four variables. For better visualization, the 3D plot (Hunter 2007) was used for every two variables, while the other two variables are kept constant. In the following figures, Fig. 5.13 illustrates the control performance results of the PR-TS-ANN method. Figures 5.14 and 5.15 illustrate the performance result of the PRC-TS-ANN and the PRH-TS-ANN methods, respectively. From the figures of optimization results, the neural network size and penalty parameter α have stronger influence on the model performance compared with optimization algorithm. However, the optimization algorithm affects the speed of convergence during optimization.

It is apparent that the three models, with different inputs, solvers, penalty α, and sizes of neuron network, present different control performances. Of the three TS-ANN methods, the accuracy of the PR-TS-ANN method was the highest. The accuracy rate reflects the performance of each at occupancy state prediction; thus, it was used as the indicator of the parameter selection. Table 5.7 shows the results of parameter optimization. The detailed optimized parameters are summarized and listed in Table 5.8.

In Fig. 5.16, data obtained over the course of one day are presented to illustrate the control performance of the three TS-ANN methods in with 30-s temporal resolution. As seen from the visualization results, all the three methods can predict the occupancy state. However, the PR-TS-ANN method contains more details, such as the short leave of occupant in the morning. The PRC-TS-ANN and PRH-TS-ANN methods can effectively distinguish morning from afternoon. The confusion matrix results are shown in Fig. 5.17, together with detailed data. Compared with the traditional time-delay method, the newly proposed TS-ANN methods have lower false negative rate, which means the new method could effectively reduce the "false-off" conditions. While the false positive rate of TS-ANN

Fig. 5.14 Control performance results of PRC-TS-ANN method

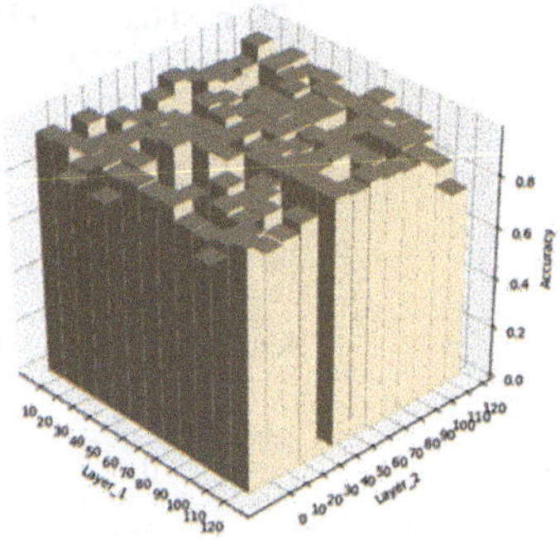

(a) Accuracy (LBFGS, α = 0.001)

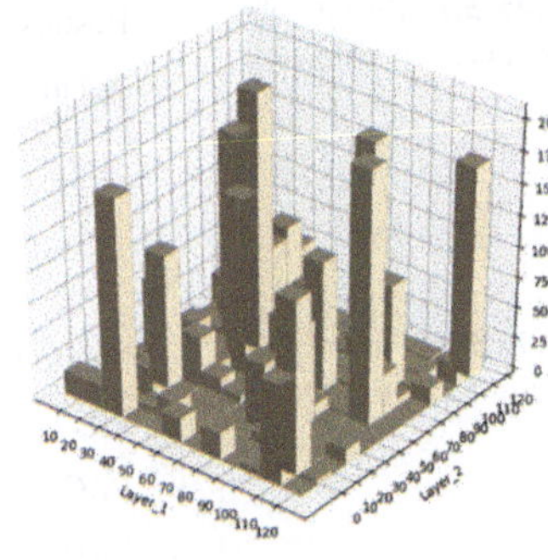

(b) False-off frequency (LBFGS, α = 0.001)

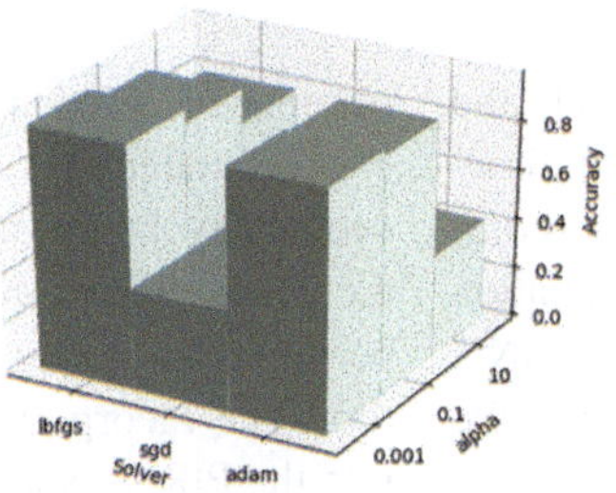

(c) Accuracy (neuron size: 100,70)

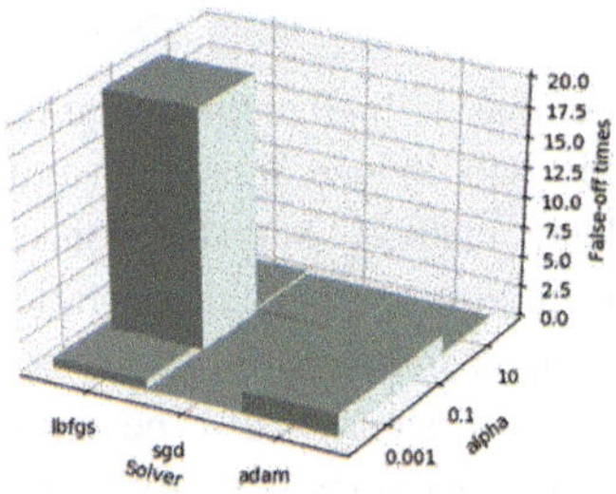

(d) False-off frequency (neuron size: 100,70)

Fig. 5.15 Control performance results of PRH-TS-ANN method

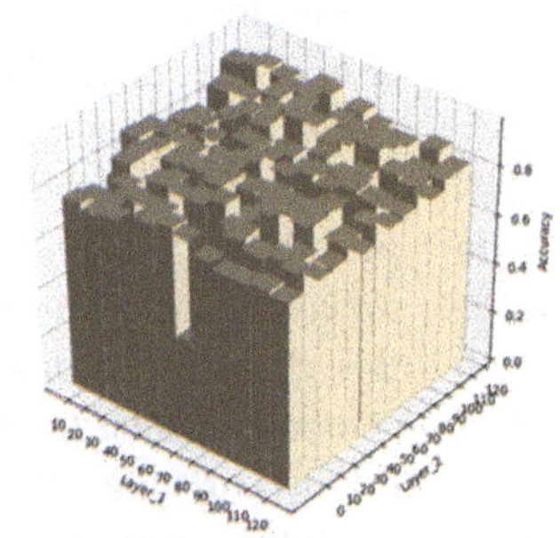

(a) Accuracy (LBFGS, α = 0.001)

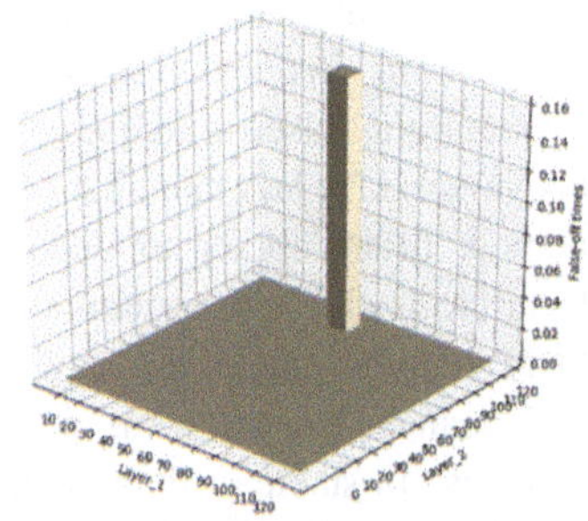

(b) False-off frequency (LBFGS, α = 0.001)

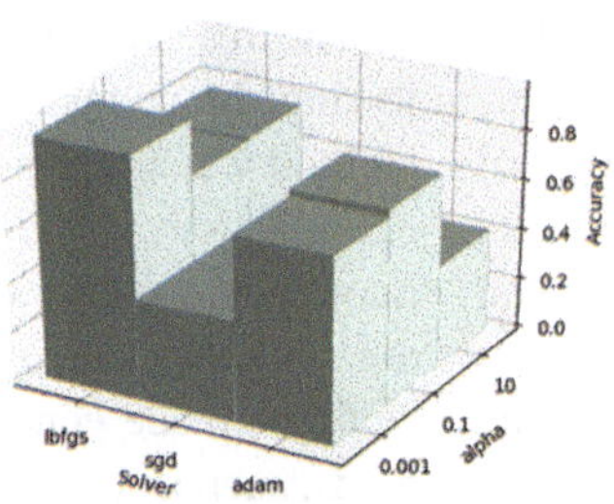

(c) Accuracy (neuron size: 60,110)

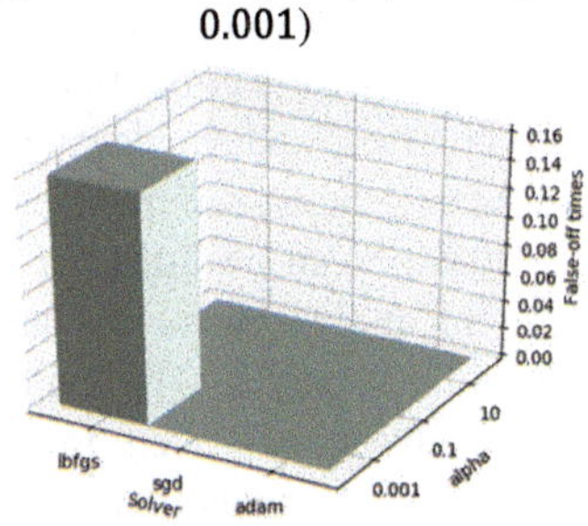

(d) False-off frequency (neuron size: 60,110)

Table 5.7 Results of parameter optimization

Results	PR method	PRC method	PRH method
Accuracy (%)	98.13	97.37	97.34
"False-off" frequency (day^{-1})	17.1	0.6	0.2

Table 5.8 Parameter setting for optimized model

Method	Solver	α	Neuron size of Layer 1	Neuron size of Layer 2
PR	LBFGS	0.001	120	120
PRC	LBFGS	0.001	100	70
PRH	LBFGS	0.001	60	110

methods is higher than time-delay method, and this will induce more consumption than time-delay method, but still conserve energy compared to static schedule operation. The evaluation of the control performance for three TS-ANN methods should also be conducted from the control accuracy, false-off frequency as well as the total duration of lighting. The evaluation results are summarized in Table 5.7 and Fig. 5.18. The PR-TS-ANN method predicts occupancy state with greater precision. However, the false-off frequency is significantly higher. Although the false-off frequency of the PRC-TS-ANN method is less than that of the PR-TS-ANN method, of its ability to distinguish short-term departure is also less pronounced. The PRH-TS-ANN method achieves remarkable prediction performance; however, in practice, its performance is not stable or reliable enough at present. In a lighting system, the accuracy and performance of the PRC-TS-ANN method is satisfactory in practice. Furthermore, it achieves less false-off times, and yields more reliable performance. In general, different temporal sequential characteristic can be fitted for different requirements of the occupancy state prediction and application scenario. The introduction of temporal analysis is proven to be useful to the occupancy state prediction.

In this research, the temporal sequential characteristic of occupancy in the office was thoroughly analyzed and incorporated into the occupancy prediction. Following the model establishment and parameter optimization, the occupancy in the office room can be well predicted. Different model inputs yield different performance under different evaluation indicators. The incorporation of occupancy prediction into the PIR sensor-based lighting system control greatly improved the control performance and accuracy of the lighting system.

In this article, we only collect an occupancy profile for a decentralized personal lighting system in one office. The occupancy profile seems simple and does not change greatly, while the detailed occupancy/non-occupancy in the whole periods is actually different, and varies with different intermediate activities. The purpose of this chapter is trying to develop the model and compare against the existing controls. Including the use of more dynamic occupancy profile in the future work is promising to further improve the model.

The research needs further work in the following directions. The parameter of the control method was from the trial experiment and model optimization; in the future study, the parameters may be adjustable with the on-site control. With the introduction of self-learning, the control could be made generalizable to other application fields, such as integrated building system control, including HVAC control. The energy saving potential will be accessed by utilizing the proposed method on building system control.

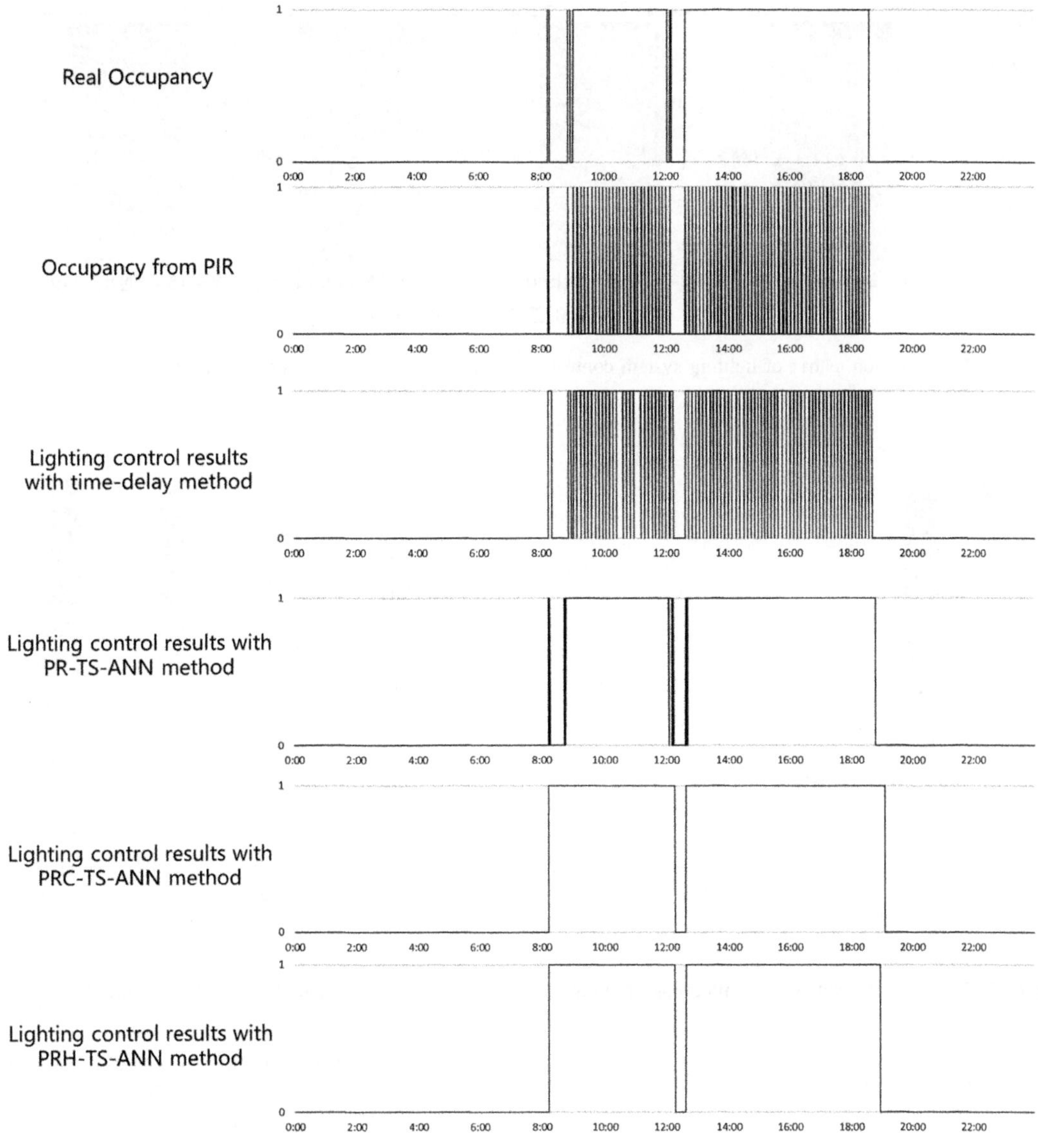

Fig. 5.16 Occupancy data of (1) real occupancy data, (2) occupancy data from PIR sensor; and (3) lighting control results with time-delay, PR-, PRC-, and PRH-TS-ANN method

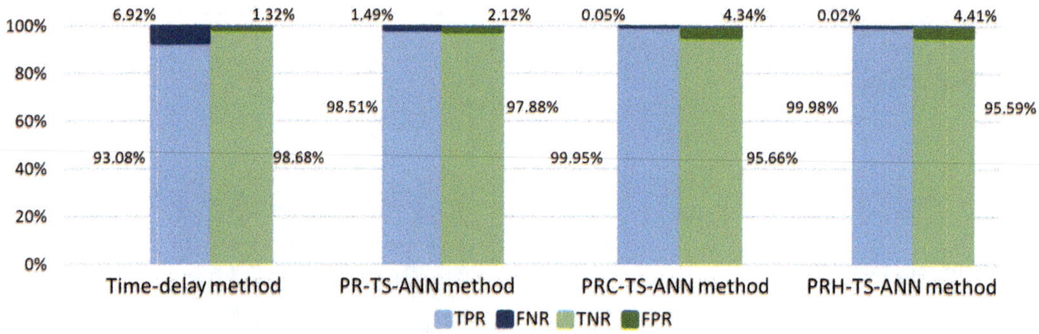

Fig. 5.17 Confusion matrix of lighting system control performance for time-delay method and three

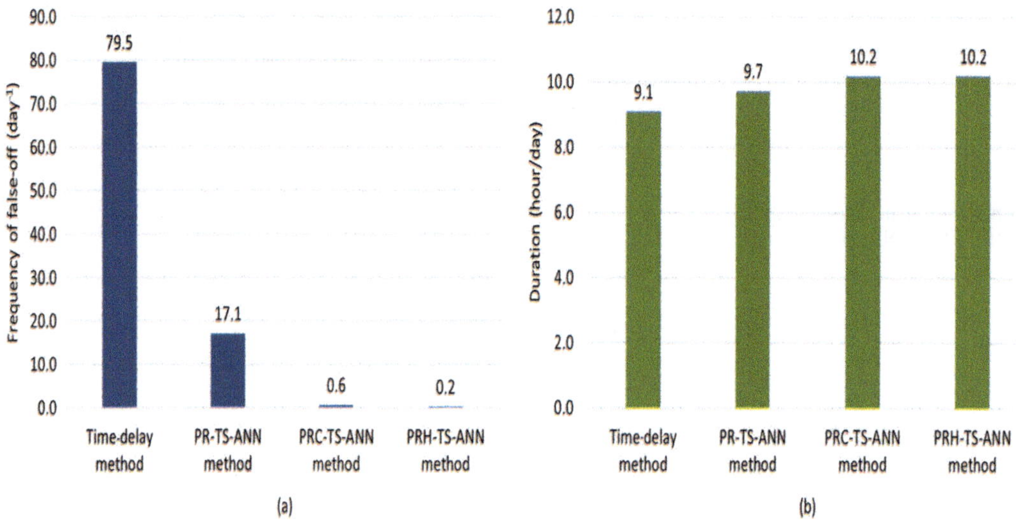

Fig. 5.18 False-off frequency and illumination duration for time-delay method and three TS-ANN methods

5.5 Summary

This research focused on enhancing the occupancy prediction accuracy of the predictive lighting system control with the goal of eliminating the discomfort arising from false offs for a room through the TS-ANN method. The proposed TS-ANN-MPC method incorporated the occupant behaviour routine analysis into on-site control, and was demonstrated to further enhance the control performance for maximum energy efficiency.

The methodology fully described how the consideration of historical occupancy contributes to occupancy prediction, establishing the feasibility of using the developed model to achieve lighting control based on the existing PIR sensors. The overall model incorporated the temporal relationship and characteristic of the occupancy patterns by utilizing the time point and historical occupancy state as inputs.

A comparison between the proposed control method and the traditional time-delay method was conducted. From the experiment, the overall accuracy of the occupancy state prediction was enhanced from 96.4% (traditional time-delay method) to 97.4% (TS-ANN method). Meanwhile, the frequency of the false-offs significantly decreased from 79.5 times per day to 0.6 times per day. Thus, it was established that the proposed control method outperformed the traditional method in terms of accuracy as well as human comfort.

A comparison among three different inputs for the proposed TS-ANN MPC method was carried out, and the parameter optimization process was conducted and illustrated. Different model inputs yielded different performances in terms of accuracy and false-off frequency.

References

Aghemo C, Blaso L, Pellegrino A (2014) Building automation and control systems: a case study to evaluate the energy and environmental performances of a lighting control system in offices. Autom Constr 43:10–22

Araji MT, Darragh SP, Boyer JL (2012) Paradigm in sustainability and environmental design: lighting utilization contributing to surplus-energy office buildings. LEUKOS 9(1):25–45

Beccali M, Bonomolo M, Lo Brano V et al (2019) Energy saving and user satisfaction for a new advanced public lighting system. Energy Convers Manage 195:943–957

Benya J (2001) Advanced lighting guidelines. New Buildings Institute (Incorporated)

Bottou L (2010) Large-scale machine learning with stochastic gradient descent. Physica-Verlag HD, Heidelberg

Bourgeois D, Reinhart C, Macdonald I (2006) Adding advanced behavioural models in whole building energy simulation: a study on the total energy impact of manual and automated lighting control. Energy Build 38(7):814–823

Buckley A, LeNir A (1985) Algorithm 630: BBVSCG–a variable-storage algorithm for function minimization. ACM Trans Math Software (TOMS) 11(2):103–119

Chen Y, Hong T, Luo X (2018) An agent-based stochastic occupancy simulator. Build Simul 11(1):37–49

Chenaru O, Hanganu CE, Popescu D et al (2019) Virtual sensor for behaviour pattern identification in a smart home application. In: 2019 8th International conference on systems and control (ICSC). IEEE, New York

Chew I, Karunatilaka D, Tan CP et al (2017) Smart lighting: the way forward? Reviewing the past to shape the future. Energy Build 149:180–191

de Bakker C, Aries M, Kort H et al (2017) Occupancy-based lighting control in open-plan office spaces: a state-of-the-art review. Build Environ 112:308–321

Dubois M-C, Blomsterberg Å (2011) Energy saving potential and strategies for electric lighting in future North European, low energy office buildings: a literature review. Energy Build 43(10):2572–2582

EIA, ANNUAL ENERGY OUTLOOK 2019 (2019) EIA

Fawcett T (2006) An introduction to ROC analysis. Pattern Recogn Lett 27(8):861–874

Galasiu AD, Newsham GR, Suvagau C et al (2007) Energy saving lighting control systems for open-plan offices: a field study. LEUKOS 4(1):7–29

Gardner MW, Dorling SR (1998) Artificial neural networks (the multilayer perceptron)—a review of applications in the atmospheric sciences. Atmos Environ 32(14):2627–2636

Goyal S, Ingley HA, Barooah P (2013) Occupancy-based zone-climate control for energy-efficient buildings: complexity vs. performance. Appl Energy 106:209–221

Guo X (2007) Occupancy sensor networks for improved lighting system control. The University of Nebraska-Lincoln

Guo X, Tiller D, Henze G et al (2010) The performance of occupancy-based lighting control systems: a review. Light Res Technol 42(4):415–431

Hong TZ, Yan D, D'Oca S et al (2017) Ten questions concerning occupant behaviour in buildings: the big picture. Build Environ 114:518–530

Hong T, Chen Y, Belafi Z et al (2018) Occupant behaviour models: a critical review of implementation and representation approaches in building performance simulation programs. Build Simul 11(1):1–14

Hunter JD (2007) Matplotlib: a 2D graphics environment. Comput Sci Eng 9(3):90–95

IEA, Energy Efficiency (2017) https://www.iea.org/efficiency2017/

IEA, Energy Technology Perspectives (2017)

Jain AK, Mao J, Mohiuddin KM (1996) Artificial neural networks: a tutorial. Computer 29(3):31–44

Kim SH, Moon HJ (2018) Case study of an advanced integrated comfort control algorithm with cooling, ventilation, and humidification systems based on occupancy status. Build Environ 133:246–264

Kim S, Kang S, Ryu KR et al (2019) Real-time occupancy prediction in a large exhibition hall using deep learning approach. Energy Build 199:216–222

Kingma D, Ba L (2015) Adam: a method for stochastic optimization

Kumar R (2017) Occupancy sensor with improved functionality. Google Patents

Kwok SSK, Lee EWM (2011) A study of the importance of occupancy to building cooling load in prediction by intelligent approach. Energy Convers Manage 52(7):2555–2564

Le QV, Ngiam J, Coates A et al (2011) On optimization methods for deep learning. In: Proceedings of the 28th International conference on international conference on machine learning. 2011 of Conference. Omnipress, Bellevue, Washington, USA

Li Z, Jiang Y (2006) Investigation methods of air conditioning energy consumption in residential buildings in summer. HVAC 36(9):35–37

Liu DC, Nocedal J (1989) On the limited memory BFGS method for large scale optimization. Math Program 45 (1–3):503–528

Magno M, Polonelli T, Benini L et al (2015) A low cost, highly scalable wireless sensor network solution to achieve smart LED light control for green buildings. IEEE Sens J 15(5):2963–2973

MANDIL C (2006) Light's labour's lost: policies for energy-efficient lighting. Energy world 343:14–15

Maniccia D, Tweed A, Bierman A et al (2001) The effects of changing occupancy sensor time-out setting on energy savings, lamp cycling and maintenance costs. J Illum Eng Soc 30(2):97–110

Manzoor F, Linton D, Loughlin M (2012) Occupancy monitoring using passive RFID technology for efficient building lighting control. In: 2012 Fourth international EURASIP workshop on RFID technology. IEEE, New York

Mirakhorli A, Dong B (2016) Occupancy behaviour based model predictive control for building indoor climate—a critical review. Energy Build 129:499–513

Nagy Z, Yong FY, Frei M et al (2015) Occupant centered lighting control for comfort and energy efficient building operation. Energy Build 94:100–108

Oldewurtel F, Sturzenegger D, Morani M (2013) Importance of occupancy information for building climate control. Appl Energy 101:521–532

Pedregosa F, Varoquaux G, Gramfort A et al (2011) Scikit-learn: machine learning in Python. J Mach Learn Res 12:2825–2830

Peng Y, Rysanek A, Nagy Z et al (2018) Using machine learning techniques for occupancy-prediction-based cooling control in office buildings. Appl Energy 211:1343–1358

Pham GN, Nguyen PH, Kwon K-R (2019) Brief paper: combining object detection and hand gesture recognition for automatic lighting system control. J Multimedia Inform Syst 6(4):329–332

Powers DM (2011) Evaluation: from precision, recall and F-measure to ROC, informedness, markedness and correlation. J Mach Learn Technol 2(1):37–63

Qiu S, Feng F, Li Z et al (2019) Data mining based framework to identify rule based operation strategies for buildings with power metering system. Build Simul 12(2):195–205

Richman E, Dittmer A, Keller J (1996) Field analysis of occupancy sensor operation: parameters affecting lighting energy savings. J Illum Eng Soc 25(1):83–92

Rubinstein F, Enscoe A (2010) Saving energy with highly-controlled lighting in an open-plan office. LEUKOS 7(1):21–36

Rummelhart DE (1986) Learning internal representations by error propagation. Parallel Distrib Process: I. Found pp 318–362

Schalkoff R (1992) Pattern classification: statistical, structural and neural approaches. Wiley, New York

Tabak V, de Vries B (2010) Methods for the prediction of intermediate activities by office occupants. Build Environ 45(6):1366–1372

Tetlow RM, Beaman CP, Elmualim AA et al (2014) Simple prompts reduce inadvertent energy consumption from lighting in office buildings. Build Environ 81:234–242

Wanjiru EM, Zhang LJ, Xia XH (2016) Model predictive control strategy of energy-water management in urban households. Appl Energy 179:821–831

Yan D, O'Brien W, Hong TZ et al (2015) Occupant behaviour modeling for building performance simulation: current state and future challenges. Energy Build 107:264–278

Yan D, Hong T, Dong B et al (2017) IEA EBC Annex 66: definition and simulation of occupant behaviour in buildings. Energy Build 156:258–270

Yin C, Dadras S, Huang X et al (2017) Energy-saving control strategy for lighting system based on multivariate extremum seeking with Newton algorithm. Energy Convers Manage 142:504–522

Zhou X, Yan D, Hong TZ et al (2015) Data analysis and stochastic modeling of lighting energy use in large office buildings in China. Energy Build 86:275–287

Tailoring Future Climate Data for Building Energy Simulation

6

Jiale Chai, Pei Huang, Jingchun Shen, and Xingxing Zhang

Abstract

Net-zero energy building (NZEB) is widely considered as a promising solution to the current energy problem. The existing NZEBs are designed using the historical weather data (e.g. typical meteorological year-TMY). Nevertheless, due to climate change, the actual weather data during a NZEB's lifecycle may differ considerably from the historical weather data. Consequently, the designed NZEBs using the historical weather data may not achieve the desired performances in their lifecycles. Therefore, this study investigates the climate change impacts on NZEB lifecycle performance (i.e., energy balance, thermal comfort and grid interaction) in different climate regions, and also evaluates different measures' effectiveness in mitigating the associated impacts of climate change. In the study, the multi-year future weather data in different Chinese climate regions are firstly generated using the morphing method. Then, using the generated future weather data, the lifecycle performances of the NZEBs, designed using the TMY data, are assessed. Next, to mitigate the climate change impacts, different measures are adopted and their effectiveness is evaluated. The study results can improve understanding of the climate change impacts on NZEB lifecycle performance in different climate regions. They can also help select proper measures to mitigate the climate change impacts in the associated climate regions.

Keywords

Net-zero energy building · Climate change · Lifecycle performance · Mitigation measures

J. Chai (✉)
Institute of Textiles and Clothing, The Hong Kong Polytechnic University, 999077 Hong Kong SAR, People's Republic of China
e-mail: jiale.chai@polyu.edu.hk

P. Huang · J. Shen · X. Zhang
Department of Energy and Community Buildings, Dalarna University, 79188 Falun, Sweden
e-mail: phn@du.se

J. Shen
e-mail: jih@du.se

X. Zhang
e-mail: xza@du.se

6.1 Introduction

Building sector accounts for 40% of primary energy use worldwide. Thus, it plays an important role in addressing the problem of energy-saving. Utilizing on-site renewable energy to meet its own energy demand, net-zero energy building (NZEB) is a promising solution to the energy problem. Many countries have

X. Zhang (ed.), *Data-driven Analytics for Sustainable Buildings and Cities*,
Sustainable Development Goals Series, https://doi.org/10.1007/978-981-16-2778-1_6

established clear targets to promote the practical applications of NZEBs (Recast 2010; Lu et al. 2015a; Crawley et al. 2009). For instance, The European Union has set up nearly zero-energy targets for all the new residential buildings from 2020 (Recast 2010). The U.S. set a target that all commercial buildings should achieve zero-energy by 2050 (Crawley et al. 2009).

In recent decade, existing NZEB studies mainly focused on: NZEB definitions (Marszal et al. 2011; Sartori et al. 2012), energy-efficiency technologies in NZEBs (Sharma et al. 2009; Alizadeh and Sadrameli 2016), NZEB system design and optimization (Zhang et al. 2016; Sun et al. 2015), NZEB system control and optimization (Lu et al. 2015a, b), and NZEB lifecycle performance assessment and improvement (Huang et al. 2018; Deng et al. 2014). In particular, NZEB lifecycle performance attracted increasing attention due to its significance and complexity. Within a NZEB's lifecycle, many different factors can affect its performance, including human behaviour, affluence and affordability, building and system efficiency, and climate change, etc. Human behaviour, such as window opening, temperature set-point reset, and occupant rate change, has been proved to have significant impacts on building energy use (Heinonen and Junnila 2014). The factors of affluence and affordability could cause the change of the operation time of the air-conditioning system and the indoor temperature set-point, and thus change associated building cooling load (Sivak 2009). In a NZEB's lifecycle, the degradation of building and system efficiencies may lead to a large difference between its actual energy use and the expected one based on original design efficiency. A latest study has explored the impacts of system performance degradation on NZEB from a lifecycle perspective (Huang et al. 2018). Meanwhile, climate change can directly influence a NZEB's performance in terms of energy use, thermal comfort and renewable generation (Lucena et al. 2009; Wan et al. 2012; Wang and Chen 2014; Zhai and Helman 2019).

In existing studies, the NZEB system is mainly sized using the historical weather data (e.g. typical meteorological year-TMY) (Shen and Lior 2016); while the weather data during the NZEB system operation can be different from the historical weather data. As the existing NZEBs are expected to serve for more than fifty years (Deng et al. 2014; Cui et al. 2015, 2017), climate change can have adverse impacts on their life-cycle performance in terms of energy balance, thermal comfort and grid interaction. Energy balance refers to how the local energy generation meet the building electrical load (Voss et al. 2010); thermal comfort indicates the user's satisfaction with the indoor thermal environment (Huang et al. 2015); grid interaction measures the energy exchange between a NZEB and a power grid, which is caused by the unstable and intermittent features of renewable energy (Salom et al. 2011).

Regarding the energy balance, climate change can affect both the building energy demand and renewable energy supply. Olonscheck et al. (2011) found that due to climate change, the heating energy demands of residential buildings in Germany decreased by 44–75% while the associated cooling energy demands increased by 28–59% as the period 2031–2060 was compared with the period 1961–1990. Santamouris et al. (2015) reported building energy use increased at a rate of 0.5–8.5%/°C in response to outdoor temperature rise under climate change. Robert and Kummert (2012) showed that in comparison with 1961–1990, the average wind speed increased by 7.4% in winter and decreased by 9.2% in summer in 2050s, thereby leading to surplus wind energy in winter but insufficient wind energy in summer. Lucena et al. (2009) pointed out both biodiesel and hydropower decreased by 5–10% in the northeast of Brazil under climate change, thus resulting in insufficient renewable energy generation. As climate change affects both NZEB energy demand and supply, NZEB may not achieve the energy-balance target in future years. Investigation of climate change impacts on NZEB energy demand and supply could help improve the energy balance.

Regarding the thermal comfort, climate change can have significant impacts on NZEB's indoor thermal comfort due to the variations of future extreme weather conditions (e.g. increase of extreme hot days in future summers). Kikumoto et al. (2016) predicted that the extremely hot days in summer dramatically increased in the future of Japan. As the existing heating, ventilation and air-conditioning (HVAC) system failed to provide sufficient cooling, the total hours with unsatisfactory thermal comfort largely increased. Gupta and Gregg (2012) found that due to outdoor temperature rise in summer, the unmet hours increased from 12 under current climate to 474 under 2030s' climate in a residential building of UK. Coley et al. (2012) concluded that due to climate change, the indoor overheating hours increased by 70 when the maximum ambient temperature raised by 1 °C in summer. Climate change causes the variations of extreme weather conditions in future and thus affects the indoor thermal comfort. This study will explore the climate change impacts on thermal comfort.

Regarding the grid interaction, climate change can influence NZEB's energy consumption and generation, and thus worsen the fluctuations and increase peak value of the energy exchange between a NZEB and a grid. Consequently, it poses more stress on the grid power balance and even damages the power-supply quality. Ref. (Zhang et al. 2016) showed that increased solar radiation led to more irregular renewable energy exported to the grid, which increased the disturbance of grid power supply and posed stress on grid power balance. Ref. (Shen and Sun 2016) illustrated that due to the increased building energy use under climate change, the grid interaction index could increase by 18%. A larger grid interaction index meant increased fluctuations of energy exchange between a NZEB and a grid, thereby resulting in poor power-supply quality. Meanwhile, Salom et al. (2014) proved that the peak power delivered to the grid increased by 10–13% under climate change. A larger peak power increased the grid stress and damaged the power-supply quality. Climate change affects the NZEB's energy consumption and generation and thus leads to poor grid friendliness. To gain a better understanding of climate change impacts, this study will investigate the NZEB's grid interaction under climate change.

Meanwhile, climate change impacts on NZEB performance vary in different climate regions. Wang and Chen (2014) summarized that due to climate change, the building energy use increased in subtropical region while it decreased in cold region. Lam et al. (2005) concluded that the frequency of extreme hot weather conditions in summer in inland area was less than that in coastal area under climate change, which caused more unmet hours in air-conditioned buildings in inland area. Ref. (Salom et al. 2014) indicated that due to climate change, the peak power delivered to the grid increased in heating-dominated region while it decreased in cooling-dominated region. Climate change showed diverse impacts on NZEB performance in different climate regions. Neglecting the diversity of climate change impacts, the decision-makers may take non-effective measures to improve the NZEB performance in different climate regions. The NZEB performance in different climate regions will also be investigated under climate change.

Existing studies have utilized various measures to mitigate the climate change impacts. Wong et al. (2010) proposed that raising room temperature set point by 1.5 °C effectively reduced the building energy use under climate change. Thus, it can help improve the energy balance. Gupta and Gregg (2012) found that under climate change the unmet hours can be decreased significantly by adding thermal insulation (i.e., 65 mm) to the external walls. Chow et al. (2013) concluded that under climate change the cooling energy use can be decreased by 40% through reducing the U-value of building envelope (e.g. changing the window U-value from 5.7 to 2.5 W/m^{2}K). It was helpful to offset the renewable energy shortage and improve the grid power balance. Nevertheless, these studies merely evaluated a measure's effectiveness with a specific case while neglecting the effectiveness variations of the measure under climate change. For instance, the effectiveness of thermal insulation was evaluated with 65 mm (Gupta and

Gregg 2012) but it was not further evaluated within 65–150 mm. Meanwhile, these studies merely evaluated the effectiveness of mitigation measures under future average weather conditions (e.g. future TMY) while neglecting the measures' effectiveness under year-to-year variations. Last, as different climate regions have their own climatic features, the effectiveness of a specified measure is not systematically evaluated in improving NZEB lifecycle performance in diverse climate regions.

Limited studies consider the climate change impacts on NZEB lifecycle performance in typical climate regions, and the evaluation of mitigation measures is also lacking. Therefore, this study will investigate the climate change impacts on NZEB lifecycle performance in typical climate regions. Meanwhile, the effectiveness of different measures in mitigating the climate change impacts is also evaluated. This chapter is organized as follows. First, the morphing method is employed to generate the future hourly weather data. Second, the NZEB lifecycle performance is assessed using the generated weather data. Third, the effectiveness of various mitigation measures is evaluated as climate change considered. Conclusive remarks are drawn in the last part.

6.2 Methodology

6.2.1 Overview

This study aims at investigating the climate change impacts on NZEB lifecycle performance in typical climate regions and evaluating the effectiveness of mitigation measures on NZEB lifecycle performance under climate change. Figure 6.1 shows the basic idea of this study. The proposed study contains three steps.

The first step is to generate the future weather data using the morphing method (Belcher et al. 2005). Due to its simplicity and flexibility, morphing method is widely adopted to evaluate the building energy performance under climate change worldwide (Wang and Chen 2014; Mourshed 2011). The principle of the method is to combine a "baseline" hourly weather data file with the future monthly weather variables predicted by a global climate model (GCM). In the second step, the NZEB lifecycle performance is investigated using the future weather data regarding energy balance, thermal comfort and grid interaction. Note that the existing NZEB systems (i.e., HVAC system, renewable energy system) are sized by the TMY weather data. In the third step, the effectiveness of different measures is evaluated in mitigating the climate change impacts on NZEB performance. Two mitigation measures are taken for each NZEB performance aspect. For instance, the renewable system size increase and free cooling use are adopted to mitigate the climate change impacts on energy balance. At last, conclusive remarks on the effectiveness of mitigation measures are drawn for each performance aspect.

6.2.2 Generation of Future Weather Using the Morphing Method

6.2.2.1 Typical Meteorological year (TMY)

In the morphing method, a "baseline climate" is prerequisite, which is defined as the present-day weather sequence average over a number of years. As TMY weather data is considered as the average weather conditions over a long-term period (Hall et al. 1978), it is adopted in the morphing process as the baseline climate. Due to the accessibility and completeness, the TMY weather data files can be downloaded from the website (https://energyplus.net/weather) for all the selected cities (i.e., Harbin, Beijing, Shanghai, Kunming, Hong Kong) in this study. The TMY data of Harbin, Beijing, Shanghai and Kunming were derived from the datasets of International Weather for Energy Calculations from the ASHRAE Research Project 1015 (Thevenard and Brunger 2001). The TMY data

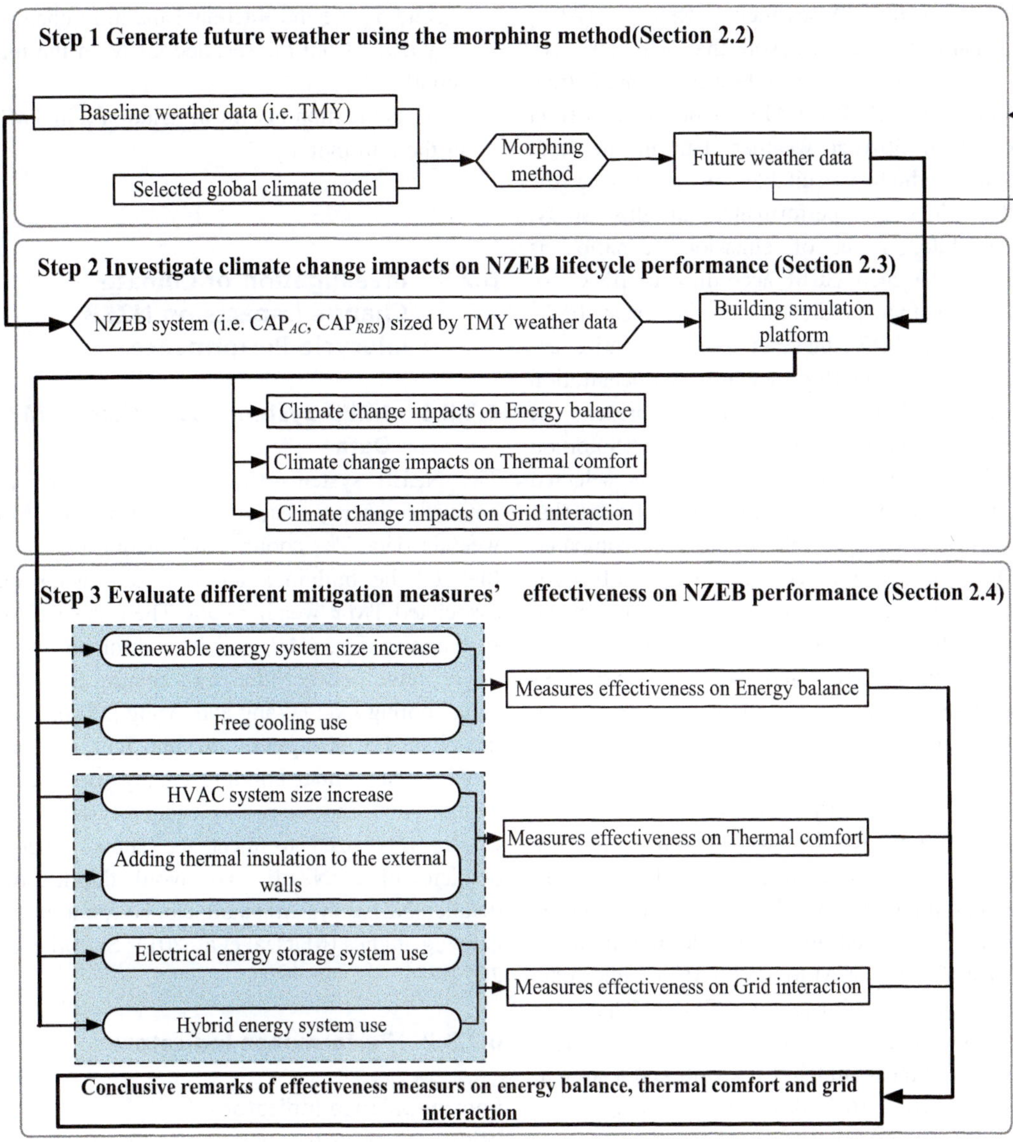

Fig. 6.1 Basic idea of the NZEB lifecycle performance evaluation under climate change

of Hong Kong was derived from the datasets of the Hong Kong Observatory (Chan et al. 2006). Note that the datasets used in the study have also been adopted by the commercial software including EnergyPlus (Crawley et al. 2001) and Trnsys (Klein 2007). Meanwhile, it should be mentioned that the latest weather data can help improve future weather prediction results and the researchers are recommended to use their available latest data for the climate change modelling.

6.2.2.2 Global Climate Model (GCM) Selection

The GCMs are combined with emission scenarios to obtain a prediction of the future climate. In this study, the CNRM-CM5 model (Voldoire et al. 2013) is employed to generate the future weather data files, and it is chosen here for two reasons. First, it has a much higher spatial resolution of 1.4° × 1.4° (latitude by longitude) over land grids than most of other GCMs in the

IPCC's Fifth Assessment Report (AR5) (Pachauri et al. 2014). Note that a higher resolution leads to more accurate weather prediction. Second, the CNRM-CM5 model can predict more than 50-year weather data in the 21th century, which is suitable for evaluating the NZEB lifecycle performance in this study. Meanwhile, a set of emission scenarios is assigned to each GCM according to IPCC for future weather prediction. In the latest IPCC report (i.e., AR5), these emission scenarios are described by the Representative Concentration Pathways (RCPs), which mainly represent the greenhouse gas emissions in the future (Solomon 2007). In this study, RCP 4.5 was selected mainly because the existing studies indicated that it was the most possible scenario as compared with others due to its consideration of a broader set of anthropogenic emissions, the usage of the updated historical data and the employment of a more sophisticated land-use and land-cover model (Kikumoto et al. 2016; Thomson et al. 2100; Zhu et al. 2016).

6.2.2.3 Morphing Method

As the future weather data obtained from a selected GCM are usually in monthly interval, the morphing method (Belcher et al. 2005) is commonly applied to downscale the monthly changes from a GCM to the hourly weather data in the baseline climate (i.e., TMY). There are three operations in the morphing process: (1) a shift; (2) a linear stretch; (3) a shift and a stretch. Equations (6.1)–(6.3) describes the three operations for different weather variables.

$$x = x_0 + \Delta x_m \tag{6.1}$$

$$y = \alpha_m y_0 \tag{6.2}$$

$$z = z_0 + \Delta z_m + \beta_m \left(z_0 - (z_0)_m \right) \tag{6.3}$$

where,

- x, y and z are the future hourly weather data;
- x_0, y_0 and z_0 are the TMY weather data;
- Δx_m *and* Δz_m are the absolute monthly changes of x and z from the selected GCM in the mth month;
- α_m and β_m are the fractional monthly changes of y and z from the selected GCM in the mth month;
- $(z_0)_m$ is the monthly mean value of z in TMY in the *m*th month.

6.2.3 Investigation of Climate Change Impacts on NZEB Lifecycle Performance

6.2.3.1 NZEB System Sizing Using TMY Data

The NZEB systems (i.e., HVAC system and renewable energy system) are sized by the TMY weather data. The cooling and heating load profiles of the buildings are obtained using the associated TMY weather data. Then, the HVAC system size (i.e., CAP_{HVAC}) is determined by the larger value between the peak heating load and peak cooling load (Wang and Wang 2000). With the given HVAC system size and building load profiles, the annual building energy use is estimated. Next, the renewable energy system (i.e., CAP_{RES}) is sized to meet the annual energy demand of a NZEB. To avoid duplication, detailed NZEB system sizing can be found in the existing studies (Marszal et al. 2011; Sartori et al. 2012).

6.2.3.2 Performance Indicators

Energy-Balance Indicator

With respect to the energy-balance evaluation, the number of years (N_{zero}) achieved the energy-balance target (i.e.,$\Phi_{energy} \geq 0$) is used (Salom et al. 2014) and it was calculated as follows.

$$N_{zero} = \sum \alpha_i \begin{cases} \alpha_i = 1, & if\ \Phi_{energy,i} \geq 0 \\ \alpha_i = 0, & if\ \Phi_{energy,i} < 0 \end{cases} \tag{6.4}$$

$$\Phi_{energy,i} = E_{gen,i} - E_{con,i} \tag{6.5}$$

where,

- α_i represents the value of energy balance in the *i*th year;

- $\Phi_{energy,i}$ is the energy difference in the *i*th year;
- $E_{gen,i}$ and $E_{con,i}$ are the annual energy generation and consumption in the *i*th year, respectively.

Thermal-Comfort Indicator

With respect to the thermal-comfort evaluation, the comfort index developed in Huang et al. (2015) was used. The comfort index ($N_{failure}$) considers the total failure time in which the supplied heating/cooling cannot meet the actual heating/cooling load, as shown in Eqs. (6.6 and 6.7).

$$N_{failure} = \sum_{i=1}^{n} Uh_i \quad (6.6)$$

$$Uh_i = \sum \tau_j \begin{cases} \tau_j = 1, & if\ L_{sup,j} < L_j \\ \tau_j = 0, & if\ L_{sup,j} \geq L_j \end{cases} \quad (6.7)$$

where,

- n is the number of future years;
- Uh_i is the annual failure time in the *i*th year;
- τ_j represents the value of failure time at the *j*th hour in a year;
- L_{sup} and L are hourly heating/cooling supply load and actual building load, respectively.

Grid-Interaction Indicator

In recent years, smart grid has been drawing attention especially when renewable energy generations are integrated with buildings (Wang et al. 2014). Due to its unstable and intermittent nature, the renewable energy generation at particular hours could be far more than the associated building energy demand (e.g. PV generation in sunny summer noon period), and the surplus part needs to be exported to the grid. In contrast, in other particular hours, the renewable energy generation could be far less than the building energy demand, and the insufficient part will be met by the energy imported from the grid. Existing studies (Voss et al. 2010; Salom et al. 2014) have shown that in comparison with the imported energy, the exported energy has much larger impacts on the grid power balance and its power supply quality. Referring to the existing studies (Voss et al. 2010; Salom et al. 2014), this research used the total amount of the annual exported energy to evaluate the grid impacts from a NZEB. Thus, with respect to the grid-interaction evaluation, the number of years (N_{grid}) that exports less energy to the grid than that of TMY (i.e., $E_{grid\_annual} \leq E_{grid\_annual,TMY}$) is employed and it is calculated as follows.

$$N_{grid} = \sum \beta_i \begin{cases} \beta_i = 1, & if\ E_{grid\_annual,i} \leq E_{grid\_annual,TMY} \\ \beta_i = 0, & if\ E_{grid\_annual,i} > E_{grid\_annual,TMY} \end{cases} \quad (6.8)$$

$$E_{grid\_annual} = \sum E_j^+ \quad (6.9)$$

where,

- β_i is the value of grid interaction in the ith year;
- $E_{grid\_annual,i}$ and $E_{grid\_annual,TMY}$ are the annual energy exported to the grid in the ith year and TMY respectively;
- E_j is the power mismatch (i.e., difference between building energy generation and consumption) in the *j*th hour in a year; and the plus sign means only the energy exported to the grid is calculated.

6.2.4 Evaluation of Different Mitigation Measures' Effectiveness on NZEB Performance

Different measures are adopted to mitigate the climate change impacts on NZEB performance and these measures' effectiveness is also evaluated. The required levels of measures for achieving the desired performances are used for the effectiveness evaluation. For instance, when the energy-balance indicator (N_{zero}) reach the desired performance (e.g., ξ_{zero} = 50 years), the PV system sizes increase by 1% and 5% in Harbin and Beijing respectively. Thus, PV system size increase is considered more effective in

Table 6.1 Mitigation measures considered in this study

Performance	Mitigation measures	Potential benefits
Energy balance	Renewable energy system size increase	Increase the renewable energy generation to offset the renewable energy shortage
	Free cooling use	Decrease the cooling energy use to offset the renewable energy shortage
Thermal comfort	HVAC system size increase	Increase the HVAC system capacity to reduce the failure time
	Adding thermal insulation to the external walls	Decrease the building load to reduce the failure time
Grid interaction	Electrical energy storage system use	Reduce the energy exchange between NZEB and the grid to decrease the exported energy
	Hybrid energy system use	Reduce the surplus renewable energy to the grid

mitigating the climate change impacts on energy balance in Harbin.

In terms of energy balance, the renewable energy shortages could be offset by increasing the renewable energy generation or decreasing the building energy use. In terms of thermal comfort, the total failure time could be reduced by increasing the HVAC system size or decreasing the building loads. In this study the grid-interaction indicator means the number of years that the annual exported energy is smaller than TMY. In terms of grid interaction, the annual exported energy could be decreased by reducing the energy exchange between NZEB and the grid or reducing the surplus renewable energy. In this study, two commonly used measures were chosen to evaluate their effectiveness differences in mitigating climate change impacts on the considered performance in the five typical climate regions, as shown in Table 6.1. Such evaluation results will help users select their proper/effective mitigation measures according to their climate regions.

6.3 Study Platform

6.3.1 Climate Regions and Selected Cities

In China, there are five typical climate regions, namely severe cold region, cold region, hot-summer cold-winter region, warm region and hot-summer warm-winter region (China 2008). A representative city is chosen from each climate region and they are Harbin, Beijing, Shanghai, Kunming and Hong Kong (Wan et al. 2012). In this study, TRNSYS (Klein 2007) is used to build the simulation platform including a building model, an air-conditioning system model and a renewable energy system model for each city.

6.3.2 Building Model

Using the multi-zone model (i.e., Type 56) in TRNSYS, a three-story office building is built and each floor has the same size (i.e., 20 m long, 10 m wide). The building has two windows facing north and south on each floor and the window to wall ratio is 0.25. The key design data of building envelope for the five cities is shown in Table 6.2 (Ministry of Housing and Urban-Rural Development of the People's Republic of China G-Dsfeeop 2015). The room temperature heating and cooling set points are 20 °C and 24 °C, respectively. The ventilation rate is set as 1 ACH (Air Change per Hour) and the lighting load is set as 10 W/m^2. The occupant density is set as 12 m^2/person in Hong Kong and it is 10 m^2/person in other four cities (Ministry of Housing and Urban-Rural Development of the People's Republic of China G-Dsfeeop 2015; Electrical Mechanical Services Department (EMSD) of Hong Kong. Guidelines on the performance-based building energy code

Table 6.2 Key design data of the building envelope for the five cities (Ministry of Housing and Urban-Rural Development of the People's Republic ofChina G-Dsfeeop 2015)

City	Climates	Building element	U-value (W/m^{2}K)
Harbin	Severe cold	Wall	0.46
		Window	1.40
		Roof	0.38
Beijing	Cold	Wall	0.52
		Window	2.80
		Roof	0.55
Shanghai	Hot-summer cold-winter	Wall	0.89
		Window	2.80
		Roof	0.69
Kunming	Warm	Wall	1.15
		Window	2.80
		Roof	0.86
Hong Kong	Hot-summer warm-winter	Wall	2.70
		Window	5.70
		Roof	0.89

ehwegheepes). A computer with a power of 80 W is assigned to each person and it is on–off controlled according to the regular occupancy schedule (i.e., from 8:00 am to 18:00 pm). The heating is not needed in Hong Kong while heating and cooling are both needed in other four cities.

6.3.3 Air-Conditioning System Model

In Harbin, Beijing, Shanghai and Kunming, since both heating and cooling are needed, the ground source heat pump systems are adopted in the air-conditioning systems. While in Hong Kong, since only cooling is needed, chiller and cooling tower system is used in the air-conditioning system. The schematics of air-conditioning systems are shown in Fig. 6.2. A constant speed pump is adopted in the ground side water loop (or cooling water loop). In the supplied water side, a constant speed pump is equipped in the primary water loop while a variable speed pump is equipped in the secondary water loop which delivers changeable water flow rate to satisfy the varying building load. The air-conditioning terminal is an air handling unit (AHU) and a supply fan was used to deliver the heated/cooled air to the room. The following section shows the calculation process for the main energy consumers in the air-conditioning systems in the five cities (Sun et al. 2015; Klein 2007; Li et al. 2016).

In heating mode of the heat pump, its energy consumption $W_{heating,HP}$ is calculated based on the fraction of full load power ($FFLP_{H,HP}$), the nominal energy efficiency ratio of the heat pump ($EER_{nom,HP}$) and the rated heating capacity ($Q_{H,HP}$), as shown in Eq. (6.10). In cooling mode of the heat pump, its energy consumption $W_{cooling,HP}$ is calculated based on the fraction of full load power ($FFLP_{C,HP}$), the coefficient of performance of the heat pump ($COP_{nom,HP}$) and the rated cooling capacity ($Q_{C,HP}$), as shown in Eq. (6.11).

$$W_{heating,HP} = FFLP_{H,HP} \times \frac{Q_{H,HP}}{EER_{nom,HP}} \quad (6.10)$$

$$W_{cooling,HP} = FFLP_{C,HP} \times \frac{Q_{C,HP}}{COP_{nom,HP}} \quad (6.11)$$

The power consumption of the chiller W_{chiller} can be calculated according to the fraction of full load power ($FFLP$), rated cooling capacity

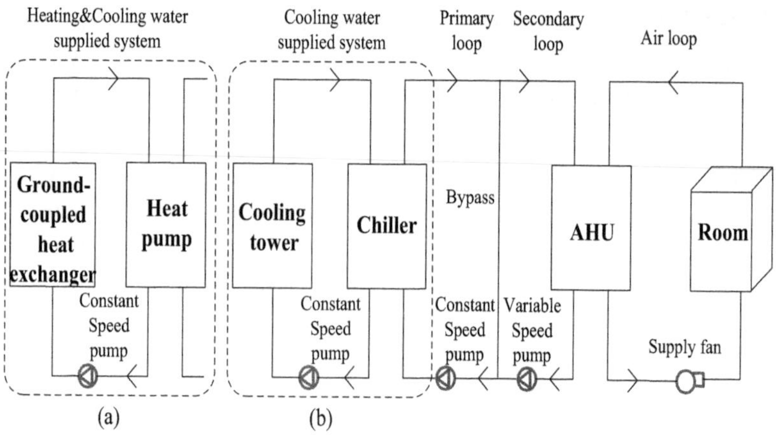

Fig. 6.2 Schematics of air-conditioning systems in **a** the four cities, i.e., Harbin, Beijing, Shanghai, Kunming; **b** Hong Kong

($Q_{\text{rated,chiller}}$) and nominal coefficient of performance of the chiller (COP_{nom}), as shown in Eq. (6.12).

$$W_{chiller} = FFLP \times \frac{Q_{rated,chiller}}{COP_{nom}} \quad (6.12)$$

The power consumption of the variable/constant speed pump can be calculated according to the water flow rate $\dot{m}_{r,w}$ and the associated water head ΔP_w, as shown in Eq. (6.13).

$$W_{pump} = \dot{m}_{r,w} \times \Delta P_w \quad (6.13)$$

The power consumption of the AHU fan can be estimated by the air flow rate $\dot{V}_{air}$ and the associated pressure drop of the air flow, as shown in Eq. (6.14).

$$W_{fan} = \dot{V}_{air} \times \Delta P_{air} \quad (6.14)$$

6.3.4 Renewable Energy System Model

As solar radiation is one of the most reliable renewable energy sources (Das et al. 2018), the PV panel is utilized to generate the renewable energy in this study. The PV panel model (i.e., Type 562) in TRNSYS is used to estimate the energy generation. Equation (6.15) shows the PV energy output (i.e., W_{PV}) with a selected size (i.e., CAP_{PV}). The parameters adopted in the PV system model were shown in Table 6.3.

$$W_{PV} = \tau \times \alpha \times IAM \times I_T \times \eta \times CAP_{PV} \quad (6.15)$$

Table 6.3 Parameters used in the PV system model

Parameter	Unit	Value
Absorptance coefficient	–	0.9
Top emissivity coefficient	–	0.9
Back emissivity coefficient	–	0.9
Refractive index	–	1.526
Extinction coefficient	1/m	4
Back resistance	m^2K/W	0.28
Cover conductivity	KJ/(hrmK)	5.04
Cover thickness	m	0.00635

where, τ and α are the transmittance and absorptance coefficients of the PV cover respectively; IAM is the overall incidence angle modifier; I_T is the total incident solar radiation and η is the overall efficiency of PV panel. The detailed parameter settings of the PV panel can be found in our previous work (Zhang et al. 2016; Sun et al. 2015).

6.4 Results and Discussions

6.4.1 Future Weather Analysis

Using the morphing method introduced in Sect. 4.2.3, the 60-year (i.e., 2021–2080) future weather files were generated for the five cities. Due to the data availability and completeness, this study took 18–25 years' historical data as baseline for climate change modelling in the five cities. In details, the TMY weather file of Hong Kong was based on the period 1979–2003 and the TMY weather files of other four cities (Harbin, Beijing, Shanghai and Kunming) were based on the period 1982–1999 (https://energyplus.net/weather). It should be mentioned that more years' historical data can improve the climate change modelling as well as future weather predictions. Under climate change, the outdoor temperature gradually increased and thus resulted in increased cooling loads and decreased heating loads. As the heating degree day (HDD) and cooling degree day (CDD) are widely used to estimate the building heating loads and cooling loads respectively (Xu et al. 2012), this study presented the changes of outdoor temperature in the forms of heating degree day changes and cooling degree day changes (compared with TMY). For simplicity, the balance point temperature was assumed to be 18 °C in the calculation of heating degree day and cooling degree day in the five cities (Büyükalaca et al. 2001).

Figure 6.3 shows the heating degree day changes (ΔHDD) and cooling degree day changes (ΔCDD) from 2021 to 2080 in the five cities. Meanwhile, the fitting curves obtained from linear regression are also presented. Note that since heating was not needed in Hong Kong, the heating degree day of Hong Kong was not calculated. As shown in Fig. 6.3, the heating degree day changes of the four cities (except Hong Kong) showed downward trends while the cooling degree day changes of the five cities showed upward trends in the future 60 years. In addition, in Harbin the decreasing rate of heating degree day changes outweighed the increasing rate of cooling degree day changes, which may lead to a downward trend of future total building load. In other three cities (i.e., Beijing, Shanghai, Kunming), the decreasing rate of heating degree day changes were smaller than the increasing rate of cooling degree day changes, which may lead to upward trends of total building loads.

To more clearly show the overall changes of the two degree days, Table 6.4 summarizes the average heating degree day changes ($\overline{\Delta HDD}$) and cooling degree day changes ($\overline{\Delta CDD}$) from 2021 to 2080 in the five cities. In Harbin and Beijing, the decrease of average heating degree day outweighed the increase of average cooling degree day, and this may lead to reduction of the total building loads in the future (Verbai et al. 2014). For instance, In Beijing, as the heating degree day decreased by 348 °C and the cooling degree day increased by 224 °C, climate change may lead to reduced total building loads. In comparison, in other three cities, the increase of average cooling degree day outweighed the decrease of average heating degree day, and this may lead to increased total building loads in the future.

In this study, the low outdoor temperatures (T_{Low}) and high outdoor temperatures (T_{High}) were introduced to represent the extreme cold and hot weather conditions, respectively. The low outdoor temperatures were the ones below the minimum outdoor temperature in TMY ($T_{\mathrm{min,TMY}}$) while the high outdoor temperatures were the ones above the maximum outdoor temperature in TMY ($T_{\mathrm{max,TMY}}$). Note that the low/high outdoor temperatures may produce the heating/cooling loads exceeding the HVAC system heating/cooling capacity, and thus caused thermal discomfort (Sun et al. 2015). To more clearly illustrate the variations of future extreme

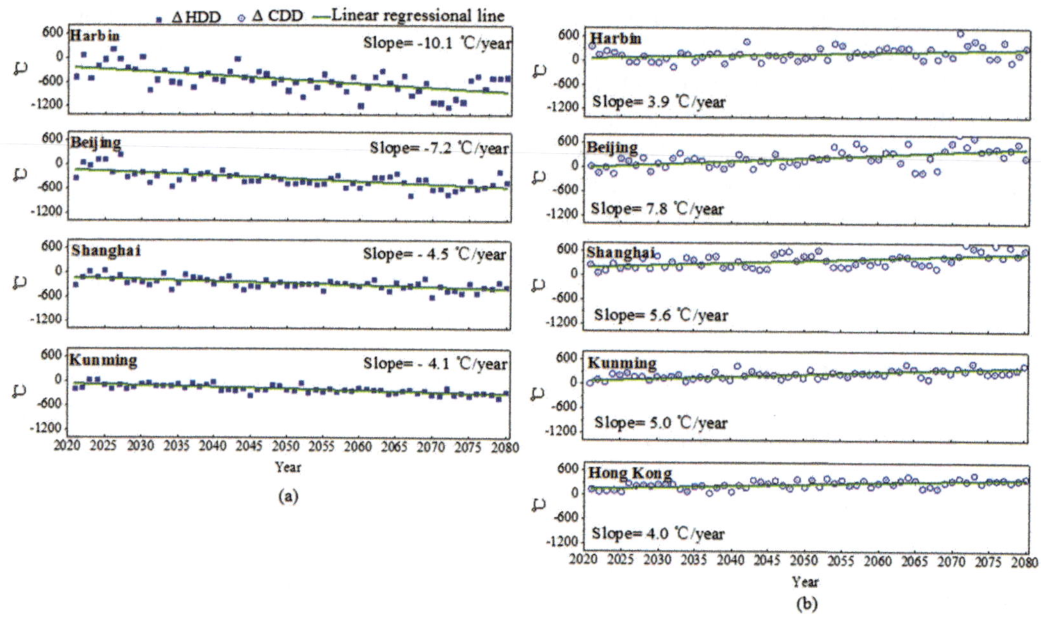

Fig. 6.3 **a** heating degree day changes (ΔHDD) and **b** cooling degree day changes (ΔCDD) from 2021 to 2080 in the five cities

Table 6.4 Average heating degree day changes ($\overline{\Delta HDD}$) and cooling degree day changes ($\overline{\Delta CDD}$) from 2021 to 2080 in the five cities

City	$\overline{\Delta HDD}$(°C)	$\overline{\Delta CDD}$(°C)
Harbin	−541	168
Beijing	−348	224
Shanghai	−267	359
Kunming	−186	241
Hong Kong	–	267

weather conditions, the low/high outdoor temperatures were compared with the minimum/maximum outdoor temperatures in TMY for each city. The deviations (i.e., $\Delta T_{Low}=T_{Low}-T_{min,\ TMY}$ and $\Delta T_{High}=T_{High}-T_{max,\ TMY}$) were presented in histograms, as shown in Fig. 6.4. The frequency of low outdoor temperatures and high outdoor temperatures were marked as n_{Low} and n_{High} respectively. A few features were observed in Fig. 6.4. Firstly, the frequency of low outdoor temperatures was much smaller than that of high outdoor temperatures in each city (except Hong Kong). It indicated that the number of heating load exceeding the HVAC system heating capacity may be smaller than the number of cooling load exceeding the HVAC system cooling capacity, which means the thermal discomfort may mainly appear in future cooling seasons. Secondly, Kunming had the largest frequency of high outdoor temperatures, and thus it may have the largest number of cooling load exceeding HVAC system cooling capacity. In comparison, Shanghai has the smallest frequency of high outdoor temperature, and thus it may have the smallest number of cooling load exceeding HVAC system cooling capacity. Thirdly, the statistics of high outdoor temperature deviations were different among these cities, and they may also result in different statistics of excessive cooling load (the

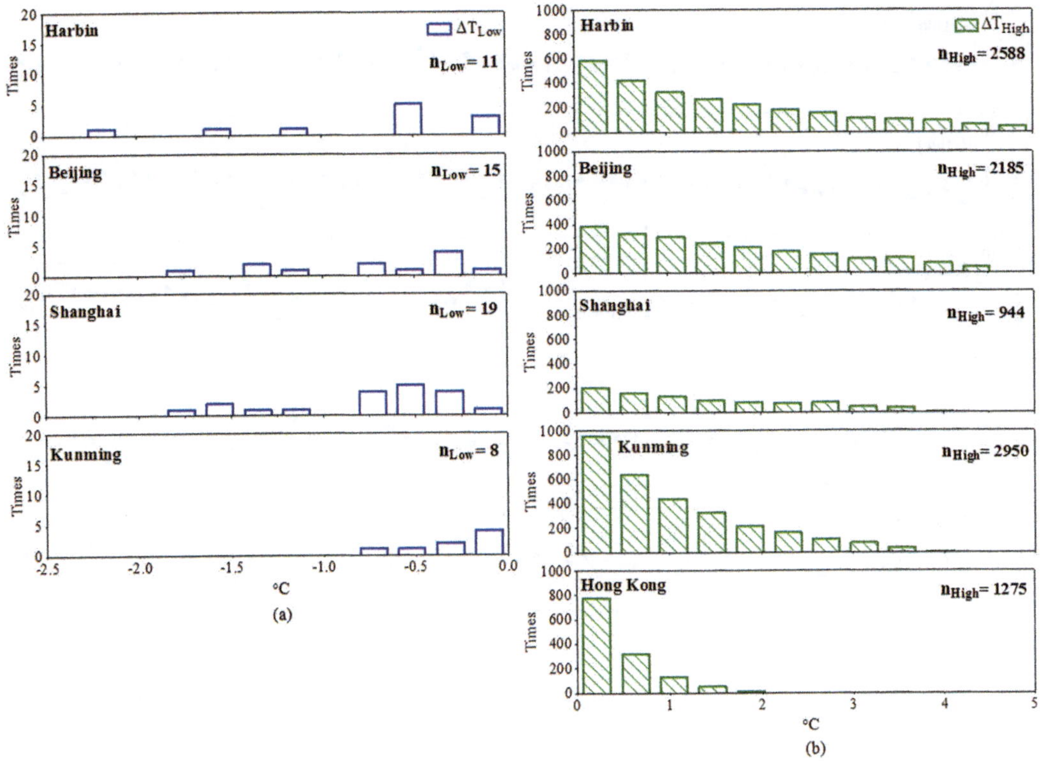

Fig. 6.4 Statistics of temperature deviations of the five cities in the future 60 years **a** low outdoor temperature deviations (ΔT_{Low}); **b** high outdoor temperature deviations (ΔT_{High})

amount of cooling load exceeding HVAC system cooling capacity). For example, in Hong Kong as more than half of the outdoor temperature deviations were close to 0 °C, most of the corresponding excessive cooling loads may be smaller than other cities under climate change.

Figure 6.5 shows the annual average solar radiation changes from 2021 to 2080 in the five cities (compared with their TMY cases). The changes of solar radiation fluctuated in a relatively small range [−20, 20] W/m^2. The solar radiation stays relatively stable under climate change, and this is consistent with findings of other studies (Chow et al. 2013; Wan et al. 2011).

6.4.2 Climate Change Impacts on NZEB Lifecycle Performance

This section mainly investigated the climate change impacts on the NZEB lifecycle performance regarding energy balance, thermal comfort and grid interaction in the five cities (i.e., Harbin, Beijing, Shanghai, Kunming and Hong Kong).

6.4.2.1 Climate Change Impacts on Energy Balance

Since the electrical loads of lighting and equipment are relatively constant during NZEB lifecycle, the total energy use changes are mainly

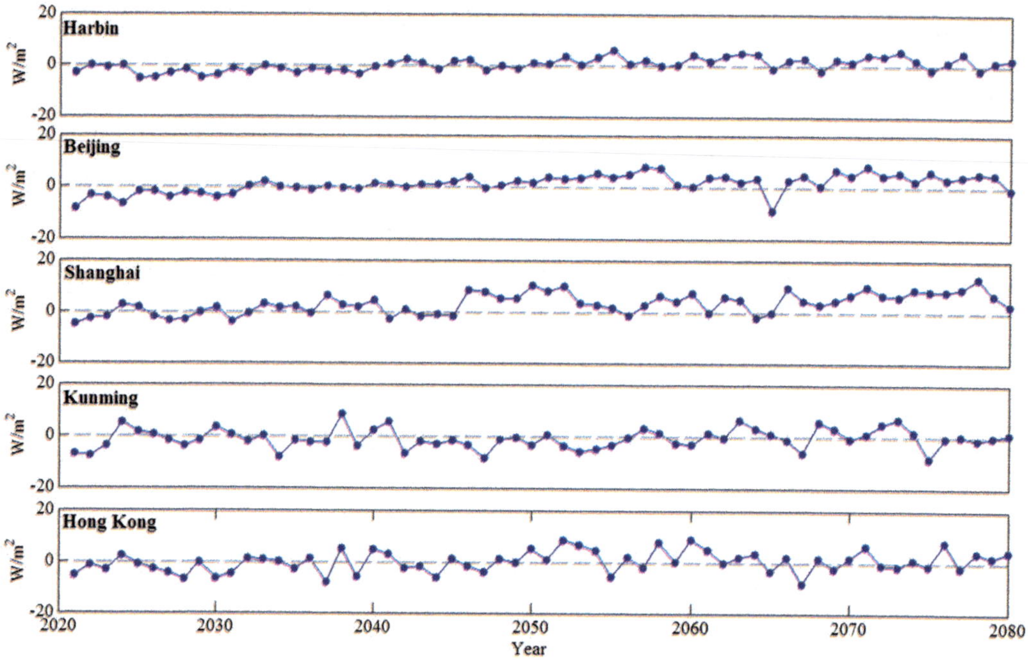

Fig. 6.5 Annual average solar radiation changes from 2021 to 2080 in the five cities (compared with their TMY cases)

attributed to the heating and cooling energy use changes. Figure 6.6 shows the heating energy use changes ($\Delta E_{heating}$), cooling energy use changes ($\Delta E_{cooling}$), and total energy use changes (ΔE_{total}) from 2021 to 2080 in the five cities, compared with TMY. The varying trends of heating/cooling energy use changes were similar to those of heating/cooling degree day changes (Fig. 6.3). In Harbin and Beijing, since the reduction of heating energy use outweighed the increase of cooling energy use, the total energy use decreased in most years. In contrast, in other three cities as the reduction of heating energy use was smaller than the increase of cooling energy use, the total energy use increased in most years.

Figure 6.7 shows the total energy use changes (ΔE_{total}), renewable energy generation changes ($\Delta E_{generation}$) and energy difference (Φ_{energy}) from 2021 to 2080 in the five cities. The energy difference was the deviation between renewable energy generation change and total energy use change. If the energy difference was non-negative (i.e.,$\Phi_{energy,i} \geq 0$) in the ith year, the NZEB was considered to achieve the energy-balance target in this year. As shown in Fig. 6.7, climate change had various impacts on energy balance (evaluated by N_{zero} in (Eq. 6.4) in the five cities. In Harbin and Beijing, the impacts were small (more than 45 years achieved energy-balance targets out of 60 years); In Shanghai, the impacts were medium (25 years achieved energy-balance targets out of 60 years); and in Kunming and Hong Kong, the impacts were large (less than 5 years achieved energy-balance targets out of 60 years). The reasons were explained below: (i) In Harbin and Beijing, the decrease of total energy use was larger than the decrease of renewable energy generation in most years. As a whole, the renewable energy generations can meet the energy demand during most years. (ii) In Shanghai, in the first 30 years, the increase of total energy use outweighed the increase of renewable energy generation in most years. As a result, the renewable energy generations cannot meet the energy demands in most years. In the later 30 years, the increase of renewable energy generation can offset the increase of total energy use, thus the energy

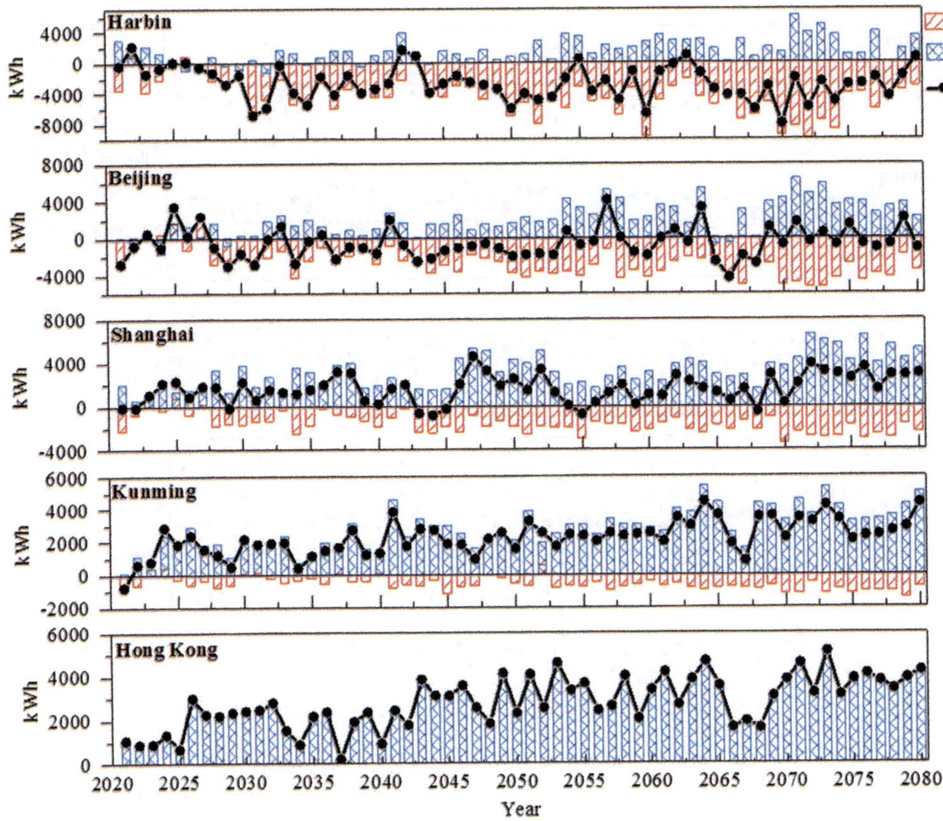

Fig. 6.6 Heating energy use changes ($\Delta E_{heating}$), cooling energy use changes ($\Delta E_{cooling}$) and total energy use changes (ΔE_{total}) from 2021 to 2080 in the five cities

balance becomes better. (iii) In Kunming and Hong Kong, the increase of total energy use was larger than the increase of renewable energy generation in most years. Thus, the renewable energy generations cannot meet the energy demands in most years.

6.4.2.2 Climate Change Impacts on Thermal Comfort

In this study, the comfort index (Eq. 6.6) represented the total failure time ($N_{failure}$) in which the HVAC system heating/cooling capacity cannot meet the actual heating/cooling loads. Here a smaller total failure time indicated better indoor thermal comfort. The failure time under climate change can be gradually reduced with the increase of HVAC system size. Figure 6.8 shows the statistics of the failure time with the increase of HVAC system capacity in the five cities. A few features were observed in Fig. 6.8. Firstly, the failure time occurred in cooling seasons in Beijing, Shanghai, Kunming and Hong Kong, while it occurred in heating seasons in Harbin. Secondly, climate change had various impacts on thermal comfort in the five cities. In Kunming and Beijing, the impacts were large (total failure time was more than 2200 h); In Shanghai and Hong Kong, the impacts were medium (total failure time was around 1100 h); and in Harbin, the impacts were small (total failure time was 11 h). This was because the total failure time was mainly determined by the low/high outdoor temperatures (the low/high outdoor temperatures may produce heating/cooling loads exceeding HVAC system heating/cooling capacity, as discussed in Sect. 4.1). Thirdly, the required HVAC

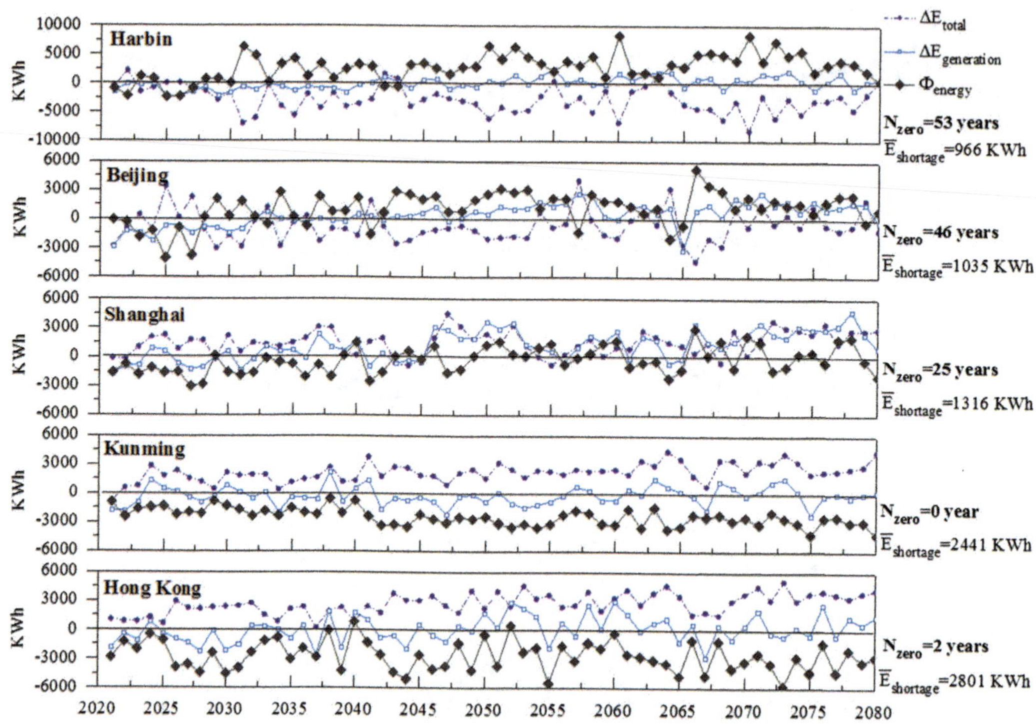

Fig. 6.7 Total energy use changes (ΔE_{total}), renewable energy generation changes ($\Delta E_{generation}$) and energy difference (Φ_{energy}) from 2021 to 2080 in the five cities. *Note* $\overline{E}_{shortage}$ was the average renewable energy shortage in the years failing to achieve energy-balance targets

system size increases ($\Delta CAP^{*}_{cooling}$) were different among the four cities (except Harbin) when the total failure time ($N^{*}_{failure}$) was reduced to the desired performance (e.g., $\xi_{failure} = 200$ h in Fig. 6.8). This was mainly attributed to the distribution of high outdoor temperature deviations under climate change (Fig. 6.4). For instance, in Hong Kong as more than half of the outdoor temperature deviations were close to 0 °C, most of the excessive cooling loads (the amount of cooling load exceeding HVAC system cooling capacity) were smaller than other cities under climate change. Thus, a smaller HVAC system size increase was required to mitigate the corresponding failure time (caused by the excessive cooling loads).

6.4.2.3 Climate Change Impacts on Grid Interaction

Figure 6.9 shows the exported energy changes in the cooling period ($\Delta E_{grid\_cooling}$), in the heating period ($\Delta E_{grid\_heating}$), and annual exported energy changes (ΔE_{grid_annual}) from 2021 to 2080 in the five cities (compared with TMY). Here the annual exported energy was the sum of exported energy in both cooling and heating periods. If the annual exported energy was smaller than TMY (i.e.,$\Delta E_{grid\_annual} \leq 0$) in *ith* year, then the NZEB was considered to achieve improved grid interaction in this year. As shown in Fig. 6.9, climate change had different impacts on grid interaction (evaluated by N_{grid} in Eq. 6.8) in the five cities. In Harbin, Beijing and Shanghai, the impacts were medium (around 30 years had improved grid interaction out of 60 years); In Kunming and Hong Kong, the impacts were small (more than 50 years had improved grid interaction out of 60 years). The reasons were explained below: (i) In Harbin, Beijing and Shanghai, the building energy use increased in cooling periods, while the increase of renewable energy generations was negligible compared with the increase of energy

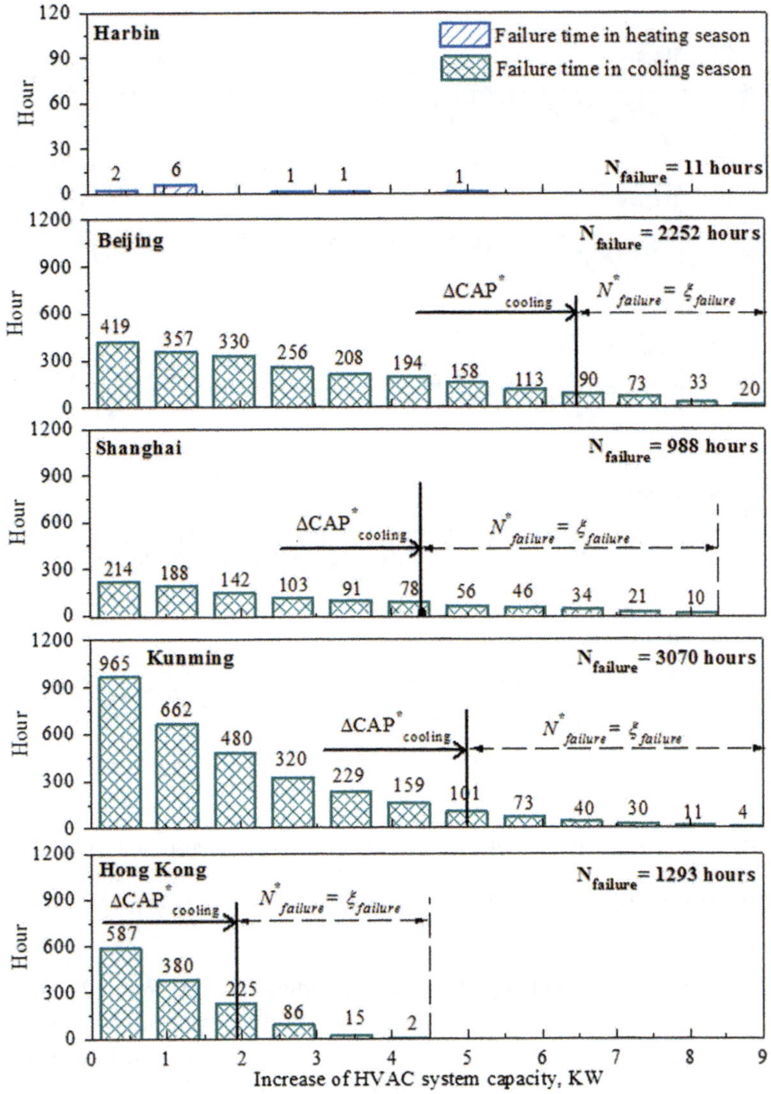

Fig. 6.8 Statistics of the failure time with the increase of HVAC system capacity in the five cities [$\Delta CAP^*_{cooling}$ was the required HVAC system cooling capacity increase for achieving the desired thermal comfort (i.e., $N^*_{failure} = \xi_{failure}$)]

demand (Fig. 6.7). Thus, more renewable energy would be consumed and less energy would be exported to the grid in most years. Meanwhile, as the building energy use decreased in the heating periods and the renewable energy generations increase slightly (Fig. 6.7). Therefore, less renewable energy would be consumed and more energy would be exported to the grid in most years. (ii) In Kunming and Hong Kong, in both heating and cooling periods the building cooling energy use increased under climate change, while the increase of renewable energy generations were negligible compared with the increase of energy demand (Fig. 6.7). Thus, more renewable energy would be consumed and less energy would be exported to the grid in most years.

6.4.3 Different Mitigation Measures' Effectiveness on NZEB Performance

In typical climate regions, climate change can worsen the NZEB performance in different magnitudes. This section evaluated different measures' effectiveness in mitigating the climate

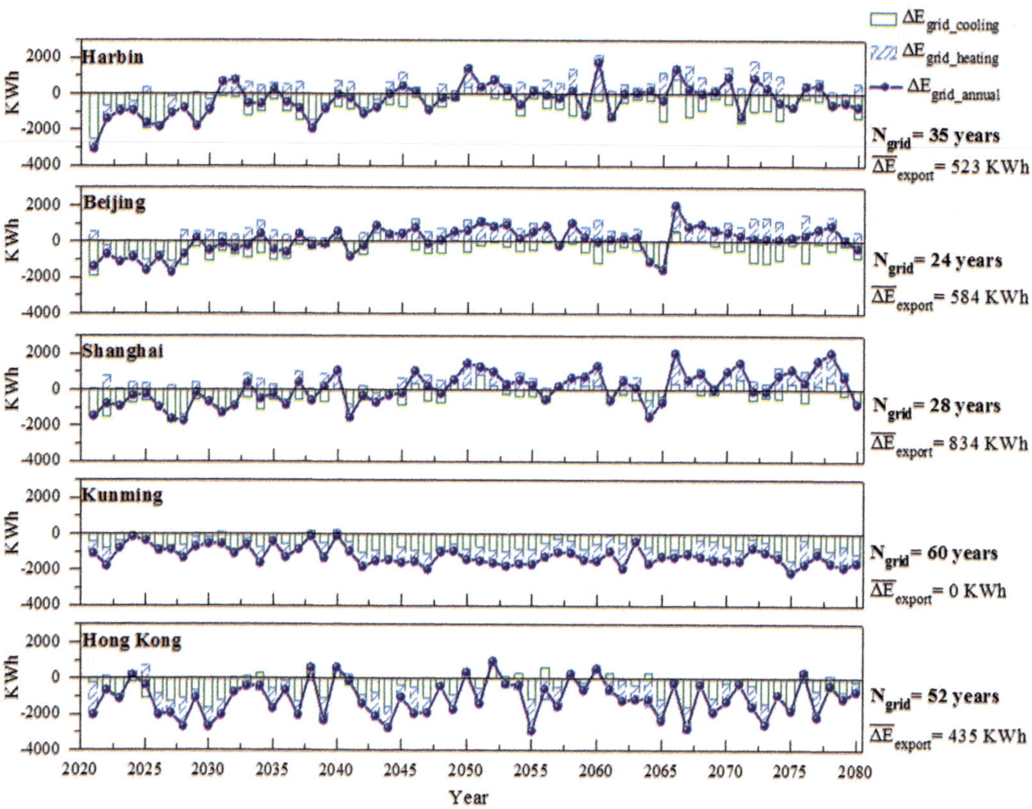

Fig. 6.9 Exported energy changes in cooling period ($\Delta E_{grid\_cooling}$) and heating period ($\Delta E_{grid\_heating}$), and annual exported energy changes (ΔE_{grid_annual}) from 2021 to 2080 in the five cities *Note* $\overline{\Delta E}_{export}$ was the average increase of exported energy in the years exporting more energy than TMY

change impacts on NZEB performance regarding energy balance, thermal comfort and grid interaction.

6.4.3.1 Effectiveness of Mitigation Measures on Energy Balance

Two measures were taken to mitigate the climate change impacts on energy balance (Shen and Lior 2016; Ascione et al. 2014), namely PV system size increase and free cooling use. In this study the energy-balance indicator (N_{zero}) meant the number of years that the annual renewable energy generation can meet the annual energy demand; and a larger N_{zero} represented better energy balance. Here, the desired energy balance (ξ_{zero}) was 54 years.

Figure 6.10a shows the effectiveness of PV system size increase on energy balance in the five cities. When the energy-balance indicator increased to 54 years, the PV system size increases of Harbin (0.5%) and Beijing (2.3%) were small; while the PV system size increases of Kunming (8.8%) and Hong Kong (9.8%) were large. In Shanghai, the increase of PV system size (4.0%) was between these two extreme regions. The differences of PV system size increase were mainly attributed to the different renewable energy shortages ($\overline{E}_{shortage}$ in Fig. 6.7) among these cities under climate change. For instance, In Harbin and Beijing, the renewable energy shortages were relatively small (smaller than 1050 KWh) and thus smaller PV system size increases were required to offset the associated renewable energy shortages.

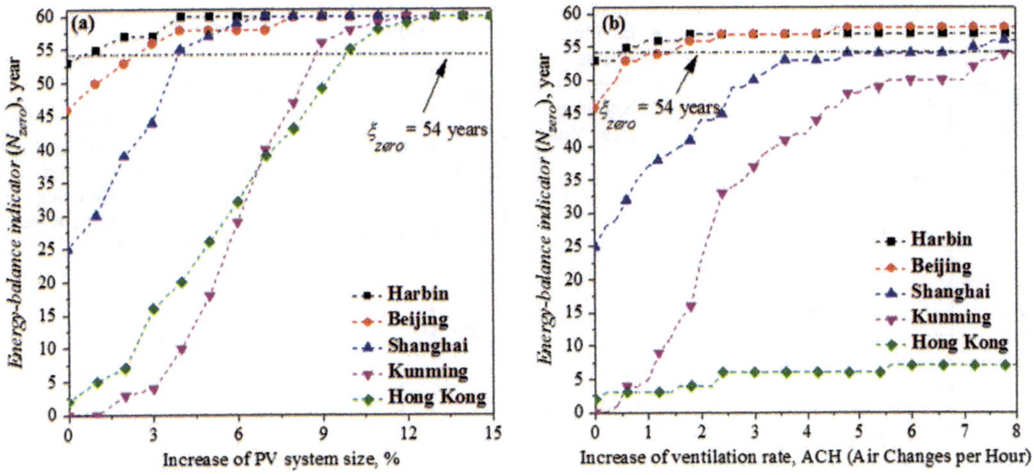

Fig. 6.10 Energy balance using the two mitigation measures in the five cities **a** PV system size increase; **b** Free cooling use

Figure 6.10b presents the effectiveness of free cooling use on energy balance in the five cities. In Hong Kong, the energy-balance indicator merely increased to 7 as the ventilation rate increased. The reason was that in the cool seasons, the outdoor temperature of Hong Kong was high (24.6 °C) and thus increasing the ventilation rate could not effectively reduce the building energy use. In contrast, in other four cities, the energy-balance indicator could reach 54 years with the increase of ventilation rate. In details, the required ventilation rate increases of Harbin (0.5 ACH) and Beijing (1.2 ACH) were much smaller than those of Shanghai (4.8 ACH) and Hong Kong (7.8 ACH). The main reason was that in the cool seasons, the outdoor temperature of Harbin (12.3 °C) and Beijing (14.7 °C) were lower than those of Shanghai (18.0 °C) and Kunming (17.5 °C). Here even the outdoor temperature of Shanghai was higher than that of Kunming, the required ventilation rate increase of Shanghai was lower than that of Kunming. The main reason was that the renewable energy shortage of Shanghai (1316 KWh) was much less than that of Kunming (2441 KWh).

6.4.3.2 Effectiveness of Mitigation Measures on Thermal Comfort

Two measures were taken to mitigate the climate change impacts on thermal comfort (Gupta and Gregg 2012; Coley et al. 2012), namely HVAC system size increase and adding thermal insulation to the exterior walls. In this study the thermal-comfort indicator ($N_{failure}$) meant the total failure time in which the HVAC system size cannot meet the actual building load; and a smaller $N_{failure}$ represented better thermal comfort. Here, the desired thermal comfort ($\xi_{failure}$) was 200 h. Note that as climate change had small impact on thermal comfort in Harbin ($N_{failure}$ = 11 h), the evaluation process was not conducted in this city.

Figure 6.11a presents the effectiveness of HVAC system size increase on thermal comfort in the five cities (Harbin was not considered). When the total failure time decreased to 200 h, the required HVAC system size increase was the largest in Beijing (6.3 KW), smallest in Hong Kong (1.9 KW), and medium in Shanghai (4.3 KW) and Kunming (5.0 KW). To avoid

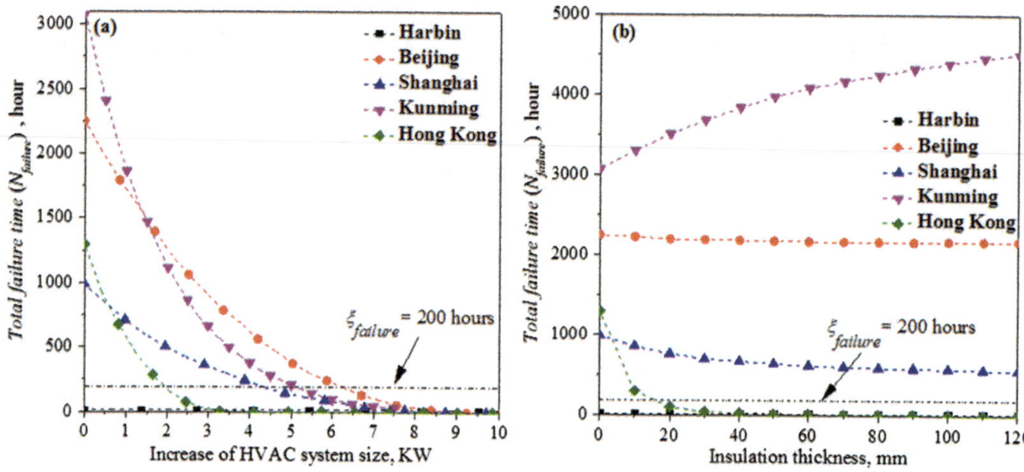

Fig. 6.11 Thermal comfort using the two mitigation measures in the five cities **a** HVAC system size increase; *b* Adding thermal insulation to the external walls

duplication, the detailed explanations can be found in Sect. 4.2.2. In addition, in the two groups of cities presenting the similar thermal comfort (Beijing and Kunming; Hong Kong and Shanghai, see Sect. 4.2.2), the total failure time of Kunming decreased faster than that of Beijing; and the total failure time of Hong Kong decreased faster than that of Shanghai. This was mainly attributed to the statistic distribution of failure time with the increase of HVAC system size (Fig. 6.8). For example, in Kunming and Beijing when the HVAC system size increased by 1.4 KW, the total failure time of Kunming decreased by 1627 h while that of Beijing decreased by 776 h. Thus, the decreasing rate of total failure time in Kunming was larger than that in Beijing.

Figure 6.11b presents the effectiveness of adding thermal insulation to the exterior walls on thermal comfort in the five cities (Harbin was not considered). In Hong Kong, when the insulation thickness increased by 17 mm, the total failure time decreased to 200 h. In Beijing and Shanghai, the total failure time decreased from 2253 to 2167 h and from 989 to 580 h respectively with the increase of insulation thickness (from 0 to 120 mm). It meant that adding thermal insulation was not so effective in Beijing and Shanghai (especially in Beijing). While in Kunming, the thermal comfort even became worse (the total failure time further increased) by adding the thermal insulation. The reasons were explained as follows. In Kunming (warm region), as the outdoor temperature was lower than the indoor temperature in most time of summer, adding thermal insulation would reduce the heat transferred from the indoor to outdoor, thereby increasing the cooling load and deteriorating the thermal comfort. In contrast, in other three cities (i.e., Hong Kong, Beijing and Shanghai), as the outdoor temperature was higher than the indoor temperature in most time of summer, adding thermal insulation would reduce the heat transferred from the outdoor to indoor, thereby decreasing the cooling load and improving thermal comfort. Furthermore, as the building envelopes were different in the diverse regions, the associated heat transfer was also different after adding thermal insulation. For instance, as the wall thickness was large in Beijing (cold region), the U-value of exterior walls decreased little by adding thermal insulation (e.g., the U-value decreased from 0.52 W/(m^2K) to 0.34 W/(m^2K) after adding a 40 mm thermal insulation). Thus, the cooling load decreased slightly and the thermal comfort was improved little. In comparison, as the wall thickness was small in Hong Kong (hot-summer warm-winter region), the U-value of exterior walls decreased sharply by adding thermal insulation (e.g., the U-value decreased from 2.70 $W/(m^2K)$ to 0.85 $W/(m^2K)$ after adding a 40 mm thermal insulation). Thus, the cooling load decreased much and the thermal comfort was improved significantly.

6.4.3.3 Effectiveness of Mitigation Measures on Grid Interaction

Two measures were taken to mitigate the climate change impacts on grid interaction (Sun et al. 2015; Yu et al. 2016), namely electrical energy storage system use and hybrid energy system use (PV and wind turbine systems combination). In this study the grid-interaction indicator (N_{grid}) meant the number of years that the annual exported energy was smaller than that of TMY; and a larger N_{grid} represented better grid interaction. Here, the desired grid interaction (ξ_{grid}) was 54 years. Note that as climate change had small impacts on grid interaction in Kunming ($N_{failure}$ = 60 years), the evaluation process was not conducted in this city.

Figure 6.12a presents the effectiveness of electrical energy storage system use on grid interaction in the five cities (Kunming was not considered). The electrical energy storage system can help mitigate the power mismatch between the energy supply and demand through charging and discharging (Chow et al. 2013). In this study, when more renewable energy was generated than the demand, the excessive part can be charged in the storage system, thereby decreasing the exported energy and improving the grid interaction. As shown in Fig. 6.12a, when the grid-interaction indicator increased to 54 years, the required electrical energy storage system size was the smallest in Hong Kong (0.6 KWh), largest in Shanghai (3.4 KWh); and medium in Harbin (1.9 KWh) and Beijing (2.3 KWh). The differences of electrical energy storage system were mainly attributed to the different increases of exported energy ($\overline{\Delta E_{export}}$ in Fig. 6.9) in these cities under climate change. For instance, in Hong Kong as the increase of exported energy (435KWh more than TMY) was relatively small among the four cities, a smaller electrical energy storage system was required to reduce the exported energy.

Figure 6.12b presents the effectiveness of hybrid energy system (PV and wind turbine (WT) systems combination) use on grid interaction in the five cities (Kunming was not considered). With the increase of WT energy generation (from 0 to 100%), the grid-interaction indicator overall experienced an increase and then a decrease in the four cities. For example, in Shanghai the grid-interaction indicator increased from 28 to 47 (indicating improved grid

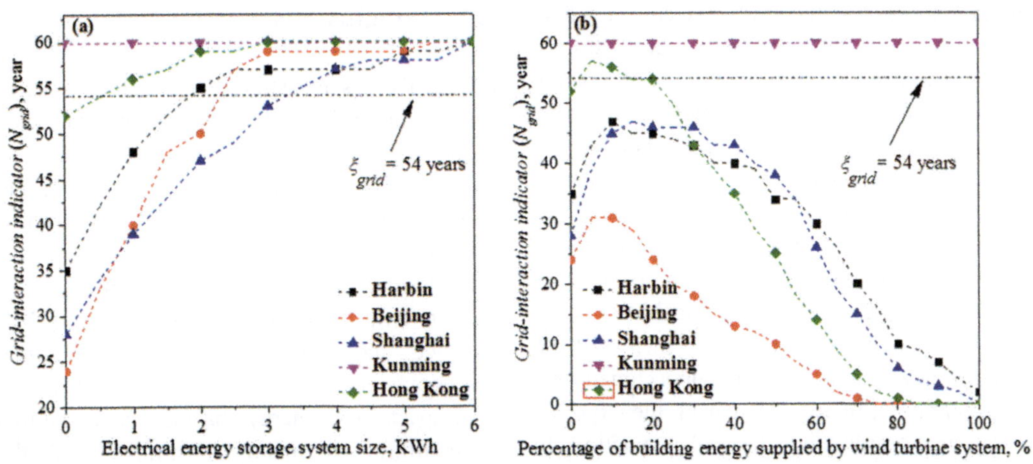

Fig. 6.12 Grid interaction using the two mitigation measures in the five cities **a** Electrical energy storage system use; **b** hybrid energy system use (PV and WT systems combination)

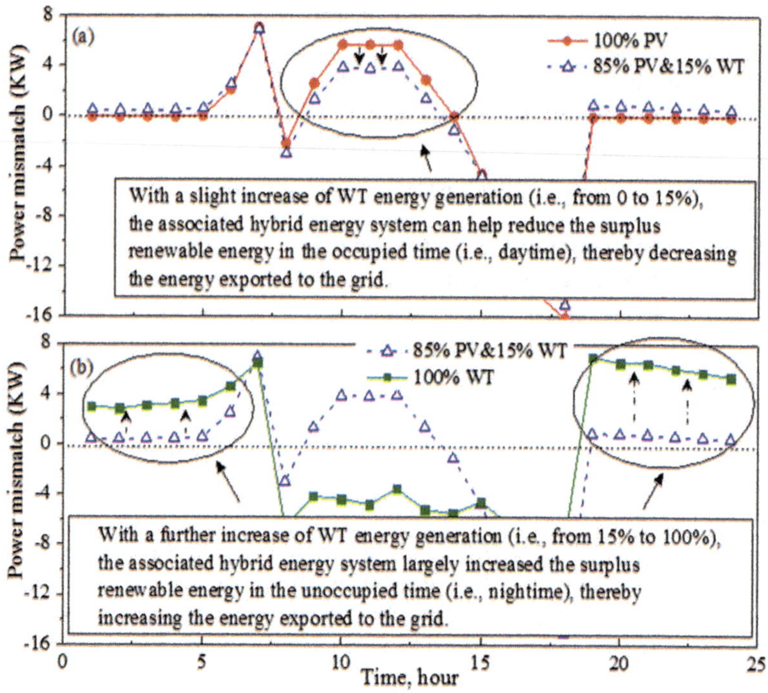

Fig. 6.13 Power mismatch using three representative energy systems (i.e., 100% PV, 85% PV and 15% WT, and 100% WT) in a typical day of Shanghai

interaction) when the WT energy generation slightly increased (from 0 to 15%); while it decreased from 47 to 0 (indicating deteriorated grid interaction) when the WT energy generation further increased (from 15 to 100%). For explanation, Fig. 6.13 presents power mismatch using three representative energy systems (i.e., 100% PV, 85% PV and 15% WT, and 100% WT) in a typical day of Shanghai. Note that only the energy exported to the grid was considered in this study (i.e., the positive power mismatch values). As shown in Fig. 6.13a, with a slight increase of WT energy generation (i.e., from 0 to 15%), the hybrid energy system can help reduce the surplus renewable energy in the occupied time (i.e., daytime), thereby decreasing the exported energy and improving the grid interaction. While in Fig. 6.13b, with a further increase of WT energy generation (i.e., from 15 to 100%), the hybrid energy system largely increased the surplus renewable energy in the unoccupied time (i.e., nighttime), thereby increasing the exported energy and deteriorating the grid interaction. In Hong Kong, the grid-interaction indicator can reach 54 years when the WT energy generation slightly increased by 4%. It was mainly because the increase of exported energy was relatively small ($\overline{\Delta E_{export}}$ in Fig. 6.9) under climate change. While in other three cities (i.e., Harbin, Beijing and Shanghai), the hybrid energy system use was not so effective as the increase of exported energy was relatively large under climate change.

6.5 Conclusions

This study has investigated the climate change impacts on NZEB lifecycle performance (i.e., energy balance, thermal comfort and grid interaction) in typical climate regions of China, and also evaluated different measures' effectiveness in mitigating the climate change impacts. The main findings are summarized as below.

(1) The climate change impacts on energy balance and thermal comfort vary largely among the different climate regions. The impacts on energy balance and thermal comfort are mainly determined by the building energy use changes and the variations of extreme hot weather conditions

(which result in cooling demands larger than the HVAC system cooling capacity), respectively. The grid interaction is relatively stable under climate change, and the associated impacts are positive as the decrease of exported energy in the cooling periods can outweigh the increase of exported energy in the heating periods.

(2) The effectiveness of mitigation measures shows distinctive features for each performance indicator. Regarding energy balance, the effectiveness of the free cooling use heavily depends on the outdoor temperature in cool seasons (e.g., the free cooling use is proved to be more effective in severe cold region). Regarding thermal comfort, adding thermal insulation to the exterior walls is less effective (compared with HVAC system size increase) in the cold region and the hot-summer cold-winter region, as it slightly reduces the heat transferred from outdoor to indoor. While adding thermal insulation is counter-effective in the warm region as it prevents the heat transferred from indoor to outdoor. The hybrid energy system use can improve the grid interaction to some limited extent; but it only mitigates the climate change impacts on grid interaction in the hot-summer warm-winter region as the corresponding increase of exported energy is smaller than other regions under climate change.

These findings will improve the understanding of the climate change impacts on NZEB lifecycle performance in different climate regions, and they will also help select proper measures to mitigate the climate change impacts in the associated climate regions. It should be mentioned that the study has not considered the impacts of the uncertainties/errors of the weather prediction results from the morphing method. The associated uncertainty analysis will be considered as a part of our future studies.

References

Alizadeh M, Sadrameli S (2016) Development of free cooling based ventilation technology for buildings: Thermal energy storage (TES) unit, performance enhancement techniques and design considerations–A review. Renew Sustain Energy Rev 58:619–645

Ascione F, Bianco N, De Masi RF, de'Rossi F, Vanoli GP (2014) Energy refurbishment of existing buildings through the use of phase change materials: Energy savings and indoor comfort in the cooling season. Appl Energy 113:990–1007

Belcher SE, Hacker JN, Powell DS (2005) Constructing design weather data for future climates. Build Serv Eng Res Technol 26(1):49–61

Büyükalaca O, Bulut H, Yılmaz T (2001) Analysis of variable-base heating and cooling degree-days for Turkey. Appl Energy 69(4):269–283

Chan AL, Chow T-T, Fong SK, Lin JZ (2006) Generation of a typical meteorological year for Hong Kong. Energy Convers Manage 47(1):87–96

China P (2008) Ministry of housing and urban-rural development. Unified Design Standard for Reliability of Engineering Structure

Chow DHC, Li Z, Darkwa J (2013) The effectiveness of retrofitting existing public buildings in face of future climate change in the hot summer cold winter region of China. Energy Build 57:176–186

Coley D, Kershaw T, Eames M (2012) A comparison of structural and behavioural adaptations to future proofing buildings against higher temperatures. Build Environ 55:159–166

Crawley DB, Lawrie LK, Winkelmann FC, Buhl WF, Huang YJ, Pedersen CO et al (2001) EnergyPlus: creating a new-generation building energy simulation program. Energy Build 33(4):319–331

Crawley D, Pless S, Torcellini P (2009) Getting to net zero. National Renewable Energy Lab.(NREL), Golden, CO (United States)

Cui B, Gao D-C, Wang S, Xue X (2015) Effectiveness and life-cycle cost-benefit analysis of active cold storages for building demand management for smart grid applications. Appl Energy 147:523–535

Cui B, Gao D-C, Xiao F, Wang S (2017) Model-based optimal design of active cool thermal energy storage for maximal life-cycle cost saving from demand management in commercial buildings. Appl Energy 201:382–396

Das UK, Tey KS, Seyedmahmoudian M, Mekhilef S, Idris MYI, Van Deventer W et al (2018) Forecasting of photovoltaic power generation and model optimization: a review. Renew Sustain Energy Rev 81:912–928

De Lucena AFP, Szklo AS, Schaeffer R, de Souza RR, Borba BSMC, da Costa IVL et al (2009) The vulnerability of renewable energy to climate change in Brazil. Energy Policy 37(3):879–889

Deng S, Wang R, Dai Y (2014) How to evaluate performance of net zero energy building–a literature research. Energy 71:1–16

Electrical & Mechanical Services Department (EMSD) of Hong Kong. Guidelines on the performance-based building energy code ehwegheepes 2007

Gupta R, Gregg M (2012) Using UK climate change projections to adapt existing English homes for a warming climate. Build Environ 55:20–42

Hall IJ, Prairie R, Anderson H, Boes E (1978) Generation of a typical meteorological year. Sandia Labs., Albuquerque, NM (USA)

Heinonen J, Junnila S (2014) Residential energy consumption patterns and the overall housing energy requirements of urban and rural households in Finland. Energy Build 76:295–303

https://energyplus.net/weather

Huang P, Huang G, Wang Y (2015) HVAC system design under peak load prediction uncertainty using multiple-criterion decision making technique. Energy Build 91:26–36

Huang P, Huang G, Sun Y (2018) Uncertainty-based life-cycle analysis of near-zero energy buildings for performance improvements. Appl Energy 213:486–498

Kikumoto H, Ooka R, Arima Y (2016) A study of urban thermal environment in Tokyo in summer of the 2030s under influence of global warming. Energy Build 114:54–61

Klein S (2007) TRNSYS 16 program manual. University of Wisconsin, Madison, USA, Solar Energy Laboratory

Lam JC, Tsang C, Yang L, Li DH (2005) Weather data analysis and design implications for different climatic zones in China. Build Environ 40(2):277–296

Li A, Xu X, Sun Y (2016) A study on pipe-embedded wall integrated with ground source-coupled heat exchanger for enhanced building energy efficiency in diverse climate regions. Energy Build 121:139–151

Lu Y, Wang S, Shan K (2015a) Design optimization and optimal control of grid-connected and standalone nearly/net zero energy buildings. Appl Energy 155:463–477

Lu Y, Wang S, Sun Y, Yan C (2015b) Optimal scheduling of buildings with energy generation and thermal energy storage under dynamic electricity pricing using mixed-integer nonlinear programming. Appl Energy 147:49–58

Marszal AJ, Heiselberg P, Bourrelle JS, Musall E, Voss K, Sartori I et al (2011) Zero energy building–a review of definitions and calculation methodologies. Energy Build 43(4):971–979

Ministry of Housing and Urban-Rural Development of the People's Republic of China G-Dsfeeop 2015

Mourshed M (2011) The impact of the projected changes in temperature on heating and cooling requirements in buildings in Dhaka Bangladesh. Appl Energy 88 (11):3737–3746

Olonscheck M, Holsten A, Kropp JP (2011) Heating and cooling energy demand and related emissions of the German residential building stock under climate change. Energy Policy 39(9):4795–4806

Pachauri RK, Allen MR, Barros VR, Broome J, Cramer W, Christ R, et al (2014) Climate change 2014: synthesis report. In: Contribution of Working Groups I, II and III to the fifth assessment report of the Intergovernmental Panel on Climate Change: IPCC

Recast E (2010) Directive 2010/31/EU of the European Parliament and of the Council of 19 May 2010 on the energy performance of buildings (recast). Off J Eur Union 18(06):2010

Robert A, Kummert M (2012) Designing net-zero energy buildings for the future climate, not for the past. Build Environ 55:150–158

Salom J, Widén J, Candanedo J, Sartori I, Voss K, Marszal A (2011) Understanding net zero energy buildings: evaluation of load matching and grid interaction indicators. In: Conference understanding net zero energy buildings: evaluation of load matching and grid interaction indicators, vol 6, pp 2514–2521

Salom J, Marszal AJ, Widén J, Candanedo J, Lindberg KB (2014) Analysis of load match and grid interaction indicators in net zero energy buildings with simulated and monitored data. Appl Energy 136:119–131

Santamouris M, Cartalis C, Synnefa A, Kolokotsa D (2015) On the impact of urban heat island and global warming on the power demand and electricity consumption of buildings—a review. Energy Build 98:119–124

Sartori I, Napolitano A, Voss K (2012) Net zero energy buildings: a consistent definition framework. Energy Build 48:220–232

Sharma A, Tyagi VV, Chen C, Buddhi D (2009) Review on thermal energy storage with phase change materials and applications. Renew Sustain Energy Rev 13 (2):318–345

Shen P, Lior N (2016) Vulnerability to climate change impacts of present renewable energy systems designed for achieving net-zero energy buildings. Energy 114:1288–1305

Shen L, Sun Y (2016) Performance comparisons of two system sizing approaches for net zero energy building clusters under uncertainties. Energy Build 127:10–21

Sivak M (2009) Potential energy demand for cooling in the 50 largest metropolitan areas of the world: implications for developing countries. Energy Policy 37(4):1382–1384

Solomon S (2007) Climate change 2007-the physical science basis: working group I contribution to the fourth assessment report of the IPCC. Cambridge university press

Sun Y, Huang P, Huang G (2015) A multi-criteria system design optimization for net zero energy buildings under uncertainties. Energy Build 97:196–204

Thevenard D, Brunger A (2001) ASHRAE research project 1015-RP, typical weather years for international locations: final report. American Society of

Heating, Refrigerating and Air-Conditioning Engineers, Atlanta, GA, USA
Thomson AM, Calvin KV, Smith SJ, Kyle GP, Volke A, Patel P, et al (2011) RCP4. 5: a pathway for stabilization of radiative forcing by 2100. Clim Change 109(1–2):77
Verbai Z, Lakatos Á, Kalmár F (2014) Prediction of energy demand for heating of residential buildings using variable degree day. Energy 76:780–787
Voldoire A, Sanchez-Gomez E, y Mélia DS, Decharme B, Cassou C, Sénési S, et al. (2013) The CNRM-CM5. 1 global climate model: description and basic evaluation. Clim Dyn 40(9):2091–2121
Voss K, Sartori I, Napolitano A, Geier S, Gonçalves H, Hall M, et al (2010) Load matching and grid interaction of net zero energy buildings. In: Conference Load matching and grid interaction of net zero energy buildings
Wan KK, Li DH, Liu D, Lam JC (2011) Future trends of building heating and cooling loads and energy consumption in different climates. Build Environ 46 (1):223–234
Wan KK, Li DH, Pan W, Lam JC (2012) Impact of climate change on building energy use in different climate zones and mitigation and adaptation implications. Appl Energy 97:274–282
Wang H, Chen Q (2014) Impact of climate change heating and cooling energy use in buildings in the United States. Energy Build 82:428–436
Wang SK, Wang SK (2000) Handbook of air conditioning and refrigeration
Wang S, Xue X, Yan C (2014) Building power demand response methods toward smart grid. HVAC&R Res 20(6):665–687
Wong SL, Wan KK, Li DH, Lam JC (2010) Impact of climate change on residential building envelope cooling loads in subtropical climates. Energy Build 42(11):2098–2103
Xu P, Huang YJ, Miller N, Schlegel N, Shen P (2012) Impacts of climate change on building heating and cooling energy patterns in California. Energy 44 (1):792–804
Yu ZJ, Chen J, Sun Y, Zhang G (2016) A GA-based system sizing method for net-zero energy buildings considering multi-criteria performance requirements under parameter uncertainties. Energy Build 129:524–534
Zhai ZJ, Helman JM (2019) Implications of climate changes to building energy and design. Sustain Cities Soc 44:511–519
Zhang S, Huang P, Sun Y (2016) A multi-criterion renewable energy system design optimization for net zero energy buildings under uncertainties. Energy 94:654–665
Zhu M, Pan Y, Huang Z, Xu P (2016) An alternative method to predict future weather data for building energy demand simulation under global climate change. Energy Build 113:74–86

A Solar Photovoltaic/Thermal (PV/T) Concentrator for Building Application in Sweden Using Monte Carlo Method

7

Yaxiu Gu and Xingxing Zhang

Abstract

The solar energy share in Sweden will grow up significantly in next a few decades. Such transition offers not only great opportunity but also uncertainties for the emerging solar photovoltaic/thermal (PV/T) technologies. This chapter therefore aims to conduct a techno-economic evaluation of a reference solar PV/T concentrator in Sweden for building application. An analytical model is developed based on the combinations of Monte Carlo simulation techniques and multi energy-balance/financial equations, which takes into account of the integrated uncertainties and risks of various variables. In the model, 11 essential input variables, i.e. average daily solar irradiance, electrical/thermal efficiency, prices of electricity/heating, operation & management (OM) cost, PV/T capital cost, debt to equity ratio, interest rate, discount rate, and inflation rate, are considered, while the economic evaluation metrics, such as levelized cost of energy (LCOE), net present value (NPV), and payback period (PP), are primarily assessed. According to the analytical results, the mean values of LCOE, NPV and PP of the reference PV/T connector are observed at 1.27 SEK/kW h (0.127 €/kW h), 18,812.55 SEK (1881.255 €) and 10 years during its 25 years lifespan, given the project size at 10.37 m^2 and capital cost at 4482–5378 SEK/m^2 (448.2–537.8 €/m^2). The positive NPV indicates that the investment on the selected PV/T concentrator will be profitable as the projected earnings exceeds the anticipated costs, depending on the NPV decision rule. The sensitivity analysis and the parametric study illustrate that the economic performance of the reference PV/T concentrator in Sweden is mostly proportional to solar irradiance, debt to equity ratio and heating price, but disproportionate to capital cost and discount rate. Together with additional market analysis of PV/T technologies in Sweden, it is expected that this chapter could clarify the economic situation of PV/T technologies in Sweden and provide a useful model for their further investment decisions, in order to achieve sustainable and low-carbon economics, with an expanded quantitative discussion of the real economic or policy scenarios that may lead to those outcomes.

Y. Gu (✉)
Department of Building Environment and Energy Engineering, Chang'an University, Xi'an 710064, China
e-mail: guyaxiu@chd.edu.cn

X. Zhang
Department of Energy and Community Buildings, Dalarna University, 79188 Falun, Sweden
e-mail: xza@du.se

X. Zhang (ed.), *Data-driven Analytics for Sustainable Buildings and Cities*, Sustainable Development Goals Series, https://doi.org/10.1007/978-981-16-2778-1_7

Keywords

PV/T · Monte Carlo · Economic · LCOE · NPV · Payback period

7.1 Market Analysis of Swedish PV/T Industry

It is likely that the economic results would be different if the study were performed in another country or even in the same country today, as costs, tariffs, and policies are changing. To make an accurate assessment, it is necessary to clarify the local policies and market structures (Poppi et al. 2018).

Currently, the Sweden's market for solar energy is very small. Nevertheless, the installation rate of PV continues to increase. In 2016, a total of 79.2 MWp were installed, showing the annual PV market grew with 63% as compared to the 48.4 MWp installed in 2015. For the PV market, the off-grid PV application accounts for a very small share, with only 1.5 MWp installed in 2016; while the market for grid-connected PV systems has grown rapidly in Sweden, with 77.7 MWp systems installed in 2016 (Lindahl 2016), as observed in Fig. 7.1. Various market segments of the yearly installed PV capacity in Sweden are illustrated. There has been a clear shift from a market dominated by off-grid systems to a grid-connected market, in which the grid-connected distributed PV systems dominates the market, mainly due to the tax law for large centralized PV systems. A new policy, published on July 2017, a 98% tax reduction on PV systems over 255 kWp has further removed the major economic barriers of PV generation systems—this will be the main source of increase of PV market. Under the new policy, the tax would be reduced from SEK 0.295 (0.0295€) per kW h to SEK 0.005 (0.0005€) per kW h (Clover 2017). The total cumulative PV capacity installed in Sweden was approximately 205.5 MWp by the end of 2016. The cumulative PV market grew with 63% under 2016, which is in line with the marked development over the last five years where the cumulative market has grown with 52%, 83%, 84%, 62% and 63%, respectively. The strong overall growth in recent years started with the introduction of the direct capital subsidy system in 2006, and has then been fuelled by the declining system prices, high popularity among the public, a growing interest from utilities and an ongoing reformation work from the Government to simplify the rules for micro producers (Lindahl 2016). According to International Energy Outlook (Lindahl 2016), solar energy in Sweden is expected to contribute 5–10% to electricity generation in 2040, which means the share of solar electricity production within the total power generation mix will rise from 0.1% in 2016 to 5–10% in 2040, which is up to 100 times of current solar power capacity if the total electricity generation remain stable as past decades. However, the PV market in Sweden is still relying much on subsidies, and the PV system prices must continue to go down, or the electricity prices to go up if PV needs to contribute to an appreciable part of the Swedish electricity generation mix. Moreover, the Swedish PV industry is becoming broader as more and more players with other core businesses, such as utilities and real estate owners, are taking an increasing interest in the PV industry. A high competition in PV industry will be expected in the near future.

The Swedish solar thermal industry is decreasing significantly from 2011. The total number of the installed solar thermal collectors in 2016 was only about 2823, nearly half of the installed capacity in 2015. The budget was about €1 million annually, while PV started out much higher, at €5 to €10 million per year. The whole solar thermal market faces increasing competition with other energy technologies, such as biomass boilers, district heating, heat pumps, solar PV. It seems as if not even the rather high national carbon tax can reinvigorate the country's solar heat market. Typical solar thermal products in Sweden are flat plate and vacuum tubes collectors, while the concentrated solar thermal plants and PV/T products are very rare. Their application includes the fields of hot water and space heating, in operation with other energy

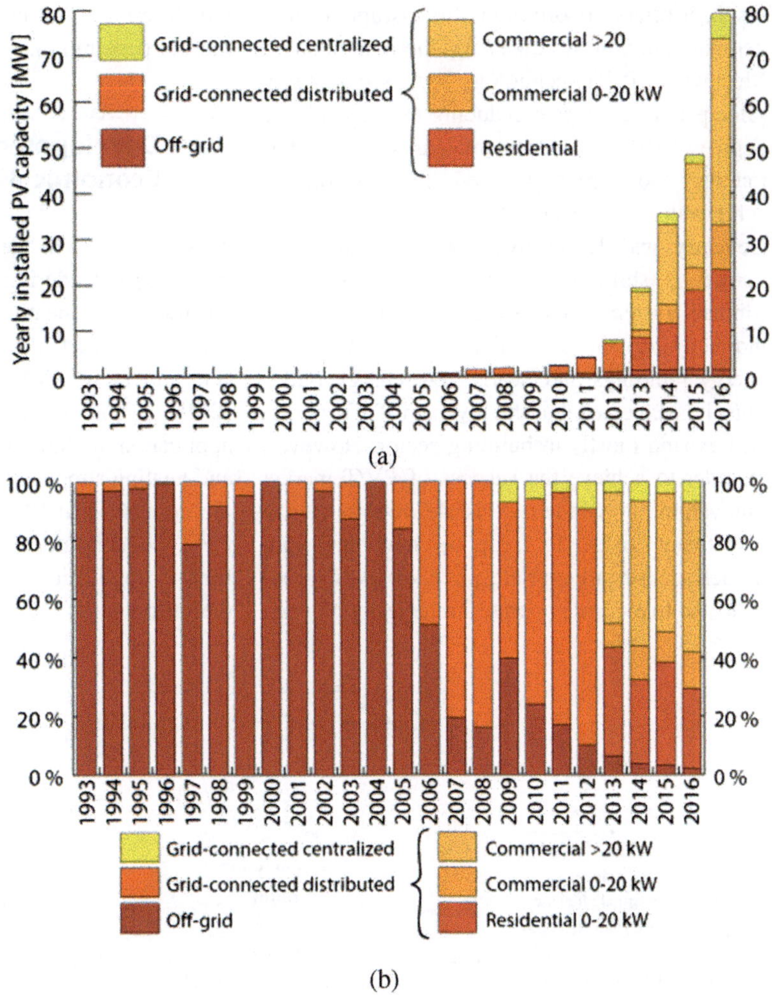

Fig. 7.1 **a** Annual installed PV capacity in Sweden from 1993 to 2016; **b** various market segments' share of the yearly installed capacity in Sweden (Lindahl 2016)

system in the meanwhile. The market price of flat plate collectors has dropped in line with the overall development of collector prices across Europe, while vacuum tube collectors have become more expensive owing to less and less players in the area. About half of the market is now being served by local manufacturers, according to the country portrait (Baerbel 2017). IEA SHC (2017) points out that the main market drivers are the national building code and indirect tax deduction for repair, conversion and extension work can be made for installing solar heating systems. In contrast to PV market, there is no direct financial support to solar thermal market. Other barriers are the lower prices of alternative energy technologies and the lack of information on solar heating systems among actors and consumers. In general, the Swedish solar thermal market is disruptive. It has to face high competition from alternative energy systems, but with limited indirect policy support. Great barriers are found for its further development.

Technology drivers offer opportunities for disruptive market innovation (Tidd et al. 2005). Innovative PV/T technology cogenerates of both electricity and heat from the same area, but with less installation cost than installing both individual PV and solar thermals collector, resulting in the lower overall cost-to-performance ratio and enabling the possible development of solar thermal market along with PV market. As a result, it offers opportunities for disruptive market innovation in Swedish solar thermal industry. Meanwhile, PV/T technology lowers the energy price per square meter, reducing the dependence degree of PV on governmental subsidy and thus making the target set by the government achievable (Tripanagnostopoulos et al. 2005; Coventry and Lovegrove 2003). Because PV market is shifting to the grid-connected distributed systems, and solar thermal market is indirectly driven by national codes for low-energy buildings and nearly zero-energy buildings, it is expected that the Swedish PV/T market will expand mostly in building sector. However, in order to achieve the success of PV/T market innovation in Sweden, it is desired to clarify investment risks, to propose new investing models, to set appropriate price of PV/T products, and to evaluate operate strategies of players, which are still not clear and are challenging the market penetration.

7.2 Development of Techno-Economic Model

This section will develop a techno-economic model by considering the present values of both electricity and heating savings from PV/T concentrator as part of the cash flow. A PV/T concentrator (X10 PVT) from a Swedish company (SMHI 2017) is used as a reference system for application in buildings, where both electricity and heating supplies are needed. As compared to the conventional PV and solar thermal devices, PV/T concentrators have plenty of advantages for the application in buildings as discussed in the introduction, i.e. higher electrical conversion

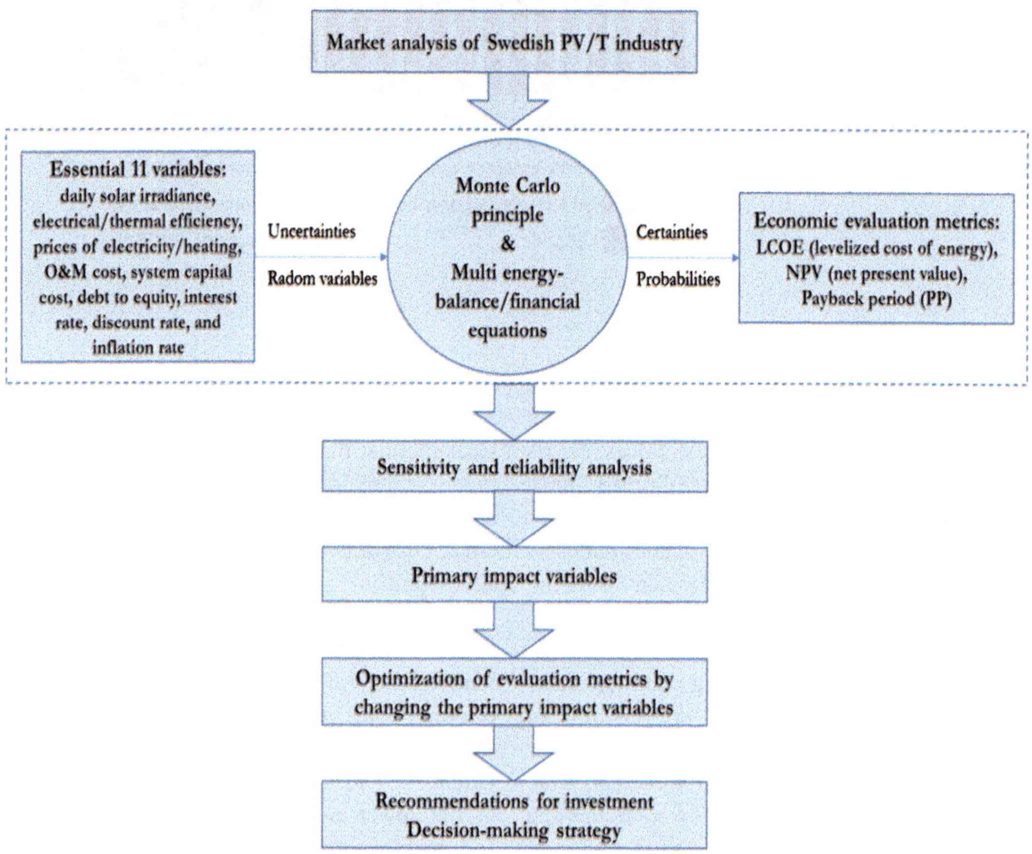

Fig. 7.2 Flow chart of techno-economic analysis method

efficiency, more energy output per unit area, efficient use of the building space, larger acceptance angle and higher optical efficiency (Xuan et al. 2017; Li et al. 2015). The selected PV/T concentrator produces more energy per unit area than typical PV and solar thermal collectors in a limited building application space, and could achieve up to 75 °C heat output in Swedish scenario (SMHI 2017), suitable for most building applications. The overall techno-economic modelling method is illustrated in Fig. 7.2.

7.2.1 Reference PV/T Concentrator

The reference PV/T technology is a concentrated type (named ×10) and Fig. 7.3 illustrates its basic configurations. The X10 PV/T consists of a cylinder-parabolic reflecting mirror, made by aluminium, that concentrates 17.8 times the solar light onto the receiver. Inside the receiver, the PV component is composed of 166 mono-crystalline solar cells in series with the size of each cell at 32 × 110 mm. The thermal component is structured in triangular with a double aluminium section bar substrate. Each bar is built in with a fluid channel for counter current flow. On the two receiver sides, opposite to the parabolic concentrator, there are laser groove buried contact solar cells on the surface; whereas the top side of the receiver is covered with a thermal absorber. The whole PV/T concentrator has a gross area of 10.91 m^2 and an aperture area of 10.37 m^2. The tracking of the sun is based on special electrical custom-designed high quality linear actuators, which are carried out by rotating the structure around an axis oriented in the east–west direction.

According to solar Keymark database and the published experimental testing results (Croce 2017; Fiorenza et al. 2016), the certified values of the concentrator's efficiencies are 9.62% (±5%) for electrical performance and 47.21–54.80% for thermal performance, under operation temperature difference between thermal medium and surrounding air from 0 to 70 °C. The stagnation temperature is high to 243 °C and the maximum operation pressure is around 1 MPa. The effective thermal capacity is almost 6.11 kJ/(m^2 K). The thermal efficiency as function of the average operation temperature (Tm) is expressed as $\eta = 0.548–0.849(t_m - t_a)/AG_w - 0.003(t_m - t_a)^2/AG_w$, with the flow mass rage at 0.01 kg/(s m^2).

7.2.2 Energy Generation Model

The energy generation model is simplified by only considering the impact from three essential variables, i.e. solar radiation and electrical/ thermal efficiencies, and the concentrator performance degradation over the operating time. The energy generated by PV/T concentrator is expressed by equations from (7.1, 7.2 and 7.3).

$$E_t = E_{et} + E_{tht} \quad (7.1)$$

$$E_{et} = (365 \times G_t \times \eta_e \times A) \times (1 - dg_e)^t \quad (7.2)$$

$$E_{tht} = (365 \times G_t \times \eta_{th} \times A) \times (1 - dg_{th})^t \quad (7.3)$$

where

E_t is the total energy generated by the PV/T concentrator in the year t, kW h/year.

E_{et} is the electricity generated by the PV/T concentrator, kW h/year.

E_{tht} is the thermal energy generated by the PV/T concentrator, kW h/year.

A is the effective collecting area of PV/T concentrator, m^2.

G_t is the average daily solar irradiation in the year t, kW h/(m^2·day).

η_e is the total concentrator electrical efficiency, %.

η_{th} is the total concentrator thermal efficiency, %.

dg_e and dg_{th} are respectively the degradations of electrical and thermal parts of the PV/T concentrator, %.

t is the index of time in the unit of number of year, starting from year '0'.

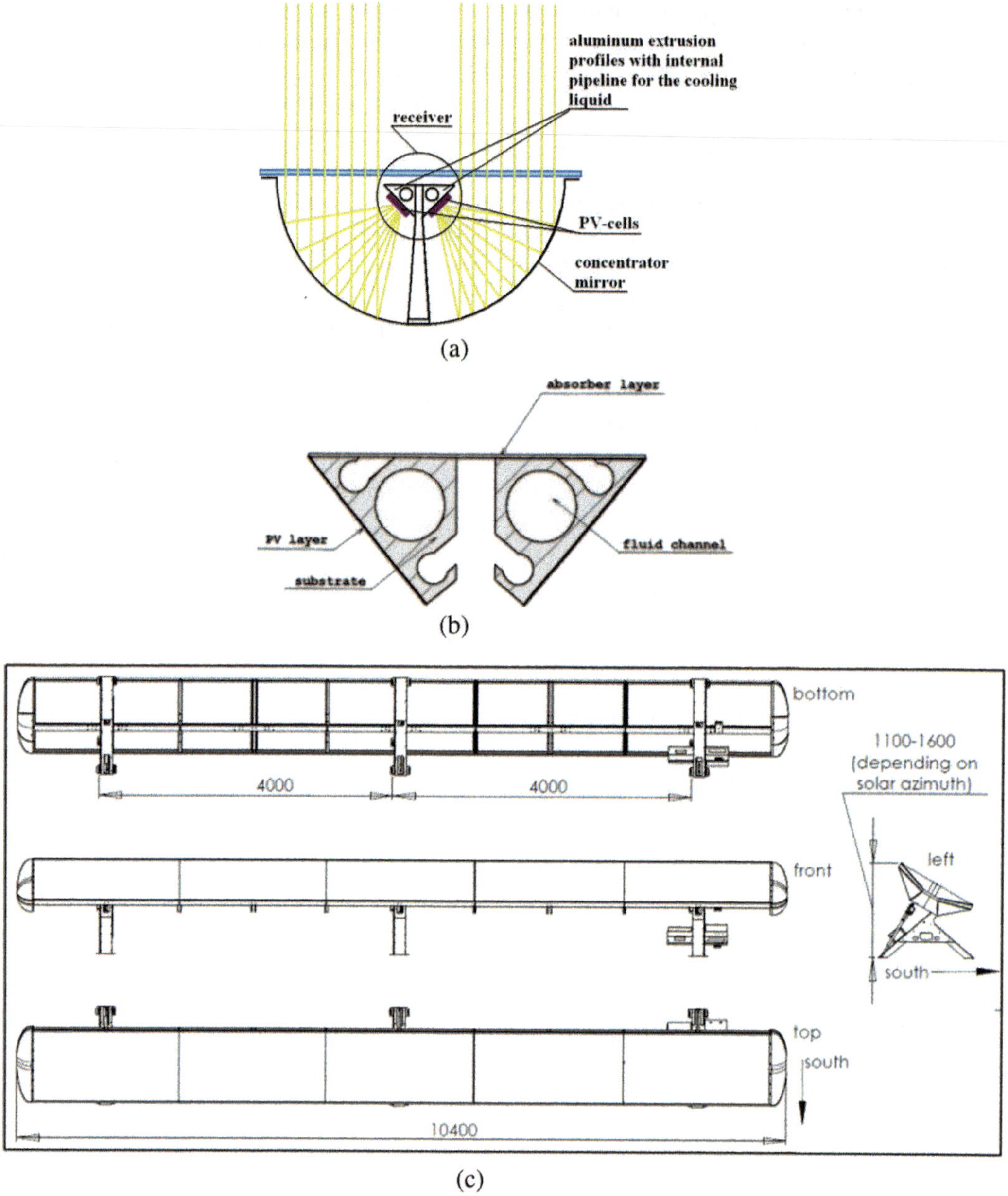

Fig. 7.3 **a** Overall structure of X10 PV/T; **b** cross section of receiver; **c** dimension of concentrator (Croce 2017; http://www.estif.org/solarkeymark/Links/Internal_links/SP/SC0987-09%202010-01-20%20Absolicon%20X10%20%2010,14m%20utg%202.pdf)

7.2.3 Economical Evaluation Metrics

LCOE weights the overall unit energy costs of establishing and operating an energy system throughout its entire life. All costs factors, such as fuel, financing, incentives, taxes, operational costs and degradation, should be considered in levelized cost calculations. Especially for the PV/T concentrator, LCOE serves a standardized metric by combining two disparate energy flows into one bottom-line metric, so that it can adequately compare energy costs of diverse generation sources. For solar energy system, LCOE values are primarily affected by capital costs (Overbye 2015). The mathematical equation for LCOE is given in Eq. (7.4).

$$LCOE = \frac{\sum_{(t=0)}^{(n-1)} \frac{C}{(1+r)^t}}{\sum_{(t=0)}^{(n-1)} \frac{E_t}{(1+r)^t}} \quad (7.4)$$

where

n is the number of year in terms of the concentrator's life time.

C_t is the total cost in the year of t, SEK.

r is the discount rate, %.

While the total cost of the PV/T concentrator in the specific year could be expressed by.

$$C_t = (I_0|t=0) + L_t + OM_t + Ta_t \quad (7.5)$$

where

I_0 is the investment expenditures in the initial year, SEK.

L_t is the financing loan cost in year t, including annual loan payment and interest SEK.

Ta_t is the tax paid for the electricity generation in the year t, SEK.

OM_t is the operating & management cost in year t, SEK.

In above equation, the investment expenditure for the first year is defined by.

$$I_0 = C_0 \times (1 - DE) \quad (7.6)$$

where

C_0 is the capital cost, SEK.

DE is the debt to equity ratio, %.

The financing loan cost in year t, L_t, is given as below equation.

$$L_t = C_0 \times DE \times \left[\frac{1}{n} + \left(1 - \frac{t}{n}\right) \times Itr_t\right] \quad (7.7)$$

where

Itr_t is the interest rate of year t, %.

And the tax is written by equation (Tripanagnostopoulos et al. 2005; Coventry and Lovegrove 2003).

$$Ta_t = TR \times E_{et} \quad (7.8)$$

where

TR is the tax rate per kW h, SEK/kWh.

NPV is defined by Eqs. (7.7–7.9), which is a measurement of l cumulative profitcalculated by subtracting the present values of cash outflows (including initial cost) from the present values of cash inflows over the PV/T concentrator's life time. The investment in a project is profitable when the NPV is positive.

$$NPV = -C_0 + \sum_{(t=0)}^{(n-1)} \frac{CF_t}{(1+r)} \quad (7.9)$$

where

CF_t is the cash flow of the year t, SEK.

The cash flow in year t is given by equation (Tripanagnostopoulos et al. 2005; Coventry and Lovegrove 2003; Xuan et al. 2017; Li et al. 2015).

$$CF_t = S_{et} + S_{tht} - C_t \quad (7.10)$$

In addition, the energy savings are respectively from electricity and heating.

$$S_{et} = EP_t \times Ifr_t \times E_{et} \quad (7.11)$$

$$S_{tht} = HP_t \times Ifr_t \times E_{tht} \quad (7.12)$$

where

EP_t is the electricity price in year t, SEK/kW h.

HP_t is the heating price in year t, SEK/kW h.

Ifr_t is the inflation rate of year t, %.

PP is the time in year for a project to break even or recover its initial investment funds, where the cash flow starts to turn positive. If PP is larger than the life-span of the reference PV/T concentrator, it is considered that PP equals 26 in this case.

$$PP = T_{(CF_t > 0)} \quad (7.13)$$

7.2.4 Key Input Parameters

The key input parameters for the analysis of the reference PV/T concentrator are categorized into technical, financial, and geographical ones. Table 7.1 displayed the basic parameters that are included in the analytical model, in which, the average values of each parameter is estimated while assuming it follows a probability distribution. For instance, the average daily irradiation within different years in a specific location of Sweden is assumed to fit with the continuous probability distribution curve. Taking Stockholm as an example, the average daily solar irradiation varies between 0.25 kW h/(m^2·day) and 5.93 kW h/(m^2·day) from January to December, and the mean of the annual average daily irradiation is 2.87 kW h/(m^2·day), and the standard deviation (SD) is 2.19 kW h/(m^2·day) in most years; however, there may be years with a higher average value and others with a lower average value within the range of deviation. As a result, it is possible to further run the probabilistic analysis using Monte Carlo simulation, conducting different combinations that follow the probability curves of the average values of different variables by determining the mean and standard deviation of the average values of those variables.

The average cost for the whole reference PV/T concentrator (including other indirect cost, such as installation, admin charges, system components etc.,) is around 4500 SEK/m^2 (450

Table 7.1 Key input parameters and their ranges for the simulation model

Description	Mean	Min	Max	Unit
Geographical parameters				
Average daily solar irradiance in Stockholm (http://solarelectricityhandbook.com/solar-irradiance.html)	2.87	0.25	5.93	kWh/m^2-day
Electricity price (Statista 2010)	1.94	1.54	2.09	SEK/kWh
Heating price (Sköldberg and Rydén 2017)	0.80	0.45	1.02	SEK/kWh
Technical parameters				
Effective PV/T area (http://www.estif.org/solarkeymark/Links/Internal_links/SP/SC0987-09%202010-01-20%20Absolicon%20X10%20%2010,14m%20utg%202.pdf)	10.37	–	–	m^2
Years of operation (X10 PV/T 2017)	25	–	–	years
STC nominal electrical efficiency (http://www.estif.org/solarkeymark/Links/Internal_links/SP/SC0987-09%202010-01-20%20Absolicon%20X10%20%2010,14m%20utg%202.pdf)	9.62	9.14	10.08	%
Nominal thermal efficiency (http://www.estif.org/solarkeymark/Links/Internal_links/SP/SC0987-09%202010-01-20%20Absolicon%20X10%20%2010,14m%20utg%202.pdf)	51.52	47.21	54.80	%
Nominal degradation rate (Overbye 2015)	1.00	–	–	%/year
Financial Parameters				
Capital product cost (include indirect cost) (X10 PV/T 2017)	4482.00	3586.00	5378.00	SEK/m^2
Debt to equity (Overbye 2015; X10 PV/T 2017)	60.00	0.00	90.00	%
Interest rate (Overbye 2015; X10 PV/T 2017)	4.00	2.00	6.00	%/year
Loan term (Overbye 2015; X10 PV/T 2017)	25.00	–	–	years
Effective tax (Clover 2017)	0.01	–	–	SEK/kWh
Nominal discount rate (Overbye 2015; Koene 2017)	8.00	6.00	10.00	%/year
OM to capital product cost (Overbye 2015; X10 PV/T 2017)	1.00	0.75	2.00	%/year
Inflation rate (Trading economics 2017)	1.20	−1.60	4.40	%/year

€/m^2), with up to 20% price-floating range. There are different ways to finance the PV/T concentrators, for instance, a loan from a bank is possible at the levels of 0%, 60%, 80% and 90% of the capital investment (Debt to equity). Meanwhile, the associated interest rate varies from 2 to 6% depending on the loan's option (flexible or fixed) and duration. The OM cost of a PV/T varies in different cases. For instance, Kalogirou and Tripanagnostopoulos (2006) and Herrando and Markides (2016) (Oracle Crystal Ball 2017) estimated at 1% of total system cost by considering parasitic costs with an increase rate of 1% per year of the system operation due to the required energy for circulate system pump. Riggs et al. (Overbye 2015) reported that OM costs for PV part and thermal part of a PV/T concentrator are respectively at 6.5 USD/kW$_{elec}$ (51.41 SEK/kW$_{elec}$) and 15 USD/kW$_{therm}$ (118.64 SEK/kW$_{elec}$) during its life time. By using these figures in the reference PV/T concentrator, the OM cost to capital cost is about 1.36%. In addition, depending on the reference PV/T concentrator's company, the OM cost is assumed in the range of 0.75% to 2% of the capital cost. In this chapter, regular labour cost for OM is not considered as the reference PV/T concentrator is a standalone system. Basic thermal OM (such as plumbing, pump power etc.) and electrical OM (such as inverter replacement, control, etc.) are taken into account. Mirror cleaning costs are retained within the OM costs but the reduced power output due to maintenance downtime is not included in the model. Further, there are a variety of business models for smaller scale installations, in which the installer covers system specific maintenance. It also increases the variation extent of OM cost. Over the past 10 years, both electricity price and oil & gas increased a lot, so it is also important to consider the inflation rate of both electricity and heating prices depending on the general inflation rate in the range of −1.6 to 4.4%. The normal depreciation of solar energy panel is estimated at 25–30 years.

By considering the tornado analysis done by Riggs et al. (Overbye 2015)and the scenarios in Sweden, there will be 11 essential input variables considered in this chapter, for further sensitivity analysis, which are: daily solar irradiance (S), electrical/thermal efficiency (η_e/η_{th}), prices of electricity/heating (EP/HP), OM cost (OM), concentrator capital cost (C_0), debt to equity ratio (DE), interest rate (Itr), discount rate (r), and inflation rate (Ifr).These variables are either affecting largely the result or are rather uncertain, and their effects on the results are expected.

The total evaluation strategy in the Monte Carlo model is expressed by input–output vector in equation (Tripanagnostopoulos et al. 2005; Coventry and Lovegrove 2003; Xuan et al. 2017; Li et al. 2015; Croce 2017; Fiorenza et al. 2016; Overbye 2015; http://www.estif.org/solarkeymark/Links/Internal_links/SP/SC0987-09%202010-01-20%20Absolicon%20X10%20%2010,14m%20utg%202.pdf).

$$\begin{bmatrix} (LCOE^1 NPV^1 PP^1) \\ (LCOE^2 NPV^2 PP^2) \\ \vdots \\ (LCOE^j NPV^j PP^j) \end{bmatrix} = MonteCarlomodel\left(\begin{bmatrix} (S^1 \eta_e^1 \eta_{th}^1 EP^1 HP^1 OM^1 C_0^1 DE^1 Itr^1 r^1 Ifr^1) \\ (S^2 \eta_e^1 \eta_{th}^1 EP^2 HP^2 OM^2 C_0^2 DE^3 Itr^4 r^5 Ifr^6) \\ \vdots \\ (S^j \eta_e^j \eta_{th}^j EP^j HP^j OM^j C_0^j DE^j Itr^j r^j Ifr^j) \end{bmatrix}\right) \tag{7.14}$$

where,
j is the number of simulation trials.

7.2.5 Simulation Process

The Monte Carlo analytical model is developed upon Crystal Ball, which is a leading spreadsheet-based application for predictive modelling, forecasting, simulation, and optimization. It offers unparalleled insight into the critical factors affecting risk so that the decision-makers can make the right tactical decisions (Oracle Crystal Ball 2017). In this model, the variables mentioned above were defined as the triangular probability distribution due to the limited data of the parameters, ranging between minimum and maximum given in Table 7.1 and the highest probability at the mean value. A range of values for assumptions was randomly generated. These inputs were then feed into formulas of evaluation metrics defined in forecast cells. This was repeated for a large number of combinations of parameters (10,000 trials in this work). After simulation, it explored ranges of outcomes, expressed as graphical forecasts, in order to exam the sensitivity/reliability of various input parameters, and to estimate the probability/certainty of different economic evaluation metrics (Oracle Crystal Ball 2017).

7.3 Sensitivity and Reliability Analysis

A sensitivity analysis was conducted to assess the sensitivity of the evaluation metrics (LCOE, NPV, and PP) to the variations in input parameters, as displayed in Fig. 7.4. Stockholm was taken as a basic example for application of the reference PV/T concentrator. The default forecast views are shown as the simples' count of values (the frequency) for each interval on the x-axis. In Fig. 7.4a, the mean value of LCOE is observed at 1.27 SEK/kW h (0.127 €/kW h). The LCOE value that occurred the most frequently is at around 0.80 SEK/kW h, which has a probability of around 6.5%, meaning that there is a 6.5% chance of LCOE values falling near this value. There is about 88.65% probability of LCOE values falling between 0.30 and 2 SEK/kW h. Comparing to the LCOE of other PV/T systems reviewed in Sect. 7.2 (0.06–0.12 €/kW h or 0.6–1.2 SEK/kW h), the mean LCOE of X10 PV/T is more closed to the upper limit owing to the much less energy generation and energy saving benefits in Sweden. Similarly in Fig. 7.4b, the reference concentrator has the mean NPV value of 18,812.55 SEK (1881.255 €) after 25 years operation, and there is nearly 74.91% probability for this PV/T concentrator to achieve positive NPV (valuable investment) when it is operated in Sweden. Comparing to the NPV values of other PV/T systems reviewed in Sect. 7.2 (20,000 € or 200,000 SEK), the mean NPV of X10 PV/T is much lower because both the energy generation and the electricity/heating prices are so low that leads to a weak financial benefits in cash flow. The mean PP, displayed in Fig. 7.4c, is nearly 10 years when the cash flow turns positive, which is equivalent to the other cases (about 11 years) reviewed in Sect. 7.2. There is around 80.92% possibility for this PV/T concentrator to have a PP less than 15 years. However, there is nearly 12% high risk for the PP over the concentrator's operation life span of 25 years.

Figure 7.5 demonstrates the sensitivity of different variables to different output values. It is found that the main parameters that determine a positive outcome of business case are average daily solar irradiance, concentrator capital cost, heating price, debt to equity ratio and discount rate. The average daily solar irradiance accounts for approximately 91.85%, 91.12% and 84.60% of the values of LCOE, NPV and PP, respectively. It is therefore considered as the most important assumption in the model since it determines energy generation and savings of the whole concentrator system. The concentrator capital cost also influences all the three evaluation metrics: 3.64% to LCOE, 1.97% to NPV and 3.00% to PP since it contributes directly to the annual savings and cash flow. The debt to equity ratio is another important factor that strongly connected with the initial investment and the associated annual cost/cash flow; so it affects the outputs too, contributing 1.85% impact to LCOE, and 7.71% to PP. Heating price also influences NPV and PP to an extent of 4.85% and 4.08% respectively. This is because the magnitude of heat generation is relative larger, comparing to electricity generation. Meanwhile, the product of heat output and heating price thus contributes largely to the cash flow. In addition, the discount rate has slight impact of LCOE while the rest variables are not so important and their impacts on the final output are so limited and that can be ignored.

Additional comparison is also carried out between two cases using different analytical methods: (1) Monte Carlo method and (2) average point value method. In the Monte Carlo case, it considers the probability distribution of each input variable and predicts the mean values of output metrics in a certain range. In the Normal case, it applies the average point value of each

Fig. 7.4 Frequency forecast chart of **a** LCOE, **b** NPV, and **c** PP

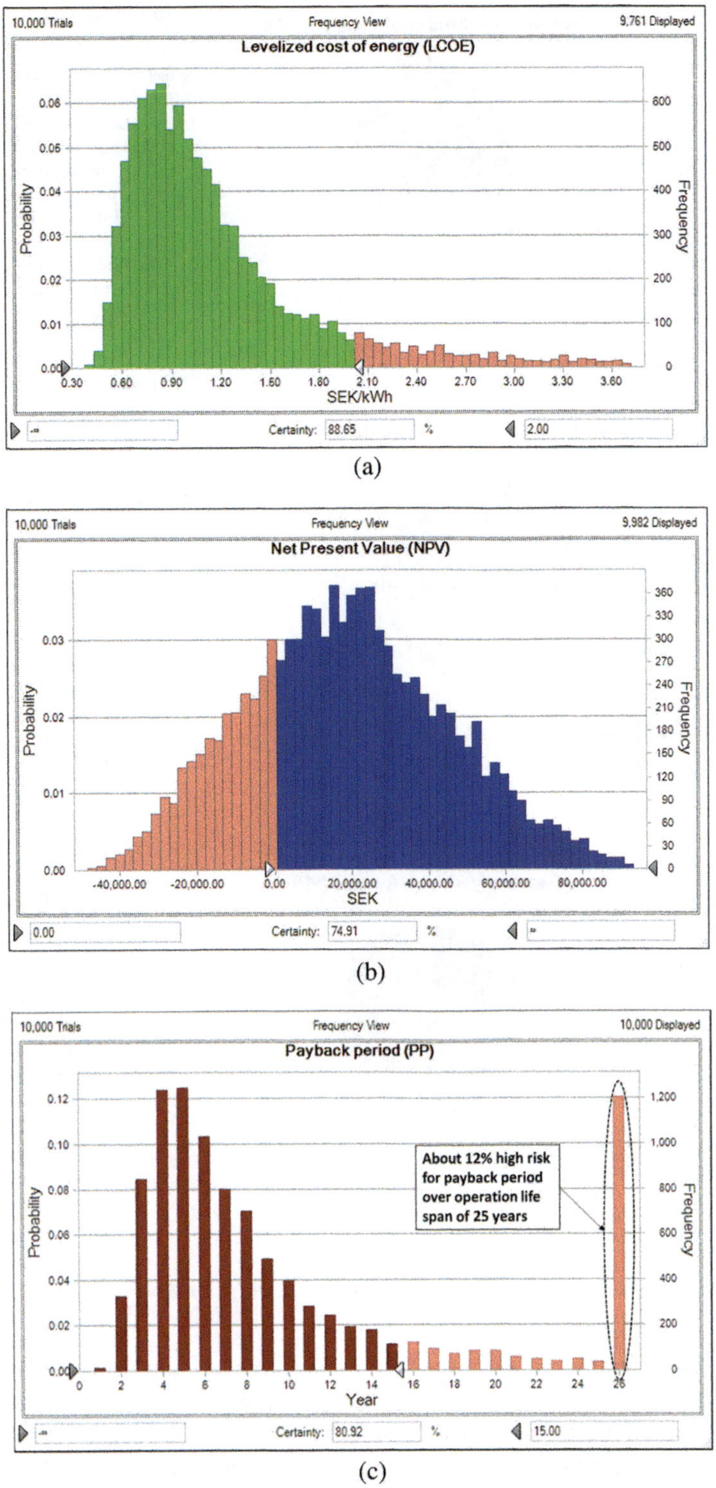

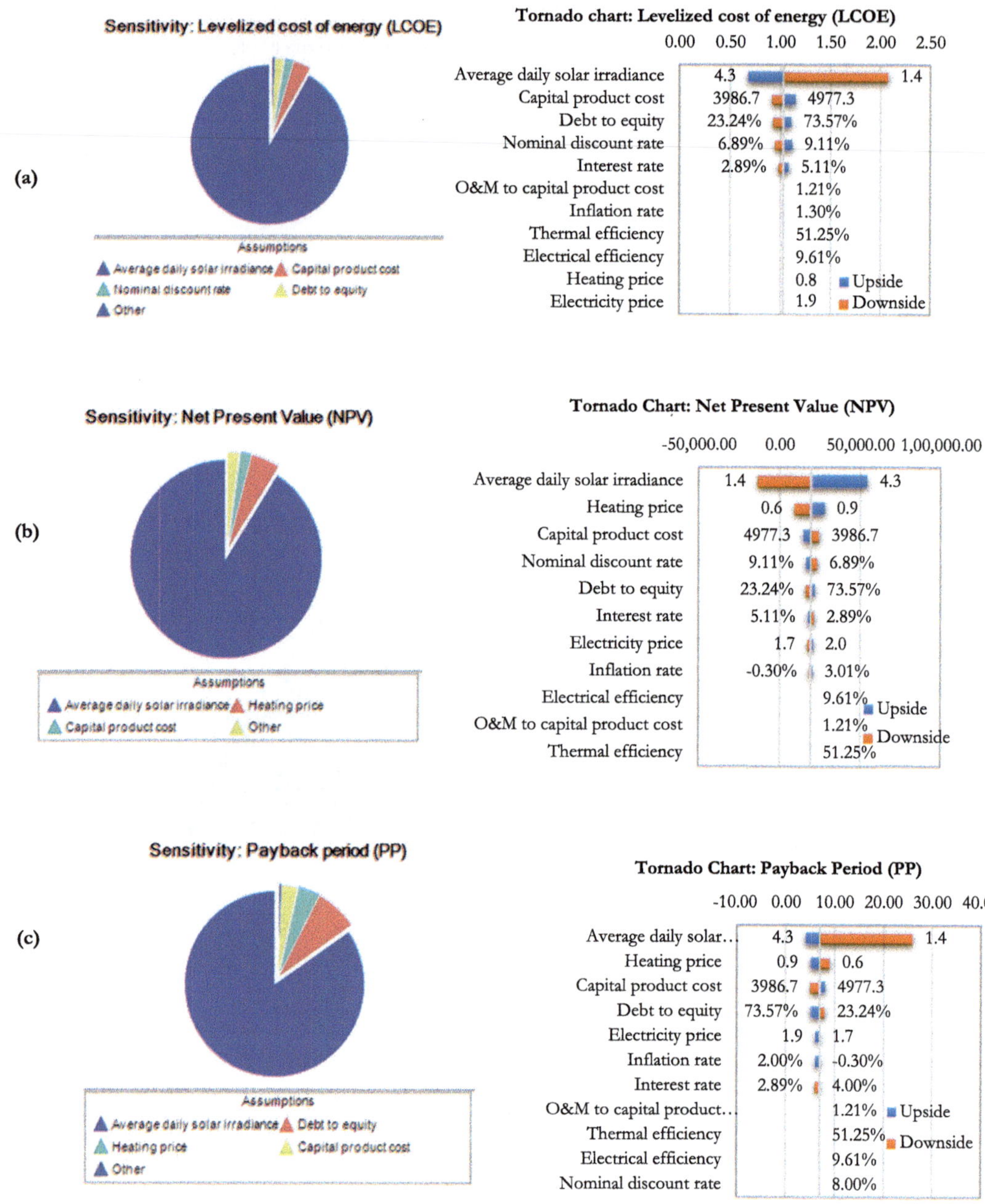

Fig. 7.5 Sensitivity and Tornado charts of **a** LCOE, **b** NPV and **c** PP

input variable and calculates the average values of output metrics. Table 7.2 displays the estimation results of Normal case, and Fig. 7.6 compares the two cases in parallel. It is seen that in Normal case, the economic performance are too optimistic, where LCOE is calculated at 1.05 SEK/kW h (0.105 €/kW h) while NPV and PP are respectively estimated at 22,973.84 SEK (2297.384 €) and 5 years. However, current Swedish PV/T market is very poor, mainly because of the weak practical economic benefits of PV/T. The results shown in the Normal case are in great disagreement to real situation. In the Normal case, the analysis neglects the inherent

Table 7.2 The estimated results of Normal case using average point value method

Year	0	1	2	3	4	5	6	7	8	9	10	11	12
Cost saving from energy generation													
Electricity generation	1030.97	1020.66	1010.46	1000.35	990.35	980.45	970.64	960.94	951.33	941.81	932.39	923.07	913.84
Heat generation	5521.39	5466.18	5411.52	5357.40	5303.83	5250.79	5198.28	5146.30	5094.84	5043.89	4993.45	4943.51	4894.08
Electricity savings	1950.33	1953.99	1934.45	1915.11	1895.96	1877.00	1858.23	1839.65	1821.25	1803.04	1785.01	1767.16	1749.49
Heating savings	4417.11	4425.42	4381.16	4337.35	4293.98	4251.04	4208.53	4166.44	4124.78	4083.53	4042.70	4002.27	3962.25
Total energy generation	6552.37	6486.84	6421.98	6357.76	6294.18	6231.24	6168.92	6107.24	6046.16	5985.70	5925.84	5866.59	5807.92
Total energy savings	6367.44	6379.41	6315.62	6252.46	6189.94	6128.04	6066.76	6006.09	5946.03	5886.57	5827.70	5769.43	5711.73
Cost of operation & maintenance and Tax													
OM cost	464.78	464.78	464.78	464.78	464.78	464.78	464.78	464.78	464.78	464.78	464.78	464.78	464.78
Tax for electricity	5.15	5.10	5.05	5.00	4.95	4.90	4.85	4.80	4.76	4.71	4.66	4.62	4.57
Total	469.94	469.89	469.84	469.79	469.74	469.69	469.64	469.59	469.54	469.49	469.45	469.40	469.35
Financial cost													
Loan payment	1,115	1,115	1,115	1,115	1,115	1,115	1,115	1,115	1,115	1,115	1,115	1,115	1,115
Loan interest	1,115	1,071	1,026	982	937	892	848	803	759	714	669	625	580
Total	2230.96	2186.34	2141.72	2097.10	2052.48	2007.86	1963.25	1918.63	1874.01	1829.39	1784.77	1740.15	1695.53
Discounted Cash Flow and Energy Generation													
Net Profit	-14924.79	3723.19	3704.06	3685.57	3667.72	3650.49	3633.88	3617.88	3602.48	3587.69	3573.49	3559.88	3546.85
Discounted Cost	21292.23	2656.23	2611.56	2566.89	2522.22	2477.55	2432.88	2388.21	2343.55	2298.88	2254.21	2209.55	2164.88
Discounted Energy Generation	6067.01	6006.34	5505.81	5046.99	4626.41	4240.88	3887.47	3563.51	3266.55	2994.34	2744.81	2516.08	2306.40
Cash flow	-14924.79	-11201.61	-7497.55	-3811.97	-144.25	3506.24	7140.11	10757.99	14360.47	17948.16	21521.65	25081.53	28628.38
Evaluation metrics													
Net Present Value	22,973.84 SEK (2,297.384 €)												
LCOE	1.05 SEK/kW h (0.105€/kW h)												
Payback period	5.00	Year											

Year	13	14	15	16	17	18	19	20	21	22	23	24
Cost saving from energy generation												
Electricity generation	904.70	895.65	886.70	877.83	869.05	860.36	851.76	843.24	834.81	826.46	818.20	810.01
Heat generation	4845.14	4796.69	4748.72	4701.23	4654.22	4607.68	4561.60	4515.99	4470.83	4426.12	4381.86	4338.04
Electricity savings	1731.99	1714.67	1697.52	1680.55	1663.74	1647.11	1630.63	1614.33	1598.19	1582.20	1566.38	1550.72
Heating savings	3922.62	388340	3844.56	3806.12	3768.06	3730.38	3693.07	3656.14	3619.58	3583.39	3547.55	3512.08
Total energy generation	5749.84	5692.34	5635.42	5579.06	5523.27	5468.04	5413.36	5359.23	5305.64	5252.58	5200.05	5148.05
Total energy savings	5654.62	5598.07	5542.09	5486.67	5431.80	5377.48	5323.71	5270.47	5217.77	5165.59	5113.93	5062.79
Cost of operation & maintenance and Tax												
OM cost	464.78	464.78	464.78	464.78	464.78	464.78	464.78	464.78	464.78	464.78	464.78	464.78
Tax for electricity	4.52	4.48	4.43	4.39	4.35	4.30	4.26	4.22	4.17	4.13	4.09	4.05
Total	469.31	469.26	469.22	469.17	469.13	469.09	469.04	469.00	468.96	468.92	468.87	468.83
Financial cost												
Loan payment	1,115	1,115	1,115	1,115	1,115	1,115	1,115	1,115	1,115	1,115	1,115	1,115
Loan interest	535	491	446	402	357	312	268	223	178	134	89	45
Total	1650.91	1606.29	1561.67	1517.05	1472.43	1427.81	1383.20	1338.58	1293.96	1249.34	1204.72	1160.10
Discounted Cash Flow and Energy Generation												
Net Profit	3534.40	3522.52	3511.20	3500.44	3490.24	3480.58	3471.47	3462.89	3454.85	3447.33	3440.34	3433.86
Discounted Cost	2120.22	2075.55	2030.89	1986.23	1941.56	1896.90	1852.24	1807.58	1762.91	1718.25	1673.59	1628.93
Discounted Energy Generation	2114.20	1938.02	1776.52	1628.48	1492.77	1368.37	1254.34	1149.81	1053.99	966.16	885.65	811.84
Cash flow	32162.78	35685.29	39196.49	42696.94	46187.17	49667.76	53139.23	56602.12	60056.97	63504.31	66944.65	70378.51

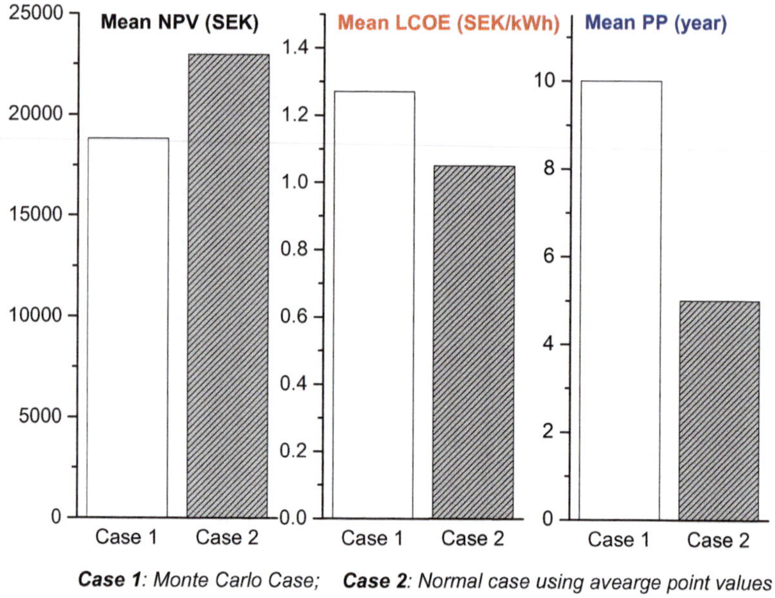

Fig. 7.6 Comparison of economic analytical results between two cases using different methods

uncertainty in the investment and thus overestimates the economic benefits from PV/T. On the contrast, the analysis in Monte Carlo case chose the input parameters randomly from pre-defined distributions for each input variable and performed the calculation for 10,000 times, whose results seem more realistic by providing more realistic information about risk and uncertainty.

7.4 Optimization Analysis

In order to measure the potential impact of each important variable mentioned above, such as average daily solar irradiance, debt to equity, heating price, concentrator capital price, and discounted rate, this section will use the existing model to predict the values of LCOE, NPV and PP, by changing the value of each important variable in a certain range when keeping the others as the same assumptions. The simulation results are expected to be useful in further marketing positioning and penetration of the reference PV/T from the points of views of marketing location, financial measures and governmental tariffs.

7.4.1 Impact of Average Daily Solar Irradiance

When keeping the other assumptions the same, increasing only the average daily solar irradiance from 0.5 to 6.5 kW h/m^2 day in the existing model, the impact of the average daily solar irradiance is illustrated in Fig. 7.7. In this case, only the average daily solar irradiance data is considered as the fixed value under each simulation running, instead of the triangular distribution, but other 10 key input variables are still assumed as the triangular distributions. Increasing the average daily solar irradiance increases NPV and reduces LCOE and PP. NPV increases from −35,068.77 SEK (−3506.877 €) to 102,639.38 SEK (10,263.938 €) with increasing solar irradiance in a purely linear way ($y = 11497x - 46{,}567$; $R^2 = 1$), and NPV starts to turn positive after the average daily solar irradiance is greater than about 2.10 kW h/m^2 day. LCOE falls down significantly at the beginning when the average daily solar irradiance starts to increase but then it drops gradually when solar irradiance becomes larger, presenting a downwards power trend ($y = 5.8109x^{-0.999}$,

$R^2 = 1$) from 5.81 SEK/kW h (0.581 €/kW h) to 0.45 SEK/kW h (0.045€/kW h). PP varies in a similar way to LCOE, also showing downwards power trend ($y = 23.057x^{-0.933}$; $R^2 = 0.9976$). PP is estimated over 25 years when the average daily solar irradiance is lower than 1.5 kW h/m^2 day, and it achieves minimum 2.52 year when the average daily solar irradiance is at 6.5 kW h/m^2 day. Such results can be interpreted that higher solar irradiance leads to more total electricity/heat generation, and thus increases the related economic savings and NPV while decreasing LCOE and PP.

The average daily solar irradiance has a significant impact on all the economic indexes; the higher solar irradiance, the better business outcomes. The appropriate places for operating the reference PV/T concentrator are suggested to have the average daily solar irradiance greater than 2.40 kW h/m^2 day (876 kW h/m^2 year) at least, in order to achieve positive NPV and low LCOE of 1.20 SEK/kW h (0.12 €/kW h) and PP of less than 10 years. Figure 7.8 illustrates the suitable places that meet such criteria in Sweden. The reference PV/T concentrators are appropriate to be applied in those representative cities, i.e. Lund, Göteborg, Norrköping, Stockholm, Borlänge and Umeä etc., (Carlund 2013). From the geographic point of view (SMHI 2017), nearly two thirds of Sweden is the suitable places to install this reference PV/T concentrator.

7.4.2 Impact of Debt to Equity Ratio

Varying the debt to equity ratio and keeping the other assumptions the same, Fig. 7.9 shows the variation trend of NPV, LCOE and PP under this circumstance. The debt to equity ratio is set as the fixed value (instead of the triangular distribution) during each simulation running. The higher debt to equity ratio results in the higher LCOE/NPV, but the lower PP. LCOE slightly increases from 1.01 SEK/kW h (0.101 €/kW h) to 1.39 SEK/kW h (0.139 €/kW h) with the increase of debt to equity from 0 to 90% in a linear way ($y = 0.0425x + 0.9493$; $R^2 = 0.994$). The impact of debt to equity ratio to NPV value is very limited, which is also demonstrated in Fig. 7.5b. NPV grows from 17,329.53 SEK (1732.953 €) to 26,418.06 SEK (2641.806 €) when the debt to equity grows in the range of 0% to 90% in a linear way ($y = 1021.2x + 16{,}537$; $R^2 = 0.9928$). PP drops against the increase of debt to equity ratio in a polynomial way ($y = -0.0391 \times 2 + 0.0309x + 10.539$; $R^2 = 0.9947$), decreasing from 10.64 year to 6.83 year. These variations are because the higher debt to equity means the higher loan from the bank, increasing the annual loan cost and therefore higher LCOE; on the other hand, high loan requires less initial investment and PP, which further increases the present values of cash flow slightly. However, the impact of debt to

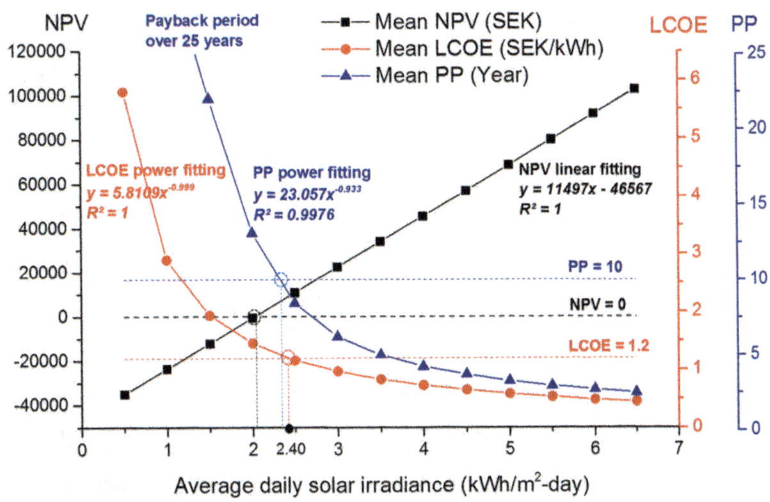

Fig. 7.7 Variations of LCOE, NPV and PP against average daily solar irradiance

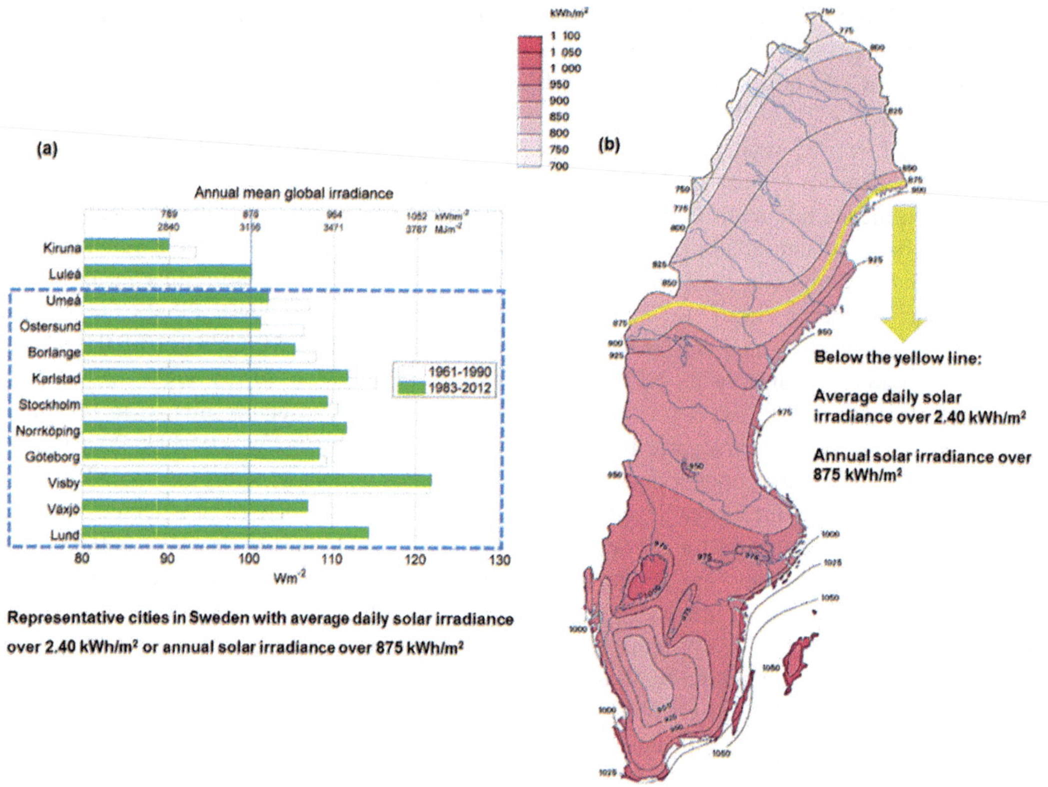

Fig. 7.8 Appropriate places for the reference PV/T concentrator in Sweden: **a** representative cities, and **b** geographic view

equity is not as much as that of the average daily solar irradiance, which is consistent with the results in sensitivity/reliability analysis. The simulation results are meaningful to encourage Swedish customers to have more loan from the bank when they decide to invest the PV/T

Fig. 7.9 Variations of LCOE, NPV and PP against debt to equity

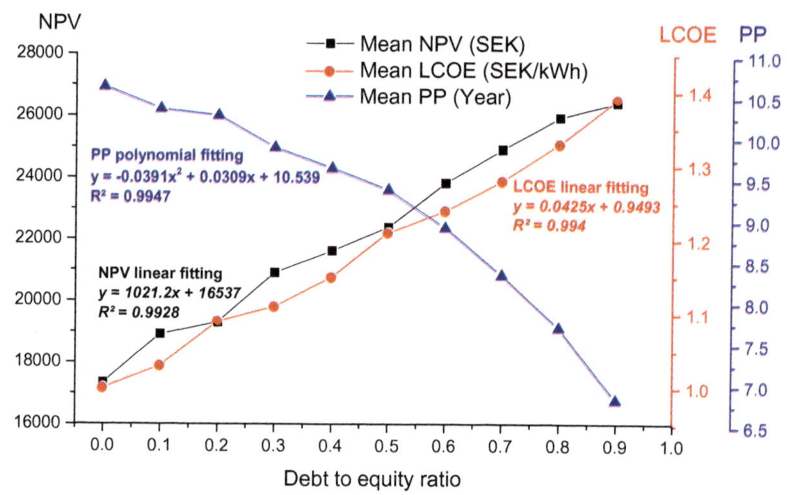

product, and also useful for development of new business models among different stakeholders.

7.4.3 Impact of Heating Price for Household

The variations of LCOE, NPV and PP against the increase of household heating price are displayed in Fig. 7.10, when the other parameters are remained as the same assumptions. The heating price is treated as the fixed value in the model (instead of the triangular distribution) during each simulation running. As concluded in above sensitivity/reliability analysis section, the impact of heating price on the LCOE can be ignored, which is also observed in this parametric study. By changing heating price from 0.50 SEK/kW h (0.05 €/kW h) to 1.00 SEK/kW h (0.1 €/kW h), LCOE almost stays constantly at around 1.20 SEK/kW h (0.12 €/kW h), while NPV increases greatly from 6805.52 SEK (680.552 €) to 38,450.23 SEK (3845.023 €) linearly ($y = 6323x + 200$; $R^2 = 0.9996$);on the contrast, PP drops from 12.47 year to 6.93 year in a linear way ($y = -1.0949x + 13.239$; $R^2 = 0.9858$). Since the energy savings are only considered into the present value of cash flow (not in the present value of system cost), the heating price only has the impact to NPV and PP but no direct influence on LCOE. The higher heating price leads to larger savings due to energy generation, contributing more to cash flows and NPV/PP. The simulation result has an indication that it would be more beneficial to apply the reference PV/T concentrator in those locations or municipalities with higher heating price in Sweden.

7.4.4 Impact of Concentrator Capital Price

Considering the change of concentrator price only and keeping the other assumptions the same, the variations of LCOE, NPV and PP against the increase of concentrator capital price are presented in Fig. 7.11. The concentrator capital price is regarded as the fixed value (instead of the triangular distribution) during each simulation running. In previous sensitivity/reliability analysis, it is observed that the impact of concentrator capital price on all the three economic indexes is remarkable, which is also observed in this parametric study. Increasing concentrator capital price from 3000 SEK/m^2 (300 €/m^2) to 5500 SEK/m^2 (550 €/m^2), LCOE goes up from 0.81 SEK/kW h (0.081 €/kW h) to 1.48 SEK/kW h (0.148 €/kW h) in a linear way ($y = 0.1351x + 0.672$; $R^2 = 0.9989$), and similarly, PP increases from 5.63 year to 11.89 year linearly too ($y = 1.2657x + 4.2567$; $R^2 = 0.9987$); on the other hand, NPV falls down from 37,989.16 SEK

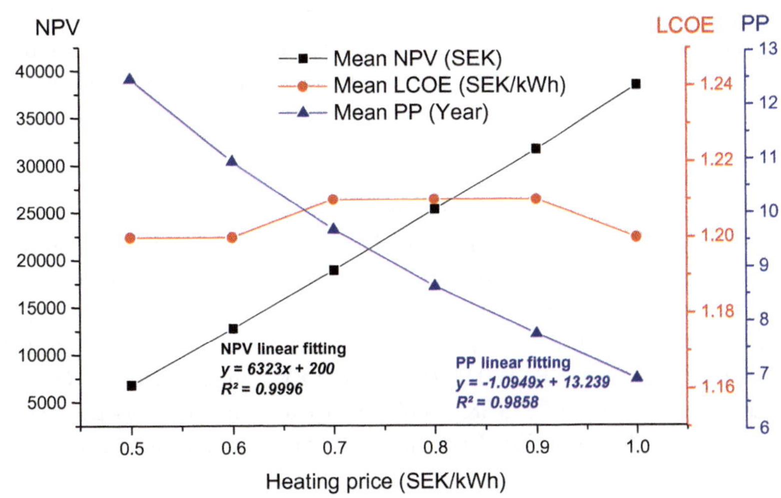

Fig. 7.10 Variations of LCOE, NPV and PP against heating price

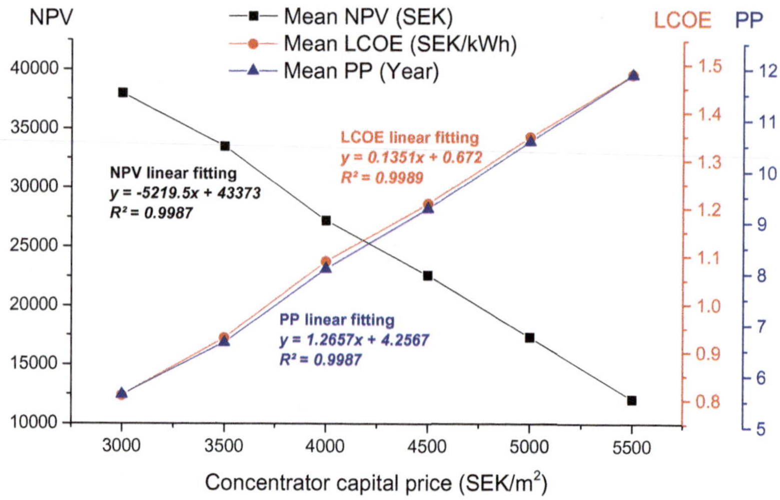

Fig. 7.11 Variations of LCOE, NPV and PP against concentrator capital price

(3798.916 €) to 12,078.33 SEK (1207.833 €) linearly (y = −5219.5x + 43,373; R^2 = 0.9987). The higher concentrator capital price results in the higher initial investment and the higher loan cost; so it has a negative impact on all the three economic indexes, lowering NPV and increasing LCOE/PP simultaneously. The simulation result indicates that the capital price of the reference PV/T concentrator is important to the final decision of investment, and it is suggested that the concentrator capital price should be controlled as low as possible for a better market penetration.

7.4.5 Impact of Discount Rate

When changing the discount rate only, its impact to the LCOE, NPV and PP is illustrated in Fig. 7.12. It needs to be noted that the discount rate is set at the fixed value, rather than triangular distribution, while the rest assumptions are defined as the same distributions. The discount rate has certain impact on both LCOE and NPV but the limited influence on PP. When increasing the discount rate from 6 to 10%, LCOE increases from 1.03 SEK/kW h (0.103 €/kW h) to 1.39 SEK/kW h (0.0139 €/kW h) linearly (y = 0.089x + 0.941; R^2 = 0.9991); while NPV reduces from 30,675.55 SEK (3067.555 €) to 16,353.94 SEK (1635.394 €) in a linear way (y = −3575.5x + 33,745; R^2 = 0.9934). The variation of PP ranges very limited from 9.15 to 9.28 years. The larger discount rate weakens the present values of cash flow and the energy generation/benefits, and it ultimately reduces NPV and enhances LCOE. According to the simulation result, the lower discount rate is beneficial to the investment of the reference PV/T concentrator in Sweden.

7.5 Future Work to Improve the Model

Since the main task in this work is to propose a methodology and develop the relevant model by using the nominal efficiencies/data, in order to find out the most influencing variables (sensitivities) on PV/T investment decisions in Sweden scenario, further work is required to improve this model's reliability in practical decision making process. The average daily solar irradiance has the highest sensitivity ranking and therefore is regarded as the most important factor. Further improvement of the model should apply the historic real-time solar irradiance data to reduce its uncertainty and increase model accuracy. As highlighted by Guarracino et al. (2016), it is important to use real climate data at high resolution instead of time-averaged data, where the

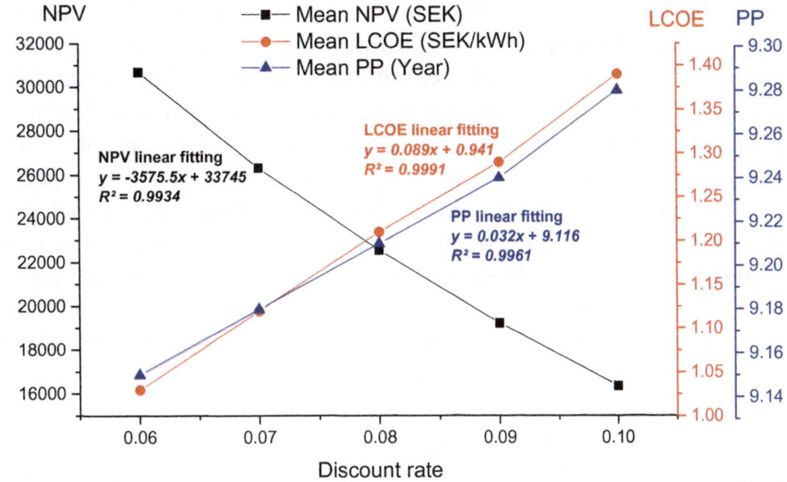

Fig. 7.12 Variations of LCOE, NPV and PP against discount rate

yearly energy yields may be discrepant by over 25%. The next important impact factors are the debt to equity ratio, the concentrator capital price, local heating prices, and the discount rate. Further simulation should be based on the actual financial plan and the location-based product & energy prices. Additional attention should be paid to the location-specific real-time PV/T efficiencies. In this chapter, only the nominal PV/T efficiencies are applied in the model, while future model should be based on real-time PV/T efficiencies that could be defined as the practical probability distribution curves in the model. Fiorenza et al. reported that the peak energy yield/efficiencies of the same PV/T concentrator were 15% lower than that in manufacturer's technical bulletin (Fiorenza et al. 2016). Figure 7.13 compared two cases by referring the PV/T efficiencies' maximum values as −15% (Fiorenza et al. 2016) and minimum values as −25% (Guarracino et al. 2016) respectively, to the nominal efficiencies according to manufacturer's technical bulletin. A normal distribution of PV/T efficiencies in case 2 was defined. It dictates that NPV decreases to 8677.80 SEK (867.780 €) while LCOE and PP only increase to

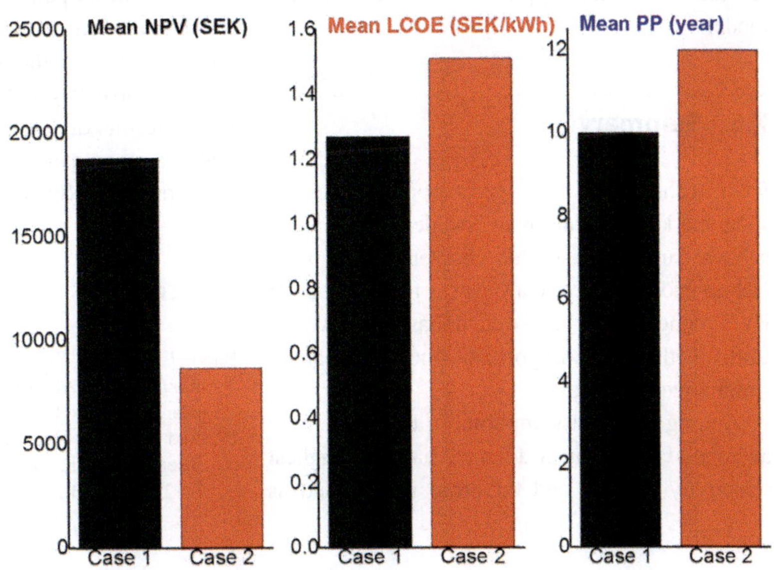

Fig. 7.13 Comparison of economic analytical results between two cases using nominal efficiencies and referenced lower practical efficiencies

1.51 SEK/kW h (0.151 €/kW h) and 12 years, respectively. Although the PV/T efficiencies only influence the overall economic performance to a certain level, they can result in a more realistic prediction. Practical PV/T efficiencies vary with many practical factors, such as seasons, locations, orientations etc. So it is not easy to simply foresee the practical efficiencies, and long-term recording of real-time efficiencies could be one solution for these data.

It is also important to introduce temperature production as a parameter in future model, together with surrounding air temperature and hot water load profile. For instance, PV electrical efficiency relies on the output temperature; the real-time thermal efficiency could be achieved though using annual temperature production and annual inlet temperature as essential parameters to estimate practical thermal output and efficiency. In addition, this chapter assumed both the thermal and power output are useful for building services (such as lighting, hot water at 40 °C level), where it would be possible when grid connection or electrical storage is available for the PV output and thermal storage is available for the thermal output. The estimation from the model may vary depending on the demand profile at different temperature levels and the nominal capacity installed in practice. Further work should also involve the integration of high-resolution demand profile with the existing model.

7.6 Summary

PV/T technologies offer great opportunities for solar market innovation in Sweden. In order to achieve such success, this chapter carries out a techno-economic evaluation of a reference solar PV/T concentrator in Sweden based on Monte Carlo method. The important conclusion can be summarized as below.

Among the 11 assumptions in the model, the average daily solar irradiance has the highest sensitivity ranking and therefore is regarded as the most important factor. The next important impact factors are debt to equity ratio, the product capital price, local heating prices, and the discount rate. Further improvement of the model should apply the historic solar irradiance data and the location-specific cell efficiency, financial plan and product & energy prices, to reducing its uncertainty and increase model accuracy.

The reference concentrator has the mean values of LCOE, NPV and PP at 1.27 SEK/kW h (0.127 €/kW h), 18,812.55 SEK (1,881.255 €) and 10 years, respectively. There is nearly 12% high risk for the PP over the concentrator's operation life span of 25 years.

The higher solar irradiance is, the better business outcomes. The appropriate places for operating the reference PV/T concentrator is suggested to have the average daily solar irradiance greater than 2.40 kW h/m^2•day (876 kW h/m^2 year) at least. The reference PV/T concentrator could be applied in those representative cities, such as Lund, Göteborg, Norrköping, Stockholm, Borlänge and Umeä etc. It is meaningful to encourage Swedish customers to have more loans from the bank when they decide to invest the PV/T product. It would be more beneficial to apply such system in those locations or municipalities with higher heating price or lower discount rate in Sweden. The concentrator capital price should be controlled as low as possible for a better market penetration.

The modelling method presenting in this chapter considers the integrated uncertainties and risks of most key input parameters, whose analytical results are expected to be useful for more realistic investment decisions of PV/T technologies applied in Swedish building sector.

References

Baerbel R (2017) Sweden's solar heat market on hold. http://www.solarthermalworld.org/content/swedens-solar-heat-market-hold. Accessed on 16 Oct 2017

Carlund T (2013) 30 years of Swedish solar radiation observations. In: EGU General Assembly 2013, held 7–12 Apr 2013 in Vienna, Austria

Clover I (2017) Sweden to effectively scrap solar energy tax: reports. https://www.pv-magazine.com/2016/11/22/sweden-to-effectively-scrap-solar-energy-tax-reports_100026987/. Accessed on 28 Nov 2017

Coventry JS, Lovegrove K (2003) Development of an approach to compare the 'value' of electrical and thermal output from a domestic PV/thermal system. Sol Energy 75:63–72

Croce SL (2017) Simulazione energetica discenari per la produzione combinataa ervizio diedificicivili in area mediterranea. Università degli Studi di Cagliari. http://veprints.unica.it/1445/1/PhD_Thesis_LaCroce.pdf. Accessed on 11 Nov 2017

Fiorenza G, Paparo G, Apicella F, Bianco N, Graditi G (2016) An innovative dynamic model for the performance analysis of a concentrating photovoltaic/thermal (CPV/T) solar collector. In: Sayigh A (ed) Renewable energy in the service of mankind Vol II. Springer, Cham, pp 337–351

Guarracino I, Mellor A, Ekins-Daukes NJ, Markides CN (2016) Dynamic coupled thermal-and-electrical modelling of sheet-and-tube hybrid photovoltaic/thermal (PVT) collectors. Appl Therm Eng 101:778–795

Herrando M, Markides CN (2016) Hybrid PV and solar-thermal systems for domestic heat and power provision in the UK: techno-economic considerations. Appl Energy 161:512–32

IEA SHC (2017) Status of solar heating/cooling and solar buildings—2017. http://www.iea-shc.org/country-report-sweden Accessed on 22 Aug 2017

Kalogirou SA, Tripanagnostopoulos Y (2006) Hybrid PV/T solar systems for domestic hot water and electricity production. Energy Convers Manage 47:3368–3382

Koene F (2017) Report on the developed techno-economic integrated concepts. Accessed on 12 Nov 2017

Li G, Pei G, Ji J, Yang M, Su Y, Xu (2015) Numerical and experimental study on a PV/T system with static miniature solar concentrator. Sol Energy 120:565–574

Lindahl J (2016) National survey report of PV power applications in Sweden—2016. Swedish Energy Agency

Oracle Crystal Ball. https://www.oracle.com/applications/crystallball/index.html. Accessed on 12 Nov 2017

Overbye O (2015) An economic valuation of solar energy potential in Nigeria. In: Power and energy conference at Illinois (PECI), IEEE, Champaign, IL, USA

Poppi S, Sommerfeldt N, Bales C et al (2018) Techno-economic review of solar heat pump systems for residential heating applications. Renew Sustain Energy Rev 81:22–32

Sköldberg H, Rydén B (2017) The heating market in Sweden: an overall picture. http://www.varmemarknad.se/pdf/The_heating_market_in_Sweden_141030.pdf. Accessed on 8 Dec 2017

SMHI (Swedish Meteorological and Hydrological Institute) Global radiation in Sweden. http://www.smhi.se/klimatdata/meteorologi/stralning/normal-globalstralning-under-ett-ar-1.2927. Accessed on 30 Nov 2017

Solar electricity handbook. Average daily solar irradiance. http://solarelectricityhandbook.com/solar-irradiance.html. Accessed on 8 Dec 2017

Statista (2017) Electricity prices for households in Sweden from 2010 to 2016, semi-annually. https://www.statista.com/statistics/418124/electricity-prices-for-households-in-sweden/. Accessed on 8 Dec 2017

Tidd J, Bessant J, Pavitt K (2005) Managing innovation: integrating technological, market and organizational change, 3rd edn. Wiley & Sons Ltd

Trading economics. https://tradingeconomics.com/sweden. Accessed on 30 Sept 2017

Tripanagnostopoulos Y, Souliotis M, Battisti R, Corrado A (2005) Energy, cost and LCA results of PV and hybrid PV/T solar systems. Progress Photovoltaics 13:235–250

X10 PV/T (2017) http://www.absolicon.se/product/absolicon-x10-pvt. Accessed on 30 Sept 2017

X10 PV/T testing report. http://www.estif.org/solarkeymark/Links/Internal_links/SP/SC0987-09%202010-01-20%20Absolicon%20X10%20%2010,14m%20utg%202.pdf

Xuan Q, Li G, Pei G, Ji J, Yuehong Su, Zhao B (2017) Optimization design and performance analysis of a novel asymmetric compound parabolic concentrator with rotation angle for building application. Sol Energy 158:808–818

Part II

Thermal Comfort and Air Quality in Buildings

Influencing Factors for Occupants' Window-Opening Behaviour in an Office Building Through Logistic Regression and Pearson Correlation Approaches

8

Song Pan, Xinru Wang, Xingxing Zhang, Li Chang, and Yiqiao Liu

Abstract

Occupants often perform many types of behaviour in buildings to adjust the indoor thermal environment. In these types, opening/closing the windows, often regarded as window-opening behaviour, is more commonly observed because of its convenience. It not only improves indoor air quality to satisfy occupants' requirement for indoor thermal comfort but also influences building energy consumption. To learn more about potential factors having effects on occupants' window-opening behaviour, a field study was carried out in an office building within a university in Beijing. Window state (open/closed) for a total of 5 windows in 5 offices on the second floor in 285 days (9.5 months) were recorded daily. Potential factors, categorized as environmental and non-environmental ones, were subsequently identified with their impact on window-opening behaviour through logistic regression and Pearson correlation approaches. The analytical results show that occupants' window-opening behaviour is more strongly correlated to environmental factors, such as indoor and outdoor air temperatures, wind speed, relative humidity, outdoor PM2.5 concentrations, solar radiation, sunshine hours, in which air temperatures dominate the influence. While the non-environmental factors, i.e. seasonal change, time of day and personal preference, also affects the patterns of window-opening probability. This chapter provides solid field data on occupant window opening behaviour in China, with high resolutions and demonstrates the way in analyzing and predicting the probability of window-opening behaviour. Its discussion into the potential impact factors shall be useful for further investigation of the relationship between building energy consumption and window-opening behaviour.

S. Pan (✉) · Y. Liu
Beijing Key Laboratory of Green Built Environment and Energy Efficient Technology, Beijing University of Technology, Beijing, China
e-mail: pansong@bjut.edu.cn

Y. Liu
e-mail: L13102158310@163.com

X. Wang
College of Mechanical enginnering, Tianjin university of commerce, Hebei, China
e-mail: xinru5263@126.com

X. Zhang
Department of Energy and Community Buildings, Dalarna University, 79188 Falun, Sweden
e-mail: xza@du.se

L. Chang
Department of Mechanical Engineering, University of Malaya, Kuala Lumpur, Malaysia
e-mail: Changli4321@163.com

X. Zhang (ed.), *Data-driven Analytics for Sustainable Buildings and Cities*, Sustainable Development Goals Series, https://doi.org/10.1007/978-981-16-2778-1_8

Keywords

Window-opening behaviour · Influential factors · Window state · Office building

8.1 Method of Study

8.1.1 Building Description

The case office building was constructed of reinforced concrete and brick, which located at a university in Beijing. The building shape and office layout were very common for office buildings in China. Around the building, there were no tall buildings and trees blocking solar gains and external noise was ignorable. As shown in Fig. 8.1a, the building has two stories, where laboratories were based on the ground floor, and a total of 9 offices with same size of 10 m^2 were placed on the second floor. The typical internal layout of all offices is shown in Fig. 8.1b. The geometry information of the measured offices is shown in Table 8.1. In these 9 offices, 5 offices were applicable and selected for the experiment.

Each office can accommodate two occupants, with a south-facing sliding window. During the experiment, there was only one occupant in each office. In winter, radiator and natural gas boiler were used as main heating generator, with municipal hot water heating as auxiliary heating occasionally. In summer, split type air conditioners were used to supply cooling. During transition seasons (the period from 1st October to 15th November and 15th March to 16th May), natural ventilation was the main strategy. To investigate the effect of regional noise on window-opening behaviour, questionnaires were developed, delivered and answered. Based on the answers, the regional noise seems no significant influence on window-opening behaviour in this case. The occupants of all 5 offices were all non-smoking, consisting of two males and three females, who had lived in Beijing for many years and adapted well to the local climate.

(a) (b)

Fig. 8.1 The case study building (**a**) and a typical office (**b**)

Table 8.1 Basic information of measured offices

The room	Size (m^2)	The orientation of windows	Window form	The number of windows	Gate number
202	11.16	South	Push–pull type	1	1
203	10.23				
205	10.23				
206	10.23				
208	10.23				

8.1.2 Measured Factors

This study covered 2 transition seasons, i.e. from 1st October to 15th November, 2014 and from 15th March to 16th May, 2015, 1 heating season, i.e. from 15th November, 2014 to 15th March, 2015 and 1 cooling season, i.e. 16th May to 15th July, 2015. During transition seasons, occupants mainly adjusted indoor air quality using natural ventilation. During heating season, circulating water radiators were used for the indoor air temperature and split type air conditioners were designed for the indoor air temperature during cooling season. In cooling season, occupants could adjust the setting value via control panel installed on the interior wall but this was not flexible in heating season as the heating was controlled by central plant. During the monitoring, infrared instruments (recording interval: 1 min; induction range: 5 m) were used to record occupancy of the monitored offices; window displacement testers were applied to detect and record the state of office windows (recording interval: 10 min; induction distance: 3 cm); indoor air temperature sensors (recording interval: 10 min; precision: ±0.5 °C) were installed to used to measure and record indoor air temperature. All the above measuring devices have been shown in Fig. 8.2a–c. A portable outdoor meteorological sensor (recording interval: 1 min) was installed at the top of the office building, as shown in Fig. 8.2d.

In the experiment, PM2.5 measurement was also considered. It refers to the atmospheric fine particulate matter with a diameter less than 2.5 μm, which has proven with a direct and harmful effect on health, especially for the respiratory system. The increase in PM2.5 concentrations is positively correlated with the death rate from respiratory diseases (Watterson et al. 2007), and such truth influence a lot on occupant's decision whether open windows or not. By considering the indirect influence of high PM2.5 concentrations on window-opening behaviour, a questionnaire survey was completed before the experiment about whether people would close windows when outdoor PM2.5 concentrations was high. The survey result showed that people tended to close windows for better indoor air quality so as to keep healthy when it comes to high outdoor PM2.5 concentrations. In consideration of serious air pollution in Beijing in recent years and the inter-linked influence of outdoor PM2.5 concentrations on window-opening behaviour, the outdoor PM2.5 concentrations were also recorded based on the data

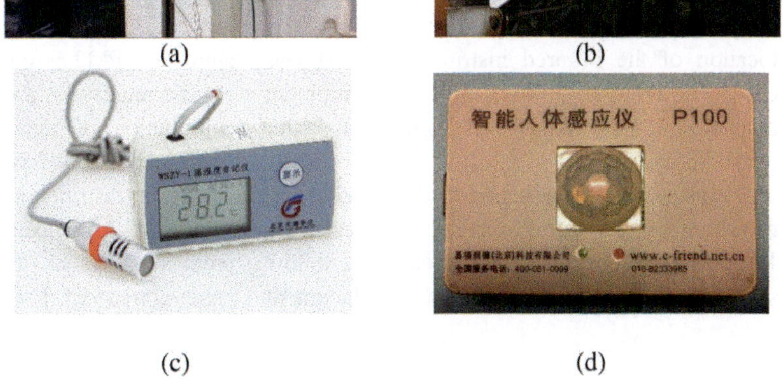

Fig. 8.2 Indoor temperature measuring device (**a**); intelligent human body inductor (**b**); the window displacement tester (**c**) and outdoor temperature measuring device (**d**)

measured in the Temple of Heaven which had a linear distance of 5 km from the investigated building. In summary, the recorded data of this study included (I) environmental factors: indoor and outdoor air temperature (°C), outdoor relative humidity (%), outdoor wind direction (°), outdoor wind speed (m/s), solar radiation (W/m^2), outdoor PM2.5 concentrations (μg/m^3), sunshine hours (the hour of daily radiation level is higher than 120 W/m^2), and (II) non-environmental factors: seasonal change, time of day, and personal preference.

8.1.3 Measuring Devices

To avoid the impact of radiators/convectors on temperature measurement, an indoor air temperature sensor TR (v1.2) produced by the China Architecture Science Institute was placed on a platform of 1.8 m from the floor, staying a distance of at least 1 m from the nearest radiator/convection inverter. By doing so, the temperature was considered the same as the one at occupant height after tests over several days before measurement. All sensors had been calibrated by the manufacturer before the measurement, an additional calibration was carried out every two months during the monitoring period.

The portable outdoor meteorological weather station was placed on the roof at 1 m height over the roof to ensure no influence from the heated generated from the building itself. An infrared instrument P-100 (v1.0) was installed on the inside walls of the office at a 2 m height and close to the office chair. Its role was to detect infrared wavelengths emitted by the human body, with a radius of approximately 5 m. The location of the infrared instrument P-100 was chosen to avoid any impact on its measurement results from devices such as computers, water fountains and radiators.

The state of the window (open/closed) was measured by a window displacement tester D-100 which automatically recorded the window state dynamically. The window displacement tester recorded the window state by means of the magnetic induction of two dry spring pipes positioned on the window. When the magnet force is approached, the dry reed pipes are connected; while magnet is weakening, the dry reed pipes are separated. So the opening state of the window are detected. The shortest time of induction switch is not less than 3 s, and the highest distance where measuring devices can be induced by the magnet is 3 cm.

8.2 Data Processing and Results Analysis

8.2.1 Environmental Factors

The outdoor and indoor air temperatures have proven as the dominating influential factors on window-opening behaviour in many studies (Haldi and Robinson 2008a, 2009a, b; Ben-David and Waring 2016; Haldi et al. 2008; Li et al. 2015; Chen 2009). In this section, the analysis is in turn to assess the impact from both outdoor and indoor air temperatures in this case building. Meanwhile, due to heavy air pollution caused by high PM2.5 concentrations in China, the influence of outdoor PM2.5 concentrations on window- opening behaviour has also been investigated.

A logistic regression approach has been used to analyze the original data by addressing the probability of the window switch against the outdoor/indoor air temperature and PM2.5 concentrations. Logistic regression analysis (Hosmer and Lemesbow 2000) is a statistic method that defines the probability of specific event happening (e.g. opening a window) based on relevant influential factors (e.g. outdoor/indoor air temperature or PM2.5 concentrations). When using logistic regression, a useful method to identify the contribution of individual factors to the event happening is called Wald statistic test, which has a chi-square distribution. Thus, a significant 2 tailed P-value of a particular predictor reflects that this predictor plays an important role in the logistic regression model. The relation is given in Eq. (8.1):

$$P = \exp(a + bx)/[1 + \exp(a + bx)] \quad (8.1)$$

where, P refers to window-opening probability, and x is relevant influential factors, a and b are constants, which represent the intercept and regression coefficients, respectively.

8.2.2 Outdoor Air Temperature

Figure 8.3 displays shows the variation of window-opening probability with increase in outdoor air temperature. It is observed that the window-opening probability increases with the increase of outdoor air temperature. The result was similar to those obtained from the European studies (O'Brien and Gunay 2014). In addition, when outdoor air temperature was lower than 10 °C, indoor occupants generally kept their windows closed, and when outdoor air temperature was higher than 10 °C, it started to influence greatly on window- opening probability. For instance, when the outdoor air temperature reached its highest level of 35.2 °C, the maximum window-opening probability was reached, which was 65.4%. In Fig. 8.3, the intercept $a = -1.45 \pm 0.04$ and the regression $b = 0.045 \pm 0.002$ in Eq. (8.1) by using the logistic regression analysis.

8.2.3 Indoor Air Temperature

Figure 8.4 illustrates the window-opening probability varying with increase in indoor air temperature. The window-opening probability seems to increase with the higher indoor air temperature, which is also in consistent with the results gained by past researchers (O'Brien and Gunay 2014; Karjalainen 2016). When the indoor air temperature was lower than 22 °C, indoor occupants generally kept their windows closed, and when indoor temperature was higher than 27 °C, the indoor air temperature turned to affect significantly on window-opening probability. The maximum probability of window-opening achieved at 62.9% when the intdoor air temperature rose to its highest level of 29.6 °C. By using the logistic regression for the data in Fig. 8.4, the intercept: $a = -4.70 \pm 0.23$, and the regression: $b = 0.171 \pm 0.009$ in Eq. (8.1).

8.2.4 *Outdoor $PM_{2.5}$ Concentrations*

Beijing has seriously suffered from outdoor air pollution because of high PM2.5 concentrations in recent years, especially in winter and transition seasons. Hence, the indirect impact of outdoor PM2.5 concentrations on window-opening

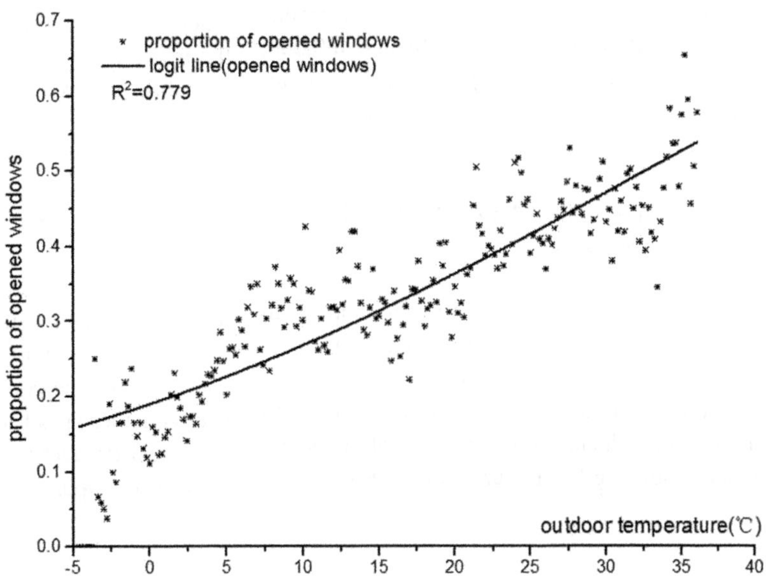

Fig. 8.3 Relationship between window-opening probability and outdoor air temperature

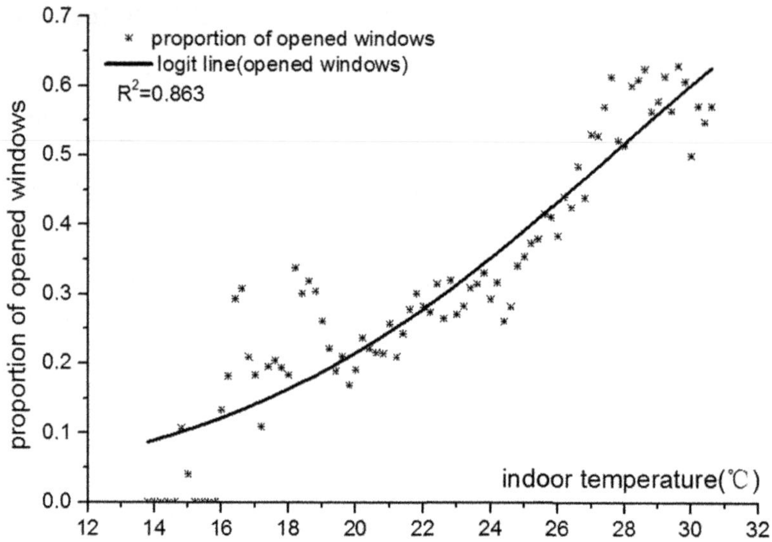

Fig. 8.4 Relationship between window-opening probability and indoor air temperature

behaviour was also investigated when considering the outdoor air temperature in parallel since it dominates the influence on window state (Herkel et al. 2008). The analysis, therefore, was a 3-dimension problem (probability against PM2.5 concentrations and outdoor air temperature) rather than a 2-dimension issue. The logistic regression analysis was applied again to identify whether outdoor PM2.5 had a significant impact on window state in the case building. Both outdoor air temperature and PM2.5 concentrations were set up as predictors of the model and the observed state of windows was regarded as the model output. The ultimate regression results gave that both predictors have a P-value of 0.000, meaning that both the two predictors had a significant influence on the observed window state.

8.2.5 Correlation Analysis for All Factors

Correlation approach has been widely applied in the analysis of the correlation between occupants' window-opening behaviour and its influential factors. This method is adopted to analyze the influence of environmental factors on window-opening behaviour in this study.

In statistics, correlation analysis refers to the analysis on two or more relevant variables. A significant feature of correlation analysis is that all variables are given the same priority in the analysis. In this chapter, the Pearson correlation coefficient, r, has been introduced to calculate the data of the interval variable, as presented in Eq. (8.2).

$$r = \sum_{i=1}^{n} (x_i - x)(y_i - y) \Big/ \sum_{i=1}^{n} (x_i - x)^2 \sum_{i=1}^{n} (y_i - y)^2 \quad (8.2)$$

where, r is the correlation coefficient; x and y are the mean value of x and y respectively; x_i and y_i are the ith observed variable for x and y respectively.

Because of existing sampling errors, the correlation coefficient between two variables not being '0' does not mean that the correlation coefficient between them in all samples not being '0'. As a result, a test needs to be performed to examine the result of the correlation coefficient. The null hypothesis of the examination is that the correlation coefficient between two variables in all samples is '0'. The tool of Statistical Product and Service Solutions (SPSS) is hereby applied used in calculating the probability of the hypothesis correctness, as defined in Eq. (8.3).

$$t = r(n - 2)/(1 - r^2) \quad (8.3)$$

where, n is the number of sample observations; $n - 2$ is degrees of freedom.

When the significance probability, i.e., t, for the correlation coefficient is less than 0.05, it is indicates that the correlation between the two variables is significant; when it's less than 0.01, the correlation between two variables becomes much more significant; when it is higher than 0.05, there is no significant correlation between the two variables, and t is only the probability value.

The results from the correlation analysis are listed in Table 8.2, illustrating that except wind direction ($t = 0.075$), the significance probabilities of indoor air temperature, outdoor air temperature, wind speed, relative humidity and outdoor PM2.5 concentrations, solar radiation, and sunshine hours were all less than 0.05. This means that these factors have statistical significance and they can be used to illustrate the relevance of parameter and the window-opening probability.

The correlation coefficient of outdoor air temperature parameters was 0.507, indicating its relatively strong correlation with the window-opening probability (correlation coefficient greater than or equal to 0.5 representing relatively strong correlation). The results of correlation analysis demonstrated that the probability of window being opened would be increasing with the higher outdoor air temperature, consistent with existing previous studies (O'Brien and Gunay 2014). The correlation coefficient of indoor air temperature parameters was 0.2, reflecting the weaker correlation comparing to outdoor air temperature parameters. Except the wind direction, the correlation coefficients of other factors had statistical significance but much weaker comparing to outdoor and indoor air temperatures. Factors, i.e. outdoor relative humidity, wind speed, outdoor PM2.5 concentrations and the sunshine hours had a rather negative correlation with window-opening probability, indicating the values of window being closed was inversely proportional to these parameters. It can be interpreted that being exposed to high humidity, high speed, high

Table 8.2 Correlation between the probability of window being opened and environmental factors

Descriptive statistics			Correlation between environmental parameters and window opening probability		
Environmental conditions	Mean value	Standard deviation	Correlation coefficient	Significance probability	Covariance
Indoor temperature (°C)	23.35	2.841	0.200**	0.000	0.268
Outdoor temperature (°C)	15.65	10.734	0.507**	0.000	1.044
Wind speed (m/s)	0.67	0.782	−0.024**	0.000	−0.009
Wind direction (°)	189.51	99.703	0.009**	0.075	0.425
Outdoor relative humidity (%)	45.91	20.626	−0.012*	0.019	−0.116
Outdoor PM2.5 concentrations (μg/m^3)	91.58	98.708	−0.056**	0.000	−2.517
Solar radiation (W/m^2)	265.05	291.955	0.063**	0.000	8.609
Sunshine time (h)	12.41	6.012	−0.082**	0.000	−0.233

**Significantly correlated on 0.01 level (bilateral); *significantly correlated on 0.05 level (bilateral)

PM2.5 concentrations, or long sunshine hours may cause discomfort to the indoor occupants.

8.3 Non-environmental Factors

This section presented the detailed analysis of the impacts of non-environmental factors to occupants' window opening behaviours, including seasonal change, time of day and personal preference.

8.3.1 Seasonal Change

Figure 8.5 illustrates the window-opening probability varying with time, the parameters including outdoor air temperature, indoor air temperature. It can be seen that the variation profile of window-opening probability was similar to that of the outdoor air temperature, but not similar to that of the indoor air temperature. In winter, when the outdoor air temperature was lowest, the window-opening probability reached the minimum nearly at the same time, reflecting an action of closing windows to prevent cold air flow going into the room. During this period, however, the indoor air temperature was almost keeping unchanged. In summer, the window-opening probability and the outdoor air temperature almost achieved their maximum values simultaneously, demonstrating the strong impact of outdoor air temperature on the operation of windows. In addition, the variation of year-round outdoor air temperature exhibited a cosine law, which presented a strong correlation with window-opening probability. Such characterized variation along with the outdoor air temperature is meaningful for the occupant behaviour modelling buildings.

Figure 8.6 gives the changes in the relationship between window-opening probability and indoor air temperature during the heating season, transition seasons and cooling season. As indoor air temperature raised, the window-opening probability became higher. When the temperature was in the range of 21–27 °C, there were different window-opening probabilities for those three seasons at the same temperature range. Besides, window were most likely to be kept opened during transition seasons. One possible reason was that natural ventilation was the only way to improve indoor thermal comfort and indoor air quality during transition seasons. In addition, there was a more comfortable outdoor air temperature when the indoor air temperature was in the range of 21–27 °C, even lower or higher during transition seasons. In that case, indoor occupants would prefer more natural ventilation to improve the indoor air environment. On the other hand, comparing to cooling season, window-opening probabilities for heating

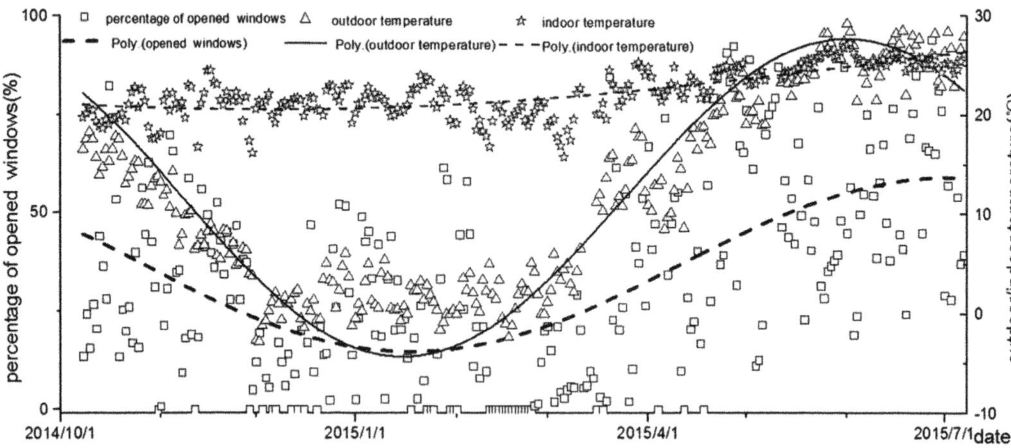

Fig. 8.5 Relationship between window-opening probability and indoor and outdoor air temperature over time

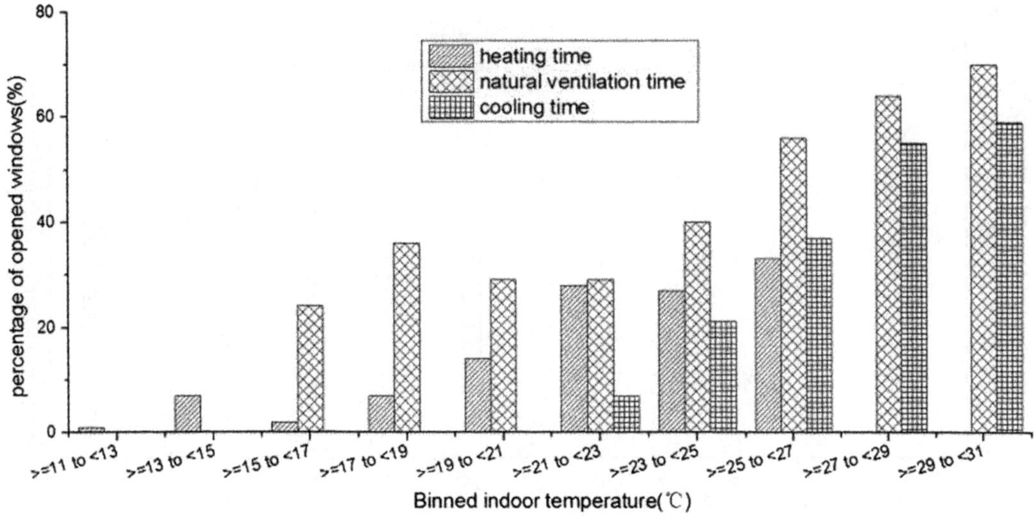

Fig. 8.6 Relationship between window-opening probability and indoor air temperature in different seasons

season were much higher at the same indoor air temperature range. One possible explanation is that the indoor heating system was operated centrally during the heating season, leading to a higher percentage of uncomfortable indoor temperature, and opening the windows were considered the only ways to enhance the indoor thermal comfort.

In Fig. 8.7, the detailed changes in the relationship between window-opening probability and outdoor air temperature are illustrated during heating season, transition seasons and cooling season. The window-opening probability changed proportionally to the outdoor air temperature. There were different window-opening probabilities for those three seasons even at the same temperature range. In comparison to cooling season, the probability of window being opened was higher during transition seasons for the same outdoor air temperature range of 17–35 °C. This is probably because only the natural ventilation was adopted to improve indoor thermal comfort and indoor air quality during transition seasons. The probability of window being opened was lower during transition seasons than that in the heating season for the same outdoor air temperature range of 2–17 °C. It is mainly the same reason as the case result in Fig. 8.6, where central heating brought forward too much uncomfortable, resulting in more frequent switch of window status. In addition, it also reflects that window-opening behaviour is the result of all kinds of all comprehensive factors.

8.3.2 Time of Day

Previous studies have shown that the change of window states often occurred at the time when occupants arrived or left their offices (Herkel et al. 2008). This is usually reflected by the factor called time of day. Figure 8.8 shows the proportion of change of the window state from opened to closed (a) and from closed to opened (b). It can be seen that indoor occupants tended to open their windows when they firstly arrived at their offices and closed their windows when they left their offices at the end of the day. Figure 8.8b reveals an interesting finding that during cooling season, people preferred to open the windows to inlet fresh air in the early morning, although there would be strong direct sunlight for south-facing windows in the late morning. This finding proves that people in this case prefer to open windows to increase the air flow for higher indoor air quality.

It is well-known that when occupants come to their offices in one day, their windows may have

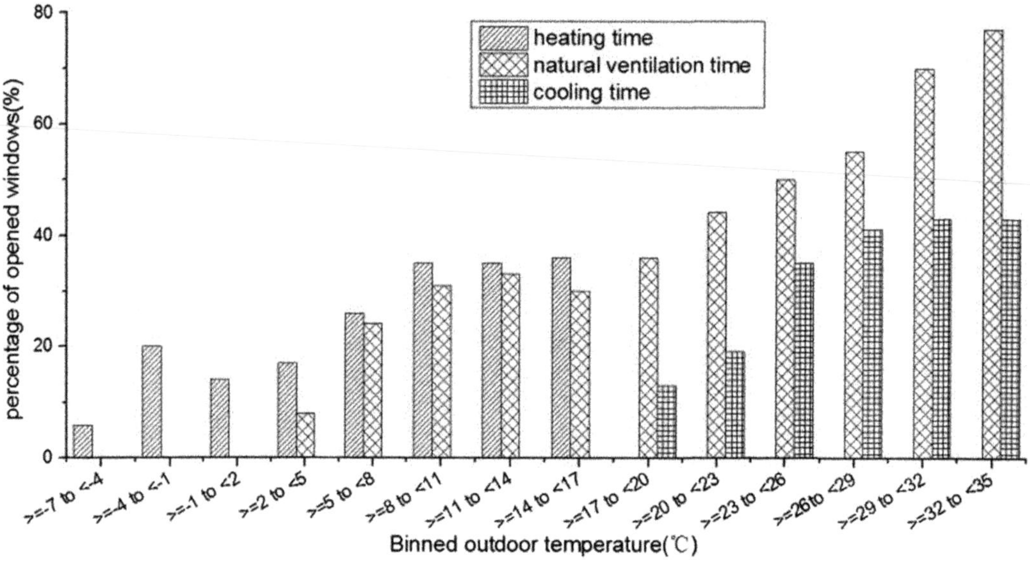

Fig. 8.7 Relationship between window-opening probability and outdoor air temperature in different seasons

Fig. 8.8 Relationship between time of day and window closure (**a**) and open (**b**)

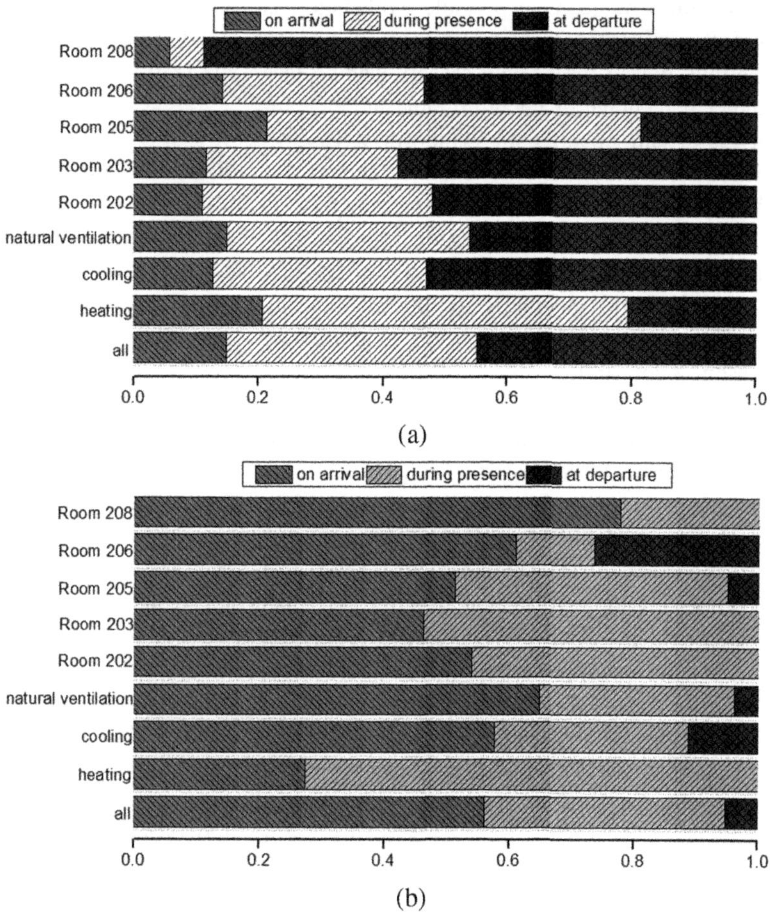

be already opened or closed. When they feel uncomfortable, they will change the window state to adjust the indoor environment. As shown in Fig. 8.9a, when occupants arrived at their offices with closed windows, the window state would not change until they felt uncomfortable. Then, the window state would been kept at that state until occupants felt uncomfortable. In most case, people often close windows when they leave their offices at the end of day. However, occupants in this case preferred to keeping window open until next arrival in order to improve indoor air quality through natural ventilation. In addition, the other situation is shown in Fig. 8.9b.

The variation of the window-opening probability in the 24 h of each season in all offices is depicted in Fig. 8.10. The window-opening probability first increased and then decreased in the day for all seasons. In addition, it can be seen that the window-opening probability climbed to the highest values in core working time till the late afternoon. One possible reason is that the solar illumination is strongest for the windows facing south at noon, and the indoor temperature is higher than morning or afternoon, when an improvement for indoor environment is needed.

8.3.3 Personal Preference

Table 8.3 lists some basic information of occupant behaviour differences for all monitored offices. The percentage of window-opening state has been used to distinguish the differences in occupant behaviour. From Table 8.3, it can be seen that the window-opening rates for all five offices were between 15.0 and 48.7%, and Rooms 202, 206 and 208 had more time with open windows. On the other hand, the average times for each occupant to open their window in a week has been used to judge the frequency of occupant behaviour. The difference in average time for each occupant can also reflect the contribution of personal preference. It was defined a low frequency as the times from 0 to 2.4, an average frequency when the times from 2.4 to 5.6, and a high frequency when the time more than 5.6. The result in the last column of Table 8.3 illustrates that Room 205, 206 have the most frequent occupant behaviour.

It is worth noting that the trend of occupant behaviour is defined as a predictable case if the window-opening probability rises with the increase of indoor and outdoor temperatures. Figure 8.11 predicts the window-opening

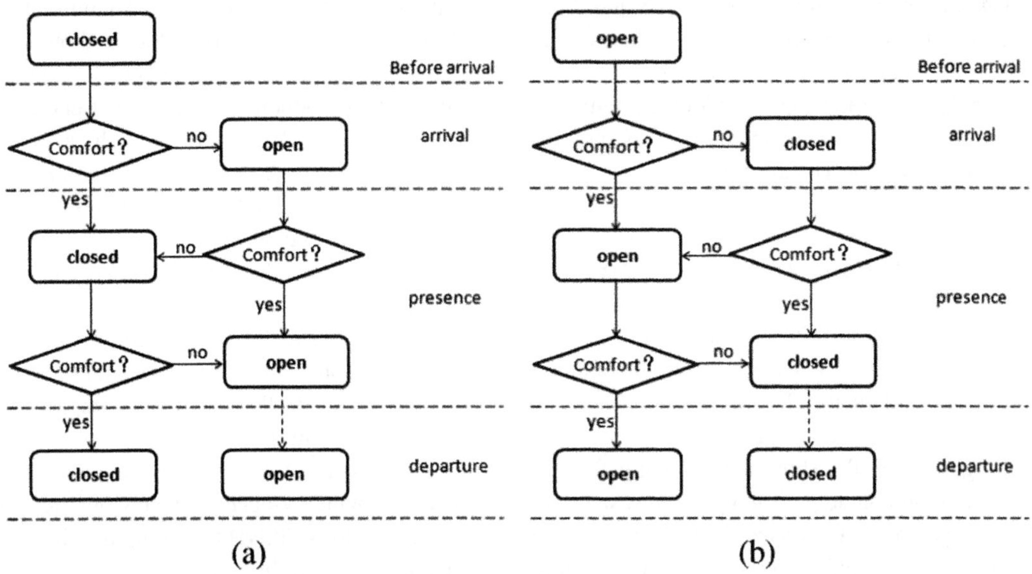

Fig. 8.9 The change of window status in different time periods when it's closed (**a**) and open (**b**) before arrival

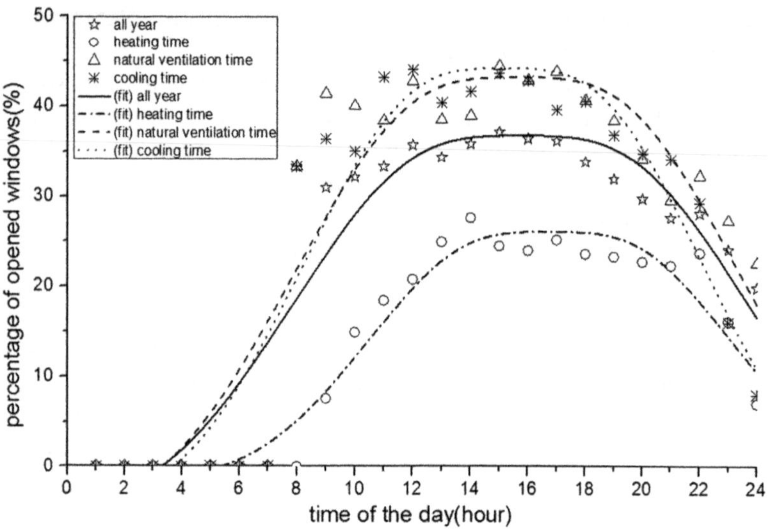

Fig. 8.10 The change of window-opening probability in different time periods

Table 8.3 The window probability description for each room

Ref	Nb. Pers	Age	Gender	Ratio open (%)	Actions	Smoking	Predictive	Overall activity
202	1	42	Male	48.7	4.68	No	No	Average
203	1	39	Female	22.0	2.54	No	No	Average
205	1	43	Male	15.0	7.80	No	No	High
206	1	50	Female	43.4	6.98	No	No	High
208	1	45	Female	31.2	1.76	No	Yes	Low

Actions: the percentage of the open window per person per week

probability against different occupants when indoor or outdoor air temperature rises. The behaviour of occupant in Room 208 has been considered be predictable. Nevertheless, in regard to a certain occupant, the trend of occupant behaviour referring to indoor air temperature or outdoor air temperature was not all consistent with the general trend.

8.4 Conclusions and Discussion

This chapter elaborates a longitudinal field measurement on occupant window-opening behaviour and their relevant influential factors for a period of 9.5 months in an office building in Beijing. The results show that all these factors, i.e. indoor and outdoor air temperatures, seasonal change, personal preference, time of day all have performed great influences on window-opening behaviour. Main conclusions are as follows:

(1) Environmental factors that affect window-opening behaviour are mainly indoor and outdoor air temperatures, against other factors. With the increase of indoor and outdoor air temperatures, the window-opening probability will gradually rises and the trend is in line with the logit model with a good goodness-of-fit: the goodness-of-fit for outdoor temperature is 0.779, and it is 0.863 for indoor air temperature.
(2) In addition to indoor and outdoor air temperatures, other factors, such as wind speed, relative humidity, outdoor PM2.5 concentrations, solar radiation, sunshine hours showed statistically significant impact on window-opening probability.
(3) Seasonal change also has an obvious influence on window-opening behaviour, reflected by various different patterns in window-opening probability in different seasons, even at similar indoor or outdoor air

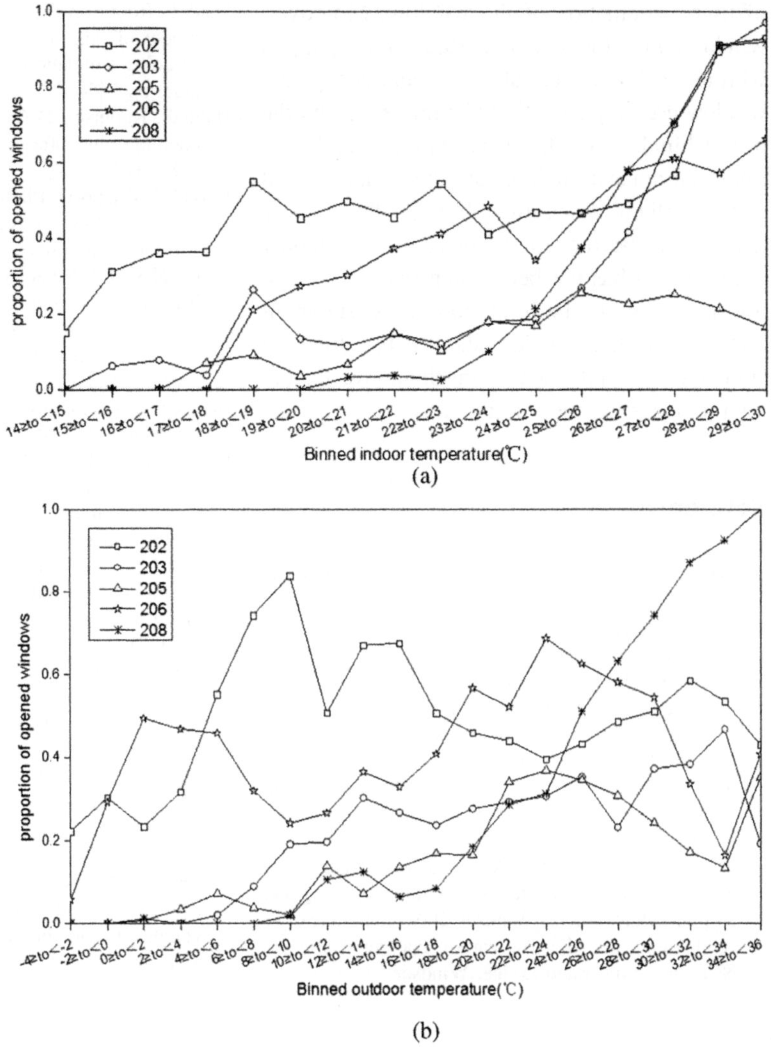

Fig. 8.11 Relationship between personal preference and indoor temperature (**a**) and outdoor temperature (**b**)

temperatures conditions. In the whole year, the window-opening probability mainly varies with outdoor air temperature.

(4) The window state also varies at different time of the day. Occupants tend to maintain the existing state of their windows until they feel uncomfortable. When they feel dissatisfied with the environment, they will change the window state to adjust the indoor comfort environment. In addition, there is the highest window-opening probability at the core working time till the late afternoon. Occupants prefer to open their windows when they firstly arrived their offices and close the windows when they left their offices at the end of the day.

(5) Different people have various personal preferences on using windows. They may prefer different window states even at similar indoor or outdoor air temperature conditions. However, the general trend keeps the same, which is general proportional to indoor and outdoor air temperatures.

Owing to the inherent limitations on monitored samples, influences from some other potential factors cannot been covered in this study, such as occupancy pattern, gender and age of users, orientation of the window. Moreover, windows monitored in this study are equipped with internal and external blinds, and occupants' window use may potentially be influenced by the use of blinds, view from the window, and daylight illuminance falling on the window and direct sunlight penetrating through the window (Zhang and Barrett 2012). In addition, window opening area has not been taken into account in this study, due to the limitation of measurement method. Further studies should be done to strengthen the influences from the above factors with a higher number of samples.

References

Ben-David T, Waring MS (2016) Impact of natural versus mechanical ventilation on simulated indoor air quality and energy consumption in offices in fourteen U.S. cities. Build Environ 104:320–336

Chen W (2009) Study on the summer thermal comfort and the occupant window behaviour for the hot summer and cold winter region. In: Hunan: Master's thesis, HVAC, Hunan University

Haldi F, Robinson D (2008a) On the behaviour and adaptation of office occupants. Build Environ 43:2163–2177

Haldi F, Robinson D (2008b) A comparison of alternative approaches for the modelling of window opening and closing behaviour. In: Windsor 2008 conference: air conditioning and the low carbon cooling challenge. NCEUB, Cumberland Lodge, Windsor

Haldi F, Robinson D (2009a) A comprehensive stochastic model of window usage: theory and validation. In: Building simulation conference 2009. Glasgow, Scotland, pp 545–552

Haldi F, Robinson D (2009b) Interactions with window openings by office occupants. Build Environ 44 (12):2378–2395

Herkel S, Knapp U, Pfafferott J (2008) Towards a model of user behaviour regarding the manual control of windows in office buildings. Build Environ 43 (4):588–600; Nicol FJ (2001) Characterising occupant behaviour in buildings: towards a stochastic model of occupant use of windows, lights, blinds, heaters and fans. In: Building simulation conference. Rio de Janeiro, Brazil

Hosmer DW, Lemesbow S (2000) Applied logistic regression, 2nd edn. Wiley, New York

Karjalainen S (2016) Should we design buildings that are less sensitive to occupant behaviour? A simulation study of effects of behaviour and design on office energy consumption. Energy Eff 1–14

Li N, Li J, Fan R et al (2015) Probability of occupant operation of windows during transition seasons in office buildings. Renew Energy 73:84–91

O'Brien W, Gunay HB (2014) The contextual factors contributing to occupants' adaptive comfort behaviours in offices—a review and proposed modeling framework. Build Environ 77:77–87

Watterson TL, Sorensen J, Martin R, Coulombe RA Jr (2007) Effects of PM2.5 collected from Cache Valley Utah on genes associated with the inflammatory response in human lung cells. J Toxicol Environ Health A 70(20):1731–1744

Zhang Y, Barrett P (2012) Factors influencing the occupants' window opening behaviour in a naturally ventilated office building. Build Environ 50(50):125–134

9 Reinforcement Learning Methodologies for Controlling Occupant Comfort in Buildings

Mengjie Han, Ross May, and Xingxing Zhang

Abstract

Classical building control systems are becoming vulnerable with increasing complexities in contemporary built environments and energy systems. Due to this, the reinforcement learning (RL) method is becoming more distinctive and applicable in control networks for buildings. This chapter, therefore, conducts a comprehensive review of RL techniques applied in control systems for occupant comfort in indoor built environments. The empirical applications of RL-based control systems are presented, depending on comfort objectives (thermal comfort, indoor air quality, and lighting) along with other objectives which invariably includes energy consumption. The class of RL algorithms and implementation details regarding how the value functions have been represented and how the policies are improved are also illustrated. This chapter shows there are limited works for which RL has been explored for controlling occupant comfort, especially in indoor air quality and lighting. Relatively few of the reviewed works incorporate occupancy patterns and/or occupant feedback into the control loop. Moreover, this chapter identifies a gap with regard to the performance of implementing cooperative multiagent RL (MARL). Based on our findings, current challenges and further opportunities are discussed. We expect to clarify the feasible theory and functions of RL for building control systems, which would promote their widespread application in built environments.

Keywords

Reinforcement learning · Building · Indoor comfort · Occupant

9.1 Overview of Comfort Control in Buildings

9.1.1 Background

Around 90% of people spend most of their time in buildings (Shaikh et al. 2013) and they spend 80–90% of the day indoors, and consequently occupant comfort becomes more and more important. Therefore, maintenance of comfort factors is crucial for improving occupant's feeling of

M. Han (✉) · R. May
Micro Data Analysis, Dalarna University, 79188 Falun, Sweden
e-mail: mea@du.se

R. May
e-mail: rom@du.se

X. Zhang
Department of Energy and Community Buildings, Dalarna University, 79188 Falun, Sweden
e-mail: xza@du.se

X. Zhang (ed.), *Data-driven Analytics for Sustainable Buildings and Cities*, Sustainable Development Goals Series, https://doi.org/10.1007/978-981-16-2778-1_9

comfort, health, morale, working efficiency as well as productivity (Li et al. 2016). Thermal comfort, visual comfort and indoor air quality (IAQ) seem to be the key parameters that jointly influence the level of comfort of a building occupant (Boodi et al. 2018; Park and Nagy 2018; Frontczak and Wargocki 2011). Thus, reaching a comfortable indoor environment is a multi-objective task and needs comprehensive cooperation between different building components.

Building design and the building management system (BMS) are direct key factors that affect the comfort level of an occupant. The design of buildings relates to the occupancy level, ventilation, use of natural resources, etc., which remains critical for a comfortable indoor climate in future building development (Wang et al. 2018). However, it is a difficult task to find sound design alternatives satisfying different conflicting criteria, such as natural ventilation against heating/cooling loss (Wang et al. 2005). Compared to the design of buildings, the BMS considers both the maintenance and the improvement of the comfort level of the buildings' occupants through a diversity of control methods. A BMS generally refers to the integrated monitoring, transmitting and control of the indoor environment based on various protocols and communication interfaces. Such a characteristic enables the BMS to have a wider application in practice.

In a BMS, the essential function is the building control system (BCS), which is usually designed to maintain indoor comfort at a certain level when responding to dynamic climate and operational conditions. The advanced control methods are able to, not only take advantage of realtime data—data available as soon as it is created—to produce the desired comfort level, but can also minimise the operational and maintenance cost, and in turn improve the building's energy performance (Marinakis et al. 2013). As a result, there is a high demand for the development of advanced control methods for future smart and economic-friendly building environments.

9.1.2 Necessity of New Methods for the Building Control System (BCS)

With the development of diverse building systems and the movement towards improving adaptive indoor comfort, buildings are becoming more and more complex to control. In practice, advanced realtime control strategies attempt to make a correct action at a prescribed point in time within defined time tolerances (Gambier 2004). In a BCS, the controller uses real-time data, which is presented as it is acquired, to make decisions, where its related impact on the indoor habitat is often delayed in such a dynamic setting. The ideal real-time control strategies can deliver the signal so as to avoid the delayed influence on the indoor surroundings. They work effectively based on the building models, the building system models, weather forecast models, and energy tariff forecast models, etc. However, these models are not as accurate in the sense of prediction, thus leading to potential inappropriate control in the future. Therefore, the existing control approaches are facing a serious challenge in real-time adaption/influence to/on occupant comfort and may fail to respond/maintain to/the indoor environment efficiently.

Reinforcement learning (RL), as one of the model-free control techniques, can be an alternative solution to such challenges when it is applied together with real-time control strategies. Model-free control techniques are able to work independently without having a priori knowledge of specific models. For instance, a recently realised Markov-based method, can work in both a model-based and model-free context where the former refers to learning a model and using this to obtain a policy and the latter to learning a policy without learning a model (Kaelbling et al. 1996). With this approach are the classic learning algorithms, such as Q-learning, $TD(\lambda)$, Dyna, and simulation-based search, that make RL much more attractive and efficient in artificial intelligence applications (Mnih et al. 2015; Silver et al.

2016, 2017; Sutton and Barto 2018). Moreover, the efforts made on solving deep RL problems open up the possibility of working on continuous large datasets (Gu et al. 2016; Lillicrap et al. 1509). The distinctive property of RL is that the learner or agent, via a trial-and-error paradigm, can make optimal actions without having a supervisor, which essentially fits the goal of a complex control problem.

In BCSs, performances of using RL for occupant comfort have not been analysed from the methodological point of view and the future tasks in this field are still rare. Relevant review works examining RL control methods has been limited (see Table 9.1 and Fig. 9.1). Unlike energy demand response (Vazquez-Canteli and Nagy 2019), this chapter considers occupant comfort as the principal optimisation target. Therefore, the aim of this chapter is to methodologically review the empirical works on how RL methods have been implemented for comfort control in buildings, and provide instructive directions for future research.

This chapter will thus highlight following issues. Firstly, it summarises the existing relevant review works in different areas of occupant comfort control, including thermal control, indoor air quality (IAQ) control, lighting control, air velocity control, and visual comfort control, etc. Secondly, under this control setting, it provides a comprehensive review of RL being implemented. Thirdly, it analyses the application of RL for comfort control in multi-agent environments. Fourthly, it highlights the potential of RL as a sustainable forerunner for truly occupant-centric building operation in the evolving smart city. Beyond these, this chapter finally identifies the current research gaps and proposes future research from both the application and methodology points of view.

9.1.3 Review Methods

In this sub-chapter, we make our search for articles (Tables 9.1 and 9.2) in the search engines, Web of Science, ScienceDirect and Google Scholar. We do not limit the publication time. For the review articles given in Table 9.1, our searching keywords are, *review* and *control* and *occupant* and *building*; *review* and *control* and "*indoor comfort*" and *building*. For the core articles given in Table 9.2, our searching keywords are

$$\left[(building(s)) \cap \begin{pmatrix} (reinforcement\ learning) \cup \\ (Markove\ decision\ processes) \cup \\ Q-learning \end{pmatrix} \cap \begin{pmatrix} comfort \cup \\ (thermal\ comfort) \cup \\ (visual\ comfort) \cup \\ (indoor\ air\ quality) \cup \\ occupant \cup \\ (indoor\ environment) \end{pmatrix} \right] \cup \begin{bmatrix} (model\ free\ control) \cup \\ (intelligent\ control) \end{bmatrix}.$$

The intersection notation, $\cap$, is the relation operation, "and", where the words on both sides are searched for simultaneously. $\cup$ is equivalent to, "or", where only one word or phrase on either side is taken for searching. For example, (*building* and "*reinforcement learning*" and *comfort*) or ("*model free control*") is one of our searching records. We also search *Markov decision processes* (MDPs) and Q-learning to guarantee that the underlying theory of RL and the most popular algorithm are covered. We also include "*model free control*" and "*intelligent control*" as alternative keywords because some articles treat RL as a special case of these control methods. We read through every search outcome and excluded irrelevant articles without direct optimisation on comfort. That is, we only included those core articles that have clearly optimized for comfort. Other joint optimisation objectives may have also been considered in the core articles but our main interest was in those articles containing at least one comfort component in the optimization objectives. Furthermore, only those papers which empirically investigated their proposed approaches, either through the use of synthetic data or real data were included. Doing so, we have identified 33 core articles that are summarised later.

Table 9.1 Relevant review works

References	Comfort type	Control methods	RL reviewed	Future indications
Galasiu and Veitch (2006)	Lighting	Lighting control; shading control	No	Satisfaction and outdoor conditions need to be considered in control system
Royapoor and Caraiscos (2009)	Thermal; visual; IAQ	Conventional control; intelligent Control; agent-based control	Yes	Future trends: balance between thermal comfort and energy usage and random neural networks with RL
Wenqi and Zhou (2009)	Thermal	ANN; fuzzy logic	No	More real-time environmental data and human activity level can be collected and applied in the system design
Guo et al. (2010)	Lighting	Occupant- based control	No	–
Roetzel et al. (2010)	Thermal; IAQ (ventilation)	Occupant-based control	No	–
Haq et al. (2014)	Lighting	Occupant-based control; daylight-linked control; scheduling control	No	Development of control algorithm can be helpful in improving the effectivity of commissioning and lead to better user satisfaction
Shaikh et al. (2014)	Thermal; visual; IAQ	Conventional control; intelligent control	No	Various other artificial intelligent techniques need to be future research objectives
Vesely and Zeiler (2014)	Thermal	Occupant based control	No	–
Song et al. (2015)	Thermal; IAQ; visual	Conventional control; computational intelligent control	No	Future trends: model-independent control strategies for general purpose use which can reduce the development time for model matching and parameters tuning
Chenari et al. (2016)	Thermal; IAQ; humidity	Model-based control; rule-based control; GA	No	Study of intelligent window-based hybrid ventilation strategies for maintaining IAQ and reducing energy consumption is missing
Merabti et al. (2016)	Thermal; IAQ	PID; fuzzy; fuzzy PID; adaptive fuzzy PD; NN; neuro-fuzzy; GA	No	Intelligent control system needs to be upgraded
Enescu (2017)	Thermal	ANN; AR-ANN; fuzzy, hybrid ANN-fuzzy	No	Refined adaptive comfort models in smart building control systems
Wang et al. (2017a)	Thermal; humidity; CO2; air velocity	Binary; iterative; PID; MPC; nonlinear; pole-placement; optimal; fuzzy; ANN; adaptive	No	Advanced control strategies combined with HVAC technologies have been currently becoming a new trend in building energy conservation and indoor environment quality research
Ye et al. (2017a)	IAQ (pollutants)	Occupant-based control	No	–

(continued)

Table 9.1 (continued)

References	Comfort type	Control methods	RL reviewed	Future indications
Boodi et al. (2018)	Thermal; lighting; IAQ; humidity	MPC; PID; fuzzy logic; RL; ANN; rule-based	Yes	Future trends: intelligent building models and adaptive building controller
Guyot et al. (2018)	IAQ	–	No	Smart ventilation is still an emerging technology
Kruisselbrink et al. (2018)	Lighting quality	–	No	–
Park and Nagy (2018)	Thermal	Rule-based; optimisation; intelligent control; MPC	No	Only 5.2 and 15.6% of thermal comfort and building control publications cited each other
Royapoor et al. (2018)	–	Classis (binary and PID); computational	Yes	More advanced computational techniques (ANN or agent-based) that have so far largely remained in demonstration stage

9.2 Summary of Relevant Review Works and Their Indications

A summary of previous review works that have examined building control methods for occupant comfort factors is shown in Table 9.1 in chronological order. In addition to examining all mentioned comfort types and control methods in those works, we have also checked if RL has been reviewed and whether there are any future indications in each work. Herein, we focus on those indications that point out some possibilities of implementing RL for comfort control. For instance, Dounis and Caraiscos suggest using random neural networks with RL to control thermal comfort and energy usage (Royapoor and Caraiscos 2009), while Song et al. indicate model-independent control strategies for general purpose use which can reduce the development time for model matching and parameter tuning (Song et al. 2015). Similarly, the development of artificial intelligence control strategies is regarded as an important future task (Shaikh et al. 2014; Merabti et al. 2016). In this sense, the neglected model-free RL technique has been regarded as a promising and attractive method for controlling occupant comfort, and we therefore have a rationale to highlight it.

In all the 19 review works, thermal comfort accounts for the majority of all topics and ANN, model predictive control (MPC) and the fuzzy method seem to be the most often reviewed control methods. RL has appeared three times and only around ten relevant empirical papers have been investigated (Li et al. 2016; Royapoor and Caraiscos 2009; Royapoor et al. 2018). To have an overview of the distribution, we divide the control methods into classic control (on/off and PID), hard control (MPC, optimal control, nonlinear control, adaptive control), soft or intelligent control (ANN, fuzzy-based), RL control, agent-based control and occupant-based control. Considering the frequency, we divide comfort factors into the four most common occurring amongst the review articles, namely, thermal comfort, IAQ, humidity and visual comfort. Each intersection of control method and comfort type is counted once if at least one paper was reviewed. We aggregate them and divide

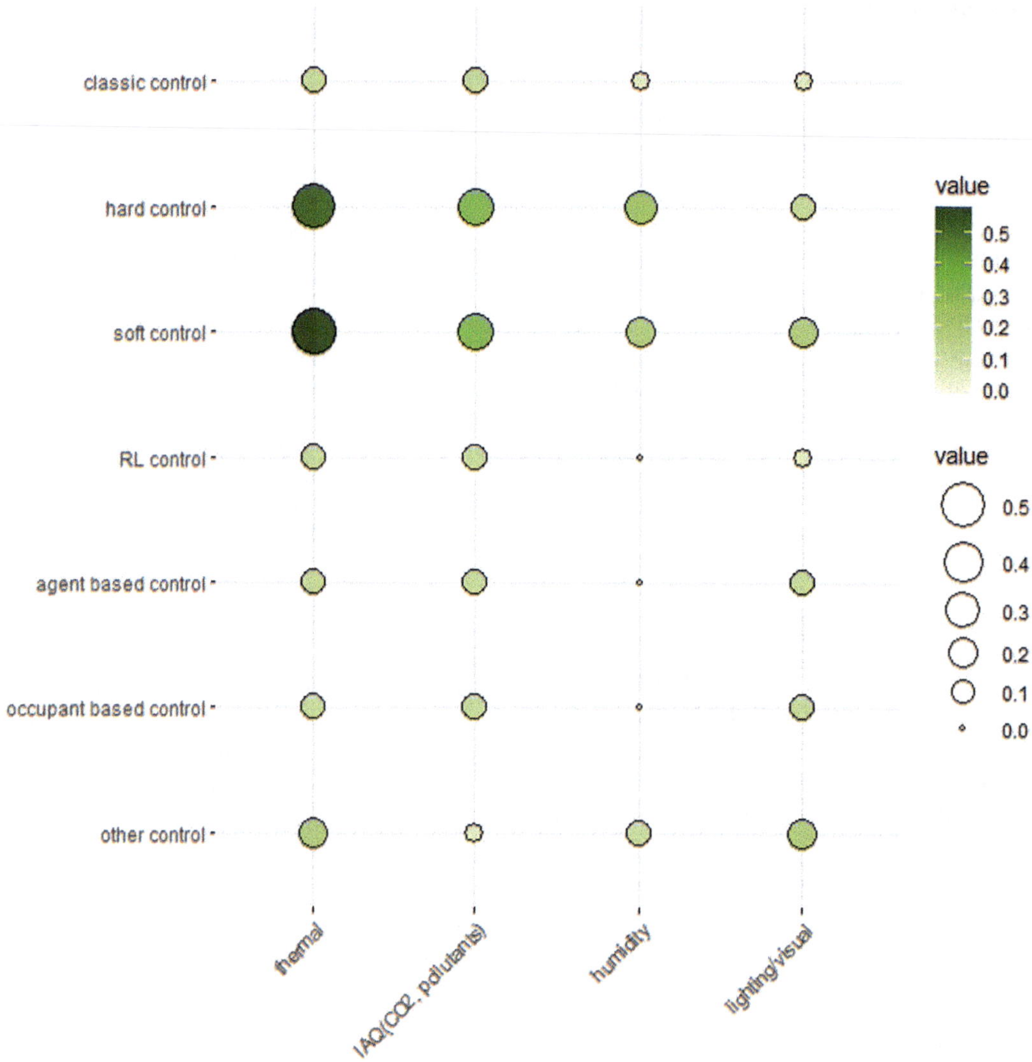

Fig. 9.1 Average number of intersections of different comfort types and control methods reviewed in each work

them by nineteen to obtain the average examining rate. Specifically, Fig. 9.1 gives the average number of intersections of different comfort types and control methods reviewed in each work. For thermal comfort, both hard and soft control appear most frequently and reach at least a 50% examining rate. For IAQ and humidity, the percentage ranges from 19 to 31%. For visual comfort, the control methods distribute relatively evenly. Looking at RL, the rates are not more than 13% among all comfort types. This low examining rate makes it appealing to investigate the framework of combining RL with occupant comfort control.

A recent work reviewed the application of reinforcement learning for demand response, which is relevant for integrating renewable energy sources into the smart grid (Vazquez-Canteli and Nagy 2019). The authors pointed out that human comfort and satisfaction in buildings

Table 9.2 Empirical applications of RL control on indoor comfort

References	Energy system	Optimisation objectives		Class	Learning control method	Value function representation	Policy improvement	Pre-training	Agent	State	Action
		Comfort	Others								
Baghaee and Ulusoy (2018)	HVAC	IAQ	Energy	VB	Q-learning	Tabular	N/A	No	SA	D*	D
Barrett and Linder (2015)	HVAC	Thermal	Energy	VB	Q-learning	Tabular	ϵ-greedy	No	SA	D	D
Bielskis et al. (2013)	HVAC, lighting	Thermal, light	N/A	AC	TD Actor-Critic	Function approximator	Policy gradient	No	SA	C**	C
Bonte et al. (2014)	Heating	Thermal	N/A	VB	Q-learning	N/A	N/A	No	SA	N/A	N/A
Chen et al. (2018)	HVAC, window	Thermal, IAQ	Energy	VB	Q-learning	N/A	ϵ-greedy	No	SA	N/A	D
Cheng et al. (2016)	Blinds, lighting	Light	Energy	VB	Q-learning	N/A	ϵ-greedy	No	SA	D	D
Dalamagkidis et al. (2007)	HVAC	Thermal, IAQ	Energy	VB	RLS-TD(λ)	Function approximator	ϵ-greedy	No	SA	C	D
Dalamagkidis and Kolokots (2008)	HVAC	Thermal	Energy	VB	RLS-TD(λ)	Function approximator	N/A	No	SA	C	D
Eller et al. (2018)	Heating, ventilation	Thermal, IAQ	Energy	VB	Q-learning	Tabular	N/A	No	SA	D	D
Fu et al. (2018)	HVAC	Thermal, IAQ	Energy	VB	SARSA	Function approximator	ϵ-greedy	No	SA	C	D
Hurtado et al. (2018)	Generic loads	Thermal, IAQ	Demand	VB	eJAL, Q-learning	N/A	N/A	N/A	MA	N/A	N/A
Jouffe (1997)	Ventilation	Thermal, IAQ	N/A	AC	TD(λ), AC	Function approximator	Policy gradient	No	SA	N/A	D
Khalili et al. (2010)	Lighting	Lighting	Energy	VB	MAXQ	Tabular	HRL epsilon-decay, Greedy	No	SA	D	D

(continued)

Table 9.2 (continued)

References	Energy system	Optimisation objectives		Class	Learning control method	Value function representation	Policy improvement	Pre-training	Agent	State	Action
		Comfort	Others								
Li et al. (2015)	HVAC	Thermal	N/A	VB	MEC, Q-learning	Function approximator, tabular	PAC, ϵ -greedy	N/A	SA	D	D
Li and Xia (2015)	HVAC	Thermal	Energy	VB	Q-Learning	Multi-grid method	Ergodic exploration	No	SA	N/A	N/A
Lu et al. (2019)	HVAC	Thermal	N/A	VB	Q-Learning	Tabular	ϵ-greedy	No	SA	D	D
Mozer (1998)	HVAC, DHW, lighting	Thermal, light	Energy	N/A	N/A	N/A	N/A	N/A	SA	N/A	N/A
Nagy et al. (2018)	Heat pump	Thermal	Energy	VB	Dyna, D-DNFQI	N/A, function approximator	ϵ-greedy	No	SA	C	D
Park et al. (2019)	Appliances	Light	Energy	VB	Value iteration	Tabular	N/A	N/A	MA	D	D
Pedro et al. (2014)	HVAC	Thermal	Energy	VB	Q-learning	Tabular, Function approximator	N/A	No	SA	D,C	D,C
Ruelens et al. (2015)	Heat pump	Thermal	Energy	VB	Fitted Q-iteration	Function approximator	Soft-max	No	SA	N/A	N/A
Sato et al. (2012)	Air Conditioner	Thermal	Energy	VB	Q-learning	Tabular	N/A	No	SA	D	D
Schmidt et al. (2017)	HVAC	Thermal	Energy	VB	Fitted Q-iteration	Function approximator	N/A	No	SA	N/A	D
Sun et al. (2013)	HVAC	Thermal, IAQ	Energy	VB	Q-learning	N/A	N/A	N/A	SA	N/A	N/A
Sun et al. (2015a)	HVAC	Thermal	Energy	VB	N/A	N/A	ϵ-greedy	No	MA	D	D
Sun et al. (2015b)	HVAC	Thermal, IAQ	Energy	VB	Q-learning	N/A	ϵ-greedy	N/A	SA	N/A	N/A

(continued)

Table 9.2 (continued)

References	Energy system	Optimisation objectives		Class	Learning control method	Value function representation	Policy improvement	Pre-training	Agent	State	Action
		Comfort	Others								
Urieli and Stone (2013)	HVAC	Thermal	Energy	VB	Simulation based tree search	N/A	N/A	No	SA	N/A	D
Wang et al. (2017b)	HVAC	Thermal	Energy	AC	A2C	Function approximator	Policy gradient	Yes	SA	N/A	D
Wei et al. (2017)	HVAC	Thermal	Energy	VB	Variant of DQN, Q-learning	Function approximator	ϵ-greedy, N/A	N/A	MA/SA	N/A	D
Yang et al. (2015)	PV/T, thermal storage, heat pump	Thermal	Energy	VB	Q-learning	Tabular, function approximator	ϵ-greedy	No	MA	C,D	D
Yu and Dexter (2010)	HVAC	Thermal	Energy	VB	Q(λ)	Tabular	ϵ-greedy	Yes	SA	D	D
Zhang et al. (2018)	HVAC	Thermal	Energy	AC	A3C	Function approximator	Policy gradient	Yes	SA	N/A	D
Zhang and Lam (2018)	HVAC	Thermal	Energy	AC	A3C	Function approximator	Policy gradient	No	SA	N/A	D

have been mostly studied in single agent systems. The performance of RL in multi-agent systems needs to be explored, which is one of our goals in our study.

9.3 The Reinforcement Learning Method

The idea of reinforcement learning originated from the term "optimal control" which emerged in the late 1950s, where a problem was formulated by designing a controller to minimise a measure of the behaviour of a system over time (Sutton and Barto 2018). Bellman (1957a) came up with the concept of MDPs or finite MDPs, a fundamental theory of RL, to formulate optimal control problems.

The learner or agent of RL learns how to map situations to actions to maximise a numerical delayed reward signal. It does not have to have a "teacher" telling it how to take an action but, rather, makes decisions via implementing a trial-and-error search, and recognizing the delayed reward from the environment that the agent interacts with (Sutton and Barto 2018). RL is neither supervised learning nor unsupervised learning; it is a third category of machine learning. Whereas supervised learning gets signals of correct actions, RL gets signals from the reward of an action without knowing if the action was correct or not. RL, in a sense, is the core of machine learning techniques. In the context of artificial intelligence, RL allows the agent to automatically determine behaviours, which cannot be achieved by supervised learning or unsupervised learning.

9.3.1 Elements of Reinforcement Learning and MDPs

In a dynamic sequential decision-making process, the state $S_t \in \mathcal{S}$ refers to a specific condition of the environment at discrete time steps $t = 0, 1, \ldots$. By realising and responding to the environment, the agent chooses a deterministic or stochastic action $A_t \in \mathcal{A}$ that tries to maximise future returns and receives an instant reward $R_{t+1} \in \mathcal{R}$ as the agent transfers to the new state S_{t+1}. The reward is usually represented by a quantitative measurement. Figure 9.2 (Sutton and Barto 2018) shows how a sequence of state, action, and reward is generated to form an MDP.

The Markov property tells us that the future is independent of the past and depends only on the present. In Fig. 9.2, S_t and R_t are the outcomes after taking an action and are considered as random variables. Thus, the joint probability density function for S_t and R_t is defined by:

$$p(s', r|s, a) = \mathbb{P}[S_t = s', R_t = r|S_{t-1} = s, A_{t-1} = a], \tag{9.1}$$

where $s, s' \in \mathcal{S}$, $r \in \mathcal{R}$, and $a \in \mathcal{A}$. It can be seen from Eq. (9.1) that the distribution of state and reward at time t depends only on the state and action one step before. Equation (9.1) implies the basic rule of how the MDP works and one can easily determine the marginal transition probabilities $p(s'|s, a)$:

$$\begin{aligned} p(s'|s, a) &= \mathbb{P}[S_t = s'|S_{t-1} = s, A_{t-1} = a] \\ &= \sum_{r \in \mathcal{R}} p(s', r|s, a). \end{aligned} \tag{9.2}$$

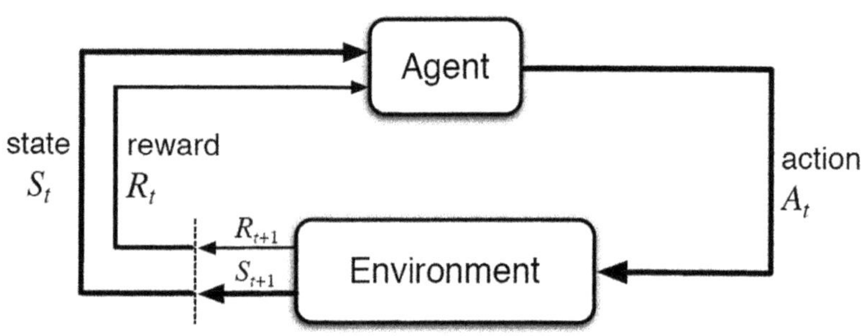

Fig. 9.2 The interaction between agent and environment in an MDP

Equation (9.3) gives the expected reward by using the marginal distribution of R_t:

$$r(s,a) = \mathbb{E}[R_t|S_{t-1} = s, A_{t-1} = a] \\ = \sum_{r\in\mathcal{R}} r \sum_{s'\in\mathcal{S}} p(s', r|s,a). \quad (9.3)$$

Both Eqs. (9.2) and (9.3) are used for solving the optimal value functions presented in Sect. 9.3.3.

9.3.2 Policies and Functions

A policy π is a distribution over actions given states. It fully defines the behaviour of an agent by telling the agent how to act when it is in different states. The policy itself is either deterministic or stochastic and the probability of taking an action, a, in state s is:

$$\pi(a|s) = \mathbb{P}[A_t = a|S_t = s]. \quad (9.4)$$

The policy can be considered as a function of actions. It acts either as a look-up table or in an approximation form (see Sect. 9.4 for the discussion). The overall goal of RL is to find the optimal policy given a state.

An optimal policy tries to maximise the expected future return from time t: $G_t = R_{t+1} + \gamma R_{t+2} + \gamma^2 R_{t+3} + \ldots$, where $0 \le \gamma \le 1$ is the discount parameter. The state-value function, $v_\pi(s)$, and the action-value function, $q_\pi(s,a)$, are two useful measures in RL that can be estimated from the data. The literature defines $v_\pi(s)$, of an MDP, under policy π, as the expectation of the return starting from state s:

$$v_\pi(s) = \mathbb{E}_\pi[G_t|S_t = s] \\ = \mathbb{E}_\pi\left[\sum_{k=0}^{\infty} \gamma^k R_{t+k+1}|S_t = s\right], \textit{for all } s \in \mathcal{S}. \quad (9.5)$$

In practical applications, $v_\pi(s)$ is more applicable for model-based problems, whereas the action-value function, $q_\pi(s,a)$, is more useful in the model-free context. When the full environment or the model is unknown, episodic simulations are often used to estimate $q_\pi(s,a)$, that is, under policy π, the expectation of the return starting from state s and taking the action a:

$$q_\pi(s,a) = \mathbb{E}_\pi[G_t|S_t = s, A_t = a] \\ = \mathbb{E}_\pi\left[\sum_{k=0}^{\infty} \gamma^k R_{t+k+1}|S_t = s, A_t = a\right], \textit{for all } \\ s \in \mathcal{S} \textit{ and } a \in \mathcal{A} \quad (9.6)$$

The task of finding the optimal policy, π_*, is achieved by evaluating either the optimal state-value function

$$v_*(s) = \max_\pi v_\pi(s), \quad (9.7)$$

or the optimal action-value function

$$q_*(s,a) = \max_\pi q_\pi(s,a). \quad (9.8)$$

9.3.3 Bellman Optimality Equation

One way of optimising Eqs. (9.7) and (9.8) is to make use of the recursive relationships between two states or actions in a sequential order. Since the procedures are similar, we only present the relationship starting from the action-values, i.e. the Bellman optimality equation for $q_*(s,a)$ (Bellman 1957b).

The backup diagrams in Fig. 9.3 show relationships between the value function and a state or state-action pairs. Figure 9.3a considers the optimal state-value function when taking an action. The agent looks at each of the possible actions it might take and selects the action with maximum action-value that tells the agent how good the state is. That is,

$$v_*(s) = \max_a q_*(s,a). \quad (9.9)$$

Similarly, Fig. 9.3b evaluates the dynamic and stochastic environment when an action is taken. Each of the states it ends up in has an optimal value. Thus, the optimal action-value counts the immediate expected reward, $r(s,a)$,

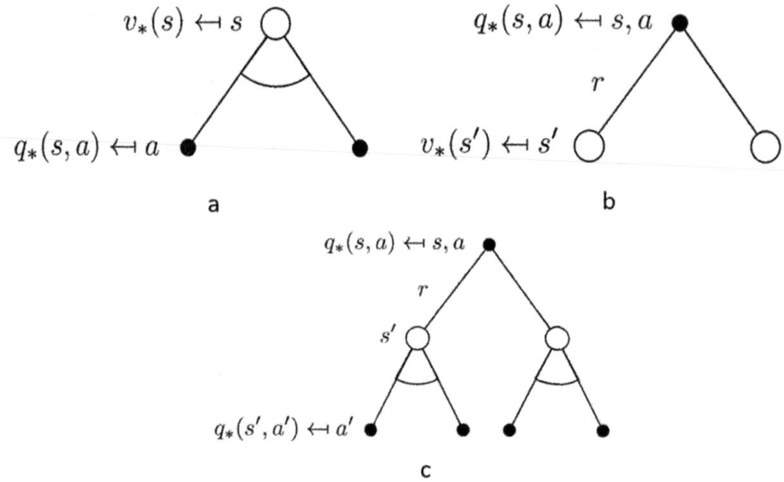

Fig. 9.3 Backup diagrams for the optimal value functions

from Eq. (9.3), and a discounted optimal state-value:

$$q_*(s,a) = r(s,a) + \gamma \sum_{s' \in \mathcal{S}} p(s'|s,a) v_*(s'). \tag{9.10}$$

Thus, as shown in Fig. 9.3c, the Bellman optimality equation for $q_*(s,a)$ is obtained by substituting Eqs. (9.9) into (9.10):

$$q_*(s,a) = r(s,a) + \gamma \sum_{s' \in \mathcal{S}} p(s'|s,a) \max_{a'} q_*(s',a'). \tag{9.11}$$

In a similar way, we can derive the Bellman optimality equation for $v_*(s)$. Both of them are the fundamental expressions for MDPs. The recursive relationship assists in splitting the current value function into the immediate reward and the value of the next action. Some of the specific learning algorithms, presented in Sect. 9.3.4, make use of the Bellman optimality equations to reach optimal policies.

9.3.4 Categorisation of RL Algorithms

The purpose of this sub-chapter is to examine how current research has methodologically explored the algorithms and to propose potential future work. There are many categorisation methods for RL algorithms. As shown in Fig. 9.4, an iteration-based classification suggests value-based methods, policy-based methods and a fusion of value-based and policy-based methods known as actor-critic methods.

Value-based methods, such as the off-policy Q-learning algorithm (Watkins 1989), start with a random value function and update to an improved value function in an iterative process until reaching the optimal value function $Q(S,A)$. The optimal policy is made by selecting the action that optimise value function at a certain state. For some value-based methods, e.g. the on-policy SARSA and SARSA(λ) algorithms (Rummery and Niranjan 1994), they evaluate policies by constructing their value functions and use these value functions to find improved policies. The distinguishing feature between off-policy and on-policy learning is that in the former the policy being learned is different to the one being followed whereas the latter approach follows the policy being learned. One of the advantages about the former way is that the optimal policy can be learned whilst following a different control strategy, for example, an MPC or rule-based control (RBC) strategy (Sutton and Barto 2018; Vazquez-Canteli et al. 2019).

In systems with small and discrete state or state-action sets, it is preferable to formulate the estimations using look-up tables with one entry

Fig. 9.4 Classification of RL algorithms

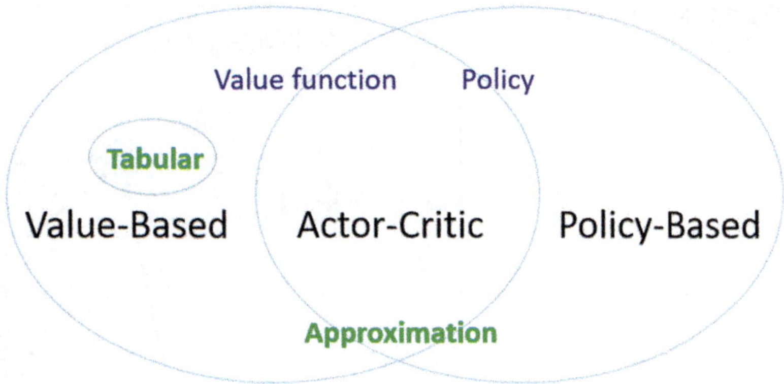

for each state- or action-value. The tabular method is straightforward to implement and guarantees convergence. For large MDP problems, however, we do not always want to separately see the trajectory of each entry of the look-up table. The parameterized value function approximation $\hat{q}(s, a; \mathbf{w}) \approx q_\pi(s, a)$ gives a mapping from the state-action to a function value, for which there are many mapping functions available, for example, linear combinations, neural networks, etc. It generates the state-actions that we cannot observe. For the incremental approximation method, $\mathbf{w}$ is updated by gradient descent:

$$\mathrm{w}_{t+1} = \mathrm{w}_t + \beta[q_\pi(S_t, A_t) - \hat{q}(S_t, A_t; \mathrm{w}_t)]\nabla\hat{q}(S_t, A_t; \mathrm{w}_t). \tag{9.12}$$

The learning target $q_\pi(S_t, A_t)$ is iteratively obtained from the Bellman Equation Eq. (9.11). Whereas the incremental method makes use of the experience once to update the estimate of the value function, and then throwing it away before going to the next step, the batch method is sample efficient and tries to find the best fit to all of the data (Ernst et al. 2005; Xu et al. 2002).

Policy-based methods use optimisation techniques to directly search for an optimal policy. Both the tabular and approximation methods work in value-based paradigms where the value functions have to be approximated and the policy is taken by greedy or ϵ-greedy strategies, whereas the policy-based method directly searches for the parametrised policy:

$$\pi_\theta(a|s; \theta) = \mathbb{P}[A_t = a|S_t = s; \theta_t = \theta]. \tag{9.13}$$

The policy-based method gives better convergence, especially for the continuous state-action space. In episodic experiments, the expected value of the start state is used as the objective function. The gradient ascent technique iteratively updates $\boldsymbol{\theta}$ for the optimisation. The action preference is usually assigned to a probability to avoid the deterministic policy.

Furthermore, a combination of value-based and policy-based methods, e.g. the Actor-Critic algorithm (Grondman et al. 2012; Mnih et al. 2016), is also appealing. The Actor makes an action when it observes a state. The Critic then marks the Actor's performance. The Actor adjusts the policy, e.g. parameters in ANN according to the score it obtained and the Critic adjusts its marking policy by evaluating the reward. In this way, both the Actor and the Critic improve themselves from random policies to better policies.

9.3.5 Multi-agent Systems

As an agent-based technology, the multi-agent system (MAS) provides promising paradigms in artificial intelligence (Sycara 1998). The decomposition of complex systems facilitates each agent to share a common environment and work independently on a specific sub-problem. MAS extends the single agent system by allowing each agent to interact with other agents—not

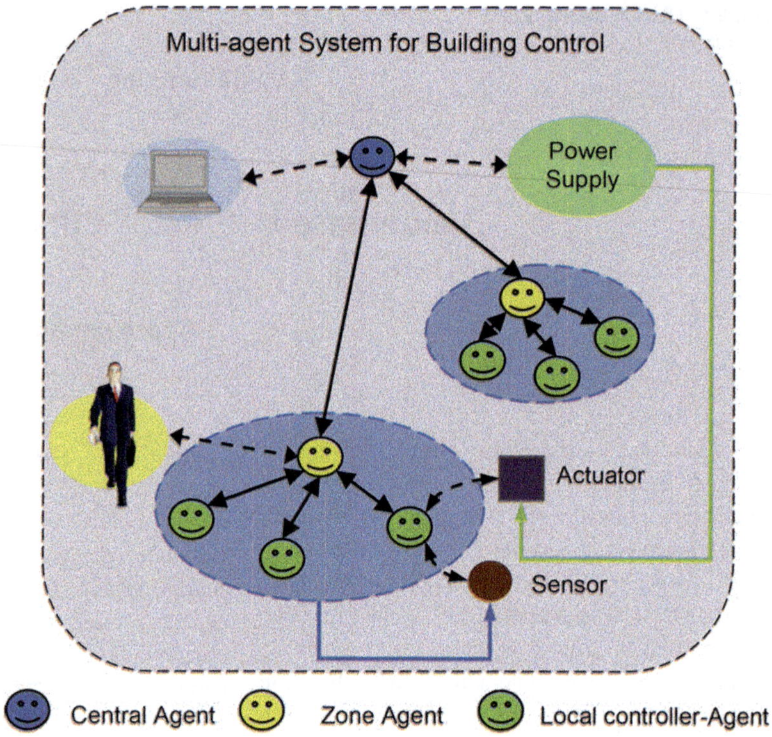

Fig. 9.5 MAS for building energy control

simply by exchanging data, but also by engaging in analogues of social activity: cooperation and negotiation.

When it comes to a multi-agent control system (MACS) in buildings, each agent implements autonomous actions in order to optimally run building models in a dynamic system. In most of the studies, the hierarchical central-local agent structure is embedded in the building models to balance the energy consumption and the occupants comfort (Yang and Wang 2012, 2013). A multi-objective optimisation technique is usually utilized for optimising intelligent management. In Fig. 9.5 (Yang and Wang 2013), for instance, a central agent communicates with building managers and zone agents to decide the optimal power distribution for each zone by considering the comfort demand. A zone agent, on the other hand, communicates with occupants and decides power demand. The local agents take care of temperature control, illumination control, and air quality control. An occupant-driven control is studied in MASs including HVAC agents, occupant agents and meeting agents (Klein et al. 2012), where the MDP based coordination tries to find the optimal policy by considering energy consumption, occupant comfort, and scheduling convenience individually.

The MDP property for multi-agent reinforcement learning (MARL) has been extensively studied in matrix game playing (Littman 1994) since both cooperative and competitive as well as a mixed environment can be modeled and simulated. The survey work (Buşoniu et al. 2010) has summarised and explored MARL theory, algorithms and applications. The benefit of MARL comes from experience sharing, information exchange, and skill learning among agents. When one or more agents fail to work in the system, the remaining agents are still able to react optimally by learning from the new environment. Generally, the transition probability for MARL extends Eq. (9.1) to a multi-action case:

$$p(s'|s,\boldsymbol{a}) = P\{S_t = s'|S_{t-1} = s, \boldsymbol{a} = (\boldsymbol{a_1},\ldots,\boldsymbol{a_n})\} \tag{9.14}$$

where n is the total number of agents, s', s belong to the set of states S and $\boldsymbol{a} \in \boldsymbol{A_1} \times \cdots \times \boldsymbol{A_n}$ is the joint action of the agents with $A_i, i = 1,\ldots,n$ is the action set for agent i. In Eq. (9.14), the stochastic transition is a probability distribution over next states s' given the current state s and joint action a of the agents. For the policy $\pi_i \in \Pi_i$, the optimal policy π_i^* for agent i fulfills the Nash equilibrium:

$$\sum_{a_1,\ldots,a_n} q_*(s,\boldsymbol{a})\pi_1^*(a_1|s)\cdot\ldots\cdot\pi_i^*(a_i|s)\cdot\ldots \cdot\pi_n^*(a_n|s) \geq \sum_{a_1,\ldots,a_n} q_*(s,\boldsymbol{a})\pi_1^*(a_1|s)\cdot\ldots \cdot\pi_i(a_i|s)\cdot\ldots\cdot\pi_n^*(a_n|s), \tag{9.15}$$

where $q_*(s,\boldsymbol{a})$ is the optimal action-value function for agent i and $\pi_i^*(a_i|s)$ is the individual probability of taking action a_i given the Nash equilibrium policy.

9.4 Applications of Reinforcement Learning Methods for Comfort Control in Buildings

The following sub-chapter reviews the applications of RL methods for occupant comfort control in buildings. We consider the algorithms implemented, their exploration versus exploitation strategy, and whether the application was from a single or multi-agent perspective. We close the sub-chapter by discussing those applications that were implemented in a physical setting.

Table 9.2 gives a summary of the reviewed literature pertaining to RL methods applied to comfort controls in buildings. We show specific learning algorithms and the classes (VB for value-based and AC for Actor-Critic) they belong to for each publication. We also investigate the representation of value functions to highlight optimisation techniques. Pre-training refers to whether or not the agents were implemented with pre-trained policies using existing data or simplified models of the physical system. We further distinguish discrete (D) state and continuous (C) state; single agent (SA) and multi-agent (MA). Unless otherwise stated, any reference to RL methods should be assumed to be model-free methods.

9.4.1 Comfort Factors

In the building literature, there are several well-established comfort factors (Dalamagkidis and Kolokots 2008). These are thermal comfort, IAQ, light, and noise. To ensure the quality of comfort of the buildings' occupants, intelligent controls are seen as an ideal strategy for maintaining the standards of these environmental factors as outlined by, for example, ASHRAE 55 (2017) and EN 15,251 (CEN 2005). We have thus divided the articles according to these factors. In the literature reviewed, we have identified three out of the four factors, outlined above, in which RL control has been applied. These are thermal comfort, IAQ, and light. Among these articles were some that also optimised over a combination of these comfort factors. Because of this, we first analyse those articles in which only a single comfort objective was considered. Following this, we then analyse those articles in which at least two comfort components were included in the objectives.

9.4.2 Thermal Comfort

Thermal comfort alone has had the most interest compared to the other comfort factors. Dalamagkidis and Kolokotsa implemented an RL control for an HVAC system with the goal of maximising both thermal comfort and energy conservation, where a heavier emphasis on thermal comfort was made (Dalamagkidis and Kolokots 2008). They compared the performance of the RL control to the performance of a fuzzy-PD and a common on/off control over a 5-year simulated time period. They found that after

4 years of simulation the RL control achieved as good as if not better performance than the other two controls. They also suggested pre-training the control before deploying it in a real environment to mitigate suboptimal performance due to policy exploration. For example, Yu and Dexter used RL to tune a fuzzy rule-based supervisory control for an HVAC system. They found that by pre-training the RL control it was able to improve the performance of a low-energy building system in an acceptable period of time (Yu and Dexter 2010). Sato et al., by considering occupants daily action plans, implemented RL to maximise both energy conservation and thermal comfort through controlling the operation of an air conditioning system (Sato et al. 2012). They found their proposed method to be effective. Similarly, Pedro et al., using RL to control an HVAC system and based on the tenant's preferences and occupancy patterns, were able to achieve energy efficiency while maintaining the tenant's thermal comfort level (Pedro et al. 2014).

A number of comparisons have been made between RL and other control strategies. Li and Xia apply RL to optimally control an HVAC system with respect to minimizing energy consumption while maintaining thermal comfort inside an acceptable range (Li and Xia 2015). The objective of the study was biased more towards energy. Compared to a blank control with a constant temperature schedule they found that the RL agent achieved better performance in operating the system with regards to energy conservation. In a similar study, Barret and Linder produced an RL control for an HVAC system and, through a comparative analysis, found their RL control outperformed, in terms of energy cost, two common strategies for controlling HVAC systems, namely, the "Always On" and "Programmable Control" methods. Improved thermal comfort was also demonstrated by offline training. Yang et al. compared the performance of RLCs with typical RBCs using numerical simulations on the thermal model of a LowEx full-building system (Yang et al. 2015). They found the RL controls outperformed the RBCs in all of the given thermal objectives. Compared with traditional methods RL control achieved better thermal comfort results with less energy consumed (Schmidt et al. 2017; Wang et al. 2017b). The comparison of a model-free method against model-based bidding strategies shows that RL gave similar performance to the model-based bidding strategy of smart thermostats (Sun et al. 2015b). Urieli and Stone applied RL to control an HVAC system in a simulated residential home, with the aim of minimising energy consumption while maintaining an acceptable temperature range (Urieli and Stone 2013). Compared with a default strategy of thermostat setback, their RL agent learned an effective setback strategy that both reduced energy costs and minimised violations of the temperature constraints. Nagy et al. applied a novel RL algorithm to the problem of controlling a heat pump in a building (Nagy et al. 2018). The objective of the controller was to maintain occupant comfort while reducing energy cost. A higher priority was given to occupant comfort. They compared their novel RL method to an RBC, an MPC with perfect information, and a model-based RL method. It was concluded that the model-based controls outperformed the model-free RL method; however, regarding computational complexity their novel RL method was much better. It also outperformed its model-based counterpart regarding changes to environmental dynamics. Furthermore, compared to the RBC, energy and cost savings were observed.

Zhang et al. developed a framework that uses a whole building energy model for optimal control of an HVAC system by deep RL (DRL) (Zhang et al. 2018). They found that the DRL control consumed 15% less heating energy with similar thermal comfort as compared to a base case. In a follow up study the same DRL control was deployed in a real-life office building over a period of 3 months (Zhang and Lam 2018). They included a thermal preference feedback app so that each occupant could state his/her thermal comfort preference. In conclusion, they found that their DRL control saved 16.6% heating energy consumption in the real setting. The thermal preference feedback system, however, had a very low participation rate and thus was not an accurate representation of the thermal comfort level. The comparison of DRL has also been implemented for maintaining zonal room temperature within a desired range, while minimising energy cost (Wei et al. 2017). Compared with a conventional RL technique (Q-learning) and a rule-based strategy, DRL was found to be the most effective in both keeping the number of temperature violations to a minimum and reducing energy cost.

9.4.2.1 IAQ

Optimising just IAQ has had the least attention. Indeed, we have identified only one article, namely, Baghaee and Ulusoy's use of RL for operating an HVAC system (Baghaee and Ulusoy 2018). Here, the objective of the control was to maintain CO_2 concentration at an acceptable range while also minimising energy consumption. In a simulation study they compared their RL control to an on/off and set point control. The RL method was found to be superior to the other two controls regarding energy consumption and CO_2 concentration.

9.4.2.2 Lighting

The earliest RL control on lighting was established about a decade ago where hierarchical RL (HRL) was developed to enable fast convergence and practical application (Khalili et al. 2010). The goal was to optimise over both lighting and energy cost. Compared to regular RL, they found HRL to be much faster in adapting to ideal light settings. After a hiatus of six years, lighting control was reconsidered in the work by, Cheng et al., who applied RL with a human feedback mechanism in order to control the blinds and lights in a single-occupant office of a building (Cheng et al. 2016). They jointly maximised over lighting comfort and energy conservation, with a heavier emphasis on comfort. With respect to the test environment, they found their results suggested an improvement in luminosity from both a comfort perspective and an energy saving perspective. With regard to the latter, compared to a manual and a traditional integrated automated control, their RL method reduced energy consumption.

In a more recent article, and the last among those just considering lighting, Park et al. developed an RL control, LightLearn, with an occupant feedback mechanism (Park et al. 2019). The aim of the control was to balance energy consumption and occupant comfort. They tested their control framework in five single-occupant offices in a building over a period of five weeks. In comparison with schedule-based and occupancy-based control strategies, LightLearn saved on energy consumption. Furthermore, LightLearn achieved good performance regarding occupant comfort. In conclusion, their RL control outperformed the other two controls in achieving a good balance between energy saving and occupant comfort.

9.4.2.3 Combinations of Factors

As can be seen, studies controlling combined factors are rare in the 1990s. For example, the earliest work in this area was by Jouffe who used RL to tune a ventilation controller for controlling temperature and relative humidity (Jouffe 1997). The policy obtained from the control was exactly to the experts' specifications. In the seminal work by Mozer, RL was used to control an HVAC and water heating system (Mozer 1998). The aim of the control framework was to minimise both discomfort (heating and lighting) and energy cost. The RL control was found to outperform alternative control strategies.

There have been more studies in the 21st century. Dalamagkidis applied RL to control an HVAC system (Dalamagkidis et al. 2007). The objective of the RL control was to minimise energy consumption and maximise user comfort where comfort was made up of two components, namely, thermal comfort and IAQ. In a computer experiment consisting of a simulated period of 4 years, they compared the performance of the RL control with an On/Off control and a Fuzzy-PD control. In terms of occupant comfort, the RL control was superior. This however came at the expense of higher energy consumption. Overall, they found the RL control to have achieved a level of performance close to that of the other control strategies. As in their later work

(Dalamagkidis and Kolokots 2008) they raised the issue of exploration during deployment in a real building setting. Their advice on this was to exhaustively train the controller beforehand, allowing little to no exploration during deployment.

Sun et al. implemented a novel RL technique based on events where an event was defined as a set of states—as opposed to time, to control an HVAC system (Sun et al. 2013). The objective of the control problem was to minimise energy cost while satisfying thermal comfort and IAQ constraints. In a numerical simulation, they compared their RL method to two other optimisation techniques, namely, backward stochastic dynamic programming (SDP) and a greedy method. They found the RL approach to be much more efficient in solving the given problem while satisfying the comfort constraints and saving energy costs. Later, Sun et al. refined their novel RL method and applied it to the aforementioned problem (Sun et al. 2015a). In a more detailed numerical simulation, they compared their RL technique to the same algorithms as before, namely, a greedy algorithm and an SDP algorithm. The same conclusions as in their former paper were reached, namely, they found their novel RL approach to be more efficient at solving the given problem as compared to the other algorithms. Similar topics can also be found in recent years (Eller et al. 2018; Fu et al. 2018), especially in MAS (Hurtado et al. 2018).

Apart from IAQ, there have also been articles studying a combination of thermal comfort along with the other factors, namely, lighting and humidity. Bielskis et al. applied RL to control an HVAC system and a Red–Green–Blue-Yellow LED lighting system (Bielskis et al. 2013). The goal of the RL controller was to maximise occupant comfort conditions, which consisted of thermal and lighting conditions. They tested their RL controller in a controlled laboratory experiment and found the performance acceptable. Chen et al. applied RL to the control problem of natural ventilation (Chen et al. 2018). In particular, they used RL to control HVAC and window systems. The aim of the control was to minimize both energy consumption and occupant discomfort. The comfort component consisted of two parts, namely, a thermal part and humidity (relative) part. In a numerical simulation over a simulated period of one year, they compared the RL control with a rule-based heuristic control under two different climates. In the two case studies, the RL control exhibited superior performance compared with the other control resulting in less energy consumption and occupant discomfort.

9.4.2.4 Algorithm Class

We can classify our learning algorithms according to whether they are value-based, policy-based, or exhibit a combination of these two classes, known in the literature as Actor-Critic algorithms. We can see from Fig. 9.6 that value-based algorithms seem to dominate the building literature followed by Actor-Critic algorithms. Furthermore, there does not (at the time of writing) seem to be any applications of policy-based algorithms. A couple of reasons for this may have something to do with the fact that less is known about this class of algorithms as well as the fact that policy evaluation is known to be less efficient with high variance (Silver 2015). In Fig. 9.6, we include the methods that are missed in Sutton and Barto's introduction (Sutton and Barto 2018) into the group of "Other" for further discussion: RLS-TD(λ), eJAL, MEC, DDNFQI, MAXQ, simulation based tree search and a variant of DQN.

9.4.2.5 Value-Based

Figure 9.6 clearly shows that the most popular value-based algorithm applied to occupant comfort optimisation is Q-learning. This is a temporal-difference (TD) learning method whereby two policies are used, one for generating the behaviour and the other being the learned policy, which eventually becomes the optimal policy. We call this process off-policy learning, and it is one of two classes, the other being on-policy learning, used to address the exploration vs exploitation dilemma. The simplicity of Q-learning leads to a majority of studies in comfort control (Baghaee and Ulusoy 2018; Barrett and Linder 2015; Bonte et al. 2014; Chen et al. 2018;

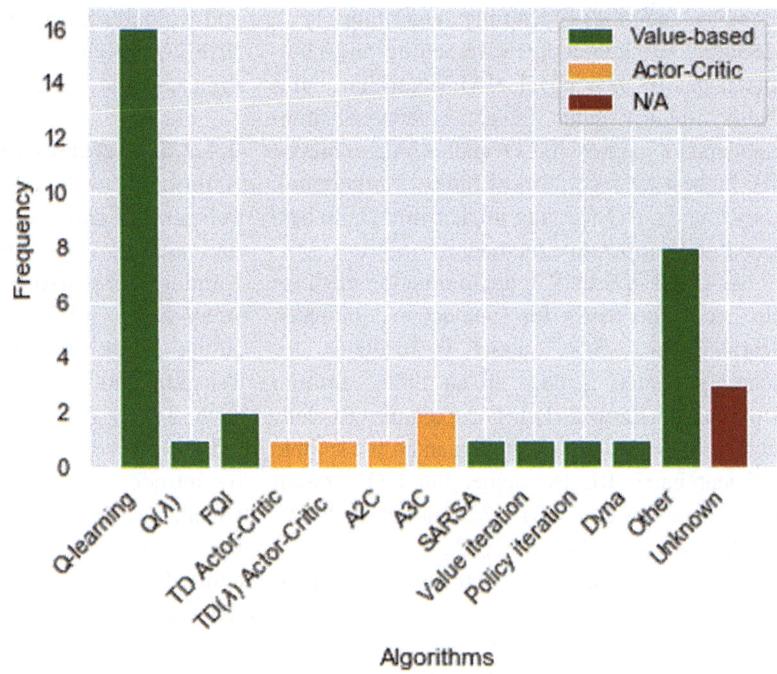

Fig. 9.6 Distribution of algorithms used for indoor comfort control

Cheng et al. 2016; Eller et al. 2018; Khalili et al. 2010; Li et al. 2015; Li and Xia 2015; Lu et al. 2019; Pedro et al. 2014; Sato et al. 2012; Sun et al. 2013, 2015a; Yang et al. 2015). Specifically, two consecutive studies applied Q-learning within a Lagrangian relaxation framework (Sun et al. 2013, 2015a). Moreover, instead of the usual time-based approach to optimisation, they took a novel event-based approach in which an action follows an event (a set of state transitions). Fazenda and Lima implemented Q-learning in two contexts, one in which action and states were discrete and thus tabular Q-learning was used; another in which the action and states were continuous. Here they combined Q-learning with a wire fitted neural network in order to approximate the Q-values (Pedro et al. 2014). Similarly, Yang et al. (2015) implemented tabular Q-learning and batch Q-learning with memory replay, where in the latter various neural network architectures were tested with the best performing being a 5–4 structure. Li and Xia implemented a novel multi-grid Q-learning algorithm. They compared its performance to standard Q-learning and found it to be more efficient (Li and Xia 2015). Yu and Dexter implemented a simplified version of Q(λ) (Yu and Dexter 2010), which extends Q-learning by including eligibility traces: these are an efficient technique that allows information to be propagated backwards over multiple time steps (Silver 2015). By applying a fuzzy discretization to the state space they were able to implement Q(λ) in tabular form, reducing the learning time. Furthermore, by incorporating a pre-trained policy into their implementation they were able to reduce the learning time even further, which they concluded to be essential for an RL control to be able to improve the operation of a building system online in an acceptable time frame.

Hurtado et al. implemented an extended joint action learning algorithm (eJAL) (Hurtado et al. 2018). The JAL method is an extension of Q-learning to a cooperative multi-agent setting whereby the actions of the other competing agents are included in the optimal policies. Hurtado et al. extended this algorithm in two ways, first by defining a conditional action space over all the other agents, and secondly by attaching a preference to each agent in a full cooperative game thereby allowing a joint reward function to be constructed. They also implemented the Q-learning algorithm in a decentralized non-cooperative setting and compared this to their novel eJAL algorithm and a Nash n-player game (a game theoretic non-cooperative approach). Overall eJAL achieved the highest average fairness index. Furthermore, they found the Q-learning implementation to be a good non-cooperative method.

As stated in Sect. 9.3, approximation methods do not create tables for state-actions. In batch learning the whole dataset is available for learning, which is done in an offline fashion (Sutton and Barto 2018). For example, Ruelens et al. and Schmidt et al. implemented the sample efficient batch RL technique, fitted Q-iteration (FQI) (Ruelens et al. 2015; Schmidt et al. 2017). In the paper by Ruelens et al., they used past observations at the end of each episode (a day in their case) to learn a policy that was then used online during the next episode with a Boltzmann exploration strategy. To reduce the state feature space they used an auto-encoder network which mapped features to a smaller dimensional subspace of the original feature space. In combination with this feature extraction technique, they used extremely randomized trees to approximate the Q-values. A novel variant of FQI, double deep neural FQI (D-DNFQI), uses deep neural networks for approximating the Q-values (Nagy et al. 2018). Here, Nagy et al. used target Q-networks and prioritised experience replay to address the problems of convergence and learning instability and, furthermore, used double Q-learning to address the upward bias problem of Q-estimates. This novel algorithm exhibited faster compute times with the added benefit of robustness under changes to system dynamics. The on-policy batch learning recursive least-squares TD(λ) (RLS-TD(λ)) is used in adaptive filtering, system identification and adaptive control. Its popularity lies in its fast convergence speed. Along with RLS-TD(λ), radial basis functions (RBFs) can be used to construct the feature vector used for the approximation of Q-values (Dalamagkidis et al. 2007; Dalamagkidis and Kolokots 2008). Similarly, Fu et al. used the SARSA algorithm, another on-policy TD learning technique (Fu et al. 2018).

9.4.2.6 Actor-Critic

Actor-only methods work with parameterized policies and critic-only methods rely exclusively on value function approximation and aim at learning an approximate solution to Eq. (9.11). Actor-Critic methods aim at combining the strong points of these techniques (Konda and Tsitsiklis 2000).

A TD(λ) Actor-Critic technique is used to reduce the variance of the policy gradient method by introducing a baseline function, called the advantage function. Bielskis et al. used the backward view of TD(λ), which uses independent eligibility traces for the actor and critic (Bielskis et al. 2013). To compute the action preferences and the state-values, they used a radial basis neural network. In a much earlier work, Jouffe also implemented the backward view of TD(λ) Actor-Critic but in a fuzzy logic network, which he called Fuzzy Actor-Critic Learning (Jouffe 1997). He defined fuzzy reinforcement functions to smooth the distinction between goal and failure states with the aim to speed up learning. To further increase the speed of learning, a priori knowledge was used.

Wang et al. implemented advantage actor-critic (A2C) (Wang et al. 2017b). This uses the same advantage function as described before but instead of using TD(λ) for different time scales it uses Monte Carlo. They combined this with two long-short-term-memory networks (LSTM), one for the actor and the other for the critic. These were used in order to mitigate against noise and the partial observability of states. Differing from A2C, asynchronous advantage actor-critic (A3C) (Mnih et al. 2016) executes multiple agents asynchronously on multiple instances of the same environment, independent of each other (Zhang et al. 2018; Zhang and Lam 2018). This approach is attractive from a practical point of view since state-of-the-art results can be achieved on, for example, a standard multi-core CPU as opposed to a GPU, which is the norm for deep learning.

9.4.2.7 Exploration Versus Exploitation Strategies

From a value-based perspective, the dilemma of exploration vs exploitation has to do with, on the one hand, exploiting current information by choosing the best (greedy) action-value, and on the other hand, exploring non-greedy actions. By exploring, the agent is able to improve on current estimates of non-greedy action-values and thus, in the long run, can lead to greater returns. Most of the works reviewed applied the naive ε-greedy approach to addressing this problem but several have taken a more sophisticated approach (Khalili et al. 2010; Li et al. 2015; Li and Xia 2015; Ruelens et al. 2015) with good results. However, none of these works systematically addressed this issue, that is, a parameter study of the given exploration strategies was not investigated. Given the sensitive nature of this topic on optimal control of buildings' energy conservation and, in particular, occupant comfort, further studies should be carried out on this topic.

Policy-gradient methods, on the other hand, approach this dilemma from another perspective. By parameterizing the policy they can learn suitable levels of exploration and asymptotically approach deterministic policies, which for action-value methods is extremely difficult to achieve and would involve placing strong assumptions on the problem. We have seen examples of policy gradient methods in Bielskis et al. (Jouffe 1997; Wang et al. 2017b; Zhang et al. 2018; Zhang and Lam 2018, 2013) with good results.

9.4.3 Agent Perspectives

Multiple agents can be treated both cooperatively and non-cooperatively. For non-cooperative agents, RL collapses to independent learnings in multiple zones or scenarios (Park et al. 2019; Wei et al. 2017; Yang et al. 2015). Single-agent algorithms may apply but the performances have seldom been compared with cooperative agents. Nevertheless, Eq. (9.14) depicts a joint action probability distribution for a Markov game—a combination of MDPs and matrix games. The Nash equilibrium is achieved via communication and interaction among agents. The reward function for multiple agents has become $r_i(s'|s,\boldsymbol{a})$, which denotes the reward for agent i in state s' given state s and joint action a (Schwartz 2014). The corresponding joint action learning method was also extended (Hurtado et al. 2018). Even though an extensive survey of MARL has been developed for many years (Buşoniu et al. 2010), there remains a great number of examples for occupant comfort control that still need to be made.

9.4.4 Physical Implementations

Only six out of the thirty-three articles reviewed involved a case study in an actual real setting rather than a simulated environment. For practitioners we believe it will be useful to have an indication of the state of the art of RL control implementation, given the relatively new adoption of this burgeoning field in building control methods, and the fact that conventional control techniques such as On/Off and PID strategies are still preferred by industry-based experts (Royapoor et al. 2018).

The earliest example of RL being applied to building control was Mozer's Neural Network House (Mozer 1998). In his seminal work, Mozer implemented a lighting control in a former school residence. It exhibited good performance of occupant comfort and conservation of energy. Much later, Bielskis et al. (2013) implemented their ACAR-Controller for HVAC and LED lighting in a laboratory setting. The performance was deemed acceptable with regard to thermal and lighting comfort constraints. Cheng et al. applied RL with a human feedback mechanism in order to control the blinds and lights in a single occupant office of a university building in Beijing (Cheng et al. 2016). It achieved good results, both from an occupant comfort point of view and an energy saving point of view. They pointed out that to achieve a more accurate comfort model, the agent would require more exploration which they saw as infeasible in a working office. Schmidt et al. deployed an RL controller on a Spanish school's heating system over two zones, in a building with a low level of thermal insulation (Schmidt et al. 2017). It was successful in improving thermal comfort with reduced energy consumption. Park et al. implemented an occupant centric lighting control with an occupant feedback mechanism at the University of Texas (Park et al. 2019). They selected five single-occupant rooms, each with a manually adjustable blind. The control strategy reduced energy consumption with good performance on occupant comfort. They pointed out that further research should be conducted to investigate the adaptability of the control to multi-occupant rooms, as well as the effect of control hyperparameters such as the discount factor and the reward structure. Zhang and Lam deployed an RL control for an HVAC system in the Intelligent Workplace (IW) at the University of Pittsburgh (Zhang and Lam 2018). The case study consisted of a multi-occupant single office. They included a thermal preference feedback phone app so that each occupant could state his/her thermal comfort preference. The control saved 16.6% heating energy consumption. However, the feedback system had a very low participation rate and thus was not an accurate representation of the thermal comfort level. They suggested investigating the effects of hyperparameter tuning on convergence of DRL since the thermal inertia of IW caused convergence problems in the DRL training. They also pointed out that the inertia of the heating system may have discouraged the occupants from using the app, which in turn may have affected the occupants' psychological feeling of comfort.

9.5 Discussions

In this sub-chapter, we discuss some challenges that may be encountered by building designers and managers. Although this sub-chapter is not an exhaustive review in building environment studies, we, nevertheless, convey ideas about how we might orient ourselves to face these challenges in the future building comfort control and management.

There have only been 33 works up to the current time for which RL has been explored for controlling occupant comfort, much less than those for building energy control. In particular, the studies including comfort factors such as indoor air quality and lighting are relatively rare in comparison to thermal comfort. Furthermore, relatively few of the reviewed works incorporate occupancy patterns and/or occupant feedback into the control loop which are crucial for occupant-centric building operation. Moreover, there is a gap with regards to the performance of implementing cooperative MARL.

The majority of our reviewed articles are found after 2010. This is not to say that people have not realised RL before 2010. Rather, it was the learning efficiency (e.g. the curse of dimensionality) that hindered people from implementing desirable experiments. Even though proper discretization of a continuous state space can make the exploration space more tractable, the scale of the problem may still be huge for high dimensional states. Even a long-term training period does not guarantee an optimal policy. However, the breakthroughs in computing power that have occurred since the 2010s have reversed the situation. The change-over to new hardware and computation platforms affords the chances to conduct complex computations, which in turn facilitates function approximation in RL and thus the ability to generate to unseen states and hence a more accurate representation of reality. In the process of designing and constructing smart buildings, considerations of incorporating real-time big data computing and learning platforms to BMSs have become one of the forthcoming challenges. This requires not only that the agent can gather real-time data, but also that building automation systems react in a timely fashion. The integration of computation infrastructures with BMSs is not explicitly formulated in most of our reviewed articles and thus requires more practical studies on this issue.

Further challenges originate from the MARL paradigm (Buşoniu et al. 2010; Ye et al. 2017b). Firstly, the agents might work independently instead of cooperatively resulting in sub-optimal returns. In an ideal framework, the cooperative agents can communicate immediate actions, rewards or learning experiences. Any interruption of the communication among agents makes the overall goal difficult to achieve. Secondly, the decision-making process for a single agent in MARL is non-stationary due to the dynamic policy changes from other agents. In systems with incomplete information share, the time may affect individual decisions. Thirdly, the exploration strategy becomes complex when the number of agents increases.

In some of our reviewed articles, the learning strategy for an MAS is still limited by applying single-agent RL algorithms to the multi-agent case. Learning depends only on the current agent's action and without being aware of the other agents. This strategy is simple to implement (Mataric 1994; Sen et al. 1994), but reaching the Nash equilibrium in Eq. (9.15) is challenging.

Defining the reward function seems to be an additional challenge. Apart from energy consumption, people with different personalities prioritise different comfort factors (Frontczak and Wargocki 2011; Zalejska-Jonsson and Wilhelmsson 2013). The magnitude for each factor is also subjective. There are a number of studies where researchers tried to optimise other objectives by setting comfort factors as the constraints. RL implementation in this circumstance is known as a constrained MDP, which is usually solved by linear programming. The dynamic programming techniques that apply in the non-constrained control problem do not hold any more and optimal policies need not exist (Altnan 1999). Thus, defining the reward function is not an easy task.

Successfully incorporating occupancy schedules and human feedback into the control loop is another challenge and which has drawn only a handful of studies (Cheng et al. 2016; Park et al. 2019; Pedro et al. 2014; Sato et al. 2012; Zhang et al. 2018). As occupants have a strong effect on energy consumption in buildings (Park and Nagy 2018; Park et al. 2019; Oca et al. 2018; Yan et al. 2017) including occupancy patterns as well as occupant feedback into the control system is crucial for efficient and occupant-centred building operation. Including human feedback can be particularly troublesome since we must be economical in our requirement of occupant feedback for it not to be too intrusive and time costly (Pedro et al. 2014; Zhang et al. 2018; Christiano et al. 2017).

Another challenge relates to the development of a toolkit, such as OpenAI Gym (Brockman et al. 2016), providing a framework whereby building simulation and advanced RL control using one's favorite machine learning library, such as, TensorFlow, PyTorch, etc. can be seamlessly tried and tested in a controlled and reproducible manner. A move in this direction can be seen in a recent work (Vazquez-Canteli et al. 2019).

9.6 Chapter Summary

The indoor environment affects not only working efficiency and living standards but also influences the occupants' health. Apart from building design, efficient control methods for the indoor environment not only improve the occupants comfort, but can also mitigate CO_2 emissions (Vazquez-Canteli et al. 2019). This chapter briefly examines and analyses empirical articles regarding the reinforcement learning control method for occupant comfort in buildings. Based on our analysis, we conclude our findings and formulate future works.

Firstly, the cutting-edge RL technique, which not only can adapt to the dynamic indoor environment of a building, but can also simultaneously adapt to the buildings' occupants, has drawn only limited attention regarding indoor climate oriented smart building controls, even though some studies have empirically tested its feasibility and comparability to other methods. The promising results lead us to a new frontier of occupant-centric building control. We have identified thirty-three empirical articles in this field, which is much less than the studies in building energy control and needs to be extended. The value-based Q-learning method is easy and straightforward to implement and it dominates among learning algorithms. This leaves a question of how policy-based or Actor-Critic algorithms perform in a practical building environment. Secondly, the computation platform and the ways of interaction with the BMS are important for conducting real-time control. Especially in the works with physical tests, the working paradigms are still vague. For example, policy-based and Actor-Critic algorithms require more function approximations and thus the power of computing resources should be updated accordingly. Thirdly, while maintaining proper indoor temperature is the foremost objective to be considered in BMSs, automation of smart buildings is an integrated system and studies about comfort factors like IAQ and lighting are relatively rare in comparison. Forthly, the empirical study of MARL for controlling the indoor environment has been modest. For example, the performance of implementing cooperative MARL still needs to be examined and confirmed through large studies. Fifthly, as occupants have a significant impact on energy consumption of buildings, it is important to include the occupant dimension into the control system. However, only five of the reviewed works considered occupancy patterns and/or human feedback in the control loop. Finally, model-based RL has had little attention in this area, appearing only twice among the reviewed works. The Dyna architecture and simulation-based search should be further explored in building applications as these have already been shown to be effective in other AI applications.

Looking ahead to the future, it is valuable to identify some future works. These include engaging in designing and training non-linear approximation approaches such as deep learning for RL; exploring algorithms for cooperative agents in MARL; setting up pre-training paradigms; stressing strategies for the exploration–exploitation dilemma; carrying out more studies for including occupancy patterns and/or human feedback into the control loop; exploring model-based RL; finally, further efforts in the direction of creating a seamless framework combining building simulation, advanced RL and other control strategies which can be compared in a standard and reproducible way. We also anticipate some promising practical works in standardising the measurement of indoor comfort, and integrating computation platforms and the ways of interacting with the BMS into smart building systems.

References

Altnan E (1999) Constrained Markov decision processes. Chapman & Hall/CRC

ASHRAE Standard 55 (2017) Thermal environmental conditions for human occupancy. ASHRAE Inc.

Baghaee S, Ulusoy I (2018) User comfort and energy efficiency in HVAC systems by Q-learning. In: 2018 26th signal processing and communications applications conference (SIU), pp 1–4

Barrett E, Linder S (2015) Autonomous HVAC control, a reinforcement learning approach. In: Bifet A, May M, Zadrozny B, Gavalda R, Pedreschi D, Bonchi F et al (eds) Machine learning and knowledge discovery in databases. Springer International Publishing, pp 3–19

Bellman R (1957a) A Markovian decision process. Indiana Univ Math J 6(4):679–684

Bellman R (1957b) Dynamic programming. Princeton Univ. Press, Princeton, NJ

Bielskis AA, Guseinoviene E, Drungilas D, Gricius G, Zulkas E (2013) Modelling of ambient comfort affect reward based adaptive laboratory climate controller. Elektronika Ir Elektrotechnika 19(8):79–82

Bonte M, Perles A, Lartigue B, Thellier F (2014) An occupant behaviour model based on artificial intelligence for energy building simulation. In: Proceedings of the 13th international IBPSA conference

Boodi A, Beddiar K, Benamour M, Amirat Y, Benbouzid M (2018) Intelligent systems for building energy and occupant comfort optimization: a state of the art review and recommendations. Energies 11(10):2604

Brockman G, Cheung V, Pettersson L, Schneider J, Schulman J, Tang J et al (2016) OpenAI Gym. arXiv: 1606.01540 [Cs]

Buşoniu L, Babuška R, De Schutter B (2010) Multi-agent reinforcement learning: an overview. In: Srinivasan D, Jain LC (eds) Innovations in multi-agent systems and applications—1, vol 310, pp 183–221. Springer Berlin Heidelberg

CEN prEN15251 (2005) Criteria for the indoor environment including thermal, indoor air quality, light and noise

Chen Y, Norford LK, Samuelson HW, Malkawi A (2018) Optimal control of HVAC and window systems for natural ventilation through reinforcement learning. Energy Build 169:195–205

Chenari B, Dias Carrilho J, Gameiro da Silva M (2016) Towards sustainable, energy-efficient and healthy ventilation strategies in buildings: a review. Renew Sustain Energy Rev 59:1426–1447

Cheng Z, Zhao Q, Wang F, Jiang Y, Xia L, Ding J (2016) Satisfaction based Qlearning for integrated lighting and blind control. Energy Build 127:43–55

Christiano P, Leike J, Brown TB, Martic M, Legg S, Amodei D (2017) Deep reinforcement learning from human preferences. arXiv:1706.03741 [Cs, Stat]

Dalamagkidis K, Kolokots D (2008) Reinforcement learning for building environmental control. In: Weber C, Elshaw M, Michael N (eds) Reinforcement learning. ITech Education and Publishing

Dalamagkidis K, Kolokotsa D, Kalaitzakis K, Stavrakakis GS (2007) Reinforcement learning for energy conservation and comfort in buildings. Build Environ 42(7):2686–2698

Eller L, Siafara LC, Sauter T (2018) Adaptive control for building energy management using reinforcement learning. IEEE Int Conf Industr Technol (ICIT) 2018:1562–1567

Enescu D (2017) A review of thermal comfort models and indicators for indoor environments. Renew Sustain Energy Rev 79:1353–1379

Ernst D, Geurts P, Wehenkel L (2005) Tree-based batch mode reinforcement learning. J Mach Learn Res 6:503–556

Frontczak M, Wargocki P (2011) Literature survey on how different factors influence human comfort in indoor environments. Build Environ 46(4):922–937

Fu Q, Hu L, Wu H, Hu F, Hu W, Chen J (2018) A Sarsa-based adaptive controller for building energy conservation. J Comput Methods Sci Eng 18(2):329–338

Galasiu AD, Veitch JA (2006) Occupant preferences and satisfaction with the luminous environment and control systems in daylit offices: a literature review. Energy Build 38(7):728–742

Gambier A (2004). Real-time control systems: a tutorial. In: Presented at the 5th Asian control conference (IEEE Cat. No. 04EX904), pp 1024–1031

Grondman I, Busoniu L, Lopes GA, Babuska R (2012) A survey of actor-critic reinforcement learning: standard and natural policy gradients. IEEE Trans Syst Man Cybern Part C 42(6):1291–1307

Gu S, Lillicrap T, Sutskever I, Levine S (2016) Continuous deep Q-learning with model-based acceleration. In: Presented at the conference on machine learning, vol 48

Guo X, Tiller D, Henze G, Waters C (2010) The performance of occupancy-based lighting control systems: a review. Light Res Technol 42(4):415–431

Guyot G, Sherman MH, Walker IS (2018) Smart ventilation energy and indoor air quality performance in residential buildings: a review. Energy Build 165:416–430

Haq MA, Hassan MY, Abdullah H, Rahman HA, Abdullah MP, Hussin F et al (2014) A review on lighting control technologies in commercial buildings, their performance and affecting factors. Renew Sustain Energy Rev 33:268–279

Hurtado LA, Mocanu E, Nguyen PH, Gibescu M, Kamphuis RIG (2018) Enabling cooperative behaviour for building demand response based on extended joint action learning. IEEE Trans Industr Inf 14(1):127–136

Jouffe L (1997) Ventilation control learning with FACL. In: Proceedings of 6th international fuzzy systems conference, vol 3, pp 1719–1724

Kaelbling LP, Littman ML, Moore AW (1996) Reinforcement learning: a survey. J Artif Intell Res 4: 237–285

Khalili AH, Wu C, Aghajan H (2010) Hierarchical preference learning for light control from user feedback. In: IEEE computer society conference on computer vision and pattern recognition—workshops, pp 56–62

Klein L, Kwak J, Kavulya G, Jazizadeh F, Becerik-Gerber B, Varakantham P et al (2012) Coordinating occupant behaviour for building energy and comfort management using multi-agent systems. Autom Constr 22:525–536

Konda VR, Tsitsiklis JN (2000) Actor-critic algorithms. Presented Adv Neural Inform Process Syst 12:1008–1014

Kruisselbrink T, Dangol R, Rosemann A (2018) Photometric measurements of lighting quality: an overview. Build Environ 138:42–52

Li B, Xia L (2015) A multi-grid reinforcement learning method for energy conservation and comfort of HVAC in buildings. IEEE Int Conf Autom Sci Eng (CASE) 2015:444–449

Li D, Zhao D, Zhu Y, Xia Z (2015) Thermal comfort control based on MEC algorithm for HVAC systems. Int Joint Conf Neural Netw (IJCNN) 2015:1–6

Li N, Cui H, Zhu C, Zhang X, Su L (2016) Grey preference analysis of indoor environmental factors using sub-indexes based on Weber/Fechner's law and predicted mean vote. Indoor Built Environ 25 (8):1197–1208

Lillicrap TP, Hunt JJ, Pritzel A, Heess N, Erez T, Tassa Y et al (2016) Continuous control with deep reinforcement learning. arXiv:1509.02971 [Cs, Stat]

Littman ML (1994) Markov games as a framework for multi-agent reinforcement learning. In: Presented at the conference on machine learning, pp 157–163

Lu S, Wang W, Lin C, Hameen E (2019) Data-driven simulation of a thermal comfort-based temperature set-point control with ASHRAE RP884. Build Environ

Marinakis V, Karakosta C, Doukas H, Androulaki S, Psarras J (2013) A building automation and control tool for remote and real time monitoring of energy consumption. Sustain Cities Soc 6:11–15

Mataric MJ (1994) Reward functions for accelerated learning. In: Presented at the proceedings 11th international conference on machine learning (ICML-94), pp 181–189

Merabti S, Draoui B, Bounaama F (2016) A review of control systems for energy and comfort management in buildings. In: 2016 8th international conference on modelling, identification and control (ICMIC), pp 478–486
Mnih V, Kavukcuoglu K, Silver D, Rusu AA, Veness J, Bellemare MG et al (2015) Human-level control through deep reinforcement learning. Nature 518 (7540):529–533
Mnih V, Badia AP, Mirza M, Graves A, Lillicrap TP, Harley T et al (2016) Asynchronous methods for deep reinforcement learning. arXiv:1602.01783 [Cs]
Mozer MC (1998) The neural network house: an environment hat adapts to its inhabitants, vol 5
Nagy A, Kazmi H, Cheaib F, Driesen J, Leuven K (2018) Deep reinforcement learning for optimal control of space heating. arXiv:1805.03777 [Stat.AP]
Oca S, Hong T, Langevin J (2018) The human dimensions of energy use in buildings: a review. Renew Sustain Energy Rev 81:731–742
Park JY, Nagy Z (2018) Comprehensive analysis of the relationship between thermal comfort and building control research—a data-driven literature review. Renew Sustain Energy Rev 82:2664–2679
Park JY, Dougherty T, Fritz H, Nagy Z (2019) LightLearn: an adaptive and occupant centered controller for lighting based on reinforcement learning. Build Environ 147:397–414
Pedro F, Kalyan V, Pedro L, Una-May O (2014) Using reinforcement learning to optimize occupant comfort and energy usage in HVAC systems. J Ambient Intell Smart Environ 6:675–690
Roetzel A, Tsangrassoulis A, Dietrich U, Busching S (2010) A review of occupant control on natural ventilation. Renew Sustain Energy Rev 14(3):1001–1013
Royapoor AI, Caraiscos C (2009) Advanced control systems engineering for energy and comfort management in a building environment—a review. Renew Sustain Energy Rev 13(6–7):1246–1261
Royapoor M, Antony A, Roskilly T (2018) A review of building climate and plant controls, and a survey of industry perspectives. Energy Build 158:453–465
Ruelens F, Iacovella S, Claessens BJ, Belmans R (2015) Learning agent for a heatpump thermostat with a setback strategy using model-free reinforcement learning. Energies 8(8):8300–8318
Rummery G, Niranjan M (1994) On-line Q-learning using connectionist systems. Cambridge University
Sato K, Samejima M, Akiyoshi M, Komoda N (2012) A scheduling method of air conditioner operation using workers daily action plan towards energy saving and comfort at office. In: Proceedings of 2012 IEEE 17th international conference on emerging technologies & factory automation (ETFA 2012), pp 1–6
Schmidt M, Moreno MV, Schulke A, Macek K, Mařik K, Pastor AG (2017) Optimizing legacy building operation: the evolution into data-driven predictive cyber-physical systems. Energy Build 148:257–279
Schwartz HM (2014) Multi-agent machine learning. a reinforcement approach, 1st ed. Wiley
Sen S, Sekaran M, Hale J (1994) Learning to coordinate without sharing information. In: Presented at the 12th national conference on artificial intelligence (AAAI-94), pp 426–431
Shaikh PH, Nor NBM, Nallagownden P, Elamvazuthi I, Ibrahim T (2013) Robust stochastic control model for energy and comfort management of buildings. Aust J Basic Appl Sci 7(10):137–144
Shaikh PH, Nor NBM, Nallagownden P, Elamvazuthi I, Ibrahim T (2014) A review on optimized control systems for building energy and comfort management of smart sustainable buildings. Renew Sustain Energy Rev 34:409–429
Silver D (2015) RL course by David Silver. UCL. Retrieved from http://www0.cs.ucl.ac.uk/staff/d.silver/web/Teaching.html
Silver D, Huang A, Maddison CJ, Guez A, Sifre L, van den Driessche G et al (2016) Mastering the game of go with deep neural networks and tree search. Nature 529 (7587):484–489
Silver D, Schrittwieser J, Simonyan K, Antonoglou I, Huang A, Guez A et al (2017) Mastering the game of go without human knowledge. Nature 550(7676):354–359
Song Y, Wu S, Yan YY (2015) Control strategies for indoor environment quality and energy efficiency—a review. Int J Low-Carbon Technol 10(3):305–312
Sun B, Luh PB, Jia Q, Yan B (2013) Event-based optimization with non-stationary uncertainties to save energy costs of HVAC systems in buildings. IEEE Int Conf Autom Sci Eng (CASE) 2013:436–441
Sun B, Luh PB, Jia Q, Yan B (2015a) Event-based optimization within the lagrangian relaxation framework for energy savings in HVAC systems. IEEE Trans Autom Sci Eng 12(4):1396–1406
Sun Y, Somani A, Carroll TE (2015b) Learning based bidding strategy for HVAC systems in double auction retail energy markets. Am Control Conf (ACC) 2015:2912–2917
Sutton RS, Barto AG (2018) Reinforcement learning: an introduction, 2nd edn. The MIT Press, Cambridge, Massachusetts
Sycara KP (1998) Multiagent systems. AI Mag 19:79–92
Urieli D, Stone P (2013) A learning agent for heat-pump thermostat control, p 8
Vazquez-Canteli JR, Nagy Z (2019) Reinforcement learning for demand response: a review of algorithms and modeling techniques. Appl Energy 235:1072–1089
Vazquez-Canteli JR, Ulyanin S, Kampf J, Nagy Z (2019) Fusing tensorflow with building energy simulation for intelligent energy management in smart cities. Sustain Cities Soc 45:243–257
Vesely M, Zeiler W (2014) Personalized conditioning and its impact on thermal comfort and energy performance—a review. Renew Sustain Energy Rev 34:401–408
Wang W, Zmeureanu R, Rivard H (2005) Applying multi-objective genetic algorithms in green building design optimization. Build Environ 40(11):1512–1525

Wang Y, Kuckelkorn J, Liu Y (2017a) A state of art review on methodologies for control strategies in low energy buildings in the period from 2006 to 2016. Energy Build 147:27–40

Wang Y, Velswamy K, Huang B (2017b) A long-short term memory recurrent neural network based reinforcement learning controller for office heating ventilation and air conditioning systems. Processes 5(3):46

Wang N, Phelan PE, Harris C, Langevin J, Nelson B, Sawyer K (2018) Past visions, current trends, and future context: a review of building energy, carbon, and sustainability. Renew Sustain Energy Rev 82:976–993

Watkins CJCH (1989) Learning from delayed rewards PhD thesis. University of Cambridge

Wei T, Wang Y, Zhu Q (2017) Deep reinforcement learning for building HVAC control. In: Proceedings of the 54th annual design automation conference 2017 on—DAC'17, pp 1–6

Wenqi G, Zhou M (2009) Technologies toward thermal comfort-based and energyefficient HVAC systems: a review. In: 2009 IEEE international conference on systems, man and cybernetics, pp 3883–3888

Xu X, He H, Hu D (2002) Efficient reinforcement learning using recursive leastsquares methods. J Artif Intell Res 16:259–292. https://doi.org/10.1613/jair.946

Yan D, Hong T, Dong B, Mahdavi A, D'Oca S, Gaetani I et al (2017) IEA EBC Annex 66: definition and simulation of occupant behaviour in buildings. Energy Build 156:258–270

Yang R, Wang L (2012) Multi-objective optimization for decision-making of energy and comfort management in building automation and control. Sustain Cities Soc 2(1):1–7

Yang R, Wang L (2013) Multi-zone building energy management using intelligent control and optimization. Sustain Cities Soc 6:16–21

Yang L, Nagy Z, Goffin P, Schlueter A (2015) Reinforcement learning for optimal control of low exergy buildings. Appl Energy 156:577–586

Ye W, Zhang X, Gao J, Cao G, Zhou X, Su X (2017a) Indoor air pollutants, ventilation rate determinants and potential control strategies in Chinese dwellings: a literature review. Science Total Environ 586:696–729

Ye D, Zhang M, Vasilakos AV (2017b) A survey of self-organisation mechanisms in multi-agent systems. IEEE Trans Syst Man Cybern Syst 47(3):441–461

Yu Z, Dexter A (2010) Online tuning of a supervisory fuzzy controller for low-energy building system using reinforcement learning. Control Eng Pract 18(5):532–539

Zalejska-Jonsson A, Wilhelmsson M (2013) Impact of perceived indoor environment quality on overall satisfaction in Swedish dwellings. Build Environ 63:134–144

Zhang Z, Lam KP (2018) Practical implementation and evaluation of deep reinforcement learning control for a radiant heating system. In: Proceedings of the 5th conference on systems for built environments—BuildSys' 18, pp 148–157

Zhang Z, Chong A, Pan Y, Zhang C, Lu S, Lam KP (2018) A deep reinforcement learning approach to using whole building energy model for HVAC optimal control. In: Presented at the 2018 building performance modeling conference and simbuild co-organized by ASHRAE and IBPSA-USA

A Novel Reinforcement Learning Method for Improving Occupant Comfort via Window Opening and Closing

10

Ross May, Mengjie Han, and Xingxing Zhang

Abstract

An occupant's window opening and closing behaviour can significantly influence the level of comfort in the indoor environment. Such behaviour is, however, complex to predict and control conventionally. This chapter, therefore, proposes a novel reinforcement learning (RL) method for the advanced control of window opening and closing. The RL control aims at optimising the time point for window opening/closing through observing and learning from the environment. The theory of model-free RL control is developed with the objective of improving occupant comfort, which is applied to historical field measurement data taken from an office building in Beijing. Preliminary testing of RL control is conducted by evaluating the control method's actions. The results show that the RL control strategy improves thermal and indoor air quality by more than 90% when compared with the actual historically observed occupant data. This methodology establishes a prototype for optimally controlling window opening and closing behaviour. It can be further extended by including more environmental parameters and more objectives such as energy consumption. The model-free characteristic of RL avoids the disadvantage of implementing inaccurate or complex models for the environment, thereby enabling a great potential in the application of intelligent control for buildings.

Keywords

Markov decision processes · Reinforcement learning · Control · Indoor comfort · Occupant

10.1 Introduction

Indoor comfort, for example, thermal comfort and air quality, have become major concerns for building designers and operators (Roulet 2006). The maintenance of these factors is important for improving the level of an occupant's comfort, health, morale, working efficiency, and productivity (Shaikh et al. 2013; Singh 1996). A survey shows that more than 80% of occupants are satisfied with their thermal comfort in only 11% of the buildings. For indoor air quality (IAQ), only 26% of the buildings have 80% or more

R. May (✉) · M. Han
Micro Data Analysis, Dalarna University, SE-79188 Falun, Sweden
e-mail: rom@du.se

M. Han
e-mail: mea@du.se

X. Zhang
Department of Energy and Community Buildings, Dalarna University, SE-79188 Falun, Sweden
e-mail: xza@du.se

X. Zhang (ed.), *Data-driven Analytics for Sustainable Buildings and Cities*,
Sustainable Development Goals Series, https://doi.org/10.1007/978-981-16-2778-1_10

satisfied occupants (Huizenga et al. 2006). In Denmark, 54% of a group of surveyed inhabitants claimed that they have at least one problem related to indoor comfort. A majority of those respondents did not try to search for information on how to solve the problem (Frontczak et al. 2012). Improving the comfort level of occupants is therefore urgent for a sustainable society, and the realisation of it seems to be a joint task between the occupants, the building designers, and the building management system (BMS).

Thermal comfort is used to manifest the thermal state of a human within a given environment (Enescu 2017). According to ASHRAE-55, the ambient parameter, temperature, is considered as the most relevant one for thermal comfort (ASHRAE Standard 2017). For IAQ, air quality index (AQI)—which measures the levels of pollutants in the air—is often used. Kyrkilis et al. (2007) reported a combination of CO, SO_2, NO_2, O_3, and PM10 as the five components of AQI. Cheng et al. (2007) included particulate matter with less than 2.5 μm diameter (PM2.5) since it can trigger cardiovascular disease-related mortality and nonfatal events. Control strategies for maintaining thermal comfort and IAQ at a desired level have been mostly implemented on heating, ventilation, and air conditioning (HVAC) systems since these have a direct influence on both the indoor environment and energy consumption. In a building with natural ventilation, however, indoor comfort depends largely on the control of window opening and closing. Compared to HVAC systems, the control of windows changes the indoor environment through naturally exchanging the air with the outdoor environment and therefore does not demand additional energy. Nevertheless, arbitrary and customary window control by an occupant does not guarantee the improvement of the indoor environment. For example, keeping an open window when the outdoor air quality becomes poor may increase the discomfort level. The occupant can easily fail to sense this slow deterioration of their surroundings. Thus, intelligent automation for window control has substantial potential to increase the level of comfort of an occupant.

Machine learning, a subfield of artificial intelligence (AI), has been used in buildings research for many years, and has demonstrated its potential to enhance building performance (Hong et al. 2020). Indeed, a number of previous studies have applied logistic regression as a prediction method for the control of window opening and closing behaviour. In recent times, a prevalent machine learning technique known as model-free reinforcement learning (RL) has made breakthroughs in intelligent controls and decision making (Mnih et al. 2013; Silver 2016, 2017). An RL agent learns how to optimally act given the environment it interacts with. An early work in this area for buildings is Mozer's Neural Network House (Mozer 1998). In this groundbreaking piece of work, a residential building's environmental parameters and observations of the occupants' actions are used by an RL agent for optimally controlling the building system. Other applications of RL to building energy systems have followed since this work, for example, in systems such as HVAC (Chen et al. 2019; Fazenda et al. 2014), lighting (Park et al. 2019), heat pump (Nagy et al. 1805; Ruelens et al. 2015), water heaters (Ruelens et al. 2014), and battery and photovoltaic systems (Shi et al. 2017). There are many more such examples of RL being applied in the building literature, however, as the scope of this chapter specifically concerns the application of RL used for window control, our focus therefore lies in considering current RL approaches to this building system.

Although there are existing studies using RL in window operating, for example, in the control of HVAC systems, particularly for ventilation purposes (Chen et al. 2018; Dalamagkidis et al. 2007), as well as in a holistic setting among the four subsystems, HVAC, lighting, blind, and window systems (Ding et al. 2019), there exists no research regarding the application of RL in window opening and closing from the aspect of occupant behaviour. This chapter, therefore, aims to fill this research gap.

In this study, we train two RL agents to learn when to open or close a window in an office building in Beijing, so as to maximise the comfort level of the indoor environment as measured by a combination of thermal comfort and IAQ, where, respectively, ambient temperature and AQI have been used as proxies. A recurrent neural network (RNN) is used to predict the indoor temperature as a result of an action taken on the window system. This enables the comparison of the agent's window opening/closing behaviour, with that of the actual observed historical occupant window/opening behaviour, under the same environmental conditions as experienced by the occupant at that time. As shown in Fig. 10.1, the contributions of this chapter are as follows. We propose a model-free reinforcement learning method for controlling windows in office buildings. We optimise the opening and closing of a window system with regard to maximising a combination of thermal comfort and air quality—using air temperature and AQI, respectively, as proxies—where a data-driven approach is used for simulating the environmental changes. A comparison is made between the window opening/closing policies of the RL agents and the occupant under identical conditions using a dataset containing the occupant's window behaviour and environmental measurements at that time. A theoretical basis is established for the future live deployment of the developed control method in the physical office, and thus for further incorporation of occupant feedback into its control logic.

The rest of the chapter is organised as follows. Section 10.2 examines drivers for window opening and closing and respective control methods used for controlling the indoor environment via window systems. Section 10.3 then briefly introduces the RL method and algorithms used in this chapter. In Sect. 10.4, we summarise the data and implementation details, and in Sect. 10.5 the results are discussed. Lastly, we conclude the chapter in the final section.

10.2 Behaviours of Window Opening and Closing

Occupant behaviour is a complex process and there are many drivers for an occupant to interact with building control. Apart from the contextual, psychological, physiological, and social factors of a building occupant, physical environmental factors have been considered as the most direct driver (Fabi et al. 2012). Since the behaviour of window opening and closing has a significant impact on both the indoor environment and energy consumption, understanding the underlying drivers and modelling methods will contribute to implementing efficient control techniques.

10.2.1 Drivers of Window Opening and Closing

Investigation of window opening and closing behaviour can be conducted in many ways. One approach is to use surveys in the form of questionnaires. This makes it flexible for the investigator to raise desired questions (Jeong et al. 2016; Freitas and Guedes 2015). These surveys reveal that most of the time, such behaviour aims

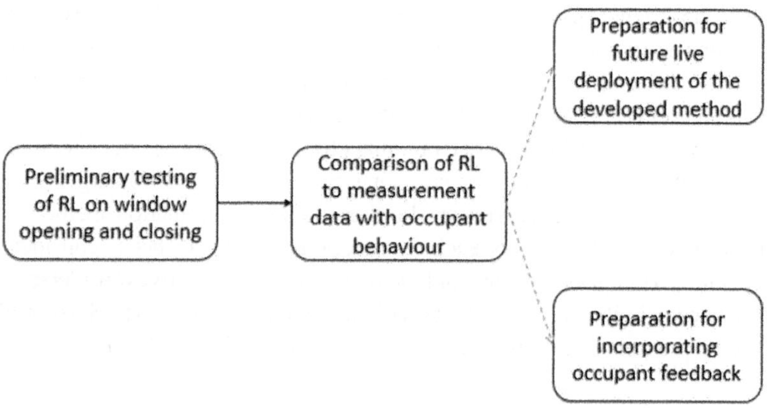

Fig. 10.1 Flowchart of the contributions

at improving physical feelings and that it varies for different seasons. For example, opening a window in winter is more explained by air quality, whereas closing a window in summer may be due to outside noise (Freitas and Guedes 2015). And a drop in indoor temperature can explain more the act of window closing than the outdoor temperature in winter (Jeong et al. 2016).

Another approach is to use statistical modeling to explain occupant behaviour (Haldi and Robinson 2009; Pan 2019). The probabilistic paradigm further allows us to monitor the distribution of behaviour in simulation studies. In an earlier work (Fritsch et al. 1990), it was stated that the probability of finding a window position depends on the preceding position of windows. A popular method for studying such behaviour is logistic regression analysis. It studies the binary dependent variables by fitting linearly independent variables and has been comprehensively used to model window opening and closing (Rijal et al. 2018; Andersen et al. 2013; Yun and Steemers 2008). It can both identify influential factors and predict window opening probabilities. For example, in residential buildings in Denmark, the probability can be explained by indoor CO_2 concentration and outdoor temperature as the common patterns (Andersen et al. 2013). Large variations, however, are found between the patterns in naturally ventilated buildings and mechanically ventilated buildings (Fabi et al. 2013). For office buildings in a natural ventilation season, the outdoor temperature was identified as the primary driver among other environmental factors. Another study has found that the trigger point for occupants' window opening is to get better thermal comfort and air quality (Li et al. 2015). Simulation results indicate that during transition seasons, the probability of window opening in office buildings follows a normal distribution and increases linearly with the outdoor temperature growth. Further studies have revealed that factors such as indoor temperature, occupant arrival and leaving time, presence, window positions, solar radiation, wind speed, seasons and time of the day also contribute to explain window operation for office buildings (D'Oca and Hong 2014; Pan 2018). In a study of two general hospital wards conducted in Nanjing, China (Shi 2018), the effects of air quality (i.e. indoor CO_2 concentration and outdoor PM2.5 concentration) and climatic parameters (i.e. indoor/outdoor temperature, relative humidity, outdoor wind speed, wind direction, and rainfall) on window opening/closing behaviour were analysed. Indoor air temperature and relative humidity were found to be dominant factors for window opening behaviour. Outdoor temperature was found to be associated with the probability of window opening negatively during the cooling season, but positively during the transition and heating seasons. Indoor relative humidity positively affects the probability of window opening during the transition season while a negative impact appears during the cooling and heating seasons.

Logistic regression models have been successfully developed to predict window opening/closing and have been verified to be promisingly adaptable for an accurate result. Similarly, a Probit analysis also models the probability of window operations (Yun and Steemers 2008). Yun and Steemers monitored data and gave evidence that there is a statistically significant relationship between window-opening behaviour patterns and indoor stimulus in summer. For example, a window in an office that featured a night cooling strategy was always open upon departure whenever the room temperature was over 23.6 °C.

10.2.2 Occupant Comfort and Intelligent Controllers

Having understood the drivers, we can single out the control targets and focus on specific control or automation methods. Thermal comfort and IAQ have been considered as the most pertinent objectives (Andersen et al. 2013; Li et al. 2015; Jin et al. 2015; Stazi et al. 2017; Tanner and Henze 2014). The measurement of these largely depends on the operation of windows via changing the airflow rate. Thus, an advanced automatic window control method leads to a consequential change in the indoor environment and hence optimises the occupant's overall comfort level in terms of these aforementioned objectives.

Rijal et al. demonstrated that the adaptive Humphreys algorithm could assist in achieving more comfortable, lower energy buildings while avoiding overheating (2008). This algorithm can also be used to adjust CO_2 concentration to the desired level while keeping the operative temperature at a constant (Stazi et al. 2017).

Four algorithms were compared for reducing energy consumption and improving comfort with regard to smart windows in commercial buildings (Dussault et al. 2016). The ruled-based controller and the quasi-optimal controller obtained by the genetic algorithm showed the best real-time control. It was also pointed out that genetic algorithms and model predictive control (MPC) are powerful tools that can easily accept more complex objective functions or scenarios.

When the occupant is involved in the control system, a stochastic process for occupant behaviour can be modeled following a known distribution (Tanner and Henze 2014). Tanner and Henze demonstrated this by implementing a stochastic MPC. Compared to deterministic optimal control, stochastic optimisation is more conservative but offers better performance. In a survey work by Han et al., the benefits of model-free control in such settings are illustrated from the methodology point of view (2019).

10.2.3 Building Environment

Both descriptive and analytical methods for finding drivers for window opening and closing aim to efficiently operate windows so that the occupants are satisfied. By defining an objective function, we can design a controller for solving this sequential decision-making problem. A key issue for a computational agent to develop a control method is the ability to sense the change in the state of the environment. A common strategy for achieving this is to use building simulation programs (Li et al. 2015; Wang and Greenberg 2015). This approach is fast and flexible for obtaining data, but cannot guarantee accuracy as occupant presence presents a significant influence on building performance. In this chapter, we propose a data-driven approach to predict the change in the indoor environment due to the operation of windows. Distinct from building simulation programs, a data-driven learning process can gradually improve accuracy. As the model-free RL control method for improving indoor comfort has not been studied in this way, a prototype of the prediction-based implementation and its achievement is demonstrated.

10.3 RL and Algorithms

Reinforcement learning (RL) essentially looks for best policies in the process of decision-making over time. The RL agent optimises its actions through interacting with and learning from the environment. It learns how to map situations to actions so as to maximise a numerical delayed reward signal. It doesn't need to have a "teacher" telling it how to take an action, rather, it makes decisions via implementing a trial-and-error search and recognising the delayed reward from the environment that the agent interacts with (Sutton and Barto 2018).

The environment gives stochastic feedback to the agent. In most cases, the environment cannot be modeled accurately and thus model-free RL techniques such as Q-learning and SARSA are employed as learning algorithms. Richard Bellman came up with the concept of Markov decision processes (MDPs) or finite MDPs, a fundamental theory of RL, to formulate the underlying framework for solving such problems (Bellman 1957a).

10.3.1 Markov Decision Processes

In a dynamic sequential decision-making process, the state $S_t \in \mathcal{S}$ refers to a specific condition of the environment at discrete time steps $t = 0, 1, \ldots$. By realising and responding to the

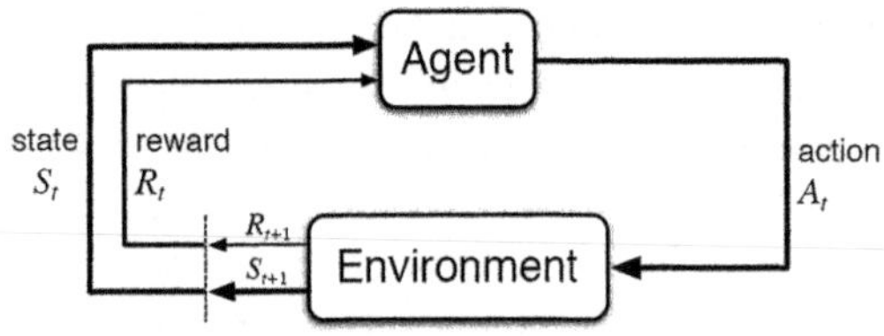

Fig. 10.2 The interaction between agent and environment in an MDP

environment, the agent chooses a deterministic or stochastic action $A_t \in \mathcal{A}$ that tries to maximise future returns and receives an immediate reward $R_{t+1} \in \mathcal{R}$ as the agent transfers to the new state S_{t+1}. The reward is usually represented by a quantitative measurement. Figure 10.2 (Sutton and Barto 2018) shows how a sequence of state, action, and reward are generated to form an MDP.

The Markov property tells us that the future is independent of the past and depends only on the present. In Fig. 10.2, S_t and R_t are the outcomes after taking an action and are considered as random variables. Thus, the joint probability density function for S_t and R_t is defined by,

$$p(s', r|s, a) = \mathbb{P}[S_t = s', R_t = r|S_{t-1} = s, A_{t-1} = a], \tag{10.1}$$

where $s, s' \in \mathcal{S}$, $r \in \mathcal{R}$ and $a \in \mathcal{A}$. It can be seen from Eq. (10.1) that the distribution of state and reward at time t depends only on the state and action one step before. Equation (10.1) implies the basic rule (or dynamics) of how the MDP works and one can easily determine the marginal transition probabilities $p(s'|s, a)$,

$$\begin{aligned} p(s'|s, a) &= \mathbb{P}[S_t = s'|S_{t-1} = s, A_{t-1} = a] \\ &= \sum_{r \in \mathcal{R}} p(s', r|s, a). \end{aligned} \tag{10.2}$$

Equation (10.3) gives the expected reward by using the marginal distribution of R_t:

$$\begin{aligned} r(s, a) &= \mathbb{E}[R_t|S_{t-1} = s, A_{t-1} = a] \\ &= \sum_{r \in \mathcal{R}} r \sum_{s' \in \mathcal{S}} p(s', r|s, a). \end{aligned} \tag{10.3}$$

Both Eqs. (10.2) and (10.3) are used for solving the optimal value functions presented in Sect. 10.3.2.

10.3.2 Policies and Value Functions

A *policy* π is a distribution over actions given states. It fully defines the behaviour of an agent by telling the agent how to act when it is in different states. The policy itself is either deterministic or stochastic (Sutton and Barto 2018) and the probability of taking an action, a, in state s is:

$$\pi(a|s) = \mathbb{P}[A_t = a|S_t = s]. \tag{10.4}$$

The policy can be considered as a function of actions. The selection of actions can be achieved by either creating a look-up table (see Sect. 10.3.3) or building an approximation model. The overall goal of RL is to find the optimal policy given a state.

An optimal policy tries to maximise the expected future return from time t: $G_t = R_{t+1} + \gamma R_{t+2} + \gamma^2 R_{t+3} + \ldots$, where $0 \leq \gamma \leq 1$ is the discount parameter. The *state-value function,* $v_\pi(s)$, and the *action-value function,* $q_\pi(s, a)$, are two useful measures in RL that can be estimated from the data. We define $v_\pi(s)$, of an MDP, under policy π, as the expectation of the return starting from state s:

$$\begin{aligned} v_\pi(s) &= \mathbb{E}_\pi[G_t|S_t = s] \\ &= \mathbb{E}_\pi\left[\sum_{k=0}^{\infty} \gamma^k R_{t+k+1}|S_t = s\right], \end{aligned} \tag{10.5}$$

for all $s \in \mathcal{S}$.

In practical applications, $v_\pi(s)$ is more applicable for model-based problems, that is, problems where a model of the dynamics is known a priori. Whereas the action-value function, $q_\pi(s,a)$, is more useful in the model-free context when the dynamics is not known. Episodic simulations are often used to estimate $q_\pi(s,a)$, where,

$$\begin{aligned} q_\pi(s,a) &= \mathbb{E}_\pi[G_t|S_t = s, A_t = a] \\ &= \mathbb{E}_\pi\left[\sum_{k=0}^{\infty}\gamma^k R_{t+k+1}|S_t = s, A_t = a\right], \\ &\text{for all } s \in \mathcal{S} \text{ and } a \in \mathcal{A}. \end{aligned} \tag{10.6}$$

The task of finding the optimal policy, π_*, is achieved by evaluating either the optimal state-value function

$$v_*(s) = \max_\pi v_\pi(s) \tag{10.7}$$

or the optimal action-value function

$$q_*(s,a) = \max_\pi q_\pi(s,a). \tag{10.8}$$

The way to optimise Eq. (10.7) or (10.8) is to make use of the recursive relationships between two states in sequential order, known as the Bellman optimality equation for $q_*(s,a)$ (Bellman 1957b), which is obtained by summating the following,

$$q_*(s,a) = r(s,a) + \gamma\sum_{s' \in \mathcal{S}} p(s'|s,a)\max_{a'} q_*(s',a'). \tag{10.9}$$

10.3.3 Q-learning and SARSA

A straightforward method to find $q_*(s,a)$ given a policy π is to iteratively update the values of $q_\pi(s,a)$ by maximising the sum of the discounted future returns and the immediate reward, known as the learning target. In a general iteration process, the new estimate of a target is updated by summing the old estimate with an error induced by the incremental observation, namely:

$$\begin{aligned} &\text{New Estimate} \leftarrow \text{Old Estimate} \\ &+ \text{Step Size(Target-Old Estimate)}. \end{aligned}$$

Different strategies regarding Q-value updates and action selections yield off-policy and on-policy algorithms. Among those theoretically convergent-guaranteed algorithms, the off-policy Q-learning and on-policy SARSA algorithms learn policies efficiently when the state-action pairs are discrete and the size is moderate. Thus, in this chapter, we consider Q-learning and SARSA as our testing algorithms and examine their performances in adaptive window controls.

10.3.3.1 Q-learning

Q-learning (Watkins 1989) is a value-based tabular method. A look-up table is built to store all state-action pairs and the corresponding action-values. When the agent is in a specific state and an action is selected, i.e. (s,a), the update for this state-action pair evaluates the transited state-action pair, i.e. (s',a'). The subsequent action, a', is taken such that $q(s',a')$ is maximised. As seen in Algorithm 1, the update to a new action-value is achieved by adding a so-called TD-error, $\alpha\left[R + \gamma\max_{a'} Q(S',a') - Q(S,A)\right]$, to the old action-values. The value function $Q(S, A)$[1] asymptotically converges to $q_*(s,a)$. An ϵ-greedy exploration indicates that the agent chooses an action that has maximal estimated action-value with probability $1-\epsilon$, but with probability ϵ the agent selects an action at random with equal probability.

[1]We use $Q(S, A)$ to represent an approximate value function from the data and q(S,A) to represent the target of the approximation.

Algorithm 1. Tabular Q-learning

Input: discount parameter γ; step size parameter α; $\{s, a\} \in \{\mathcal{S}, \mathcal{A}\}$; $\epsilon > 0$;

initialised $Q(s, a)$.

Loop for each episode
  Initialise S
    Loop for each time step
    Choose A from S by e.g., ϵ-greedy policy
    Take action A and observe R and S'
    $Q(S,A) \leftarrow Q(S,A) + \alpha\left[R + \gamma \max_{a'} Q(S',a') - Q(S,A)\right]$
    $S \leftarrow S'$
    Until S is terminal

Output: Q-table

10.3.3.2 SARSA

Compared to Q-learning, SARSA is more conservative and sensitive to errors. When the agent of SARSA updates its Q-table, it observes the successor state and takes an action according to, for example, an ϵ-greedy policy (or another exploration method) derived from Q, whereas Q-learning always looks for the maximum Q-value by evaluating those possible successor actions. Moreover, Q-learning re-selects the successor actions after updating the Q-table (due to exploration), which makes the policy of the learning agent distinct from the policy for updating the Q-table and thus behaves off-policy. The incremental update in the SARSA algorithm uses all of the elements in (S, A, R, S', A') to obtain the action-value,

$$Q(S,A) \leftarrow Q(S,A) + \alpha[R + \gamma Q(S',A') - Q(S,A)],$$

where A' is derived from an ϵ-greedy or another exploration method. Algorithm 2 gives the implementation details of SARSA.

Algorithm 2. Tabular SARSA

Input: discount parameter γ; step size parameter α; $\{s, a\} \in \{\mathcal{S}, \mathcal{A}\}$; $\epsilon > 0$;

initialised $Q(s, a)$.

1: Loop for each episode

2: Initialise S

3: Choose A from S by e.g. ϵ-greedy policy

4: Loop for each time step

5: Take action A and observe R and S'

6: Choose A' from S' by e.g. ϵ-greedy policy

7: $Q(S,A) \leftarrow Q(S,A) + \alpha[R + \gamma Q(S',A') - Q(S,A)]$,

8: $S \leftarrow S'$ and $A \leftarrow A'$

9: Until S is terminal

Output: Q-table

10.4 Data and Methods

10.4.1 Data

In this study, data from an office building constructed of reinforced concrete and brick at a university in Beijing are used. The construction material of the building is composed of 370 mm common brick with thermal conductivity 0.6 $W/(m \cdot K)$ and 200 mm polystyrene foam with thermal conductivity 0.033 $W/(m \cdot K)$. All the offices, shown in Fig. 10.3a, are located on the second floor in the building, whereas the floor spaces are used for laboratories (Pan 2018). There is one door and one push–pull south pointing window for the experimental room. As

(a) (b)

Fig. 10.3 Office building **a** and room **b**

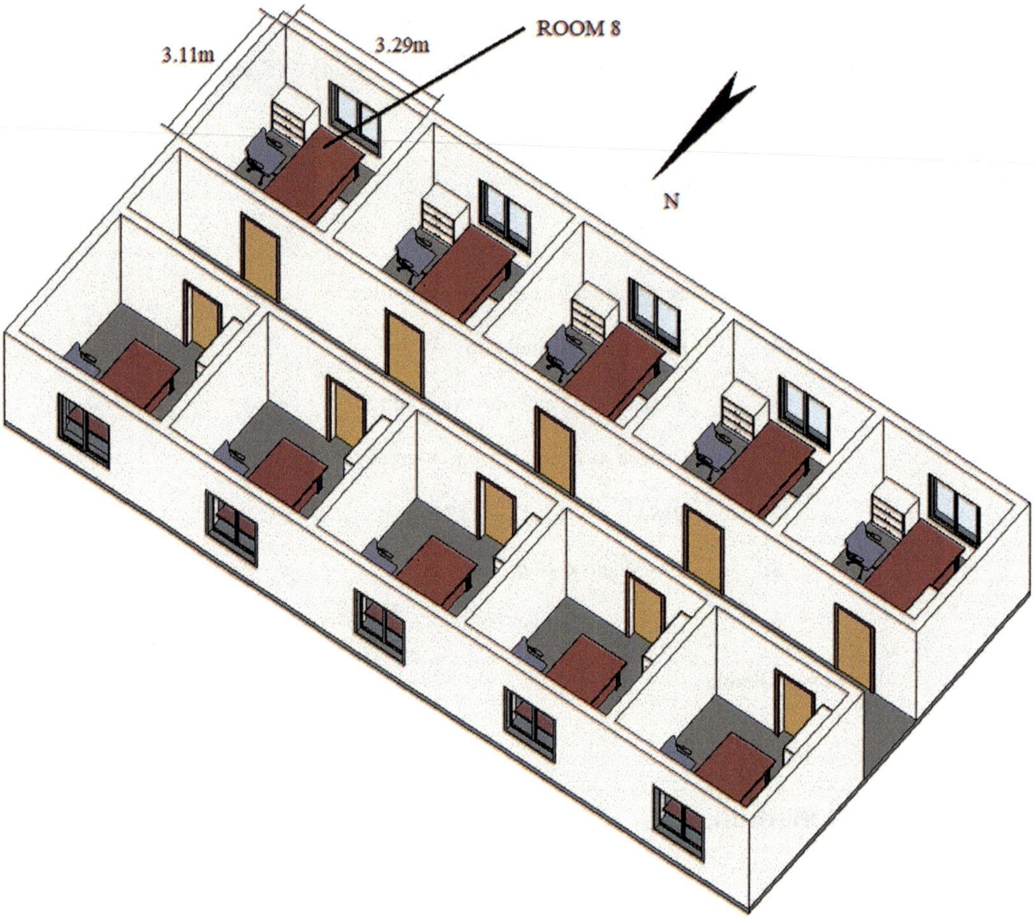

Fig. 10.4 The selected office room

shown in Fig. 10.3b, the window size is 1.5 m 1.6 m and can be half-open. One out of ten offices is selected for our experiment. As illustrated in Fig. 10.4, the experimental room (No. 8 according to the serial number of the building) is located at the southeast corner with a size of 3.29 m × 3.11 m.

The data collection took place between March 16, 2015 and May 15, 2015. This period in Beijing is the transition season with moderate outdoor temperature and so natural ventilation is highly preferred. To facilitate the comparison, the same occupant following the working routine in the university was in the room during the data collection period. An earlier work (Pan 2018) gives a detailed description of the variables and the settings of the sensors. The devices that were used for collecting the data are highlighted in Fig. 10.5. An indoor air temperature sensor TR (v1.2) was placed inside the room to avoid direct sunshine and local heating sources. A portable outdoor meteorological weather station was put over the roof where outdoor temperature, solar radiation, AQI, and wind speed and directions are measured. Moreover, an intelligent human body inductor P100 was used to detect the wavelength of the human body. For the days when the room was occupied for at least 30 min, the daily occupied time ranges from 50 min to 11 h and has a mean value of 5.5 h. To monitor the window, a displacement tester was applied to detect and record the position of the window. This was achieved by the magnetic induction of two dry spring pipes positioned on the window.

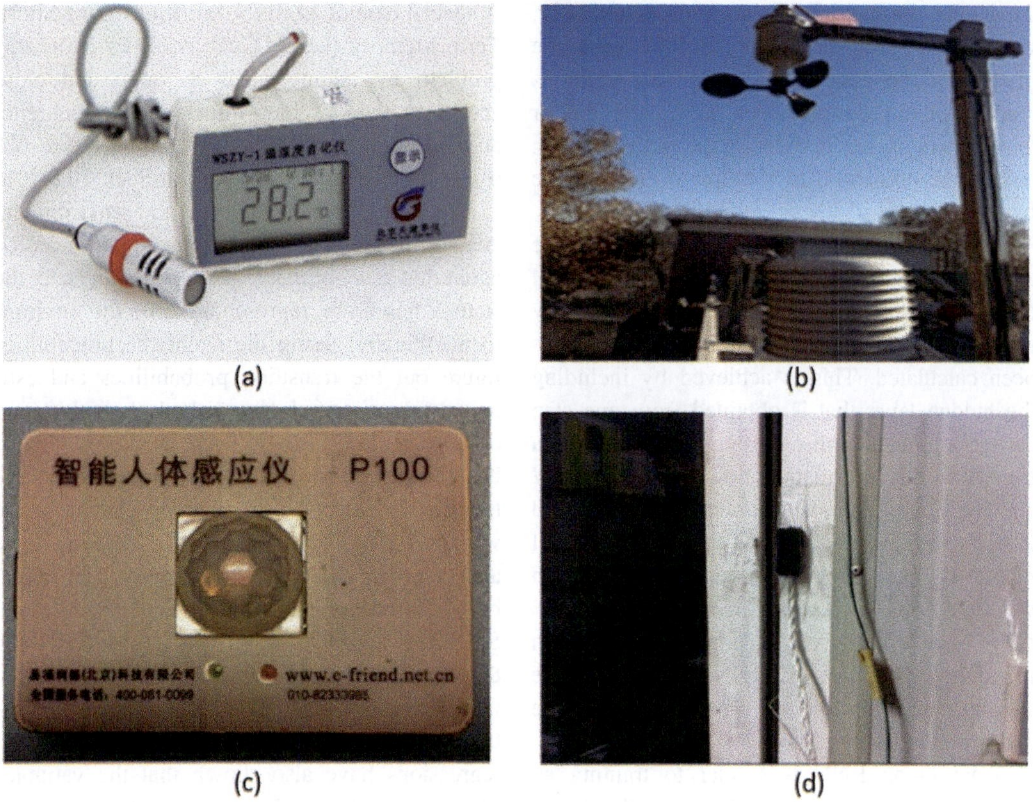

Fig. 10.5 Data collection devices: **a** indoor temperature sensor; **b** portable outdoor meteorological weather station; **c** human body inductor; **d** window tester

The action of opening was recorded when the window was opened more than 3 cm and the opening period exceeded 3 s.

Indoor temperature (T^{in}), outdoor temperature (T^{out}), solar radiation (*SR*), wind speed (*WS*), wind direction (*WD*), and outdoor air quality index (*AQI*) were selected as our environment variables. The position of the window (P_{window}, open/closed) and the occupancy information of the room (occupied/unoccupied) were also tracked. Given that the comfort factors T^{in} and AQI do not change drastically, each data record was collected at a time resolution of 10 min—which are accessible from a data logger. Indoor temperature is the main component for thermal comfort. Due to the limitation of the devices and time, we did not measure other thermal comfort factors, for example, metabolic rate, clothing insulation, radiant temperature, air speed and humidity. Given this, we simplify the factors of thermal comfort by only considering the most representative factor, temperature. For full factors, the adjustment of Eq. (10.10) is straightforward according to the method from ASHRAE standard 55 (ASHRAE Standard 2017). Since simplified thermal comfort also generates an interval or zone to indicate discomfort that affects reward in Eq. (10.10), the simplification has no impact on our training method compared to comfort with full factors. Hence, thermal comfort mentioned in this study refers to the concept with simplified factor.

10.4.2 Methods

The RL agent optimises its behaviour by exploring a number of different trajectories as time goes. The observed state at a time point has a strong correlation to the following states, because the change of the environment is time-dependent. Instead of simulating the trajectories, we conduct a data-driven approach to mimic the impact on the environment when an action is made. Among the variables in $\mathcal{S}$, the indoor temperature is susceptible to the actions taken and so the prediction of it helps to evaluate the future return.

The recurrent neural network (RNN) (Mandic and Chambers 2001) makes use of sequential data to make predictions. An RNN has a memory that stores previous information about what has been calculated. This is achieved by including the hidden layer that is obtained from one step earlier as input to the current hidden layer. In Fig. 10.6, the information flows of the input I and the output O have been stored and passed into the hidden layer a. An additional weight matrix W_a connects the hidden layers between two time points by computing the function $a_t = f(W_1 I_t + W_a a_{t-1})$ given a nonlinear activation function f. The unrolled RNNs share the same weight parameters W_1, W_2 and W_a across the entire prediction steps.

Training an RNN is similar to training an ordinary neural network. By using the back-propagation Through Time (BPTT) method (Werbos 1990), the gradient for an RNN at each output depends on both the current input and previous output. The gradient at time t needs to sum up all previous $t-1$ gradients. In practice, the long-term dependency makes BPTT unable to work due to the problem of vanishing or exploding gradients (Pascanu et al. 2013). A special case of an RNN, known as Long Short-Term Memory (LSTM), can perfectly avoid this problem by adding gates to open and close access to the previous information (Hochreiter and Schmidhuber 1997). We analyse the results of our RNN predictive model in Sect. 10.5.

Predicting the change in the environment enables the agent to start learning. To train an RL agent that acts optimally given a certain state, the state S has to be representative of the environment, thereby giving the agent the potential to figure out the transition probabilities and estimated future reward. Hence, both the indoor and outdoor temperatures as well as the AQI have been identified as direct environmental factors that have impacts on the position change of the window. Wind speed and solar radiation have also been included as factors since they affect both the airflow rate and ambient temperature. Since the current position of the window forms the baseline for the agent, this too has been included as part of the state components. Furthermore, Gauss distribution models and logistic regressions have also shown that the variables $T^{in}, T^{out}, SR, WS, AQI$ significantly influence window operations (Pan 2019; Pan 2018). We therefore formulate a single state as a sextuple: $\mathrm{S} = T^{in}, T^{out}, SR, WS, AQI, P_{window}$. Dynamic variables such as airflow rate or variables that have indirect effects to comfort parameters such as wind direction are excluded from S. As some of the components of the state are continuous and imperative for making policies, we have thus

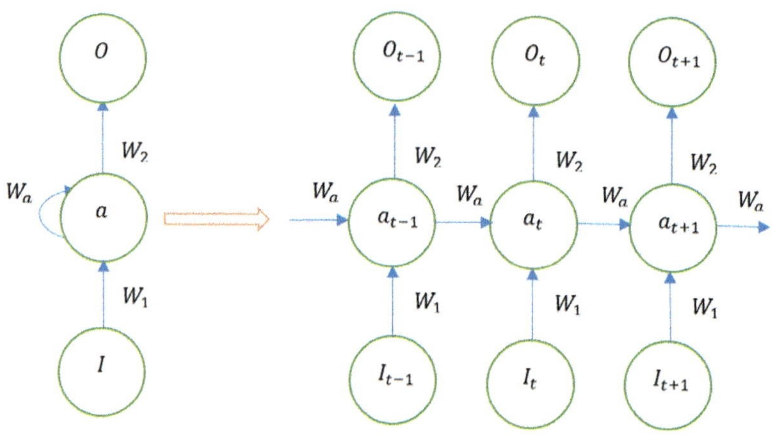

Fig. 10.6 Structure of RNN

discretised these so that the tabular RL algorithms can work.

Given the observed state, the action set, $\mathcal{A} = \{Switch, Inaction\}$, consists of only two elements, since we do not measure the degree of opening. *Switch* refers to either closing or opening the window depending on the current state. *Inaction* means keeping the position of the window unchanged under either scenario. The rationality to formulate the actions in this way is so that we are able to keep an eye on the current position of the window as well as track the actual position change. Given an open window, for example, it is more natural to say "keep it open" than to say "open the window".

The reward R reflects the comfort and is composed of both thermal comfort (as measured by the indoor temperature) and air quality (as measured by the proxy variable, AQI). We first define the thermal discomfort, τ, at time t by evaluating the squared difference between T^{in} and given thresholds,

$$\tau_t = \begin{cases} \left[\min\left(\left|T_t^{in} - T_{UB}\right|, \left|T_t^{in} - T_{LB}\right|\right)\right]^2, & T_t^{in} \notin [T_{LB}, T_{UB}] \\ 0, & otherwise \end{cases}, \tag{10.10}$$

where T_{UB} and T_{LB} are, respectively, the upper and lower thresholds of the comfort temperature. Higher outdoor AQI can also bring discomfort to the occupant. A survey shows that people have an incentive to close windows for better indoor air quality when high outdoor AQI is detected (Pan 2018). Hence, an additional component of discomfort, σ, is considered when the window is open and the AQI is higher than a given threshold, where,

$$\sigma_t = \begin{cases} AQI_t - AQI_{LB}, & AQI_t \in (AQI_{LB}, \infty) \\ 0, & otherwise \end{cases}. \tag{10.11}$$

A combination of the normalised thermal and air quality discomfort components yields the following reward for discomfort,

$$R_t = \omega_1 \frac{\tau_t - \min \tau_t}{\max \tau_t - \min \tau_t} + \omega_2 P_{window} \frac{\sigma_t - \min \sigma_t}{\max \sigma_t - \min \sigma_t}, \tag{10.12}$$

where the indicator variable, P_{window}, takes zero for a closed window and one for an open window, and the weight parameters, $0 \le \omega_i \le 1$, allocate the importance between the components. An inverse transformation, $R_t' = 1/(R_t + \xi)$, where $\xi > 0$ is some small real number, allows us to solve the maximisation problem defined in Eq. (10.9). It should be noted that unlike the discretised components making up a state, T_t^{in} and AQI_t in Eqs. (10.11 and 10.12) are, respectively, the RNN predictions and the numerical values from the observations.

10.5 Results

The results for both indoor temperature modeling given by a trained RNN, as well as the control performance of the trained RL agents are discussed in this section. All experiments are conducted in python 3.6.5 using TensorFlow's (v1.12.0) high-level API, Keras (v2.2.4). These were implemented on a Single Intel(R) 64-Bit Core(TM) i5-7300U 2.70 GHz CPU with 16 GB RAM.

10.5.1 RNN Predictions

An RNN-LSTM with a single hidden layer of 50 LSTM units and a dense output layer for predicting inside temperature was trained for the experimental room. Since opening and closing the window has a direct effect on the airflow rate and ventilation speed, the rate of change of indoor temperature may increase after an immediate switch of the window position. This effect of a change in window position will gradually wear off until the next switch occurs. Therefore, we incorporate the following lagged—by one time-step—as input variables: T^{in}, T^{out}, SR, WS, P_{window}, WD, time of day, time since a switch to

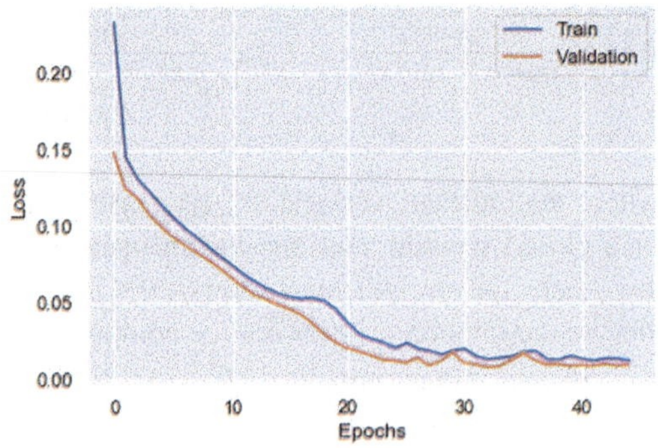

Fig. 10.7 Loss for RNN training

the window was made, the presence of the occupant, and outdoor humidity.

We use the 70–30% rule to divide the data into training and validation sets under the period between March 20 and May 7. Since the sequential order matters in RNNs, we strictly follow the time series observations and do not shuffle the data. We further select observations between May 12 and May 15 as the testing set. The training stops when the average losses are not significantly reduced. As can be seen in Fig. 10.7, the number of epochs needed for getting a stable loss is about 40. In our experiments we do not shuffle the data and so the sequential feature may produce higher loss for unusual observations in the early stage of training. Nevertheless, this unusual high loss diminishes as the number of epochs increases. Moreover, the predicted T^{in} s of a continuous 3-day period are compared against the actual values in the validation set in Fig. 10.8a. For almost all time points, the predicted values are close to the actual values.

Once the RNN-LSTM is trained, we further test its accuracy on a test set distinct from both the training or validation datasets. The comparisons between the actual vs the predicted values are given in Fig. 10.8b, where it can be seen that no significant deviations are found, and hence the RNN-LSTM predictive model thus trained generalises well to new inputs. The root mean squared error (RMSE) for the experimental room is 0.2 °C that is too trivial for the occupant's sensory-receptors to sense. Thus, the trained RNN-LSTM is verified as a suitable predictive model for simulating the environment.

10.5.2 Performance of RL Agents

Due to a lack of computational resources, the learning outcomes are illustrated for a single day (April 8) as a prototype. The occupant spent eight hours in the room on this day, which is a typical office routine in China. This same procedure can be applied to any other day, but we do not see any reason why our agents would behave differently.

Discretisation to the continuous states for our experimental room and date are shown in Table 10.1. We restrict $T^{in} \in [20\,°\mathrm{C}, 27\,°\mathrm{C}]$ and $T^{out} \in [6\,°\mathrm{C}, 15\,°\mathrm{C}]$ to be within the bounds of the actual extreme values. Considering the lengths between the bounds in the intervals, a step-size of 1 °C is set for both T^{in} and T^{out}. In running the tabular algorithms, both the actual and predicted states are numerically rounded to the nearest endpoints. The distribution of SR is skewed and so we consider ordinal indicators for representing uneven intervals. The AQI and P_{window} are binary variables where we only distinguish if AQI is greater than AQI_{LB} or not. We take $AQI_{LB} = 150$ as vulnerable groups of people start to have aggravation of symptoms of heart and respiratory diseases when $AQI_{LB} > 150$, furthermore, outdoor activities are not recommended when values of AQI are greater than the given threshold (Jin et al. 2015; Jassim and Coskuner 2017; Pu et al. 2017). Further, in Eq. (10.10) we take T_{LB} = 21 °C and T_{UB} = 21 °C (ASHRAE Standard 2017; Chen et al. 2018; Zhang et al. 2011). There is no obvious best temperature within this interval. Since we try to establish a general framework of agent training, we consider the most representative thresholds for the majority of occupants. Although individual comfort preference can vary, the consequence of the occupant's behaviour is emphasized in this study.

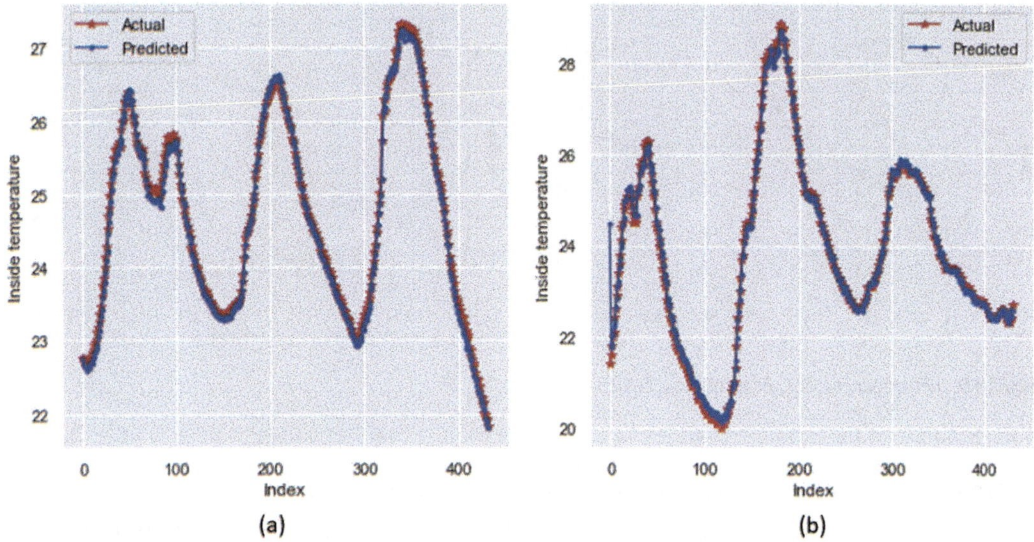

Fig. 10.8 3-day period comparisons: **a** predicted validation sets and actual values; **b** predicted test sets and actual values

Table 10.1 Discretization of states

Variables	Minimum	Maximum	Interval
T^{in}	20 °C	27 °C	1 °C
T^{out}	6 °C	15 °C	1 °C
WS	0m/s	2.5m/s	0.5m/s

We take 144 time steps as one episode for both Q-learning and SARSA. To evaluate the learning performances of our agents, we not only monitor the reward function but we also examine the accumulative number of penalty actions (defined in Table 10.2) for each epoch. When the agent is in a specific state, we consider six different scenarios made by the values of AQI and T_t^{in}. For example, the agents should have learnt to close the window when $AQI_t > 150$ and $T_t^{in} < 21°C$; opening under this scenario would result in a penalty. Clearly, the reward function and the accumulative number of penalties are inversely related—an increase in the former will result in a decrease in the latter. By the end of the 20th epoch in Fig. 10.9, both Q-learning and SARSA are able to improve their reward functions and reduce their penalties. As stated in theory, Q-learning has higher variance than SARSA due to following a different policy to its behaviour policy. As we can see, by the end of training, SARSA slightly outperforms Q-learning.

Table 10.2 Penalty actions

	$AQI_t > 150$	$AQI_t \leq 150$
$T_t^{in} < 21$ °C	Open	Open
21 °C $\leq T_t^{in} \leq$ 27 °C	Open	–
$T_t^{in} > 27$ °C	Close when,[1]$\alpha > \beta$	Close
	Open when,$\alpha < \beta$	

[1]$\alpha = \frac{\tau_t - \min \tau_t}{\max \tau_t - \min \tau_t}$, $\beta = \frac{\sigma_t - \min \sigma_t}{\max \sigma_t - \min \sigma_t}$

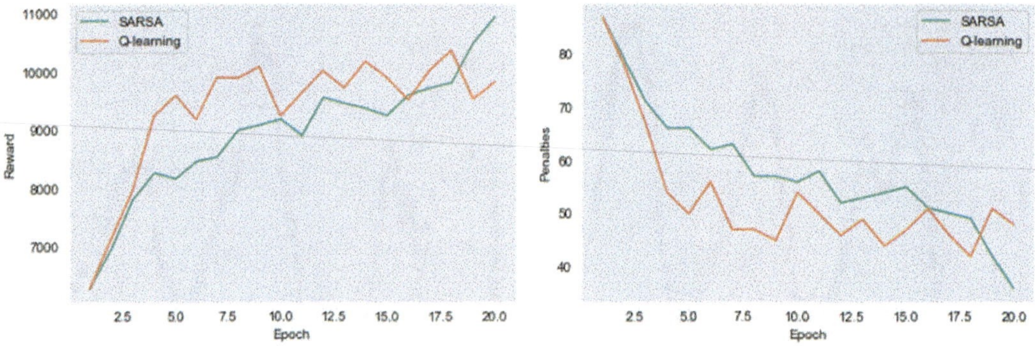

Fig. 10.9 Performances of Q-learning and SARSA

Table 10.3 Average reward and penalty for the agents and occupant per time step

		24 h	8:25–12:45	13:25–15:15	15:45–17:35
QL[2]	Avg.R	70.12	7.60	55.26	47.20
	Avg.P	0.31	0.96	0.27	0.45
S	Avg.R	76.07	7.60	64.54	73.88
	Avg.P	0.26	0.96	0.27	0.18
OC	Avg.R	23.57	7.82	28.89	3.10
	Avg.P	0.85	0.96	0.55	0.90

[2]QL is short for Q-learning; S is short SARSA; OC is short for Occupant; Avg.R is short for Average Reward; Avg.P is short for Avergage Penalty

We further evaluate the performances of the agents against the actual occupant. Specifically, we compare the average reward and penalty per time step for the whole day and three periods when the occupant was in the room, namely, 8:25–12:45, 13:25–15:15, and 15:45–17:35. As shown in Table 10.3, the trained agents give on average more than 70 in reward for a 24 h period, whereas they reduce to 7.60 in the morning and increase in the afternoon. For the three occupied periods, the agents give lower average rewards than the 24 h period. This is because the agents have to compromise the gain when $AQI_t > 150$ and $T_t^{in} > 27$ °C.

For the morning period, negligible differences in reward can be seen between the agents and the occupant, indicating that the agents can perform at least as good as the occupant. This tiny difference in reward is due to the prediction of indoor temperatures. If we scrutinize the actual actions given by both the agents and the occupant in Fig. 10.10, we see that the actions in the morning coincide with each other. In the afternoon, however, the longer time steps with an open window for the occupant make the reward higher for the agents. An explanation for the occupant's irrational behaviour is the inertial thinking from the morning, failing to sense the gradually increased AQI. The occupant may have been concentrating on his work and so easily forgot to close the window. The RL agent, however, is always able to learn from the environment and keep a level that is close to the 24 h average.

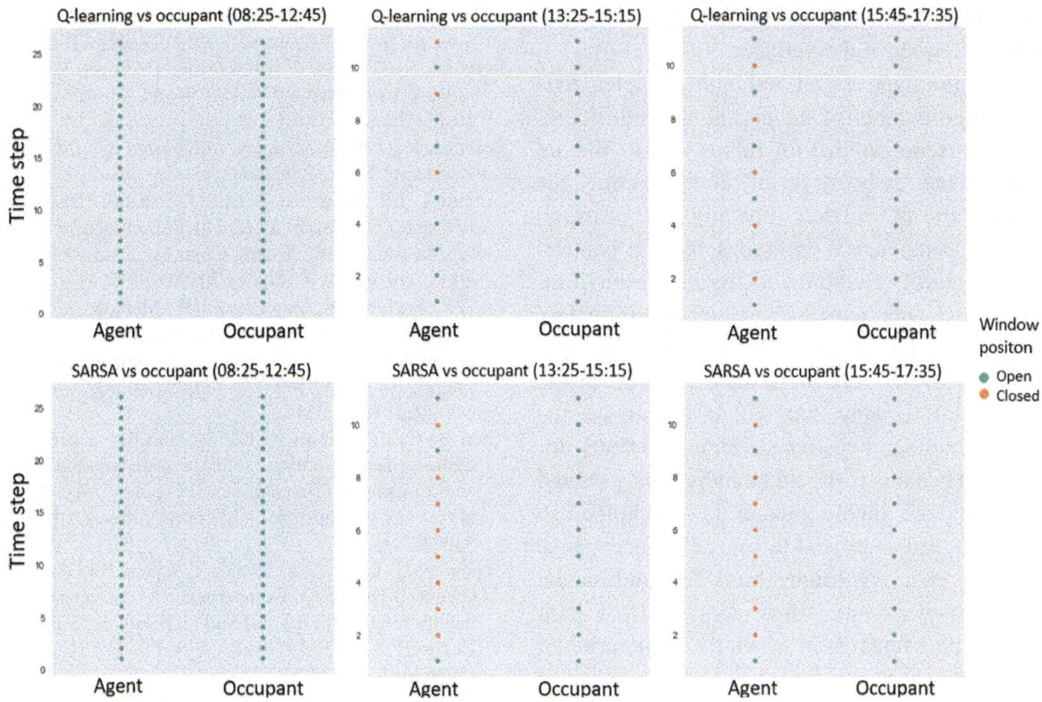

Fig. 10.10 Comparison of actions for different time periods

10.6 Conclusions

The control of windows in naturally ventilated buildings have a large impact on occupant comfort. Among the comfort factors, better thermal comfort and IAQ are of most concern for occupants. In typical Chinese office buildings, occupants may not behave optimally due to the complex climate and weather, and therefore intelligent control methods aiming at improving thermal comfort and IAQ become indispensable for smart and sustainable buildings. Previous intelligent control methods applied to adaptive window control have been based on models and their performances are therefore heavily dependent on the accuracy of these models. Furthermore, these models need to be corrected and re-identified as a consequence of a change in the dynamics of the building caused by, for example, retrofits. Thus, as an alternative solution which addresses such challenges, we have developed an automatic control prototype based on reinforcement learning for improving occupant comfort and tested it in a data-driven simulated environment.

An RNN-LSTM predictive model was used for predicting the indoor temperature given environmental variables and was verified by a test set. The high accuracy of the predictive model enabled us to simulate the actions of an agent in a flexible setting. Two tabular algorithms, Q-learning and SARSA, were used to train two RL agents whose learned behaviours were evaluated against the occupant's historically observed behaviour. The agents achieved much better policies than that of the historically observed occupant's policy measured in terms of both accumulative reward and penalties. An RL agent aims to maximise cumulative future return instead of the immediate reward at a single time point. Even though the performance of our trained agents failed to surpass the average level of the complete learning period for some specific sub-periods when the room was occupied, the agents still behaved close to the average level.

This therefore means that tabular algorithms can inherently reduce the variance.

The prototype established in this chapter leads to a large number of novel and valuable topics that are recommended for future works. We are still at the early stage of understanding the behaviours of window opening and closing. Human behaviour is indispensable for controlling the level of comfort in the indoor environment, and with regard to occupant-centric RL, we believe that occupant feedback will not only continuously correct the reward function in the process of learning, but will also increase the actual learning experience. Human effects for different occupants should be individually treated and they are highly related to psychological, physiological and social factors of the occupants. To this end, algorithms built for multi-agent cooperative systems (where agents have a joint action-value function in which the exploration of new states becomes complex as the number of agents increases) are valuable to explore and adapt accordingly in order to make them feasible in practice. The comfort level of an occupant is made up of the four factors, thermal comfort, indoor air quality, lighting, and noise, and therefore holistic approaches to measuring the comfort level of occupants should be explored and integrated in the design of an intelligent agent. While discretising the state space allows for the application of tabular RL methods, in so doing errors may arise. Hence, solutions for training an agent with a continuous state space are therefore promising and thus approximation techniques need to be developed. With advanced computational power, as well as advances in algorithm design that address the problem of sample efficiency in RL (Botvinick et al. 2019), high performance is expected.

References

Andersen R, Fabi V, Toftum J, Corgnati SP, Olesen BW (2013) Window opening behaviour modelled from measurements in Danish dwellings. Build Environ 69:101–113. https://doi.org/10.1016/j.buildenv.2013.07.005

ASHRAE standard 55—thermal environmental conditions for human occupancy (2017). ASHRAE Inc

Bellman R (1957a) A Markovian Decision Process. Indiana Univ Mathe J 6(4):679–684. https://doi.org/10.1512/iumj.1957.6.56038

Bellman R (1957b) Dynamic programming. Princeton Univercity Press, Princeton, NJ

Botvinick M, Ritter S, Wang JX, Kurth-Nelson Z, Blundell C, Hassabis D (2019) Reinforcement learning, fast and slow. Trends Cogn Sci 23(5):408–422. https://doi.org/10.1016/j.tics.2019.02.006

Chen Y, Norford LK, Samuelson HW, Malkawi A (2018) Optimal control of HVAC and window systems for natural ventilation through reinforcement learning. Energy Build 169:195–205. https://doi.org/10.1016/j.enbuild.2018.03.051

Chen B, Cai Z, Berges M (2019) Gnu-RL: a precocial reinforcement learning solution for building hvac control using a differentiable MPC policy. New York, NY, USA, pp 316–325.https://doi.org/10.1145/3360322.3360849

Cheng W-L, Chen Y-S, Zhang J, Lyons TJ, Pai J-L, Chang S-H (2007) Comparison of the revised air quality index with the PSI and AQI indices. Sci Total Environ 382(2–3):191–198. https://doi.org/10.1016/j.scitotenv.2007.04.036

D'Oca S, Hong T (2014) A data-mining approach to discover patterns of window opening and closing behaviour in offices. Build Environ 82:726–739. https://doi.org/10.1016/j.buildenv.2014.10.021

Dalamagkidis K, Kolokotsa D, Kalaitzakis K, Stavrakakis GS (2007) Reinforcement learning for energy conservation and comfort in buildings. Build Environ 42(7):2686–2698. https://doi.org/10.1016/j.buildenv.2006.07.010

Ding X, Du W, Cerpa A (2019) OCTOPUS: deep reinforcement learning for holistic smart building control. New York, NY, USA, pp 326–335. https://doi.org/10.1145/3360322.3360857

Dussault J-M, Sourbron M, Gosselin L (2016) Reduced energy consumption and enhanced comfort with smart windows: comparison between quasi-optimal, predictive and rule-based control strategies. Energy Build 127:680–691. https://doi.org/10.1016/j.enbuild.2016.06.024

Enescu D (2017) A review of thermal comfort models and indicators for indoor environments. Renew Sustain Energy Rev 79:1353–1379. https://doi.org/10.1016/j.rser.2017.05.175

Fabi V, Andersen RV, Corgnati S, Olesen BW (2012) Occupants' window opening behaviour: a literature review of factors influencing occupant behaviour and models. Build Environ 58:188–198. https://doi.org/10.1016/j.buildenv.2012.07.009

Fabi V, Andersen RV, Corgnati SP, Olesen BW (2013) A methodology for modelling energy-related human behaviour: Application to window opening behaviour in residential buildings. Build Simul 6(4):415–427. https://doi.org/10.1007/s12273-013-0119-6

Fazenda P, Veeramachaneni K, Lima P, O'Reilly U-M (2014) Using reinforcement learning to optimize occupant comfort and energy usage in HVAC systems. J Ambient Intell Smart Environ 6(6):675–690. https://doi.org/10.3233/AIS-140288

Fritsch R, Kohler A, Nygård-Ferguson M, Scartezzini J-L (1990) A stochastic model of user behaviour regarding ventilation. Build Environ 25(2):173–181. https://doi.org/10.1016/0360-1323(90)90030-U

Frontczak M, Andersen RV, Wargocki P (2012) Questionnaire survey on factors influencing comfort with indoor environmental quality in Danish housing. Build Environ 50:56–64. https://doi.org/10.1016/j.buildenv.2011.10.012

Haldi F, Robinson D (2009) Interactions with window openings by office occupants. Build Environ 44(12):2378–2395. https://doi.org/10.1016/j.buildenv.2009.03.025

Han M et al (2019) A review of reinforcement learning methodologies for controlling occupant comfort in buildings. Sustain Cities Soc 51:101748. https://doi.org/10.1016/j.scs.2019.101748

Hochreiter S, Schmidhuber J (1997) Long short-term memory. Neural Comput 9(8):1735–1780. https://doi.org/10.1162/neco.1997.9.8.1735

Hong T, Wang Z, Luo X, Zhang W (2020) State-of-the-art on research and applications of machine learning in the building life cycle. Energy Build 212(109831):1–15

Huizenga C, Abbaszadeh S, Zagreus L, Arens EA (2006) Air quality and thermal comfort in office buildings: results of a large indoor environmental quality survey. Healthy Build Lisbon 3:393–397

Jassim MS, Coskuner G (2017) Assessment of spatial variations of particulate matter (PM10 and PM2.5) in Bahrain identified by air quality index (AQI). Arab J Geosci 10(19). https://doi.org/10.1007/s12517-016-2808-9

Jeong B, Jeong J-W, Park JS (2016) Occupant behaviour regarding the manual control of windows in residential buildings. Energy Build 127:206–216. https://doi.org/10.1016/j.enbuild.2016.05.097

Jin W, Zhang N, He J (2015) Experimental study on the influence of a ventilated window for indoor air quality and indoor thermal environment. Procedia Eng 121:217–224. https://doi.org/10.1016/j.proeng.2015.08.1058

Kyrkilis G, Chaloulakou A, Kassomenos PA (2007) Development of an aggregate air quality Index for an urban Mediterranean agglomeration: Relation to potential health effects. Environ Int 33(5):670–676. https://doi.org/10.1016/j.envint.2007.01.010

Li N, Li J, Fan R, Jia H (2015) Probability of occupant operation of windows during transition seasons in office buildings. Renew Energy 73:84–91. https://doi.org/10.1016/j.renene.2014.05.065

Mandic DP, Chambers JA (2001) Recurrent neural networks for prediction: learning algorithms, architectures, and stability. John Wiley, Chichester; New York

Mnih V et al (2013) Playing Atari with Deep Reinforcement learning. arXiv:1312.5602 *[cs]*, Accessed: 26 Jan 2019. [Online]. Available: http://arxiv.org/abs/1312.5602

Mnih V et al (2015) Human-level control through deep reinforcement learning. Nature 518(7540):529–533. https://doi.org/10.1038/nature14236

Mozer MC (1998) The neural network house: An environment that adapts to its inhabitants. AAAI Spring Symp Intell Environ 58:110–114

Nagy A, Kazmi H, Cheaib F, Driesen J (2018) Deep reinforcement learning for optimal control of space heating. arXiv:1805.03777

Nunes de Freitas P, Guedes MC (2015) The use of windows as environmental control in 'Baixa Pombalina's' heritage buildings. Renew Energy 73:92–98. https://doi.org/10.1016/j.renene.2014.08.029

Pan S et al (2018) A study on influential factors of occupant window-opening behaviour in an office building in China. Build Environ 133:41–50. https://doi.org/10.1016/j.buildenv.2018.02.008

Pan S et al (2019) A model based on Gauss Distribution for predicting window behaviour in building. Build Environ 149:210–219. https://doi.org/10.1016/j.buildenv.2018.12.008

Park JY, Dougherty T, Fritz H, Nagy Z (2019) LightLearn: an adaptive and occupant centered controller for lighting based on reinforcement learning. Build Environ 147:397–414. https://doi.org/10.1016/j.buildenv.2018.10.028

Pascanu R, Mikolov T, Bengio Y (2013) On the difficulty of training recurrent neural networks. In: International conference on machine learning, pp 1310–1318

Pu H, Luo K, Wang P, Wang S, Kang S (2017) Spatial variation of air quality index and urban driving factors linkages: evidence from Chinese cities. Environ Sci Pollut Res 24(5):4457–4468. https://doi.org/10.1007/s11356-016-8181-0

Rijal HB, Tuohy P, Nicol F, Humphreys MA, Samuel A, Clarke J (2008) Development of an adaptive window-opening algorithm to predict the thermal comfort, energy use and overheating in buildings. J Build Perform Simul 1(1):17–30. https://doi.org/10.1080/19401490701868448

Rijal HB, Humphreys MA, Nicol JF (2018) Development of a window opening algorithm based on adaptive thermal comfort to predict occupant behaviour in Japanese dwellings. Jpn Architectural Rev 1(3):310–321. https://doi.org/10.1002/2475-8876.12043

Roulet C-A et al (2006) Perceived health and comfort in relation to energy use and building characteristics. Build Res Inf 34(5):467–474. https://doi.org/10.1080/09613210600822279

Ruelens F, Claessens BJ, Vandael S, Iacovella S, Vingerhoets P, Belmans R (2014) Demand response of a heterogeneous cluster of electric water heaters using batch reinforcement learning. Wroclaw, Poland, pp 1–7

Ruelens F, Iacovella S, Claessens BJ, Belmans R (2015) Learning agent for a heat-pump thermostat with a set-

back strategy using model-free reinforcement learning. Energies 8:8300–8318. https://doi.org/10.3390/en8088300
Shaikh PH, Nor NBM, Nallagownden P, Elamvazuthi I, Ibrahim T (2013) Robust stochastic control model for energy and comfort management of buildings. Aust J Basic Appl Sci 7(10):137–144
Shi G, Liu D, Wei Q (2017) Echo state network-based Q-learning method for optimal battery control of offices combined with renewable energy. IET Control Theory Appl 11(7):915–922
Shi Z et al (2018) Seasonal variation of window opening behaviours in two naturally ventilated hospital wards. Build Environ 130:85–93. https://doi.org/10.1016/j.buildenv.2017.12.019
Silver D et al (2016) Mastering the game of Go with deep neural networks and tree search. Nature 529 (7587):484–489. https://doi.org/10.1038/nature16961
Silver D et al (2017) Mastering the game of go without human knowledge. Nature 550(7676):354–359. https://doi.org/10.1038/nature24270
Singh J (1996) Review: health, comfort and productivity in the indoor environment. Indoor and Built Environ 5 (1):22–33. https://doi.org/10.1177/1420326X9600500105
Stazi F, Naspi F, Ulpiani G, Di Perna C (2017) Indoor air quality and thermal comfort optimization in classrooms developing an automatic system for windows opening and closing. Energy Build 139:732–746. https://doi.org/10.1016/j.enbuild.2017.01.017
Sutton RS, Barto AG (2018) Reinforcement learning: an introduction, 2nd edn. The MIT Press, Cambridge, MA
Tanner RA, Henze GP (2014) Stochastic control optimization for a mixed mode building considering occupant window opening behaviour. J Build Perform Simul 7(6):427–444. https://doi.org/10.1080/19401493.2013.863384
Wang L, Greenberg S (2015) Window operation and impacts on building energy consumption. Energy Build 92:313–321. https://doi.org/10.1016/j.enbuild.2015.01.060
Watkins CJCH (1989) Learning from delayed rewards. Ph.D. thesis, University of Cambridge
Werbos PJ (1990) Backpropagation through time: what it does and how to do it. Proc IEEE 78(10):1550–1560. https://doi.org/10.1109/5.58337
Yun GY, Steemers K (2008) Time-dependent occupant behaviour models of window control in summer. Build Environ 43(9):1471–1482. https://doi.org/10.1016/j.buildenv.2007.08.001
Zhang H, Arens E, Pasut W (2011) Air temperature thresholds for indoor comfort and perceived air quality. Build Res Inf 39(2):134–144. https://doi.org/10.1080/09613218.2011.552703

11 Development of an Adaptation Table to Enhance the Accuracy of the Predicted Mean Vote Model

Yu Li, Yacine Rezgui, Annie Guerriero, Xingxing Zhang, Mengjie Han, Sylvain Kubicki, and Yan Da

Abstract

The Predicted Mean Vote (PMV) model is extensively used by current thermal comfort standards, such as ASHRAE 55 and ISO 7730, despite its discrepancy in predicting Thermal Sensation (TS). The implicit assumption is that PMV can be applied for predicting TS of a large population. Our statistical analysis of a subset of ASHRAE global database of thermal comfort field study shows that occupants' expectations towards TS are affected by factors that are not accounted for in the classic PMV model, such as climate, building type, age group, season and gender. The influences of the climate and building type are more determinant. An adapted PMV (PMVa) model and an adaptation table were developed based on the selected samples to reduce this discrepancy. After adaptation, the medians of each category corresponding to the discrepancy are zero or near zero. The results also show that the adapted PMV outperforms the classic PMV in predicting TS, while increasing the overall accuracy from 36 to 39%.

Keywords

Predictive mean vote · Thermal sensation · Discrepancy · Adapted model · Adaptation table · Adaptive thermal comfort

11.1 Introduction

Buildings are responsible for 40% of the energy consumption in the EU (European Union), with Heating, Ventilation and Air Conditioning (HVAC) equipment used to regulate the indoor

Y. Li (✉)
School of Environmental Science and Engineering, Donghua University, Shanghai, China
e-mail: liyu@dhu.edu.cn

Y. Li · A. Guerriero · S. Kubicki
Luxembourg Institute of Science and Technology LIST, 5, avenue des Hauts-Fourneaux, L-4362 Esch-sur-Alzette, Luxembourg
e-mail: annie.guerriero@list.lu

S. Kubicki
e-mail: sylvain.kubicki@list.lu

Y. Rezgui
BRE Trust Centre for Sustainable Engineering, Cardiff University, Cardiff C24 3AA, UK
e-mail: RezguiY@cardiff.ac.uk

X. Zhang
Department of Energy and Community Buildings, Dalarna University, 79188 Falun, Sweden
e-mail: xza@du.se

M. Han
Micro Data Analysis, Dalarna University, 79188 Falun, Sweden
e-mail: mea@du.se

Y. Da
Building Energy Conservation Research Center, Tsinghua University, Beijing 100084, China
e-mail: yanda@tsinghua.edu.cn

X. Zhang (ed.), *Data-driven Analytics for Sustainable Buildings and Cities*, Sustainable Development Goals Series, https://doi.org/10.1007/978-981-16-2778-1_11

climate accounting for approximatively 50% of the building energy consumption (European Commission 2016; Li et al. 2017; Li et al. 2019). Despite the significant energy footprint spent in controlling the indoor climate, user satisfaction with the indoor comfort is often not yet met (Frontczak et al. 2012). A large scale survey in North America showed that only 2% of the commercial buildings achieved 80% thermal satisfaction, which is the prescribed minimum requirement by most standards for occupants' thermal comfort (Karmann et al. 2018). Citizens in industrialized countries spend around 90% of their time indoor (Höppe et al. 1988), and therefore the indoor conditions are important to human health and wellbeing. Thermal comfort has been identified as the most important indicator influencing the overall satisfaction in terms of indoor environmental quality (Al et al. 2016; Frontczak and Wargocki 2011). Several studies have proven that a dissatisfied thermal environment would result in an increased number of problems, such as complaints, absenteeism, and reduced productivity at work (Al et al. 2016; Wang et al. 2019). From an energy conservation point of view, indoor comfort can be used to understand the specific demand and requirements of occupants. Such information can inform the design and control of building operation systems to optimize energy efficiency and reduce carbon emission.

Thermal comfort is defined as the concept of mind to express satisfaction towards the thermal environment, and thus it should be evaluated through the direct feedback of the occupants (Ghahramani et al. 2016). Human response to Thermal Sensation (TS) is normally measured by asking the subjects to complete a 'comfort vote' on a descriptive scale ranging from '−3' to '3', either a 7-point ordered or continuous scale. '0' is the best condition representing thermal neutrality. Statistical methods are then applied to analyse the results. However, it is not practical for occupants to answer questionnaires on a continuous basis. As a result, other techniques have been developed to correlate TS with the built environment (Cheng et al. 2019; Yang et al. 2019).

PMV (Predicted Mean Vote) is the most widely used model to mathematically predict the average TS of a large group of individuals, which is the basis for multiple indoor thermal comfort standards, such as ASHRAE 55 (2017) and ISO 7730 (2005). It was developed based on extensive experiments conducted in well-controlled environments of European and North American subjects (Fanger 1970; Van Hoof 2008; Fanger and Toftum 2002). It rests on steady state heat transfer conditions between a human body and its surrounding environment. The static heat balance model predicts the mean TS of the occupants exposed to their thermal environment as a function of four thermal environmental parameters (indoor air temperature, radiant temperature, air speed and relative humidity), and two occupant's personal data (metabolic rate and clothing insulation).

Previous studies have revealed that human responses to indoor thermal comfort are affected by non-thermal factors that are not accounted for in the classic PMV model (Cao et al. 2011). As a result, a discrepancy exists between the TS votes and PMV model. The evidenced discrepancy triggered an investigation into improving the model's credibility in predicting indoor thermal comfort. Adaptive models have been proposed which assumed that people are able to adapt to their thermal environment through behavioural adjustment, acclimatization to the thermal environment or relaxation of expectations (Frontczak and Wargocki 2011). While the classic PMV model is capable of accounting for some degrees of behavioural adaptation such as adjusting local temperature and changing one's clothes, it overlooks psychological adaptation and acclimatization. Fanger and Toftum (2002) extended the PMV model by introducing an expectancy factor for non-air-conditioned buildings in warm climates. They argued that the occupants' TS expectation is lower, and that occupants would slow down their activities to adjust their metabolic rates under warm conditions. Nicol and Humphreys (2002) claimed that people are able to adjust themselves to suit the environment, and developed an adaptive thermal comfort model to estimate the acceptable indoor temperature in relation to the outdoor monthly average temperature in free-running buildings. Fanger and Toftum (2002) pointed out that the limitation of the adaptive model proposed by Nicol and Humphreys is that the model does not consider activity level, clothing insulation and the indoor thermal environment, which are believed to have a significant impact on human thermal comfort. Yao et al. (2009) proposed an adaptive PMV model based on a "black box" theory to explore the logical and statistical relationships between the variables involved. The established relationship was used to predict thermal comfort, accounting for behavioural and psychological adaptation. They claimed that their model might be important in the context of human interaction with the environment. Humphreys and Nicol (2002) discussed the variables affecting the accuracy of PMV model. They advocated that using PMV to predict thermal comfort votes could be misleading and proposed a modified PMV model based on the classic PMV and its discrepancy from the TS votes. The biases were reduced after modification.

The PMV model was developed based on the assumption that if the indoor climate meets the critical requirements for a thermally acceptable comfort condition, the TS is deemed the same for occupants with the same level of clothing insulation and carrying out similar activities, regardless of the demographic and contextual factors. The fact is that the influences of psychological adaptation and physiological acclimatization are well documented (Pantavou et al. 2018; Cheung et al. 2019). To address this research gap, the proposed study aims to explore the impacts of the variables contributing to this discrepancy. Currently, most of the field and experiment studies are based on a limited number of samples, which cannot be generalized to represent a large population. Thus, our study is derived from the ASHRAE Global Thermal Comfort Database II (Földváry et al. 2018) to investigate the influence of a large sample size. Following this introduction, Sect. 11.2 discusses the factors contributing to the PMV discrepancy. The chapter then elaborates on the methodology used in this study (Sect. 11.3), followed by a presentation of the results (Sect. 11.4). An adapted model and an adaptation table are presented in Sect. 11.5 with the aim to compensate the influences caused by the variables identified in Sect. 11.2. The results are then discussed in Sect. 11.6, followed by concluding remarks.

11.2 Factors Contributing to the Discrepancies Between PMV and Thermal Sensation

The PMV model was primarily developed from predefined thermal environmental and personal variables, hence overlooking other factors that may potentially affect the accuracy of the results. However, thermal comfort is a subjective factor, which is closely associated with occupants' thermal expectations and capacity of adaptation. Different occupants may have different perceptions of thermal comfort even when exposed to the same environment (Humphreys and Nicol 2002). Such differences are associated with the influence of seasonal variations (Cao et al. 2011; Fountain et al. 1996; Rijal et al. n.d.) and the general climate (Rijal et al. n.d.; Nicol and Humphreys et al. 2007; Wenzel 1989; Hancock et al. 2011), as well as lifestyle and socio-cultural factors, including the use of different clothing materials (Humphreys and Nicol 2002), expectations (influenced by the season (Zhang et al. 2016; Li et al. 2017), climate (Humphreys and Nicol 2002; Dear and Brager 1998), age (Natsume et al. 1992; Yang and Olofsson 2017), and gender (Wang et al. 2018; Karjalainen 2012), and the ability to control the thermal conditions in the actual buildings (Teli et al. 2012; Wang et al. 2019; Alwetaishi 2017). This, in turn, involves variations in (a) perceived neutral temperatures, (b) interpretation of the ASHRAE scale categories, and (c) personal judgements (Humphreys and Nicol 2002). As such, five determinant variables are identified from the literature, namely: season, climate, building type, age group, and gender; and elaborated below.

11.2.1 Season

Many previous studies have reported the effect of seasonal variations on personal perceived TS (Cao et al. 2011; Fountain et al. 1996; Rijal et al. n.d.; Zhang et al. 2016; Li et al. 2017). Li et al. (2017) conducted a large-scale thermal comfort survey in a hot summer and cold winter zone of China. Different TS and adaptive responses were detected in different seasons. The results revealed that the significant seasonal variations were due to the individuals' thermal experience and thermal expectations in difference seasons. A higher neutral temperature was expected in warm seasons. Thus, maintaining the same indoor temperature in winter as in summer results in a waste of energy (Cao et al. 2011). In addition to the difference in perceived TS, the occupants' physiological reactions are also different. The differences have been recorded from climate chamber experiments by measuring physiological reaction of the participants (Nakamura and Okamura 2009; Umemiya 2006; Lee et al. 2012). Noriko (Umemiya 2006) investigated seasonal metabolic rate variation of 6 subjects under identical thermal conditions over the course of one year. Results showed that metabolic rates were higher in winter than in summer. Lee et al. (Lee et al. 2012) compared sweating responses of 15 male participants in summer and winter. They concluded that sweat volume and evaporative rate were significantly less in winter than in summer.

11.2.2 Climate

Research studies with regard to thermal comfort have revealed that occupants' thermal adaptation is affected by climate and social custom, mainly reflected by their perceived neutral temperature and their interpretation of ASHRAE comfort vote (Humphreys and Nicol 2002; Dear and Brager 1998). Generally, these studies found that neutral indoor temperature in sub-tropical climate is higher than temperate climate, and lowest in cool climate (Rijal et al. n.d.; Nicol and Humphreys 2007). Although the clothing insulation and metabolic rate are taken into consideration in the classic PMV model, the values are normally assessed from a standard checklist. Clothing in the tropical area may allow great diffusion of moisture and air (Humphreys and Nicol 2002). The metabolic rate may be lower in the warm climate while conducting the same listed tasks (Humphreys and Nicol 2002). Thus, the identical clothing level and tasks providing identical assumption of clothing insulation and metabolic rates may result in the same PMV while the TS is different. Meanwhile, evidence suggests that human genome imparted by natural selection is strongly correlated with climate variables (Hancock et al. 2011). People exposed to certain climate may have higher tolerance for higher temperature and humidity (Wenzel et al. 1989; Yang and Wang 2018). For example, people living in warm climate prefer a warmer indoor temperature (Dear and Brager 1998).

11.2.3 Building Type

The impact of building type on thermal comfort is much less discussed when compared with other factors (Zhang and Dear 2019). Architects design buildings that address functional requirements while complying with the energy (and other) regulatory landscape (Daher et al. 2019). Moreover, the interior design as well as indoor facilities vary from one building type to another. An office building is normally equipped with personal computers and servers, which are responsible for the extra heat gain of the building. Furthermore, the occupants are not passive recipients of their thermal environment regime as they actively interact with the control systems in place (e.g. thermostats and radiator valves) to make themselves comfortable. Different types of buildings affect people's ability to adapt to clothing insulation and control of the environment (Teli et al. 2012; Wang et al. 2019). For example, occupants in residential buildings tend to adapt their clothing level, and impact on the indoor environment by controlling their HVAC system. Thus, the acceptable temperature range in residential buildings is wider than other types of buildings (Alwetaishi 2017). Oseland et al. (2010) compared 30 subjects' thermal comfort from home, office and a climate chamber. The author concluded that under the same clothing and activity, the participants felt warmer in their home than in their office, and warmer in their office than in the climate chamber. The result was validated by a follow on study from Karjalainen et al. (2009), who conducted an investigation to examine the thermal comfort of 3094 respondents. The results showed that the respondents felt warmer at home than in their office.

11.2.4 Age Group

Tolerance to cold and hot environments of different age groups is generally considered to decrease with age due to the reduced thermoregulation response (Natsume et al. 1992). This reduction in thermal sensitivity is caused by aging of the skin and the superficial skin blood flow (Guergova and Dufour 2011). Natsume et al. (1992) studied 6 older men (71–76 years old) and 6 young men (21–30 years old) to investigate their preferred temperature. The subjects were healthy both physically and mentally. The results indicated that the preferred temperature range of the older people is wider than the younger people. Another study was conducted by Schellen et al. (2010) to investigate thermal comfort response of different age groups. 8 seniors (67–73 years old) and 8 young adults (22–25 years old) participated in the investigation. The results revealed that the older people preferred a higher temperature compared with the younger people and their TS was generally 0.5 scale units less than the young people. The results were challenged by some scientists who claimed no significant influence of age on TS. Soebarto et al. (2019) investigated the thermal comfort of younger (20 samples) and older (22 samples) subjects under different test conditions. The skin temperature of the subjects was measured at four body parts. No significant difference in thermal preference or thermal sensation was observed for the two different age groups.

11.2.5 Gender

A large number of studies have examined the effect of gender on TS and yielded conflicting conclusions. A substantial amount of field studies reported a weak or insignificant influence caused by gender difference (Wang et al. 2018). Amai et al. (2007) studied thermal comfort under three types of tasks. The TS difference between male and female subjects was small. Maykot et al. (2018) conducted 116 field studies to investigate the influence of gender on thermal comfort temperature. Total 584 participants were involved in the experiments. Statistical analysis of the collected results showed that the comfort temperature for female is slightly higher (≤ 1 °C) than male. Conversely, some studies revealed a noticeable gender difference in terms of thermal comfort (Alwetaishi 2017; Beshir and Ramsey 1981; Karjalainen 2012). Beshir and Ramsey (1981) investigated the TS difference between 31 male and 15 female participants. The subjects were exposed to 23.3–43.3 °C and were not allowed to adjust the temperature throughout the experiment. The results showed a significant gender difference, with the male subjects preferring a lower comfort temperature than the female subjects. The female participants felt more uncomfortable in both cold and hot temperature extremes. In a way, the results are similar to some field and laboratory studies, which showed that the female participants are more sensitive to deviation from the comfort temperature and more likely to be thermal dissatisfaction than male under the same thermal environment (Wang et al. 2018; Karjalainen 2012). Thus, it was suggested that female should be used as primary subjects when investigating indoor thermal comfort requirements (Karjalainen 2007). However, Kingma and Lichtenbelt (2015) and Wang et al. (2018) pointed out that the gender differences are derived from the metabolic rate and clothing insulation, and thus the influence might be eliminated once the clothing and metabolic rates are well controlled (Zhang and Dear 2019).

11.3 Methodology

This section describes the data source used in the study, the adopted sampling technique, as well as the analytical methods used to analyse and quantify the relationships between the categorical variables involved, as elaborated below.

11.3.1 Data Source and Sampling Technique

The ASHRAE Global Thermal Comfort Database II, which consists of thermal comfort data from subjective comfort votes and objective instrumental measurements, is used in this study. Besides the thermal comfort information, the database also includes other related information such as season, climates, building types, age and gender. The samples with PMV, thermal sensation, season, climate (Köppen–Geiger climate classification), building type, age and gender were extracted from the database. Meanwhile, the attributes under the different variables with a sample size of less than 383 were excluded to ensure a confidence level of 95%. This number is obtained by Eq. (11.1).

$$Sample\ size = \frac{\frac{Z^2 \cdot SD \cdot (1-SD)}{e^2}}{1 + \left(\frac{Z^2 \cdot SD \cdot (1-SD)}{e^2 N}\right)} \quad (11.1)$$

where, Z = 1.96 (95% confidence level), SD (standard deviation) = 0.5, e (margin of error) = 0.05. N is the population size in the database.

The calculated population size is 383. Finally, a total of 5 variables (Season, Climate, Building Type, Age group, and Gender) consisting of 17,841 observations were used in this analysis. A new variable named discrepancy representing the difference between PMV and TS was generated. The extracted age information was a numeric variable ranging from 16 to 99. In this study, the age was grouped into three attributes: Young: 16–24, Adults: 25–64 and Older adults: 65–99. Figure 11.1 summarizes the distribution of the extracted samples. A large number of the

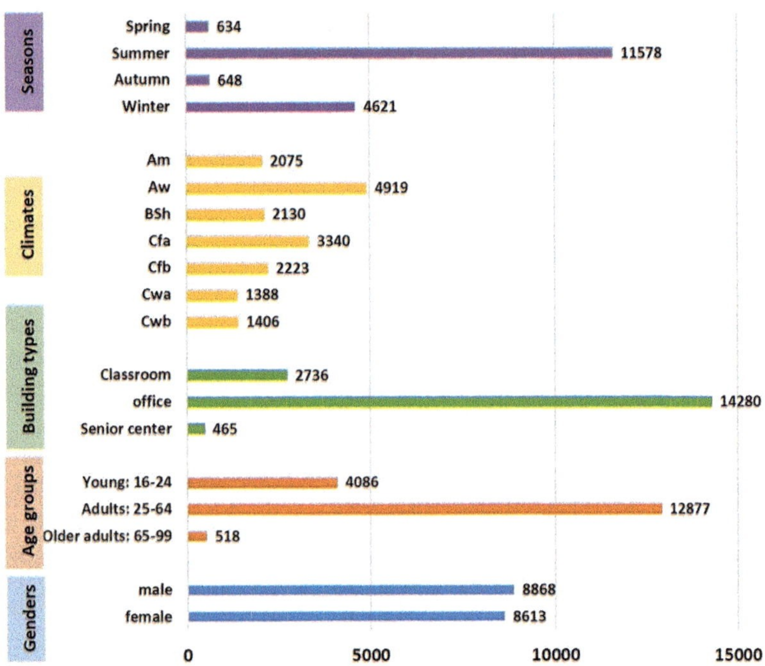

Fig. 11.1 Distribution of extracted data samples from the database. Am (Tropical monsoon climate), Aw (Tropical wet and dry climate), BSh (Hot semi-arid climate), Cfa (Humid subtropical climates), Cfb (Oceanic climate), Cwa (Dry-Winter Humid Subtropical), Cwb (Dry winter Oceanic climate)

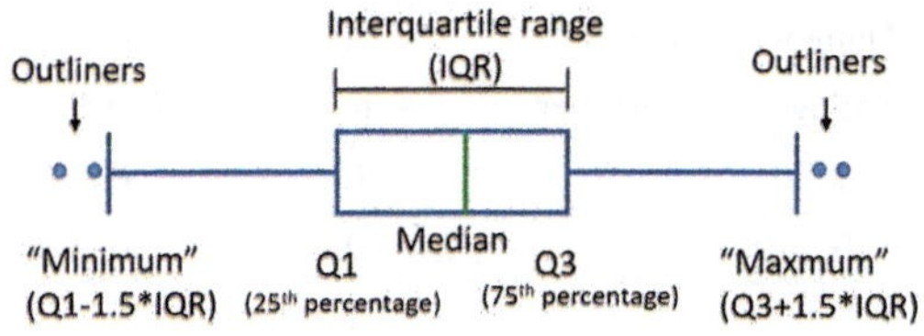

Fig. 11.2 Interpretation of the boxplot

records were collected from Adults: 25–64 in the office building during summer.

11.3.2 Analytical Methods for Categorical Variables

The factors contributing to the discrepancy in our analysis are categorical variables. Therefore, the analytical method should be able to analyse the relationships among these categorical variables. Analysis of variance (ANOVA) was applied to explore the association between the discrepancy and the different categorical variables (e.g. season, climate).

A boxplot is a straightforward graphical method to summarize the datasets, which shows whether or not a dataset is symmetric. It was used in this analysis to visually identify the dispersion of samples based on a five-number summary ("minimum", lower quartile (Q1), median, upper quartile (Q3) and "maximum"), as shown in Fig. 11.2. The difference between lower quartile and upper quartile is the length of the box. A line that divides the box into two parts represents the median of the data. For example, a median of 5 denotes that the number of data higher than 5 is the equal to the number of data lower than 5. The difference between lower quartile and upper quartile is the interquartile range (IQR). The "minimum" and the "maximum" are the Q1-1.5 * IQR and Q1 + 1.5 * IQR. Outliers are displayed as individual points.

Correspondence Analysis (CA) and Multiple Correspondence Analysis (MCA) were both used in this study to project the correlation between the discrepancy and the categorical variables in 2D map. The CA is a graphical technique designed specifically for the analysis of categorical variables, which interprets the relationship among categorical variables by identifying their differences and similarities (Greenacre 1984; Sourial et al. 2010; Cariou and Qannari 2018). This technique preserves the categorical nature of the variables and is able to accommodate any type of categorical variable (Sourial et al. 2010). It is developed from data in a contingency table, which is a two-dimensional table in matrix format showing the frequency of the variables associated attributes. The MCA is similar to CA apart from that it can be used when there are more than two categorical variables. The association distances in CA and MCA are measured by chi-square distance between the response categories. This measurement ensures that the larger population do not dominate the relative distance. Thus, CA exhibits a higher accuracy when compared with other multivariate techniques derived from the correlation coefficient (Sourial et al. 2010). The chi-square distance between row i and i' is defined by Eq. (11.2) (Saqlain et al. 2019):

$$d(i,i') = \sqrt{\sum_j \left(\frac{\left(p_{ij} - p_{i'j}\right)^2}{p_{+j}}\right)} \tag{11.2}$$

where p_{ij} and $p_{i'j}$ are relative frequencies of row i and i' in column j. p_{+j} is the marginal relative frequency for column j.

11.3.3 Analytical Methods to Quantify the Relationships

Although categorical variables are widely used in our daily life, they cannot be used directly in regression analysis to establish a statistic relationship. Dummy coding, also known as one-hot coding, is employed to incorporate categorical variables into regression analysis by converting the categorical variables into mutually exclusive binary variables. The dummy variables then can be considered as true values that consist of

Table 11.1 An example of dummy coding for season

		Dummy variables			
	Season	Spring	Summer	Autumn	Winter
Sample 1	Spring	1	0	0	0
Sample 2	Summer	0	1	0	0
Sample 3	Autumn	0	0	1	0
Sample 4	Winter	0	0	0	1

0 and 1. Table 11.1 gives an example for dummy coding of season. After coding, the attributes are converted into binary data.

Multivariate linear regression is a linear statistical model with more than one independent predictor. It was applied in this study to establish the relationship between the discrepancy and the dummy coded categorical variables. Linear regression is selected because it is the most extensively used regression model and the coefficients can be easily used to generate a compensatory table for categorical variables.

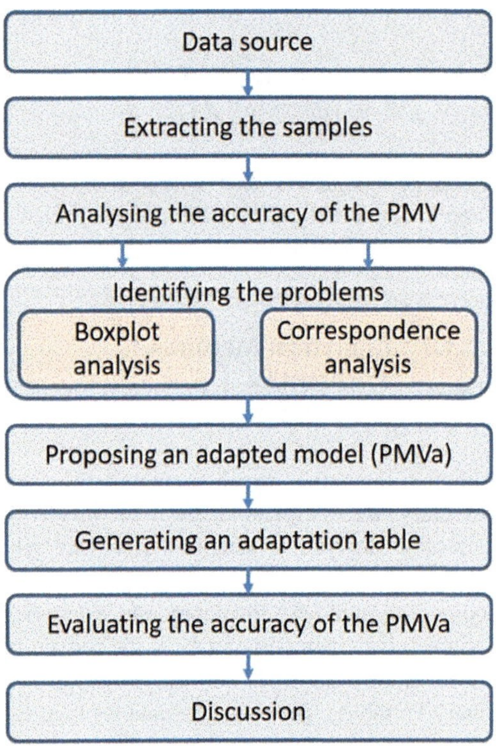

Fig. 11.3 workflow of the methodology

11.3.4 Workflow of the Methodology

The workflow of the methodology is shown in Fig. 11.3. The main concept of the chapter is to analyse the accuracy of the existing PMV model for TS prediction, identify the reasons for the discrepancy (PMV-TS), and propose an adapted model (PMVa) together with an adaptation table to enhance the accuracy of the model.

11.4 Results

This section presents the results related to the accuracy of the PMV model for TS prediction, as well as the effects of the variables involved in the discrepancy based on the extracted samples.

11.4.1 Accuracy of the PMV Model for TS Prediction

As people differ in their thermal perceptions, PMV cannot be expected to precisely predict the TS of an individual. The relationship between the PMV and its discrepancy (PMV-TS) is illustrated in Fig. 11.4. Figure 11.4a displays a scatter diagram and a linear relationship of the PMV corresponding to the recorded discrepancy. It can be seen that the discrepancy increases with the increase of the PMV. When the PMV is neutral, the absolute value of the discrepancy is smaller than when the PMV deviates from the neutral. This means that the perceived TS is not as hot or cold as it is predicted from the PMV model in hot or cold environments.

The PMV was binned into 7 categories in Fig. 11.4b: Cold (PMV ≤ -2.5), Cool

(a)
(b)

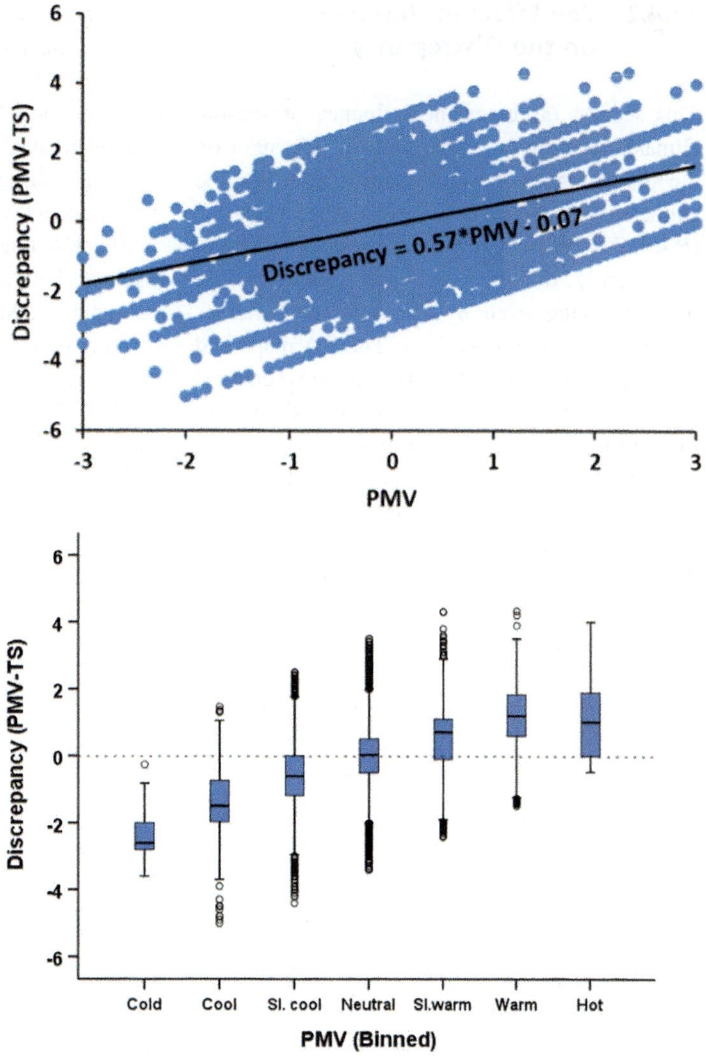

Fig. 11.4 The correlation between PMV and its discrepancy (PMV-TS) **a** Scatter plot of PMV corresponding to discrepancy, **b** Boxplots of Binned PMV corresponding to discrepancy

(−2.5 < PMV ≤ −1.5), Slightly cool (−1.5 < PMV ≤ −0.5), Neutral (−0.5 < PMV ≤ 0.5), Slightly warm (0.5 < PMV ≤ 1.5), Warm (1.5 < PMV ≤ 2.5) and Hot (PMV > 2.5). After binning the PMV, the overall sample number for each category is 58 (Cold), 467 (Cool), 3405 (Slightly cool), 7956 (Neutral), 3920 (Slightly warm), 1007 (Warm) and 668 (Hot). The magnitude of the discrepancy quantifies the success of PMV in predicting TS. The median of each boxplot is −2.6, −1.5, −0.61, 0.02, 0.7, 1.19 and 1. The medians are smaller at "Slightly cool", "Neutral" and "Slightly warm" environments. Thus, the PMV model is better for TS prediction under these three circumstances. The ISO 7730 (2005) also recommends that the PMV range should be used within ±2 to ensure a higher accuracy. The discrepancy is bigger in "Cold" and "Cool" environments than in "Hot" and "Warm" environments, which indicates that the PMV model is better in predicting "Hot" and "Warm" conditions than in that in "Cold" and "Cool" environments. This could be explained by the access to greater adaptive options for most building occupants in a cooler environment e.g. clothing modification. Thus, they do not feel as cold as predicted from the PMV model.

11.4.2 The Effect of Variables on the Discrepancy

This section discusses the influence of season, climate, building type, age group and gender on the prediction of the discrepancy based on the extracted samples. The ANOVA test results ($P < 0.001$) indicated that the discrepancy was significantly different for each category. The boxplots categorized by the different variables are displayed in Fig. 11.5a–e. The CA maps and MCA map illustrating the distance between each variable and the discrepancy are shown in Figs. 11.6a–e and 11.7. The origins of the maps correspond to the centroid of each variable. The longer distance from the attributes to the origin, the more discriminating it is. The results are discussed in the following subsections.

11.4.2.1 Season

Figure 11.5a shows the boxplots of the discrepancy observation corresponding to the four seasons. The medians of the discrepancy are −0.115 for spring, 0.09 for summer, -0.58 for autumn, and 0.2 for winter. The negative medians imply that on average the perceived TS is warmer than predicted from the classic PMV model, while the positive medians imply a cooler feedback from the ASHRAE vote when compared with the PMV model. It is apparent that the PMV underestimates the actual TS in autumn. The maximum difference of the discrepancy is 0.78, which occurs between winter and autumn.

The CA map in Fig. 11.6a demonstrates the distance between season and the categorized discrepancy. Autumn is a highly discriminating attribute indicated by its distance from the origin. It is closer to category 1, category 2 and category 3 (PMV-TS ≤ -1), leading to a larger negative median value. The results reveal that autumn has a big impact on occupants' TS. Although spring is also scattered from the origin, it is closer to category 4 and category 5 ($-1 <$ PMV-TS ≤ 1).

11.4.2.2 Climate

The boxplots of the discrepancy observation corresponding to the climate are displayed in Fig. 11.5b. The difference caused by climate is more noticeable when compared with other variables. The medians for the Am, Aw, Bsh, Cfa, Cfb, Cwa and Cwb are 0.1, 0.36, 0.5, −0.4, −0.64, 0.6 and 0.2. The maximum difference is

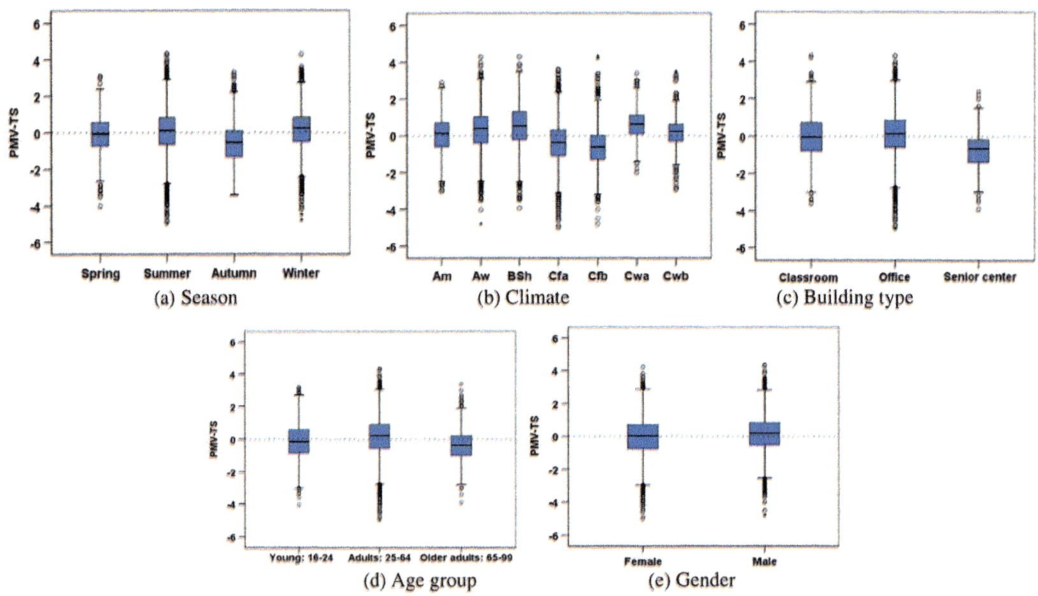

(a) Season (b) Climate (c) Building type (d) Age group (e) Gender

Fig. 11.5 Boxplots of the discrepancy (PMV-TS) categorized above a, b, c, d, e

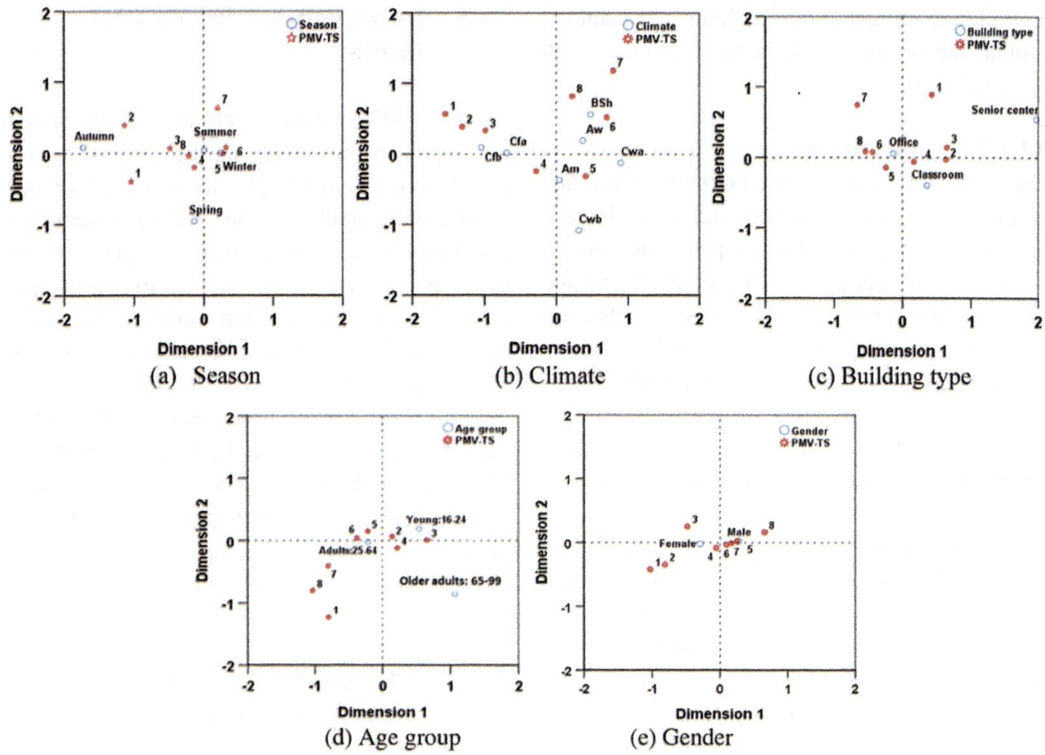

Fig. 11.6 Correspondence analysis maps of the discrepancy (PMV-TS) above a, b, c, d, e

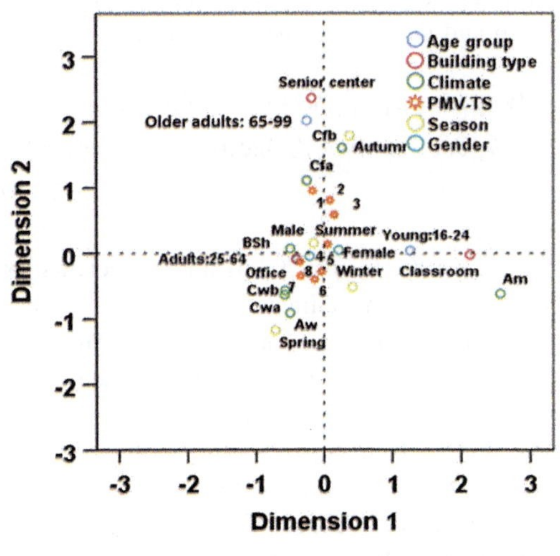

Fig. 11.7 Multiple correspondence analysis map

1.24, which occurs between Cwa and Cfb. It should be noted that the impact of the similar climates on the discrepancy is similar. For example, the tropical climates (Am and Aw) both result in a cooler TS.

On the other hand, the CA map in Fig. 11.6b indicates the different climates are significantly scattered. Cfa and Cfb are clustered around category 1, 2 and 3 (PMV-TS ≤ -1), leading to negative medians. On the contrary, Aw, Bsh and Cwa are clustered around 6, category 7 and 8 (PMV-TS > 1), and thus positive medians are observed.

11.4.2.3 Building Type

As can be seen from Fig. 11.5c, the influence of classroom and office building on the discrepancy is small, with medians of −0.1 and 0.1, respectively. However, the influence of senior center is more obvious, with a median of −0.7, which means that the perceived thermal comfort is warmer than predicted from the PMV model. The CA map in Fig. 11.6c shows that the senior center is highly differentiated, reflected in the distance between the senior center and the origin. It is close to category 1, category 2 and category 3. Therefore, a negative median is observed.

The classroom and office building are centered around the origin, which is in accordance with the small medians.

11.4.2.4 Age Group

Figure 11.5 (d) displays the boxplots of the discrepancy observation corresponding to the age group. The medians for discrepancy are −0.19, 0.19, −0.4 for Young: 16–24, Adult: 25–64 and Older adults: 65–99. The group difference between the Young: 16–24 and Adult: 25–64 is minor when compared with the Older adults: 65–99. The distance between the senior group and the origin in Fig. 11.6d also indicated that the older group is a more discriminating variable.

11.4.2.5 Gender

The gender does not have a significant influence on the discrepancy. The median values for discrepancy caused by female (0) and male (0.16) are similar. Figure 11.5e clearly demonstrates that female and male are clustered around centroid, and thus the deviation between the two attributes is small. The male occupants are clustered among point 6, point 7 and point 8. The map indicates that the male occupants tend to feel slightly warmer than PMV prediction.

11.4.2.6 Overall Analysis

The results in Figs. 11.5 and 11.6 show that the five categorical variables (season, climate, building type, age group and gender) investigated had an impact on the discrepancy. In order to examine the influence of those variables together, the five variables are mapped into one MCA in Fig. 11.7. As can be seen from the figure, the climate and building type are more scattered than the other variables, which indicates the two variables are more differentiate than the other variables. The gender is closely located to the origin. The discrepancy categories (from 1 to 8) are primarily clustered by the climate. Thus, the climate has the most significant influence on the discrepancy.

11.5 Improvement in the PMV Model

As the PMV model is an aggregated model developed to predict the average TS of a large population, unsurprisingly, its accuracy for predicting individual's thermal comfort response is not high. In fact, for occupants exposed to the same space, sharing the same environment, their thermal comfort perception varies. However, it should be able to predict the mean comfort vote of a large population. The medians shown in Fig. 11.5 are deviate from zero, and thus there are other factors contributing to the discrepancy, which are not accounted in the classic PMV model. This section attempts to quantify the influence of these factors so that the PMV can better represent the TS of a group of people.

11.5.1 Adapted PMV Model and Adaptation Table

A regression model in Eq. (11.3) was developed to account for the impact of the categorical variables (season, climate, building type, age group and gender) on the discrepancy.

$$PMV - TS = \beta_0' + \beta_1' \, Season + \beta_2' \, Climate + \beta_3' \, Buildingtype + \beta_4' \, Agegroup + \beta_5' \, Gender + \varepsilon \tag{11.3}$$

where β_0', β_1', β_2', β_3', β_4', β_5' are coefficients for constant, season, climate, building type, age group and gender. ε is the error term. The categorical variables were then dummy coded. The sample data were fitted into the model ($P < 0.001$). The target was to reduce the difference between PMV prediction and TS from the occupants' feedback. The results are shown in Table 11.2. Based on the results, we proposed an adapted PMV (PMVa) model in Eq. (11.4). As the range of TS is from −3 to 3, PMVa should also meet the requirement in Eq. (11.5).

Table 11.2 Adaptation table for PMV model

		Coefficient	Standard error	p
Constant	β_0	0.362	0.019	$P < 0.001$
Season	β_1-Spring	−0.408	0.044	$P < 0.001$
	β_1-Autumn	−0.586	0.052	$P < 0.001$
	β_1 Winter	−0.093	0.019	$P < 0.001$
Climate	β_2-Am	−0.594	0.068	$P < 0.001$
	β_2-BSh	0.170	0.028	$P < 0.001$
	β_2-Cfa	−0.756	0.025	$P < 0.001$
	β_2-Cfb	−1.032	0.032	$P < 0.001$
	β_2-Cwa	0.294	0.032	$P < 0.001$
	β_2-Cwb	−0.222	0.032	$P < 0.001$
Buildingtype	β_3-Classroom	0.391	0.057	$P < 0.001$
	β_3-Senior center	−0.358	0.068	$P < 0.001$
Agegroup	β_4-Older adults: 65–99	0.249	0.063	$P < 0.001$
	β_4-Young: 16–24	−0.092	0.024	$P < 0.001$
Gender	β_5-Male	0.087	0.016	$P < 0.001$

$$PMV_a = PMV - \beta_0 - \beta_1\ Season - \beta_2\ Climate - \beta_3\ Buildingtype - \beta_4\ Agegroup - \beta_5\ Gender \quad (11.4)$$

$$-3 \leq \mathrm{PMV}_a \leq 3 \quad (11.5)$$

where β_0, β_1, β_2, β_3, β_4, and β_5 are coefficients for constant, season, climate, building type, age group and gender. The corresponding values are displayed in Table 11.1. For example, in Summer, in Am climate, in classroom for Young: 16–24 male occupants, the PMVa = PMV −0.362−0−(−0.594)−0.391−(−0.092) −0.087 = PMV−0.154. The standard error of the coefficient is the standard deviation of the coefficient, which measures how precise the coefficient is. Compared with the coefficient, the standard error is small, which indicates the accuracy of the model is high. PMV is the value calculated from Fanger's PMV model. As it can be seen from Table 11.2, the influence of climate and building type on TS are more significant while the influence of gender is minor. The maximum difference for climate and building type are 1.324 and 0.749.

The reference categories are Adult: 25–64; Aw; Female, Summer, Office, which are equal to 0 in the equation.

11.5.2 Evaluation of the Adapted Discrepancy

The extracted samples in Fig. 11.8 together with Eqs. (11.4) and (11.5) and the results from Table 11.2 were used to obtain the adapted discrepancy (PMVa-TS). Boxplots in Fig. 11.8 show the adapted discrepancy with respect to season, climate, building type, age group and gender. As shown in the boxplots, the median of each attribute is '0' or near '0', which indicates that, on average, the PMVa is free from serious bias. Thus, it is concluded that the PMVa can be used to predict the mean TS of a large population.

After adaptation, it is also important to ensure the accuracy of PMVa is the same or even better than the classic PMV model when used for individual prediction. To compare the overall accuracy of the PMVa with the classic PMV, the adapted discrepancies and original discrepancies

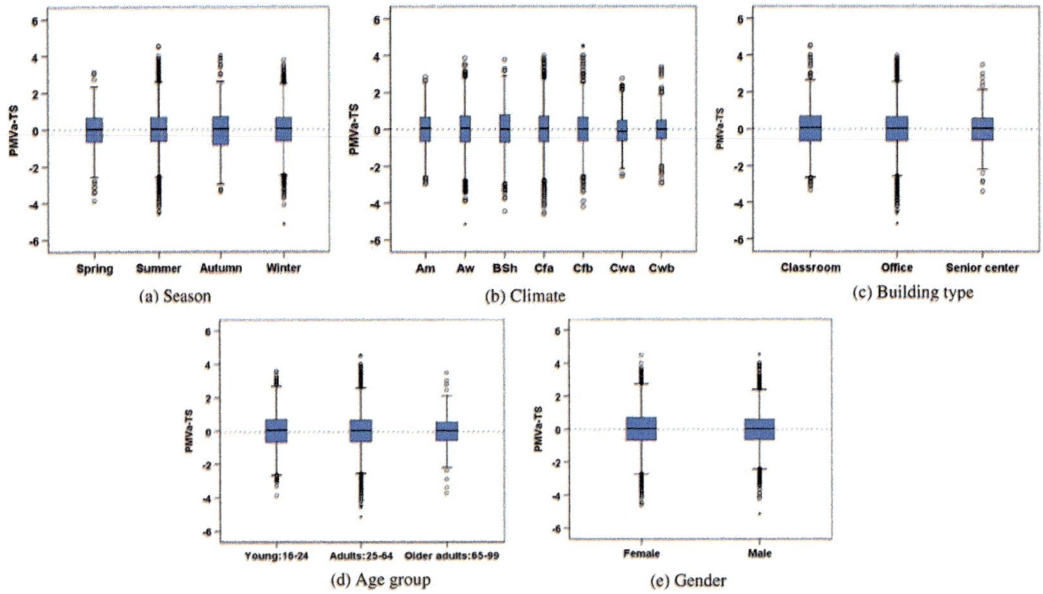

Fig. 11.8 Boxplot of the discrepancy (PMVa-TS) categorized **a** Season, **b** Climate, **c** Building type, **d** Age group, **e** Gender

were pooled into distributions in Fig. 11.8a, b, representing 17,841 discrepancies for each. The original discrepancy in Fig. 11.9b follows a normal distribution, with a mean of 0.06 scale units and a standard deviation of 1.13 scale units. The mean value demonstrates that the PMV as a whole is slightly higher than the actual ASHRAE vote by 0.06 scale units, which indicates that on average the discrepancy is small. The distribution of the adapted discrepancy is displayed in Fig. 11.8a, which also follows a normal distribution, with a mean of 0 and a standard deviation of 1.04 scale units. Both the mean value and the standard deviation are decreased. Therefore, the overall accuracy of the adapted model used for thermal comfort vote prediction has been improved.

The PMVa and TS are binned in the same way as the PMV. The PMVa was binned into 7 categories: Cold (PMVa ≤ -2.5), Cool ($-2.5 <$ PMVa ≤ -1.5), Slightly cool ($-1.5 <$ PMVa -0.5), Neutral ($-0.5 <$ PMVa ≤ 0.5), Slightly warm ($0.5 <$ PMVa ≤ 1.5), Warm ($1.5 <$ PMVa ≤ 2.5) and Hot (PMVa > 2.5). The TS was also binned into 7 categories: Cold (TS -2.5), Cool ($-2.5 <$ TS ≤ -1.5), Slightly cool ($-1.5 <$ TS ≤ -0.5), Neutral ($-0.5 <$ TS $\leq$ 0.5), Slightly warm ($0.5 <$ TS ≤ 1.5), Warm ($1.5 <$ TS ≤ 2.5) and Hot (TS > 2.5). After binning the TS, the sample number for each binned TS category is 156 (Cold), 756 (Cool), 3144 (Slightly cool), 8216 (Neutral), 3387 (Slightly warm), 1373 (Warm) and 449 (Hot). The binned PMVa and binned PMV with respect to the binned TS are illustrated in Fig. 11.10, with the ratio of correct prediction for each category and overall prediction accuracy shown in the top of the figure. When comparing Fig. 11.10a with Fig. 11.10b, the accuracy for PMVa in "Cool", "Sl. cool", "Neutral", "Sl. warm", "Warm" is higher than the PMV model. More specifically, in "Neutral" environment, the accuracy has increased by 2%. The most significant increases are observed in "Sl. cool" and "Warm" environment, increasing from 23 and 12% to 33% and 31% respectively. In addition, the overall accuracy for PMVa is 3% higher than PMV model, which indicates the PMVa model is better in predicting TS.

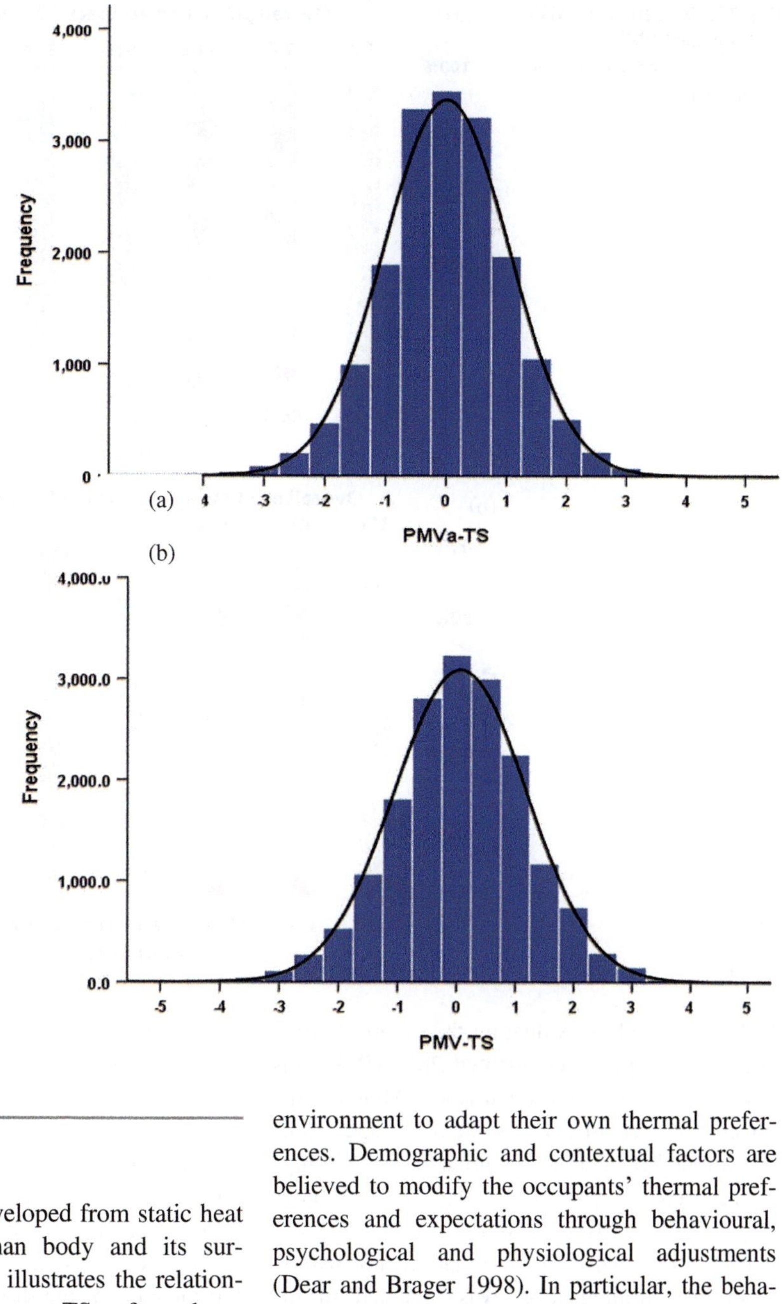

Fig. 11.9 Frequency distribution of **a** adapted discrepancy and **b** original discrepancy

11.6 Discussion

The PMV model was developed from static heat balance between a human body and its surrounding environment. It illustrates the relationship between the average TS of a large population and the surrounding indoor environment, which assumed that the TS is exclusively affected by four environmental and two personal factors (Yao et al. 2007). The occupants are regarded as passive recipients of their thermal environment (Dear and Brager 1998). In reality, they actively interact with their thermal environment to adapt their own thermal preferences. Demographic and contextual factors are believed to modify the occupants' thermal preferences and expectations through behavioural, psychological and physiological adjustments (Dear and Brager 1998). In particular, the behavioural adaptation offers the biggest opportunity for the occupants to play an active role in maintaining thermal comfort.

The literature review in Sect. 11.2 and the analysis in Sect. 11.4 discussed the effect of season, climate, building type, age group and gender on the discrepancy between PMV and TS. These variables are not used as predictors in the

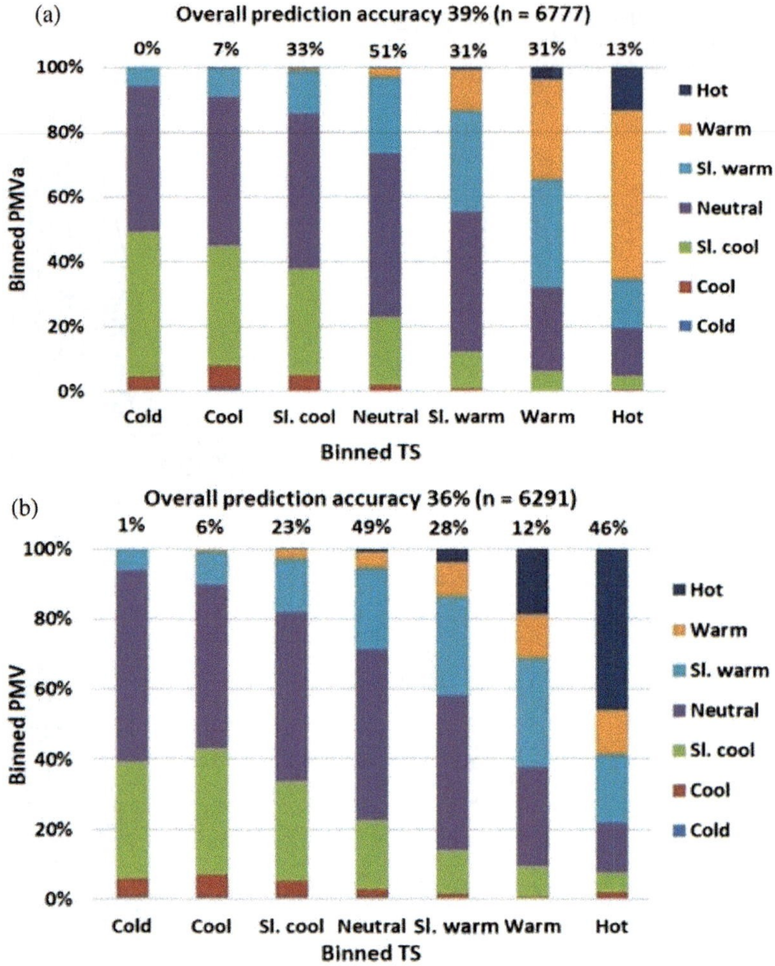

Fig. 11.10 a Binned PMVa and b binned PMV distribution corresponding to binned TS

PMV model while evaluating the indoor thermal comfort. The proposed adapted PMV (PMVa) is able to account for these variables. Although the effect of ventilation system usage (air-conditioned or naturally ventilated) in buildings is recorded (Fanger and Toftum 2002; Halawa and Van Hoof 2015), it is not used as an indicator in this study. The ventilation information from the building level was included in the database, but it is difficult to identify whether the device was on when the subjects were filling the surveys and how often it was used. Figure 11.11 shows the original discrepancy of the samples collected from the summer season with respect to the cooling strategies. The medians for the different control strategies are very close to "0". We believe more information is required to identify the effect of air conditioning on PMV accuracy. Furthermore, the use of HVAC systems can be inferred from the climate together with season and building type. For example, the office buildings in tropical areas are normally equipped with air conditioning facilities while in cold areas they are equipped with heating devices.

Boxplots in Figs. 11.5 and 11.12 compared the medians of the discrepancies with respect to one specific variable. Figures 11.12 and 11.13 discuss the interaction between the variables. Figure 11.12 compared the medians of the adapted discrepancy with the original discrepancy for different building types and seasons. After adaptation, the medians are closer to '0'. The boxplots with significant deviation from '0' for the adapted discrepancy are spring senior

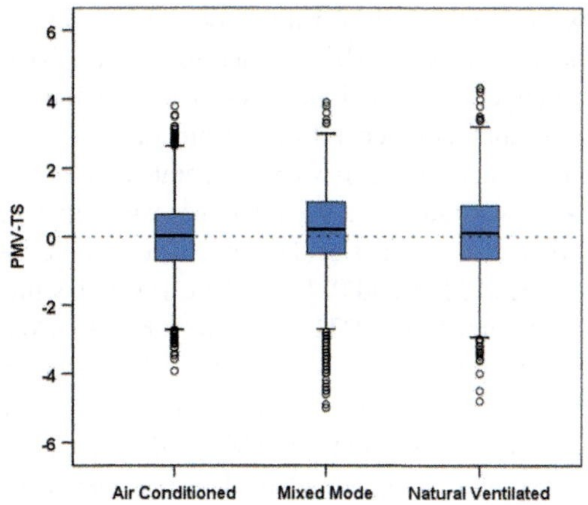

Fig. 11.11 The effect of cooling strategies at building level on original discrepancy

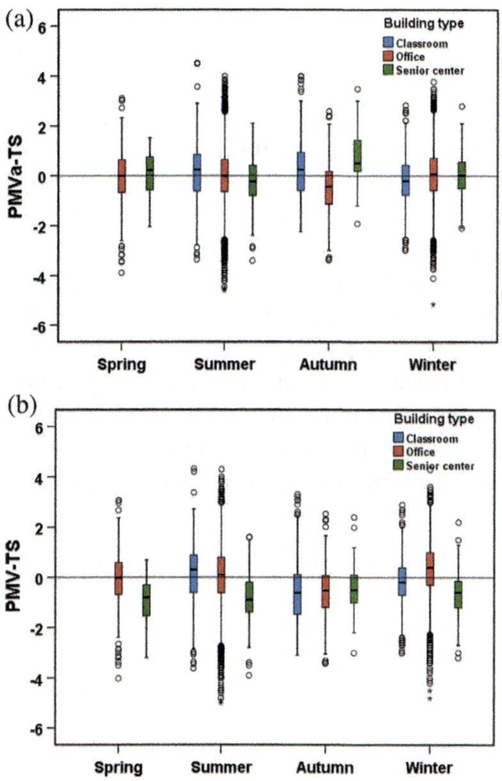

Fig. 11.12 Boxplots of adapted discrepancy **a** and original discrepancy **b** in different types of buildings and seasons

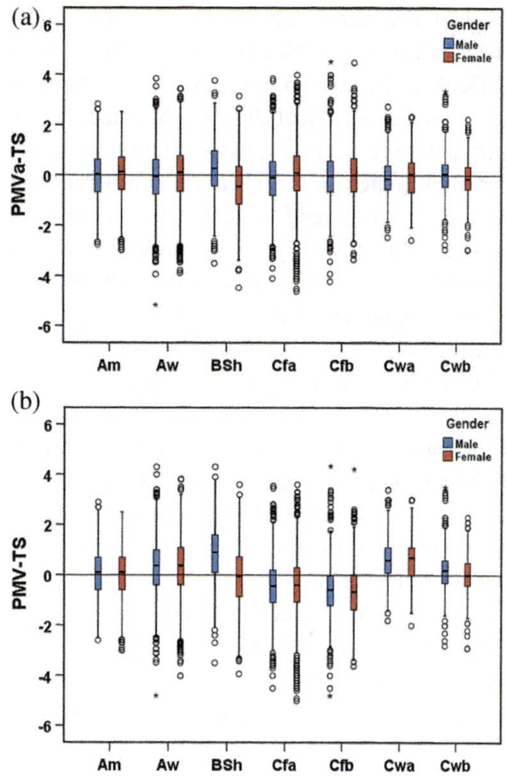

Fig. 11.13 Boxplots of adapted discrepancy **a** and original discrepancy **b** in different climates and gender

center (41 samples), summer senior center (181 samples), autumn office (233 samples) and autumn senior center (41 samples). As discussed in Sect. 11.3.1, in order to achieve a 95% confidence, the sample size should be larger than 383. Thus, the significant deviations may be caused by individual differences, while the limited number of observations cannot be scaled to a large population.

Boxplots in Fig. 11.13 compared the medians of adapted discrepancy with the original discrepancy for different climates and gender. The sample size of each attribute is larger than 383 except for female subjects from climate Cwb (345 samples). When compared with the original discrepancy, the medians of the adapted discrepancy are much closer to '0' for male and female participants from the 7 different climates.

It can be concluded from Figs. 11.12 and 11.13 that the universal application of the Fanger's model without any modification is deemed as inappropriate. Researchers proposed multiple adaptive thermal comfort models to improve the prediction accuracy. The models are criticised for their inherent complexity which makes them difficult to be applied again by others (McCartney and Fergus Nicol 2002). This study is the first one to attempt to quantify the effect of different attributes on the TS of a large population. The model proposed is an extension of the PMV model, which is easy to be utilized by other researchers. We believe this model will be useful to investigate the indoor comfort temperature, which facilitates building energy optimization. The authors also argue that the adaptation table should be incorporated into the current standards to account for the influence of season, climate, building type, age group and gender on TS. Conversely, standards should be adapted to different climates to account for the effect of the categorical variables.

The Fanger's model was developed without considering the influences of the categorical variables. Nevertheless, when it was used for predicting TS of the extracted 17,481 samples, the discrepancy is small, which is 0.06 scale unit larger than ASHRAE vote. Based on the results of this study, the authors provide a potential explanation; according to van Hoof (2008): "Fanger derived his comfort equation based on college-age students exposed to steady-state conditions in a climate chamber for a 3 h period in winter at sea level (1013 hPa) while wearing standardized clothing and performing standardized activities". According to Alfano et al. (2017), extensive experimental studies were carried out in Kansas State University (KSU), which formed the basis of Fanger's finding. Later, a substantial amount of data (including data from Danmarks Tekniske Universitet (DTU)) were integrated into the datasets. Figure 11.14 shows the calculation of PMVa for Fanger's datasets. The difference between PMVa and PMV for KSU is small, while the difference for DTU is slightly higher. Considering that Fanger's PMV was developed in well-controlled environments at steady-state without local discomfort (PMV = 0) (Van Hoof 2008), and the potential of data collected from other climate, it is reasonable that even when data from DTU were merged to the dataset, the discrepancy for Fanger's model is still small.

There are also limitations for the use of the adapted model. Due to a lot of missing information in the database, some demographic and contextual factors, such as educational background, ethnicity, body mass and social status, which may influence thermal sensation are not taken into consideration in the adapted model. With current advances in smart devices and Internet of Things (IoT) in the built environment, understanding the causes of individual difference towards perceived thermal comfort has gained increased popularity. However, this study investigates the occupants' TS as an aggregated model of a group of people, which does not differentiate individual differences. The prediction performance is poor when applied to

$$PMV_a = PMV - \beta_0 - \beta_1 Season - \beta_2 Climate - \beta_3 Buildingtype - \beta_4 Agegroup - \beta_5 Gender$$

	β_0	β_1	β_2	β_3	β_4	β_5
KSU	Constant 0.362	Winter -0.093	Cfa -0.756	Classroom 0.391	College-age -0.092	Male/female 0.087/0
DTU	Constant 0.362	Winter/summer -0.093/0	Cfb -1.032	Classroom 0.391	College-age -0.092	Male/female 0.087/0

KSU:

$$PMV_a = PMV - 0.362 - (-0.093) - (-0.756) - 0.391 - (-0.092) - 0.087/0$$

Male: PMV_a=PMV+0.188

Female: PMV_a=PMV+0.101

DTU:

$$PMV_a = PMV - 0.362 - (-0.093) - (-1.032) - 0.391 - (-0.092) - 0.087/0$$

Male in winter: PMV_a=PMV+0.377 **Male in summer:** PMV_a=PMV+0.284

Female in winter: PMV_a=PMV+0.464 **Female in summer:** PMV_a=PMV+0.371

Fig. 11.14 Calculation of PMVa for Fanger's experiments

individuals due to large variations among the occupants. Thus, the model cannot be applied to understand the specific comfort requirements of an individual occupant and characterize a set of conditions to meet personalised conditioning in a given space. The standards should be adapted to different climates.

11.7 Conclusions

The PMV model has been widely used to predict occupants' thermal comfort. The discrepancy between PMV and TS has been noted and discussed since the model was developed. Extensive studies have investigated the influence of different factors on TS through surveys or field experiments. Understanding the impacts not only contributes to our knowledge on how occupants interact with the built environment, it also provides guidance on how to operate and manage buildings to ensure comfort and health considerations are met, while optimizing energy usage. This study leverages on a global thermal comfort database to quantify the influences of season, climate, building type, age group and gender on the discrepancy. Results indicate that the impacts of climate and building type on the discrepancy are more noticeable than the other variables. An adapted model was proposed to reduce the discrepancy by the five variables and an adaptation table was generated. The maximum difference for climate and building type are 1.324 and 0.749, respectively. After adaptation, the median of each attribute is '0' or near '0', which indicates that, on average, the PMVa is free from serious bias. The prediction accuracy of the model used for individual TS was improved from 36 to 39%.

References

Alwetaishi MS (2017) Impact of building function on thermal comfort: a review paper. Am J Eng Appl Sci 9:928–945. https://doi.org/10.3844/ajeassp.2016.928.945

Al horr Y, Arif M, Katafygiotou M, Mazroei A, Kaushik A, Elsarrag E (2016) Impact of indoor environmental quality on occupant well-being and comfort: a review of the literature. Int J Sustain Built Environ 5:1–11. https://doi.org/10.1016/J.IJSBE.2016.03.006.

ASHRAE Standard 55 (2017) Thermal environmental conditions for human occupancy

Amai H, Tanabe S, Akimoto T, Genma T (2007) Thermal sensation and comfort with different task conditioning systems. Build Environ 423955–3964. https://doi.org/10.1016/J.BUILDENV.2006.07.043

d'Ambrosio Alfano FR, Olesen BW, Palella BI (2017) Povl ole fanger's impact ten years later. Energy Build 152:243–249. https://doi.org/10.1016/j.enbuild.2017.07.052

Beshir MY, Ramsey JD (1981) Comparison between male and female subjective estimates of thermal effects and sensations. Appl Ergon 12:29–33. https://doi.org/10.1016/0003-6870(81)90091-0

Cao B, Zhu Y, Ouyang Q, Zhou X, Huang L (2011) Field study of human thermal comfort and thermal adaptability during the summer and winter in Beijing. Energy Build 1051–1056. https://doi.org/10.1016/j.enbuild.2010.09.025

Cariou V, Qannari EM (2018) Statistical treatment of free sorting data by means of correspondence and cluster analyses. Food Qual Prefer 68:1–11. https://doi.org/10.1016/j.foodqual.2018.01.011

Cheng X, Yang B, Hedman A, Olofsson T, Li H, Van Gool L (2019) NIDL: A pilot study of contactless measurement of skin temperature for intelligent building. Energy Build 198:340–352. https://doi.org/10.1016/J.ENBUILD.2019.06.007.

Cheung T, Schiavon S, Parkinson T, Li P, Brager G (2019) Analysis of the accuracy on PMV – PPD model using the ASHRAE global thermal comfort database II. Build Environ 153:205–217. https://doi.org/10.1016/J.BUILDENV.2019.01.055

Daher E, Kubicki S, Guerriero A (209) Post-occupancy evaluation parameters in multi-objective optimization–based design process. Adv Inform Comput Civ Constr Eng. Springer International Publishing, Cham, pp. 463–470. https://doi.org/10.1007/978-3-030-00220-6_55

Dear RJD, Brager GS (1998) Developing an adaptive model of thermal comfort and preference, ASHRAE Trans 104

European Commission (2016) An EU strategy on heating and cooling

Fanger PO (1970) Thermal comfort: analysis and applications in environmental engineering. Copenhagen, Danish Technical Press. https://doi.org/10.1016/s0003-6870(72)80074-7

Fanger PO, Toftum J (2002) Extension of the PMV model to non-air-conditioned buildings in warm climate. Energy Build, Elsevier, pp. 533–536. https://doi.org/10.1016/S0378-7788(02)00003-8.

Fountain M, Brager G, De Dear R (1996) Expectations of indoor climate control. Energy Build 24:179–182. https://doi.org/10.1016/S0378-7788(96)00988-7.

Frontczak M, Wargocki P (2011) Literature survey on how different factors influence human comfort in indoor environments. Build Environ 46:922–937. https://doi.org/10.1016/j.buildenv.2010.10.021

Frontczak M, Schiavon S, Goins J, Arens E, Zhang H, Wargocki P(2012) Quantitative relationships between occupant satisfaction and satisfaction aspects of indoor environmental quality and building design. Indoor Air 22:119–131. https://doi.org/10.1111/j.1600-0668.2011.00745.x

Földváry Ličina V, Cheung T, Zhang H, de Dear R, Parkinson T, Arens E, Chun C, Schiavon S, Luo M, Brager G, Li P, Kaam S, Adebamowo MA, Andamon MM, Babich F, Bouden C, Bukovianska H, Candido C, Cao B, Carlucci S, Cheong DKW, Choi JH, Cook M, Cropper P, Deuble M, Heidari S, Indraganti M, Jin Q, Kim H, Kim J, Konis K, Singh MK, Kwok A, Lamberts R, Loveday D, Langevin J, Manu S, Moosmann C, Nicol F, Ooka R, Oseland NA, Pagliano L, Petráš D, Rawal R, Romero R, Rijal HB, Sekhar C, Schweiker M, Tartarini F, ichi Tanabe S, Tham KW, Teli D, Toftum J, Toledo L, Tsuzuki K, De Vecchi R, Wagner A, Wang Z, Wallbaum H, Webb L, Yang L, Zhu Y, Zhai Y, Zhang Y, Zhou X (2018) Development of the ASHRAE global thermal comfort database II. Build Environ 142:502–512. https://doi.org/10.1016/j.buildenv.2018.06.022

Ghahramani A, Castro G, Becerik-Gerberm B, Yu X (2016) Infrared thermography of human face for monitoring thermoregulation performance and estimating personal thermal comfort. Build Environ 109:1–11. https://doi.org/10.1016/j.buildenv.2016.09.005.

Guergova S, Dufour A (2011) Thermal sensitivity in the elderly: a review. Ageing Res Rev 10:80–92. https://doi.org/10.1016/j.arr.2010.04.009

Greenacre M (1984) Correspondence analysis

Halawa E, Van Hoof J (0212) The adaptive approach to thermal comfort: a critical overview. Energy Build 51:101–110. https://doi.org/10.1016/j.enbuild.2012.04.011

Hancock AM, Witonsky DB, Alkorta-Aranburu G, Beall CM, Gebremedhin A, Sukernik R, Utermann G, Pritchard JK, Coop G, Di Rienzo A (2011) Adaptations to climate-mediated selective pressures in humans. PLoS Genet 7:e1001375. https://doi.org/10.1371/journal.pgen.1001375

Höppe P, Martinac I (1998) Indoor climate and air quality. Review of current and future topics in the field of ISB study group 10. Int J Biometeorol 42:1–7. https://doi.org/10.1007/s004840050075

Humphreys MA, Nicol JF (2002) The validity of ISO-PMV for predicting comfort votes in every-day thermal environments. Elsevier. https://doi.org/10.1016/S0378-7788(02)00018-X.

Humphreys MA, Nicol JF (2002) Adaptive thermal comfort and sustainable thermal standards for buildings. Energy Build 34:563–572. https://doi.org/10.1016/S0378-7788(02)00006-3

ISO 7730 (2005) Ergonomics of the Thermal environment — analytical determination and interpretation of thermal comfort using calculation of the PMV and PPD indices and local thermal comfort criteria

Karmann C, Schiavon S, Arens E (2018) Percentage of commercial buildings showing at least 80% occupant satisfied with their thermal comfort. 10th WinD Con. Rethink Comf Cumberl Lodg

Karjalainen S (2007) Gender differences in thermal comfort and use of thermostats in everyday thermal environments. Build Environ 42:1594–1603. https://doi.org/10.1016/j.buildenv.2006.01.009.

Karjalainen S (2009) Thermal comfort and use of thermostats in Finnish homes and office. Build Environ 44:1237–1245. https://doi.org/10.1016/j.buildenv.2008.09.002.

Karjalainen S (2012) Thermal comfort and gender: a literature review. Indoor Air 22:96–109. https://doi.org/10.1111/j.1600-0668.2011.00747.x

Kingma B, van Marken Lichtenbelt W (2015) Energy consumption in buildings and female thermal demand. Nat Clim Chang 5:1054–1056. https://doi.org/10.1038/nclimate2741

Lee J-B, Kim T-W, Min Y-K, Yang H-M (2015) Seasonal acclimatization in summer versus winter to changes in the sweating response during passive heating in Korean young adult men. Korean J Physiol Pharmacol 19:9. https://doi.org/10.4196/kjpp.2015.19.1.9.

Liu H, Wu Y, Li B, Cheng Y, Yao R (2017) Seasonal variation of thermal sensations in residential buildings in the hot summer and cold winter zone of China. Energy Build 140:9–18. https://doi.org/10.1016/j.enbuild.2017.01.066

Li Y, Rezgui Y, Zhu H (2017) District heating and cooling optimization and enhancement – Towards integration of renewables, storage and smart grid. Renew Sustain Energy Rev 72:281–294. https://doi.org/10.1016/j.rser.2017.01.061

Li Y, Kubicki S, Guerriero A, Rezgui Y(2019) Review of building energy performance certification schemes towards future improvement, Renew. Sustain. Energy Rev 113:109244. https://doi.org/10.1016/J.RSER.2019.109244

Maykot JK, Rupp RF, Ghisi E (2018) A field study about gender and thermal comfort temperatures in office buildings. Energy Build 178:254–264. https://doi.org/10.1016/j.enbuild.2018.08.033

McCartney KJ, Fergus Nicol J (2002) Developing an adaptive control algorithm for Europe. Energy Build, Elsevier, 623–635. https://doi.org/10.1016/S0378-7788(02)00013-0

Nakamura Y, Okamura K (1998) Seasonal variation of sweating responses under identical heat stress. Appl Hum Sci J Physiol Anthropol 17:167–172. https://doi.org/10.2114/jpa.17.167

Natsume K, Ogawa T, Sugenoya J, Ohnishi N, Imai K (1992) Preferred ambient temperature for old and young men in summer and winter. Int J Biometeorol 36:1–4. https://doi.org/10.1007/BF01208726

Nicol JF, Humphreys MA (2002) Adaptive thermal comfort and sustainable thermal standards for buildings. Energy Build 34:563–572. https://doi.org/10.1016/S0378-7788(02)00006-3

Nicol F, Humphreys M (2007) Maximum temperatures in European office buildings to avoid heat discomfort. Sol Energy 81:295–304. https://doi.org/10.1016/j.solener.2006.07.007

Oseland NA (1995) Predicted and reported thermal sensation in climate chambers, offices and homes. Energy Build 23:105–115. https://doi.org/10.1016/0378-7788(95)00934-5

Pantavou K, Lykoudis S, Nikolopoulou M, Tsiros IX (2018) Thermal sensation and climate: a comparison of UTCI and PET thresholds in different climates. Int J Biometeorol 62:1695–1708. https://doi.org/10.1007/s00484-018-1569-4

Rijal HB, Yoshida H, Umemiya N (n.d.) Seasonal and regional differences in neutral temperatures in Nepalese traditional vernacular houses. https://doi.org/10.1016/j.buildenv.2010.06.002.

Schellen L, Van Marken Lichtenbelt WD, Loomans MGLC, Toftum J, De Wit MH (2010) Differences between young adults and elderly in thermal comfort, productivity, and thermal physiology in response to a moderate temperature drift and a steady-state condition. Indoor Air 20:273–283. https://doi.org/10.1111/j.1600-0668.2010.00657.x.

Soebarto V, Zhang H, Schiavon S (2019) A thermal comfort environmental chamber study of older and younger people. Build Environ 155:1–14. https://doi.org/10.1016/j.buildenv.2019.03.032

Sourial N, Wolfson C, Zhu B, Quail J, Fletcher J, Karunananthan S, Bandeen-Roche K, Béland F, Bergman H (2010) Correspondence analysis is a useful tool to uncover the relationships among categorical variables J Clin Epidemiol 63:638–646. https://doi.org/10.1016/j.jclinepi.2009.08.008

Teli D, Jentsch MF, James PAB (2012) Naturally ventilated classrooms: an assessment of existing comfort models for predicting the thermal sensation and preference of primary school children. Energy Build 53:166–182. https://doi.org/10.1016/j.enbuild.2012.06.022

Umemiya N (2006) Seasonal variations of physiological characteristics and thermal sensation under identical thermal conditions. J Physiol Anthropol 25:29–39. https://doi.org/10.2114/jpa2.25.29

Van Hoof J (2008) Forty years of fanger's model of thermal comfort: comfort for all? Indoor Air 18:182–201. https://doi.org/10.1111/j.1600-0668.2007.00516.x

Wang Z, de Dear R, Luo M, Lin B, He Y, Ghahramani A, Zhu V (2018) Individual difference in thermal comfort: a literature review. Build Environ 138:181–193. https://doi.org/10.1016/J.BUILDENV.2018.04.040

Wang L, Kim J, Xiong J, Yin H (2019) Optimal clothing insulation in naturally ventilated buildings. Build Environ 154:200–210. https://doi.org/10.1016/j.buildenv.2019.03.029

Wang X, Li D, Menassa CC, Kamat VR (2019) Investigating the effect of indoor thermal environment on occupants' mental workload and task performance using electroencephalogram. Build Environ 158:120–132. https://doi.org/10.1016/j.buildenv.2019.05.012

Wenzel HG, Mehnert C, Schwarzenau S (1989) Evaluation of tolerance limits for humans under heat stress and the problems involved. Scand J Work Environ Health 15(1):7–14

Yang B, Olofsson T (2017) A questionnaire survey on sleep environment conditioned by different cooling modes in multistorey residential buildings of Singapore. Indoor Built Environ 26:21–31. https://doi.org/10.1177/1420326X15604206

Yang B, Wang F (2018) Supplementary opinions on alternative cooling technologies in hot climate. Int J Biometeorol 62:1927–1928. https://doi.org/10.1007/s00484-018-1588-1

Yang B, Cheng X, Dai D, Olofsson T, Li H, Meier A (2019) Real-time and contactless measurements of thermal discomfort based on human poses for energy efficient control of buildings. Build Environ 162:106284. https://doi.org/10.1016/J.BUILDENV.2019.106284.

Yao Y, Lian Z, Liu W, Shen Q (2007) Experimental study on skin temperature and thermal comfort of the human body in a recumbent posture under uniform thermal environments. Indoor Built Environ 16:505–518. https://doi.org/10.1177/1420326X07084291

Yao R, Li B, Liu J (2009) A theoretical adaptive model of thermal comfort - adaptive predicted mean vote (aPMV). Build Environ 44:2089–2096. https://doi.org/10.1016/j.buildenv.2009.02.014

Zhang Y, Chen H, Wang J, Meng Q (2016) Thermal comfort of people in the hot and humid area of China—impacts of season, climate, and thermal history. Indoor Air 26:820–830. https://doi.org/10.1111/ina.12256

Zhang F, de Dear R (2019) Impacts of demographic, contextual and interaction effects on thermal sensation—Evidence from a global database. Build Environ 162:106286. https://doi.org/10.1016/j.buildenv.2019.106286

12 A Prediction Accuracy Weighted Voting Ensemble Method for Thermal Sensation Evaluation

Yu Li, Yacine Rezgui, Sylvain Kubicki, Annie Guerriero, and Xingxing Zhang

Abstract

PMV (Predicted Mean Vote) model is currently the most extensively used method for thermal sensation (TS) evaluation. However, the model is criticized for not being able to account for human thermal preferences and expectations. In response to these limitations, the adaptive model was developed to factor in behavioural, psychological and physiological adjustments, but it overlooks important factors such as clothing insulation, activity level and the indoor thermal environment. This chapter therefore proposes a prediction accuracy weighted voting ensemble (PAWVE) method for TS evaluation. Feature selection was used to identify the important features contributing most to TS. Data resampling was applied to improve the classification performance of the imbalanced data extracted from the ASHRAE global database. Five classifiers were eventually selected for the ensemble method based on their prediction accuracies. Results indicate that resampling is important for improving the performance of the minority classes. PAWVE with data resampling outperformed the other models, with an overall accuracy and F1-score both at 0.67. When compared with the traditional PMV model, the performance has been improved by 72%.

Keywords

Predicted mean vote · Thermal sensation · Thermal comfort · Prediction performance · Ensemble models

12.1 Introduction

Building design involves ensuring that the indoor environment meets occupants' comfort aspirations (Höppe and Martinac 1998; Li et al. 2019a, 2020). Occupants are becoming active agents in buildings, and as such, continuously interact with a wide range of appliances, including heating or cooling control systems, to optimize their indoor

Y. Li (✉)
School of Environmental Science and Engineering, Donghua University, Shanghai, China
e-mail: liyu@dhu.edu.cn

Y. Li · S. Kubicki · A. Guerriero
Luxembourg Institute of Science and Technology LIST, 5, avenue des Hauts-Fourneaux, L-4362 Esch-sur-Alzette, Luxembourg
e-mail: sylvain.kubicki@list.lu

A. Guerriero
e-mail: annie.guerriero@list.lu

Y. Rezgui
BRE Trust Centre for Sustainable Engineering, Cardiff University, Cardiff C24 3AA, UK
e-mail: RezguiY@cardiff.ac.uk

X. Zhang
Department of Energy and Community Buildings, Dalarna University, 79188 Falun, Sweden
e-mail: xza@du.se

X. Zhang (ed.), *Data-driven Analytics for Sustainable Buildings and Cities*, Sustainable Development Goals Series, https://doi.org/10.1007/978-981-16-2778-1_12

climates to meet their thermal comfort requirements (Ahmad et al. 2016). Given that these interactions influence building energy consumption, a large number of studies have delivered methods and algorithms to optimize local HVAC (Heating, Ventilation and Air Conditioning) appliances to strengthen occupants' thermal comfort (Daum et al. 2011; Kim et al. 2018a; Xin et al. 2019). Results revealed that accurate evaluation of thermal comfort is indispensable for building energy system management as poor prediction often leads to overcooling, overheating and energy wastage (Sekhar 2016). For instance, Attia and Carlucci (Attia and Carlucci 2015) stated that choosing different thermal comfort evaluation criteria for zero energy buildings would result in energy consumption variation from 16 to 24.7%. Therefore, understanding occupants' thermal comfort is critical to improve their wellbeing as well as to optimize building energy consumption.

Extensive field studies and chamber experiments have been conducted worldwide to investigate the mechanism of thermal comfort. Many thermal comfort models have associated Thermal Sensation (TS) with their thermal environment. An overwhelming majority of models to date assumed TS as a direct indicator for thermal evaluation using categories such as "Cold (−3)", "Cool (−2)", "Slightly cool (−1)", "Neutral (0)", "Slightly warm (1)", "Warm (2)" and "Hot (3)". The thermal comfort responses are prevailingly collected through questionnaires by asking occupants to assess their TS. Two major approaches are widely used to evaluate TS: (1) the Fanger's PMV (Predicted Mean Vote) model (Fanger 1970) and (2) the adaptive model (Nicol and Humphreys 2002). The PMV is a typical heat balance model, which predicts indoor TS as a function of four environment predictors (air temperature, radiant temperature, air velocity and relative humidity) and two personal predictors (clothing insulation and metabolic rate). It was developed based on chamber experiments involving over one thousand European and North American subjects exposed to well-controlled laboratory climates (Fanger 1970). This approach seeks to capture the average TS of a large population premised on steady-state heat balance of human body. The PMV model formed the theoretical basis of thermal comfort standards such as ASHRAE 55 (ASHRAE Standard 2017) and ISO 7730 (ISO 7730 2005). The heat balance model was challenged by the adaptive model, which claims that occupants actively interact with their thermal environment by modifying their thermal preferences and expectations through behavioural, physiological and psychological adjustments (Dear and Brager 1998a; Jing et al. 2018; Kim et al. 2015). It was also argued that the rigorous restrictions in the climate chamber are quite different from the actual situations in real buildings (Humphreys and Nicol 2002). Thus, the adaptive model was derived from field surveys. The adaptive hypothesis is that one's thermal satisfaction is achieved through adjusting one's thermal expectation according to the actual thermal environment (Brager and Dear 1998). The surveys revealed that the comfort indoor temperature is greatly influenced by the outdoor temperature. Therefore, it proposed a linear relationship between the indoor thermal comfort temperature and the outdoor monthly air temperature. The adaptive model is currently used as an alternative for naturally ventilated buildings.

The universal applicability of the adaptive model and the PMV model has been debated for a long time (Li et al. 2019b). A number of field studies showed that the adaptive model cannot be used to accurately predict TS as it ignores the effect of factors such as clothing insulation, activity level and the indoor thermal environment that have a huge impact on human thermal comfort (Jing et al. 2019; López-Pérez et al. 2019; Fanger and Toftum 2002). Although these variables are taken into account by the PMV model, the latter ignores demographic and contextual factors such as age, gender, climate, building type and ventilation mode, which are also believed to contribute to occupants' thermal preferences and expectations (Li et al. 2019b; Zhang and Dear 2019). It has been documented that people exposed to different environments develop different limits of tolerance to their thermal environment, and thus PMV model turns out to be a poor predictor for TS (Li et al. 2019b; Wenzel et al. 1989). Cheung et al. (2019) analysed the accuracy of using PMV for TS prediction focusing on 56,771 samples extracted from ASHRAE Global Thermal Comfort Database II (Földváry Ličina et al. 2018). Results showed that PMV is not reliable and it can explain only 34% of the TS votes. Li et al. (2019b) developed an adaptation table, based on a subset extracted from the same database, to enhance the performance of using PMV for average TS prediction by taking into account the effect of climate, season, building type, age group and gender. After adaptation, the PMV model was free from serious bias in predicting average TS of a large population. More recently, Zhang and Lin (Zhang and Lin 2020) proposed an extended PMV model to account for thermal adaptation by analysing the same database. The extended model was obtained by multiplying a factor to the original PMV model and the accuracy was improved by 65%.

TS is a subjective response, which is influenced by a multitude of intricate factors. Previous research has shown that current tools are insufficient for individual's TS prediction. Compared with the regression based approaches (PMV and adaptive model) that use a function to approximate the relationship between TS and environmental, physiological and psychologic factors, attempts have been initiated to develop more sophisticated approaches that are more realistic and powerful in describing thermal comfort. In recent years, a number of studies have attempted to develop thermal comfort models by employing various machine learning algorithms such as neural networks (Katić et al. 2018), Support Vector Machine (SVM) (Kim et al. 2018a), Random Forest (RF) (Kim et al. 2018a), Logistic Regression (LR) (Daum et al. 2011), Bayesian network (Ghahramani et al. 2015) and KNN (K-Nearest Neighbours) (Lu et al. 2019) for models development. Compared with conventional comfort models, these data-driven models have mitigated the difficulty in capturing hidden relationships while securing significant prediction accuracy improvement (17–40%). Kim et al. (Kim et al. 2018b) presented a review of the machine learning based thermal comfort models and proposed a unified modelling framework in predicting individuals' thermal comfort responses. The general process involves data collection and preparation, model selection and evaluation, and continuous learning. The data sources can be derived from existing databases or field collected data. Lu et al. (Lu et al. 2019) established TS models with KNN, SVM and RF (random forest) by using ASHRAE RP884 and KNN achieved the best performance with a recall of 49.3% (Dear and Brager 1998). However, the classifiers failed in predicting the minority classes and a majority of the samples were classified as neutral due to the imbalanced datasets. Grabe (Grabe 2016) developed an ANN (Artificial Neural Network) to predict TS votes based on ASHRAE RP884 and claimed that the prediction results outperformed the classical PMV model, with an accuracy of 53%. Wang et al. (2019a) developed two RF classification models to investigate TS of older people. The field and lab study models both exhibited higher accuracy than the PMV model, producing an overall accuracy of 56.6% and 76.7% respectively. Wu et al. (Wu et al. 2018) compared an ensemble machine learning (Bagging) method against ANN and SVM for TS votes prediction grounded on field data collected from naturally ventilated and air conditioned buildings. Results indicated that the ensemble machine learning method performed better than ANN and SVM models.

The above findings have evidenced the role of machine learning in predicting occupants' TS. Machine learning is a significant departure from the traditional models such as PMV and the adaptive model that are mainly derived from statistical regression, focusing on personal TS. Though prior studies provide valuable insights into machine learning in improving predictive performance of individual's thermal comfort, there is still a paucity of research to investigate TS prediction of an individual. A key challenge is to choose the optimal learning model given the extensive number of algorithms in the field of machine learning (Wang et al. 2020). This forms the gap addressed by the present chapter, which provides an in-depth study to re-establish the relationship between a widespread range of influential parameters and TS through machine learning, based on samples derived from ASHRAE Global Thermal Comfort Database II. Following this introduction, Sect. 12.2 provides an overview of the methodology applied in this study and proposes a prediction accuracy weighted voting ensemble (PAWVE) method for TS prediction. The PAWVE is compared with the soft voting and hard voting methods. The results are demonstrated in Sect. 12.3, followed with a discussion. The last section presents concluding remarks and directions for future research.

12.2 Methodology

The primary concept of the methodology is to identify the important features contributing to TS and then formulate a PAWVE learning method for TS classification. The data source as well as the feature selection methods, resampling method, machine learning technique and the methodology workflow are elaborated below. Python programming language is used in this study for data analysis.

12.2.1 Data Source and Samples

For a data-driven based study, a sufficient quality and quantity of datasets are required for training and validation. In this chapter, such datasets are provided by the ASHRAE Global Thermal Comfort Database II, which integrates and harmonizes the abundant data from worldwide thermal comfort related studies (Ličina et al. 2018). This combined datasets consist of 68 attributes, covering subjective comfort votes, demographic information of subjects, local climate/weather conditions, building characteristics and objective instrumental measurements (Földváry Ličina et al. 2018). Identifying a number of important features is critical before the formulation of a learning model. Clearly, the number of potential features is subject to the information provided in the database. The predictors were selected according to the most extensively used thermal comfort models and literature publications on potential influential parameters for TS. Firstly, air temperature (Air_T), relative humidity (RH), air velocity (Air_Vel), clothing (Clo) and metabolic rate (Met) were identified as important factors as they are the key inputs for Fanger's PMV model. Although radiant temperature is also used as an indicator in the PMV model, it is not commonly recorded in the database. The mean outdoor monthly temperature (Out_T) was selected as it is a prevailing parameter for the adaptive model. The samples with features of climate (Köppen climate classification), building type (BT), season, sex, ventilation and age were also extracted to ascertain possible links between TS and environmental, psychological and demographic parameters (Li et al. 2019b). Around 97% of the samples' TS adopted a seven-point scale (−3, −2, −1, 0, 1, 2, 3 corresponding to cold, cool, slightly cool, neutral, slightly warm, warm and hot). Therefore, only the observations with integer votes were included to form a classification problem. Eventually, 10,288 samples were extracted from the database. The statistic information of the numerical variables and the information with regard to the categorical variables are shown in Tables 12.1 and 12.2. The numerical data are then standardized into a consistent range between −1 and 1 for machine learning.

12.2.2 Feature Selection

Given that there are generally more features than required to perform an analysis, it is often advantageous to reduce the number of features before applying machine learning, focusing on the variables of great interest. Feature selection is a technique utilized to determine the attributes that contribute most to the prediction result. This process can not only effectively identify the important features that are critical to achieve good classification, it also removes irrelevant features that may potentially negatively impact model's performance. Pearson correlation and

Table 12.1 Statistic information of the numerical variables

	Age	Clo	Met	Air_T	RH	Air_Vel	Out_T	PMV
Mean	36.67	0.68	1.22	25.65	55.25	0.26	24.64	0.44
Std	15.18	0.32	0.19	3.81	13.81	0.40	7.09	1.03
Min	16.00	0.23	0.70	13.40	14.50	0.00	5.30	−3.00
50%	35.00	0.62	1.20	25.20	56.90	0.11	25.00	0.00
Max	95.00	2.87	3.50	45.30	88.80	4.71	38.10	3.00

Table 12.2 Information of the extracted categorical variables

Categories	Subcategories	Description	Sample size
Climate	Am	Tropical monsoon climate	2075
	Aw	Tropical wet and dry climate	2008
	Bsh	Hot semi-arid climate	1372
	Cfa	Humid subtropical climates	1589
	Cfb	Oceanic climate	1396
	Cwa	Dry-winter humid Subtropical	1344
	Cwb	Dry winter oceanic climate	504
BT	Office	/	7766
	Classroom	/	2075
	Senior center	/	447
Season	Spring	/	44
	Summer	/	6402
	Autumn	/	41
	Winter	/	3801
Sex	Female	/	4532
	Male	/	5756
Ventilation	NV	Natural ventilation	4455
	MM	Mix mode	4019
	AC	Air conditioned	1814
TS	−3	Cold	37
	−2	Cool	264
	−1	Slightly cool	1419
	0	Neutral	5703
	1	Slightly warm	1825
	2	Warm	710
	3	Hot	330

Extra-Trees classifier were both employed in this study to analyse the strength of relationships. The Pearson correlation coefficient $(\rho_{X,Y})$ measures the linear correlation between two variable X and Y, which is given by Benesty et al. (2009):

$$\rho_{(X,Y)} = \frac{cov(X,Y)}{\sigma_X \sigma_Y} \tag{12.1}$$

where cov is the covariance. σ_X and σ_Y denote the standard deviation of X and Y. The value of the correlation coefficient $\rho_{X,Y}$ varies from −1 to 1. A value close to −1 or 1 implies a strong negative or positive correlation while a value of 0 means there is no association between the two variables. If the value is greater than 0, it indicates a positive association, and vice versa. Features that are highly correlated with each other have almost the same effect on the dependent variable. Thus, the redundant attributes should be eliminated. Though categorical variables are popular in our daily life, they cannot be used directly in regression analysis to establish a statistical relationship. One-hot encoding was applied to convert each categorical feature into a binary feature, which then allows the use of correlation analysis.

Since Pearson correlation can only identify feature importance according to linear relationship, Extra-Trees classifier is also used in this study to select the features of importance with non-linear relationship (Geurts et al. 2006). Extra-Trees classifier is an ensemble that aggregates the results of multiple decision trees collected in a 'forest' to output a classification result. Each decision tree is constructed from the original samples. At each test node, the tree is provided with a random of k features subset from which each decision tree selects the best feature to split the data according to some mathematical criteria (e.g. Gini index). To perform feature selection using the forest structure, the normalized total reduction in the mathematic criteria is computed and the values represent the features' importance.

12.2.3 Resampling Imbalanced Data

Imbalanced data refers to unequally distributed sampling values in a dataset. As observed from Table 12.2, the sample sizes for different classes of TS are severely imbalanced. For example, '0' accounts for 55.4% of the sampling size, while '−3' merely takes up 0.4% of the sample size. Classification algorithms normally generate poor performance for imbalanced data as they favour the majority class, leading to a highly inaccurate prediction of the minority classes (Feng et al. 2018; Ramentol et al. 2012). For example, a classifier could achieve 55.4% accuracy for the extracted samples by simply classifying all samples to the class of '0'. Such algorithm would be misleading and could not make any sense in the real world. To improve the performance of the classifiers, resampling is implemented to deal with the imbalanced datasets. There are three resampling strategies for imbalanced data: over-sampling, under-sampling and hybrid sampling (Ramentol et al. 2012). Over-sampling randomly synthesizes the minorities until their sample sizes are equal to the majority. This method reserves all the original samples, but it could potentially result in serious overfitting. By contrast, under-sampling selectively eliminates some samples from the majority classes while keeping the original population of the minority. Since a number of observations are removed from the majorities, some useful information might be discarded. The hybrid sampling is a method to rebalance the class sizes by combining the previous two approaches. Many researchers have successfully overcome the difficulties of the imbalanced problems in machine learning by using the hybrid resampling method (Seiffert et al. 2010; Zhang and Hu 2014). This method will be adopted in this study to solve the class imbalanced problem.

12.2.4 Machine Learning Classifiers

Previous studies have provided robust evidences for the use of machine learning in thermal comfort prediction. Classification algorithms are more often employed when the variables to be predicted are discrete rather than continuous. There are many different types of machine learning classifiers, with different degrees of complexity and ability to learn. In this study, classifiers include LR (Onan et al. 2016), Linear Discriminant Analysis (LDA) (Balakrishnama and Ganapathiraju 1998a), KNN classifier (Pereira et al. 2009), Classification and Regression Tree (CART) (Dreiseitl and Ohno-Machado

2002), Naive Bayes (NB) (Chen et al. 2020) and Support Vector Classifier (SVC) (Gunn 1998a) were selected as the base models to train the extracted samples as they are widely used in solving classification problems.

LR (Onan et al. 2016) is a simple and effective supervised classification algorithm. The model builds a generalized linear regression to predict the probability of an event as a sigmoid function of a number of predictors. The LDA (Balakrishnama and Ganapathiraju 1998b) is a commonly used dimension reduction technique for classification and machine learning. It is implemented by projecting the data points onto a line to reduce their dimensionality. Then, the projected points are classified according to their distances to a given point. LDA attempts to express the dependent variable as a linear relationship of the independent features. Thus, it is better to be used when linear relationship exists. The KNN (Pereira et al. 2009) is one of the simplest classifiers as it does not involve any explicit learning functions. It is realized by computing the distance from the test sample to the nearest training samples. Afterwards, the test sample is assigned with the label of the training samples according to majority voting or mean of the k-nearest neighbours. CART (Dreiseitl and Ohno-Machado 2002) is an important decision tree algorithm, which is an effective non-parametric machine learning technique for regression and classification. This algorithm repeatedly splits the datasets into smaller subsets according to different sorting criteria, resulting in a tree-like structure. It then finds the solutions based on sequential and hierarchical decisions. The NB classifier is a statistical classification model built upon Bayes' theory. It determines the probability of an event given that some events have occurred. A probability score is obtained by multiplying the conditional probability of each class. SVC (Gunn 1998b) is a classification method developed based on the SVM. The original datasets are transformed into a higher dimension by using a nonlinear matching method. The datasets are then partitioned into different classes using a separating hyperplane.

12.2.5 Cross Validation

In order to evaluate the performance of the selected classifiers, cross validation was applied to test the effectiveness of the selected models. K-fold cross validation (Fushiki 2011) is a simple and easy to understand technique to generate less biased models. In the K-fold cross validation, the datasets are partitioned into k subsets (folds). The algorithm uses one fold for testing while the rest k-1 folds for training until every fold has been tested. This method ensures that every original dataset has the opportunity to appear in the training and testing set. Thus, it can be used to evaluate the accuracy of predicting unseen data. The processes are elaborated as follows:

(1) The datasets are randomly split into K folds (normally k is 5 or 10), as shown in Fig. 12.1.

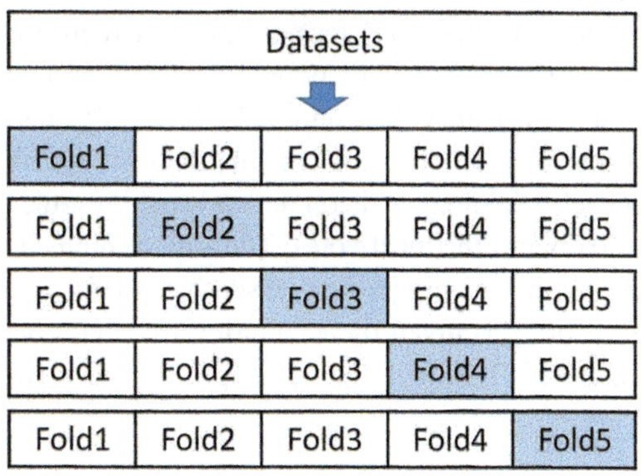

Fig. 12.1 K-fold cross validation

(2) The model is trained using k-1 folds and the fold left out is used for testing.
(3) The process is repeated for K times until every fold has been used as training and testing dataset.
(4) The average accuracy of all the tests is reported.

12.2.6 Prediction Accuracy Weighted Voting Ensemble Method

The performance of a single classifier is affected by the merits and shortcomings of its underpinning algorithm. Ensemble methods attempt to combine multiple classifiers to overcome the limitations of an individual classifier (Saqlain et al. 2019). By coupling different classifiers, the bias and variance can be mitigated. In addition, an ensemble normally outperforms any single learning method. Voting ensemble is a simple, effective and widely used method to form an ensemble from several machine learning algorithms. This idea follows the human behaviour of voting that tends to seek individual's preferences and options before making a collective decision. Therefore, the voting ensemble is in fact not a single classifier but a wrapper for several different learning methods trained and tested in parallel. There are two typical different voting strategies: hard voting and soft voting. The hard voting is also the well-known majority voting. Namely, the final decision depends on the subject who obtains the most votes. The soft voting is the weighted averaging voting ensemble. An average probability is given by accounting for each vote. Table 12.3 shows an example of the two strategies. In hard voting, V1 obtained two votes from classifier l_1 and l_2. V2 got one vote from l_3. Therefore, the probability for V1, V2 and V3 is 2/3, 1/3 and 0 respectively. Eventually, the result is classified as V1. In soft voting, the probability of l_1, l_2 and l_3 voting for V1, V2 and V3 is taken into account. As a result, the average probability for V1, V2 and V3 is 38.3%, 43.3% and 20% respectively. The result is classified as V2. It can be concluded that different voting strategies yield different results. Therefore, choosing the appropriate ensemble is important to ensure the overall performance.

The hard voting only considers the final voting result from each model, regardless of the probability of the candidates (V_i) being voted by the learning models. The soft voting remedies the disadvantage of the hard voting, but the weights of the learning models are considered the same. The PAWVE method was proposed focusing on developing an ensemble grounded on the performance of its constituent models. The proposed ensemble method is based on the soft voting strategy while the prediction accuracies of the learning methods are used as the weight, which means the algorithms with higher accuracies have bigger weights. This ensures that the models with higher accuracies contribute more to the final vote. Since the machine learning models are independent models, the probability of voting for $V_i(P(l, V_i))$ under the PAWVE strategy is given by:

$$P(l, V_i) = \frac{P(l_1, V_i) \cdot P_{l_1} + P(l_2, V_i) \cdot P_{l_2} + \ldots + P(l_m, V_i) \cdot P_{l_m}}{P_{l_1} + P_{l_2} + \ldots + P_{l_m}} \tag{12.2}$$

where m is the selected number of independent machine learning used for ensemble. $P(l_m, V_i)$ is

Table 12.3 Example of voter's preferences for hard voting and soft voting

	Hard voting			Soft voting		
	V_1	V_2	V_3	V_1 (%)	V_2 (%)	V_3 (%)
l_1	✓			50	30	20
l_2	✓			45	40	15
l_3		✓		20	60	20

the probability of learning model l_m voting for candidate V_i. P_{l_m} is the prediction accuracy of learning model l_m.

12.2.7 Results Evaluation

Evaluation criterion is a key indicator for the assessment of the classification performance. In a two-class problem, a confusion matrix is often used to record the correctly and incorrectly recognized samples, as displayed in Fig. 12.2. Traditionally, classification accuracy has been the most widely used metric to evaluate the overall ratio of correct prediction. The accuracy can be obtained from Eq. (12.3).

$$accuracy = \frac{TP + TN}{N} \times 100\% \tag{12.3}$$

where N is the total number of samples, given by

$$N = TP + TN + FP + FN \tag{12.4}$$

However, empirical evidences show that it is inappropriate to use overall prediction accuracy as the sole measure for imbalanced data (Chawla et al. 2002). Thus, precision, recall and f1 score are introduced. Precision is a measure of exactness while recall is a measure of completeness (He and Garcia 2009). More specifically, precision measures the fraction of positive observations among those predicted positive observation; recall measures the ratio of correctly classified positive observations of the total number of positive observations. Clearly, since classification intends to achieve good performance for all classes, none of the above two measures alone is sufficient for good evaluation. To make it comparable, F1-score, which is more useful than accuracy in evaluating imbalanced classification, is used to assess the weighted average of recall and precision (Fahad et al. 2015).

$$Precision = \frac{TP}{TP + FP} \tag{12.5}$$

$$Recall = \frac{TP}{TP + FN} \tag{12.6}$$

$$F1 - score = \frac{2 \times Recall \times Precision}{Recall + Precision} \tag{12.7}$$

	Predicted positive	Predicted negative
Actual positive	TP	FN
Actual negative	FP	TN

Fig. 12.2 Confusion matrix of a two-class problem (TP: the number of positive observation correctly predicted. FP: the number of positive observation wrongly predicted. FN: the number of negative observation wrongly predicted. TN: the number of negative observation correctly predicted.)

12.2.8 Overview of the Methodology

The workflow of the methodology is illustrated in Fig. 12.3. Firstly, the data samples were extracted from the database. Subsequently,

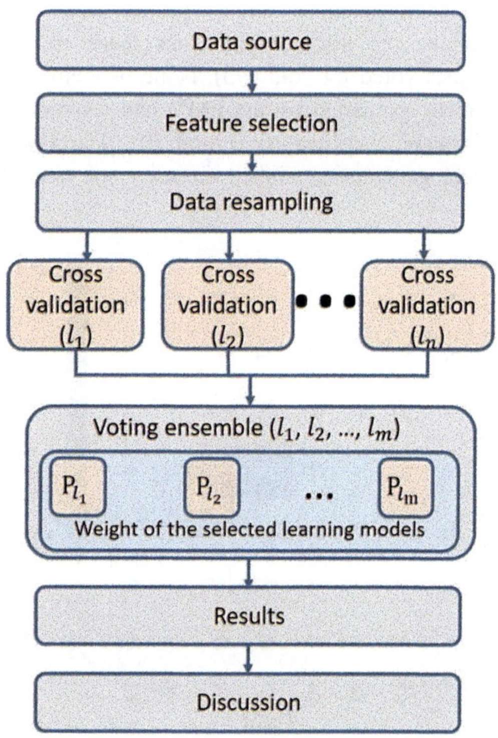

Fig. 12.3 A schematic view of the methodology

feature selection technique was utilized to identify the features of importance, and a data resampling technique was used to process the imbalanced datasets. Then, the resampled data were trained and tested by a number of classifiers. The classifiers with good performance were selected for the PAWVE prediction.

12.3 Results

12.3.1 PMV Prediction Accuracy

In order to examine the accuracy of using the PMV model for mean TS prediction, the PMV results are benchmarked with the true votes. The PMV values were binned into 7 categories: 3 (PMV > 2.5); 2 (1.5 < PMV $\leq$ 2.5); 1 (0.5 < PMV $\leq$ 1.5); 0 (−0.5 $\leq$ PMV $\leq$ 0.5); −1 (−1.5 $\leq$ PMV < −0.5); −2 (−2.5 $\leq$ PMV < -1.5); −3 (PMV > −2.5). The confusion matrix in Fig. 12.4 describes the performance of using PMV for TS classification. Each row in the confusion matrix represents the number of samples for a specific vote. For example, the first row indicates there are 37 samples voted for cold (−3). None of them was correctly predicted by the PMV and a majority (25) were classified as neutral (0). The PMV model generates a higher accuracy for neutral condition, 2852 out of 5703 were correctly predicted, which accounts for 50% of the samples voted for neutral. The accuracy of the model deteriorates in cold and warm environment. It also shows a good performance under the hot condition, with 193 out of 330 samples correctly predicted. The overall prediction accuracy of the PMV for TS is 0.39. Since the PMV was developed to predict the mean vote of a large population, it is reasonable that individual's TS deviates considerably from the PMV.

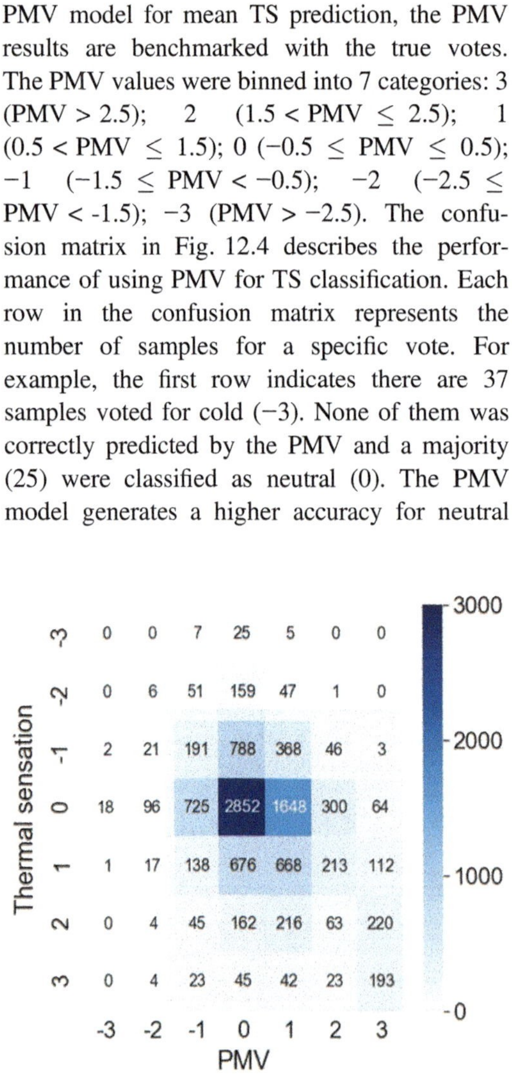

Fig. 12.4 Confusion matrix of PMV for TS prediction

12.3.2 Feature Selection

Feature selection is an effective approach to remove the irrelevant and redundant features among the inputs. Pearson correlation and Extra-Trees classifier were both used to test the dependency of input parameters on the TS votes. Figures 12.5b and 12.6b depict the order of Pearson correlation importance in which individual inputs contribute to TS. The correlation coefficient indicates if the strong dependency exists. The positive values represent positive relationships while the negative values represent negative relationships. The order of importance is determined by their absolute values. The greater the absolute value, the more important the variable is. It can be observed that indoor air temperature (Air_T) is the most important feature contributing to TS. Summer (Season_Summer), winter (Season_Winter) and clothes level (Clo) show a stronger correlation with TS, with a correlation coefficient of 0.234, −0.234 and −0.226 respectively. The features with absolute values of higher than 0.1 are considered as important variables for TS in this study. Thus, Air_Vel (0.158), Out_T (0.148) and Ventilation_AC (−0.119) were also selected though their correlation with TS is slightly lower. Closer examination of the magnitudes of the correlation index shows how individual parameters affect each other. The correlation maps in Fig. 12.5a and Fig. 12.6a present the correlation of all extracted features. The strong red and strong blue colour represent strong correlation, which should be avoided to be used as inputs simultaneously.

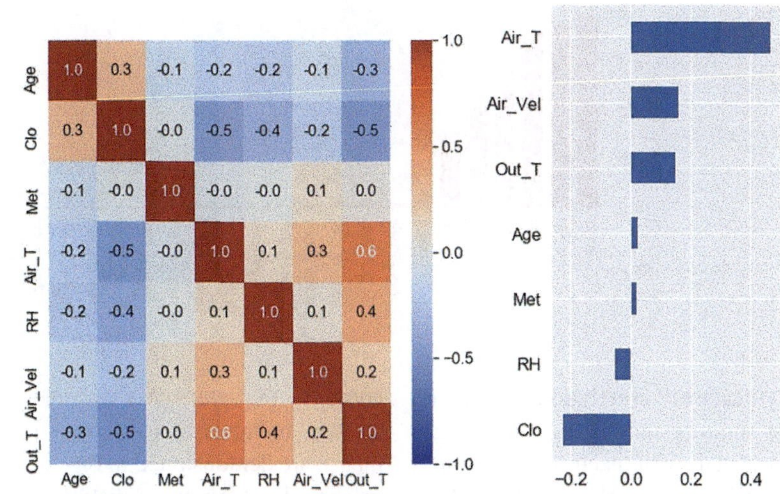

(a) Correlation matrix between continuous variables (b) Feature importance to TS

Fig. 12.5 Pearson correlation coefficients between continuous variables and TS

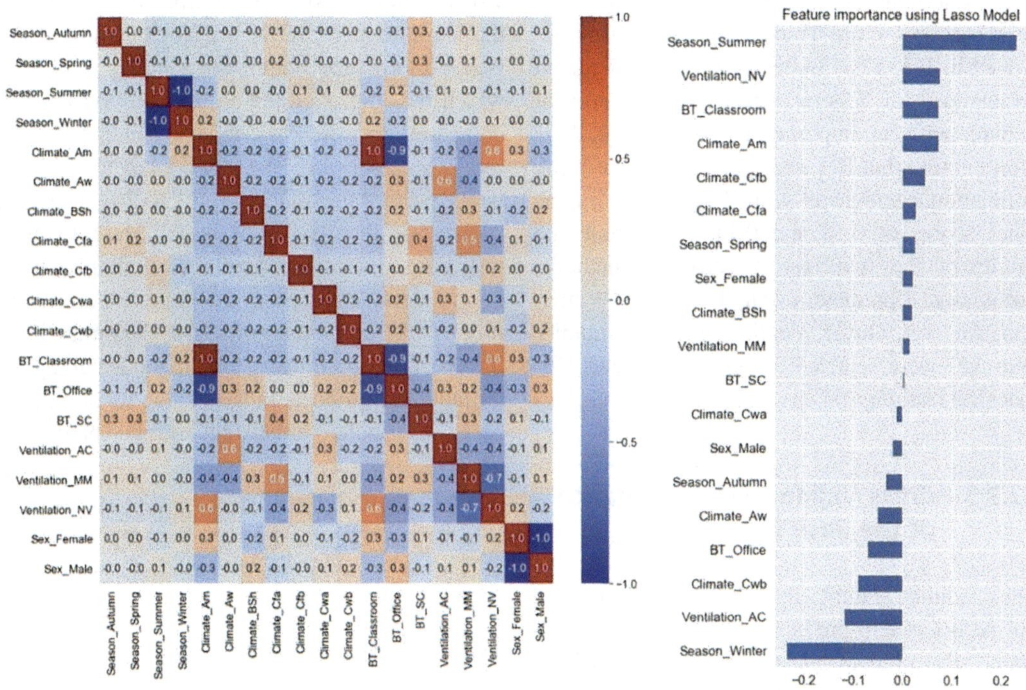

(a) Correlation matrix between categorical features (b) Feature importance to TS

Fig. 12.6 Pearson correlation coefficients between categorical variables and TS

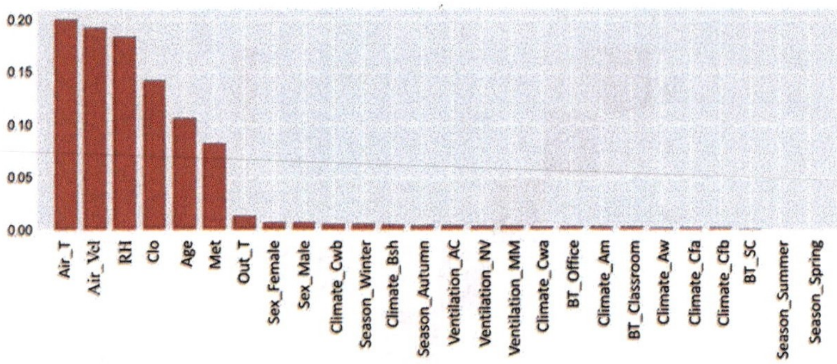

Fig. 12.7 Feature importance by Extra-Trees classifier

By comparing features' correlation, it can be concluded that summer and winter are strongly negatively correlated with each other, with a correlation coefficient of −1. Hence, one of them should be screened out from the selected features.

As Pearson correlation only filters the features with linear relationship, the Extra-Trees classifier serves as a supplement to identify features of non-linear relationship. The feature importance computed by Extra-Trees classifier is demonstrated in Fig. 12.7. In accordance with Pearson correlation, Air_T is recognized as the dominant feature, with an importance score of 0.2. It is worth noting that RH and age, which have weak correlation coefficients with TS, exhibit importance scores of 0.18 and 0.11. Combining the results of the two feature selection methods, eventually, 8 predictors (Air_T, Season_Winter, Clo, Air_Vel, Out_T Ventilation_AC, humidity and age) were selected as input parameters for machine learning.

12.3.3 Cross Validation Performance of the Base Models

The candidate feature subsets are generated based on the selected 8 predictors. 10-fold cross validation was used to test the performance of the selected six base classifiers (LR, LDA, KNN, CART, NB, SVM). This means 90% of the data were used for training and the rest 10% were used for validation. Such process was repeated 10 times until all data have been used nine times for training and once for testing. All classifiers were evaluated with the same resampled datasets. To compare the prediction accuracy of the base models, boxplots of the cross validation performance are demonstrated in Fig. 12.8. Each model generated a different prediction accuracy according to its learning abilities. There is a large variation among prediction accuracies of the individual models. Some models produce over 60% in prediction accuracy while some model is worse than the PMV model. Specifically, among the six classifiers, KNN and CART showed the best performance both with a median of 0.64, which has improved the accuracy by 64% when compared with the PMV model. This is followed by SVM (0.46), LR (0.41) and LDA (0.39). The worst performing model is NB (0.30), which is 23% worse than the PMV model. The difference may be resulted from the different complexity of the models or because the samples are more predictable by certain machine learning algorithms. The variability of prediction accuracy for all individual models in the cross validation process is small, indicating stable prediction

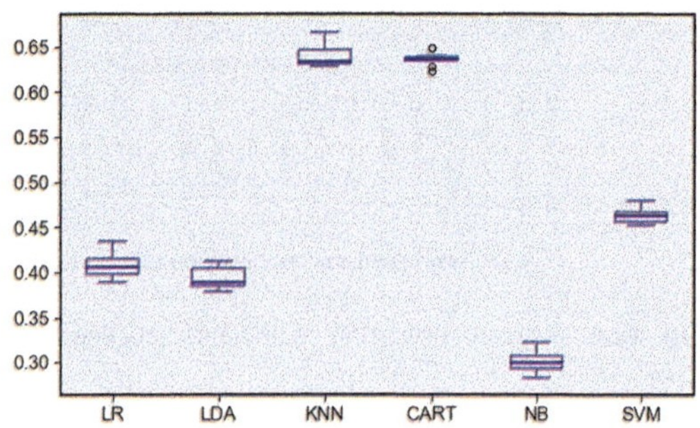

Fig. 12.8 Boxplot of cross validation performance associated with the base models

performance of the base models. Since the performance of an ensemble model depends on the accuracy of its constituent classifiers (Saqlain et al. 2019), only the models that exhibited better performance than the PMV model were selected for the ensemble. Specifically, KNN, CART, SVM, LR and LDA were employed in the ensemble.

12.3.4 Classification Performance Analysis

For the proposed PAWVE model, the prediction accuracy of each base model is assigned as the weight of the classifier. The model collects the results of the selected models and then combines them into an aggregated output. To indicate the merit of the proposed PAWVE model, the recognition performance of the models is compared with soft voting, hard voting and the PMV model. The overall classification performance of these models is presented in Table 12.4. The results show that all the ensembles outperform the PMV model, which proves that machine learning can improve individual's TS prediction performance. The proposed PAWVE with resampling performed best, as reflected in the elevated values of precision, recall, F1-score as well as accuracy. More specifically, the higher accuracy shows that more samples were correctly recognized by the PAWVE model and the higher F1-score indicates that the PAWVE with resampling produced a more balanced classification than the other models. The accuracy and F1-score for PAWVE with resampling are both 0.67, while the corresponding values for soft voting are both 0.65. This proves that weighting the best classifiers with higher weight strengthens the overall performance of the ensemble. When compared with the PMV model, the proposed PAWVE model increases the accuracy and F1-score both by 72%.

The performance of the PAWVE and soft voting with resampling is better than the corresponding models with imbalanced datasets. This reveals that resampling improves the performance and effectiveness of the imbalanced data in classification. However, an exception exists for hard voting model: the F1-score with resampling is higher than the imbalanced datasets while the accuracy of the imbalanced datasets is higher than the datasets with resampling. This lower accuracy does not mean that resampling deteriorates the performance of the hard voting model. As indicated in Sect. 12.2.3, when comparing the performance of the imbalanced datasets, F1-score is a more effective metric in indicating classification performance. A more detailed results showing the F1-scores of each class from different models are demonstrated in Table 12.5. The F1-scores of the ensembles with resampling are more evenly distributed for each class, especially for the PAWVE method. This indicates that the PAWVE with resampling performs the best at predicting all classes of the TS. For the models using the imbalanced dataset, there is a diverse difference in the F1-scores. The F1-score for 'neutral' of the ensembles using imbalanced datasets is as high as 0.75, while under the 'cold' condition, the value is 0. This means these models classify most of the imbalanced datasets to 'neutral' in order to ensure accuracy.

Table 12.4 Overall classification performance of different models

		Precision	Recall	F1-score	Accuracy
Resampling datasets	PAWVE	0.67	0.67	0.67	0.67
	Soft voting	0.66	0.65	0.65	0.65
	Hard voting	0.55	0.51	0.49	0.51
Imbalanced datasets	PAWVE	0.55	0.60	0.51	0.60
	Soft voting	0.54	0.60	0.51	0.60
	Hard voting	0.54	0.59	0.47	0.59
PMV	PMV	0.42	0.39	0.39	0.39

Table 12.5 Classification performance of each class for the different models

		F1-score (−3)	F1-score (−2)	F1-score (−1)	F1-score (0)	F1-score (1)	F1-score (2)	F1-score (3)
Resampling datasets	PAWVE	0.76	0.64	0.62	0.68	0.67	0.71	0.66
	Soft voting	0.67	0.60	0.60	0.67	0.65	0.68	0.62
	Hard voting	0.48	0.28	0.43	0.60	0.43	0.51	0.47
Imbalanced datasets	PAWVE	0	0	0.08	0.75	0.25	0.43	0.38
	Soft voting	0	0	0.09	0.75	0.27	0.42	0.39
	Hard voting	0	0.01	0.03	0.74	0.14	0.34	0.33
PMV	PMV	0	0.03	0.15	0.55	0.28	0.09	0.42

Figure 12.9 shows the normalized confusion matrices of the different models. Each row represents the proportion of predicted TS for a specific vote. From the figure, it is easy to conclude that the prediction models misclassified a majority of the imbalanced datasets to 'neutral'. This again explains the reason why the F1-scores are higher at 'neutral' condition. It should be noted that the higher prediction accuracies of the imbalanced datasets are mainly attributed to the 'neutral' samples, as reflected in Fig. 12.9d, e and f. Weighting the models with higher accuracy improves the performance of the minority classes.

12.4 Discussion

Both PMV model and the adaptive model have an inherent limitation of poor accuracy in predicting personal thermal comfort. Given that these models were developed to estimate the average TS of a large population, it is reasonable that they fail to accurately predict individual's TS. Namely, the models cannot be applied to correctly represent the thermal perception of an individual. This also reflects real situations where occupants sharing the same space claim various degrees of thermal perception. Even for the same occupants, their thermal perception differs under different contexts. For example, people have a wider tolerance of indoor thermal environmental variation if they are capable of controlling their environment (de Dear and Brager 1998b). The results are intrinsically different due to a range of reasons such as personal differences in interpretation of TS and failure in evaluating the results objectively (Wang et al. 2019b). Moreover, subjective evaluations depend on occupants' voluntary participation in the surveys in order to obtain their feedbacks. Hence, an alternative approach to inform individual's thermal comfort is needed for understanding personal thermal comfort.

Recently, the physiological measurement (skin temperature and heart rates) based techniques arouse great interest in thermal comfort evaluation (Song et al. 2016; Yao et al. 2007). A growing number of wearable devices (such as thermocouples and heart rate monitors) and thermal cameras are used to measure skin temperature and heart rates given their high accuracy, low cost and easy to manipulate. They are normally installed in certain parts of the body or nearby people, making them possible for monitoring body parameters in daily life. The collected information is then used to establish personal thermal comfort models. These models leverage machine learning and Internet of Things to understand individual's thermal comfort requirements. The strategy is based on long-term data collection to adapt the personal comfort model. Such models can provide improved prediction accuracy in capturing individual's thermal comfort, and provide a more realistic estimation of building energy consumption for thermal comfort based HVAC control (Kim et al. 2018b). In practice, there is a scale limitation in

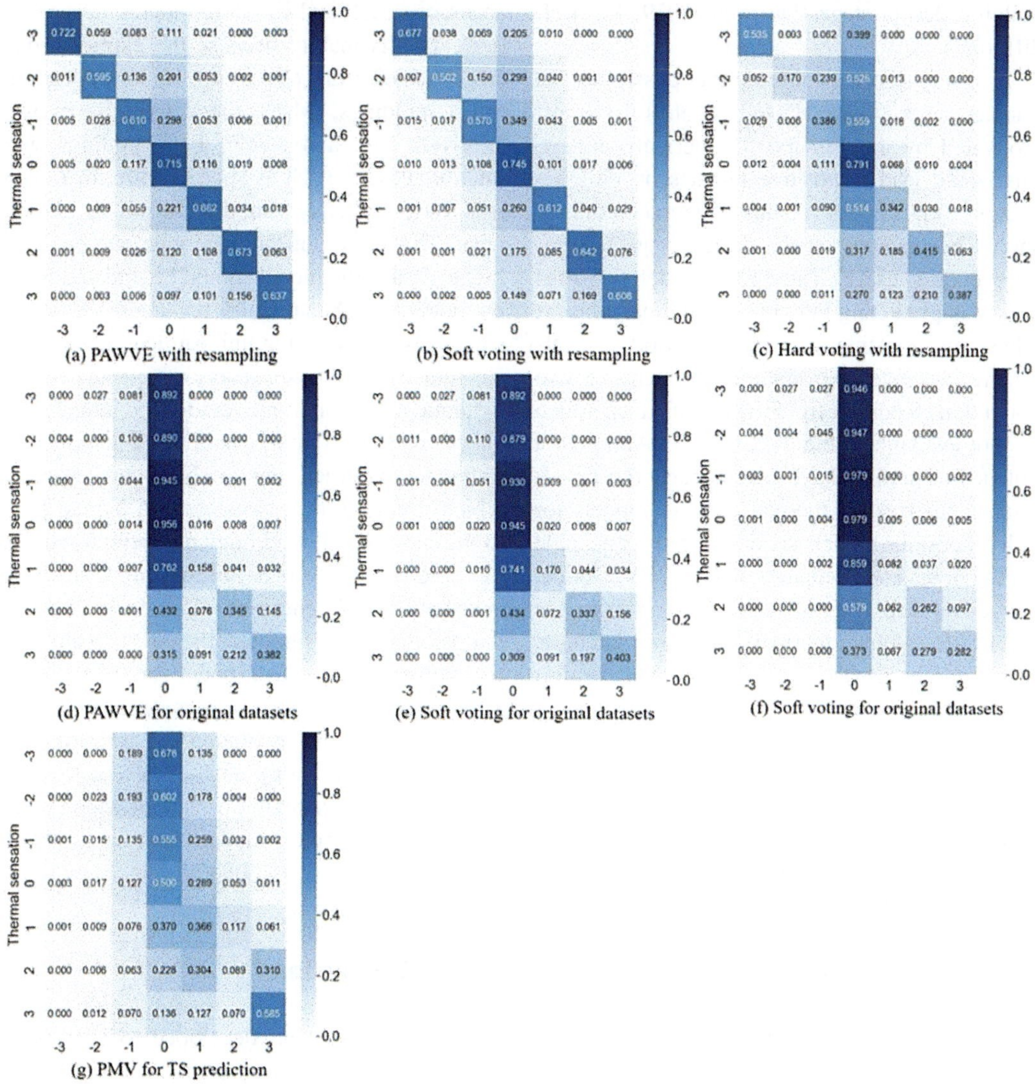

Fig. 12.9 Normalized confusion matrix of the different models

tracking sufficient comfort feedback to train a personal comfort model as repeated surveys may potentially lead to fatigue and eventually decay in participation (Rana et al. 2013). Meanwhile, due to occupants' diverse preferences and expectation, the personal thermal comfort models cannot be generalized for people who do not participant in the experiments. In addition, continuously tracking of individual's behaviour may be viewed by some participants as an invasion of privacy. It is also worth noting that for the developed personal comfort model, it may not remain the same due to temporal and spatial changes, such as seasonal variations and building type. A new learning model should be trained to update the personalized comfort model. Such kind of personal thermal comfort model requires continuous collection of individual feedback and environment sensor data, which would result in a high investment. In addition, some of the personal thermal comfort models call for more personal information (e.g. body mass index, skin temperature, body surface area and heart rate) and sensor data, and thus cannot be used at the building design stage (Du et al. 2019; Li et al. 2019c).

If occupants are offered the possibility to interact with their HVAC control systems, motivated by the demand for thermal comfort, this would lead to positive impacts on building energy consumption. Thus, thermal comfort should be predicted at the design stage to optimize energy usage. The PMV model performs poorly in predicting individual's thermal comfort while the existing personal comfort models require an extensive number of sensors and continuous feedback from the participants, which are not available at the design stage. Although a number of studies have investigated the development of personal thermal comfort models for individuals, the models are developed for a single person or several occupants. The PAWVE model was developed with a subset derived from the ASHRAE global thermal comfort database, which can be used to represent thermal comfort of the general public. The model proposed in this study may provide valuable information of thermal comfort for a broad range of population. These can serve as the foundation for thermal comfort profiles, which can be utilized to control the HVAC when a more personalized thermal comfort model is not available or is still under development. The model can also be applied to inform the design, management and retrofit of centralized heating and cooling systems in buildings, such as hospitals and schools to better characterize specific thermal comforts and to provide optimal occupant thermal comfort satisfaction.

The chapter argues that 8 predictors (Air_T, Season_Winter, Clo, Air_Vel, Out_T Ventilation_AC, humidity and age) are important variables influencing TS based on the results. The features, such as age, outdoor temperature and ventilation mode, which are not included in the PMV model and adaptive model, turn out to be important features contributing to TS. On the contrary, metabolic rate, which is considered as an important input for the PMV model is identified as a weak association variable to TS. A primary evaluation of metabolic rate by taking it as an input in the PAWVE shows that the accuracy, recall and F1-score remain the same at 0.67. This is mainly because the extracted data are collected from classroom, senior centre and office buildings, where the activity level is a relatively low and stable. For buildings with intense activity, such as sports centre, metabolic rate varies greatly with the degree of activity level. The impact of the metabolic rate on TS is bigger. Thus, the model should be updated based on field data. Meanwhile, the PAWVE model calculates TS assuming uniform temperature and air velocity within the space, which ignores the dynamic nature of air movement and temperature variation. For example, there is a temperature difference between human body and the air, which results in heat transfer from human body to the surrounding environment. Previous studies have shown that local discomfort caused by distribution differences in air velocity and air temperature may significantly affect occupants' thermal perception. The model should be modified in future work to consider such conditions. For example, the evolution of sensing and information technology enables access to more updated environmental information, which could be used by complex models for more reliable prediction.

12.5 Conclusions

The PMV model was developed to predict the average TS of a large population, but still it is widely used to evaluate the TS of individuals. It reduces thermal comfort to a steady-state heat balance equation. Results of early field studies argued that the heat balance model is inadequate to predict indoor thermal comfort accurately as it ignores individual's thermal expectation and adaptation. Actually, the occupants are actively interacting with their environment to achieve thermal comfort. Our findings indicate that reliance on the PMV model for TS prediction should be avoided and the prediction accuracy of the model can only achieve an accuracy of 0.39.

This study formulated a data-driven method for personal TS prediction under the explicit hypothesis that individuals' thermal perception is different under different thermal conditions. The proposed method attempts to associate TS with environmental variables and personal variables targeting at improving the prediction accuracy. Feature selection was conducted to identify the key variables contributing to TS votes. Eventually, eight variables (Air_T, Season_Winter, Clo, Air_Vel, Out_T, Ventilation_AC, RH and Age) were selected as input parameters for machine learning. The ensemble models were developed based on five constituent base models (LR, LDA, KNN, CART and SVC) and the prediction accuracies were assigned as the weights of the base models. Results showed that the proposed PAWVE model with resampling outperformed the other models. In particular, the overall accuracy and F1-score are both at 0.67. When compared with the PMV model, the accuracy and F1-score are both improved by 72%. In addition, the ensemble models with resampling are excellent in predicting minority classes and the F1-scores of all classes are more evenly distributed.

The proposed PAWVE model for TS evaluation is a personal thermal comfort model that can be used for the general public. It significantly improves the prediction performance by embracing the effect of microclimatic, physiological and psychological factors. This model can be used to facilitate building design, control and renovation of the heating and cooling system to meet both thermal comfort and energy saving. Future work involves modifying the PAWVE model to account for distribution differences in air velocity and air temperature in indoor environments. This will be enabled by the reliance on an enhanced indoor environmental sensing infrastructure.

References

Ahmad MW, Mourshed M, Mundow D, Sisinni M, Rezgui Y (2016) Building energy metering and environmental monitoring—a state-of-the-art review and directions for future research. Energy Build 120:85–102. https://doi.org/10.1016/j.enbuild.2016.03.059

Attia S, Carlucci S (2015) Impact of different thermal comfort models on zero energy residential buildings in hot climate. Energy Build 102:117–128. https://doi.org/10.1016/j.enbuild.2015.05.017

ASHRAE Standard 55 (2017) Thermal environmental conditions for human occupancy. https://www.techstreet.com/ashrae/standards/ashrae-55-2017?product_id=1994974&ashrae_auth_token. Accessed 27 July 2019

Balakrishnama S, Ganapathiraju A (1998) Linear discriminant analysis—a brief tutorial. http://www.music.mcgill.ca/~ich/classes/mumt611_07/classifiers/lda_theory.pdf. Accessed 12 Feb 2020

Balakrishnama S, Ganapathiraju A (1998) Linear discriminant analysis—a brief tutorial

Benesty J, Chen J, Huang Y, Cohen I (2009) Pearson correlation coefficient. In: Noise reduction in speech processing. Springer, Berlin Heidelberg, pp 1–4. https://doi.org/10.1007/978-3-642-00296-0_5

Brager GS, de Dear RJ (1998) Thermal adaptation in the built environment: a literature review. Energy Build 27:83–96. https://doi.org/10.1016/S0378-7788(97)00053-4

Chawla NV, Bowyer KW, Hall LO, Kegelmeyer WP (2002) SMOTE: synthetic minority over-sampling technique. J Artif Intell Res 16:321–357. https://doi.org/10.1613/jair.953

Chen S, Webb GI, Liu L, Ma X (2020) A novel selective naïve Bayes algorithm. Knowledge-Based Syst 192:105361. https://doi.org/10.1016/j.knosys.2019.105361

Cheung T, Schiavon S, Parkinson T, Li P, Brager G (2019) Analysis of the accuracy on PMV—PPD model using the ASHRAE global thermal comfort database II. Build Environ 153:205–217. https://doi.org/10.1016/J.BUILDENV.2019.01.055

Daum D, Haldi F, Morel N (2011) A personalized measure of thermal comfort for building controls. Build Environ 46:3–11. https://doi.org/10.1016/j.buildenv.2010.06.011

de Dear RJ, Brager GS (1998) Developing an adaptive model of thermal comfort and preference. ASHRAE Trans 145–167. https://escholarship.org/uc/item/4qq2p9c6. Accessed 11 Feb 2020

de Dear RJ, Brager GS (1998a) Developing an adaptive model of thermal comfort and preference. ASHRAE Trans 104. https://escholarship.org/content/qt4qq2p9c6/qt4qq2p9c6.pdf

de Dear RJ, Brager GS (1998b) Developing an adaptive model of thermal comfort and preference. ASHRAE Trans 104

Dreiseitl S, Ohno-Machado L (2002) Logistic regression and artificial neural network classification models: a methodology review. J Biomed Inform 35:352–359. https://doi.org/10.1016/S1532-0464(03)00034-0

Du C, Li B, Liu H, Ji Y, Yao R, Yu W (2019) Quantification of personal thermal comfort with localized airflow system based on sensitivity analysis and classification tree model. Energy Build 194:1–11. https://doi.org/10.1016/j.enbuild.2019.04.010

Fahad LG, Tahir SF, Rajarajan M (2015) Feature selection and data balancing for activity recognition

in smart homes. In: IEEE international conference communication, institute of electrical and electronics engineers Inc., pp 512–517. https://doi.org/10.1109/ICC.2015.7248373
Fanger PO (1970) Thermal comfort: analysis and applications in environmental engineering. Danish Technical Press, Copenhagen. https://doi.org/10.1016/s0003-6870(72)80074-7.
Fanger PO, Toftum J (2002) Extension of the PMV model to non-air-conditioned buildings in warm climates. Energy Build (Elsevier) 533–536. https://doi.org/10.1016/S0378-7788(02)00003-8
Feng W, Huang W, Ren J (2018) Class imbalance ensemble learning based on the margin theory. Appl Sci 8:815. https://doi.org/10.3390/app8050815
Fushiki T (2011) Estimation of prediction error by using K-fold cross-validation. Stat Comput 21:137–146. https://doi.org/10.1007/s11222-009-9153-8
Geurts P, Ernst D, Wehenkel L (2006) Extremely randomized trees. Mach Learn 63:3–42. https://doi.org/10.1007/s10994-006-6226-1
Ghahramani A, Tang C, Becerik-Gerber B (2015) An online learning approach for quantifying personalized thermal comfort via adaptive stochastic modeling. Build Environ 92:86–96. https://doi.org/10.1016/j.buildenv.2015.04.017
Gunn SR (1998a) Support vector machines for classification and regression. http://ce.sharif.ir/courses/85-86/2/ce725/resources/root/LECTURES/SVM.pdf Accessed 4 Feb 2020
Gunn SR (1998b) Support vector machines for classification and regression
He H, Garcia EA (2009) Learning from imbalanced data. IEEE Trans Knowl Data Eng 21:1263–1284. https://doi.org/10.1109/TKDE.2008.239
Höppe P, Martinac I (1998) Indoor climate and air quality. Review of current and future topics in the field of ISB study group 10. Int J Biometeorol 42:1–7. https://doi.org/10.1007/s004840050075
Humphreys MA, Nicol JF (2002) Adaptive thermal comfort and sustainable thermal standards for buildings. Energy Build 34:563–572. https://doi.org/10.1016/S0378-7788(02)00006-3
ISO 7730, 2005 Ergonomics of the Thermal Environment —analytical determination and interpretation of thermal comfort using calculation of the PMV and PPD indices and local thermal comfort criteria (2005)
Jing S, Li B, Yao R (2018) Exploring the "black box" of thermal adaptation using information entropy. Build Environ 146:166–176. https://doi.org/10.1016/j.buildenv.2018.09.038
Jing S, Lei Y, Wang H, Song C, Yan X (2019) Thermal comfort and energy-saving potential in university classrooms during the heating season. Energy Build 202:109390. https://doi.org/10.1016/j.enbuild.2019.109390
Katić K, Li R, Verhaart J, Zeiler W (2018) Neural network based predictive control of personalized heating systems. Energy Build 174:199–213. https://doi.org/10.1016/j.enbuild.2018.06.033
Kim JT, Lim JH, Cho SH, Yun GY (2015) Development of the adaptive PMV model for improving prediction performances. Energy Build 98:100–105. https://doi.org/10.1016/j.enbuild.2014.08.051
Kim J, Zhou Y, Schiavon S, Raftery P, Brager G (2018a) Personal comfort models: Predicting individuals' thermal preference using occupant heating and cooling behaviour and machine learning. Build Environ 129:96–106. https://doi.org/10.1016/j.buildenv.2017.12.011
Kim J, Schiavon S, Brager G (2018b) Personal comfort models—a new paradigm in thermal comfort for occupant-centric environmental control. Build Environ 132:114–124. https://doi.org/10.1016/j.buildenv.2018.01.023
Li Y, Kubicki S, Guerriero A, Rezgui Y (2019a) Review of building energy performance certification schemes towards future improvement. Renew Sustain Energy Rev 113:109244. https://doi.org/10.1016/J.RSER.2019.109244
Li Y, Rezgui Y, Guerriero A, Zhang X, Han M, Kubicki S, Yan D (2019b) Development of an adaptation table to enhance the accuracy of the predicted mean vote model. Build Environ 168:106504. https://doi.org/10.1016/j.buildenv.2019.106504
Li D, Menassa CC, Kamat VR (2019c) Robust non-intrusive interpretation of occupant thermal comfort in built environments with low-cost networked thermal cameras. Appl Energy 251:113336. https://doi.org/10.1016/j.apenergy.2019.113336
Li Y, Rezgui Y, Kubicki S (2020) An intelligent semantic system for real-time demand response management of a thermal grid. Sustain Cities Soc 52:101857. https://doi.org/10.1016/J.SCS.2019.101857
Ličina VF, Cheung T, Zhang H, de Dear R, Parkinson T, Arens E, Chun C, Schiavon S, Luo M, Brager G, Li P, Kaam S, Adebamowo MA, Andamon MM, Babich F, Bouden C, Bukovianska H, Candido C, Cao B, Carlucci S, Cheong DKW, Choi JH, Cook M, Cropper P, Deuble M, Heidari S, Indraganti M, Jin Q, Kim H, Kim J, Konis K, Singh MK, Kwok A, Lamberts R, Loveday D, Langevin J, Manu S, Moosmann C, Nicol F, Ooka R, Oseland NA, Pagliano L, Petráš D, Rawal R, Romero R, Rijal HB, Sekhar C, Schweiker M, Tartarini F, Tanabe S, Tham KW, Teli D, Toftum J, Toledo L, Tsuzuki K, De Vecchi R, Wagner A, Wang Z, Wallbaum H, Webb L, Yang L, Zhu Y, Zhai Y, Zhang Y, Zhou X (2018) Development of the ASHRAE global thermal comfort database II. Build Environ 142:502–512. https://doi.org/10.1016/j.buildenv.2018.06.022
López-Pérez LA, Flores-Prieto JJ, Ríos-Rojas C (2019) Adaptive thermal comfort model for educational buildings in a hot-humid climate. Build Environ 150:181–194. https://doi.org/10.1016/J.BUILDENV.2018.12.011
Lu S, Wang W, Lin C, Hameen EC (2019) Data-driven simulation of a thermal comfort-based temperature set-point control with ASHRAE RP884. Build Environ

156:137–146. https://doi.org/10.1016/j.buildenv.2019.03.010

Nicol JF, Humphreys MA (2002) Adaptive thermal comfort and sustainable thermal standards for buildings. Energy Build 34:563–572. https://doi.org/10.1016/S0378-7788(02)00006-3

Onan A, Korukoğlu S, Bulut H (2016) A multiobjective weighted voting ensemble classifier based on differential evolution algorithm for text sentiment classification. Expert Syst Appl 62:1–16. https://doi.org/10.1016/j.eswa.2016.06.005

Pereira F, Mitchell T, Botvinick M (2009) Machine learning classifiers and fMRI: a tutorial overview. Neuroimage 45. https://doi.org/10.1016/j.neuroimage.2008.11.007

Ramentol E, Caballero Y, Bello R, Herrera F (2012) SMOTE-RSB*: a hybrid preprocessing approach based on oversampling and undersampling for high imbalanced data-sets using SMOTE and rough sets theory. Knowl Inf Syst 33:245–265. https://doi.org/10.1007/s10115-011-0465-6

Rana R, Kusy B, Jurdak R, Wall J, Hu W (2013) Feasibility analysis of using humidex as an indoor thermal comfort predictor. Energy Build 64:17–25. https://doi.org/10.1016/j.enbuild.2013.04.019

Saqlain M, Jargalsaikhan B, Lee JY (2019) A voting ensemble classifier for wafer map defect patterns identification in semiconductor manufacturing. IEEE Trans Semicond Manuf 32:171–182. https://doi.org/10.1109/TSM.2019.2904306

Seiffer C, Khoshgoftaar TM, Van Hulse J, Napolitano A (2010) RUSBoost: a hybrid approach to alleviating class imbalance. IEEE Trans Syst Man Cybern Part A Syst Hum 40:185–197. https://doi.org/10.1109/TSMCA.2009.2029559

Sekhar SC (2016) Thermal comfort in air-conditioned buildings in hot and humid climates—why are we not getting it right? Indoor Air 26:138–152. https://doi.org/10.1111/ina.12184

Song WF, Zhang CJ, Lai DD, Wang FM, Kuklane K (2016) Use of a novel smart heating sleeping bag to improve wearers' local thermal comfort in the feet. Sci Rep 6:19326. https://doi.org/10.1038/srep19326

von Grabe J (2016) Potential of artificial neural networks to predict thermal sensation votes. Appl Energy 161:412–424. https://doi.org/10.1016/j.apenergy.2015.10.061

Wang Z, Yu H, Luo M, Wang Z, Zhang H, Jiao Y (2019) Predicting older people's thermal sensation in building environment through a machine learning approach: modelling, interpretation, and application. Build Environ 161. https://doi.org/10.1016/j.buildenv.2019.106231

Wang Z, Parkinson T, Li P, Lin B, Hong T (2019b) The squeaky wheel: machine learning for anomaly detection in subjective thermal comfort votes. Build Environ 151:219–227. https://doi.org/10.1016/j.buildenv.2019.01.050

Wang Z, Wang J, He Y, Liu Y, Lin B, Hong T (2020) Dimension analysis of subjective thermal comfort metrics based on ASHRAE Global Thermal Comfort Database using machine learning. J Build Eng 29. https://doi.org/10.1016/j.jobe.2019.101120

Wenzel HG, Mehnert C, Schwarzenau P (1989) Evaluation of tolerance limits for humans under heat stress and the problems involved. Scand J Work Environ Health 15(Suppl 1):7–14. http://www.ncbi.nlm.nih.gov/pubmed/2609123. Accessed 5 August 2019

Wu Z, Li N, Peng J, Cui H, Liu P, Li H, Li X (2018) Using an ensemble machine learning methodology-bagging to predict occupants' thermal comfort in buildings. Energy Build 173:117–127. https://doi.org/10.1016/j.enbuild.2018.05.031

Xin Y, Cui W, Zeng J (2019) Experimental study on thermal comfort in a confined sleeping environment heating with capillary radiation panel. Energy Build 205. https://doi.org/10.1016/j.enbuild.2019.109540

Yao Y, Lian Z, Liu W, Shen Q (2007) Experimental study on skin temperature and thermal comfort of the human body in a recumbent posture under uniform thermal environments. Indoor Built Environ. 16:505–518. https://doi.org/10.1177/1420326X07084291

Zhang F, de Dear R (2019) Impacts of demographic, contextual and interaction effects on thermal sensation—evidence from a global database. Build Environ 162:106286. https://doi.org/10.1016/j.buildenv.2019.106286

Zhang X, Hu BG (2014) A new strategy of cost-free learning in the class imbalance problem. IEEE Trans Knowl Data Eng 26:2872–2885. https://doi.org/10.1109/TKDE.2014.2312336

Zhang S, Lin Z (2020) Extending predicted mean vote using adaptive approach. Build Environ 171. https://doi.org/10.1016/j.buildenv.2020.106665

13 Analysis and Interpretation of the Particulate Matter Concentrations at the Subway Stations by General Linear Model (GLM) and Correlation Analysis

Xinru Wang, Song Pan, Xingxing Zhang, Li Chang, and Yiqiao Liu

Abstract

The particulate matters (PM10 and PM2.5) inside urban subway stations greatly influence indoor air quality and passenger comfort. This study aims to analyze and interpret the concentrations of PM10 and PM2.5, measured in several subway stations from October 9th to 22nd, 2016 in Beijing, China. The overall methodology was based on the Statistical Package for Social Science (SPSS) software while General linear model (GLM) and correlation analysis were further applied to examine the sensitivities of different variables to the particle concentrations. The data analysis showed the average overall mass ratio of PM concentrations inside subway station is about 68.7%, much lower than outdoor condition (79.6%). In the areas of the station hall and platform, the real-time PM10 and PM2.5 concentrations varied periodically. In working and operation offices, all rooms had much higher PM concentrations than the outdoor environment when its pollution level was level 3, in which the facility room reached the highest level, while the closed meeting room had the lowest. Correlation analysis results indicated that PM10 and PM2.5 concentrations were mutually correlated (average $R^2 = 0.854$), and a strong linear correlation ($R^2 = 0.897$) of the subway-station PM concentrations to the outdoor PM conditions, regardless of the outdoor atmospheric PM concentrations pollution level was. Nevertheless, the impact of passenger number and temperature and humidity on the station PM concentrations was less, when compared to the outdoor environment. This chapter is expected to provide useful information for further research and design of effective prevention measures on PM in local subway stations, towards a more sustainable and healthier built environment in the city underground.

X. Wang (✉)
College of Mechanical Engineering, Tianjin University of Commerce, Tianjin, China
e-mail: xinru5263@126.com

S. Pan · Y. Liu
Beijing Key Laboratory of Green Built Environment and Energy Efficient Technology, Beijing University of Technology, Beijing, China
e-mail: pansong@bjut.edu.cn

Y. Liu
e-mail: L13102158310@163.com

X. Zhang
Department of Energy and Community Buildings, Dalarna University, 79188 Falun, Sweden
e-mail: xza@du.se

L. Chang
Department of Mechanical Engineering, University of Malaya, Kuala Lumpur, Malaysia
e-mail: Changli4321@163.com

X. Zhang (ed.), *Data-driven Analytics for Sustainable Buildings and Cities*,
Sustainable Development Goals Series, https://doi.org/10.1007/978-981-16-2778-1_13

Keywords

PM10 · PM2.5 · Influencing factors · Correlation analysis · Subway station

13.1 Monitoring Subway Station and Measuring Sites

This chapter tested 4 subway stations totally in Beijing, as illustrated in Fig. 13.1, which included 3 most frequent subway lines, i.e. line 8 (green), line 10 (red), and line 14 (yellow). The measured subway stations were respectively Anhuaqiao station in line 8, Jinsong and Panjiayuan stations in line 10, and Beigongda-Ximen station in line 14. The reasons for selecting these stations are stated as below:

In Beijing, there are 2 different isolating door systems to partition the train tunnel and the platform/public areas for security: (1) safety door system designed at different height, one is full-height safety door and the other is half-height safety door (only line 1 in Beijing subway), and (2) full-height screen door system on the platform. In terms of full-height safety door system, as used mostly, the Beigongda-Ximen station in line 14 was selected as a typical subway station equipped with safety door system. The schematic diagram of subway structure and measuring points for the public areas (hall and platform) are shown in Fig. 13.2. From the Fig. 13.2a, the subway station is consisted of two levels: the ground level for the entrance and underground level. The underground level includes hall for tickets and security check, while the platform for ride. We set one measurement point in the entrance for ground condition, while for the underground part, although, the stations are different, including the size, depth, the system, the location and so on, the construction is the same. We put all testing points in the same picture to make it clearer for the reader to have a total understand about all testing points. In Fig. 13.2b and c, there were respectively 2 measuring points (1 and 2) for hall area and 3 measuring points (3–5) for platform area. In terms of screen door system, Anhuaqiao station in line 8 was chosen as a representative station for the testing in the tunnel (point 6) and staff's working areas. In line 10, Jinsong (phase 1) and Panjiayuan station (phase 2) were constructed at different phases. Jinsong station was opened from July at 2008 and Panjiayuan station opened at the end of 2012. These two stations were chosen in order to compare the effect of service time on concentration, where the other factors are similar, such

Fig. 13.1 Map of measured stations in Beijing subway lines

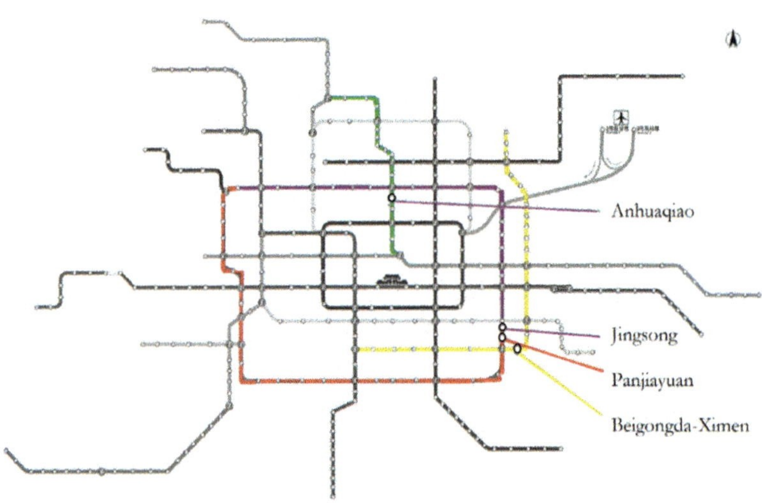

Yellow-line 14; Purple-line 10 phase 1 and Red-line 10 phase 2; Green-line 8

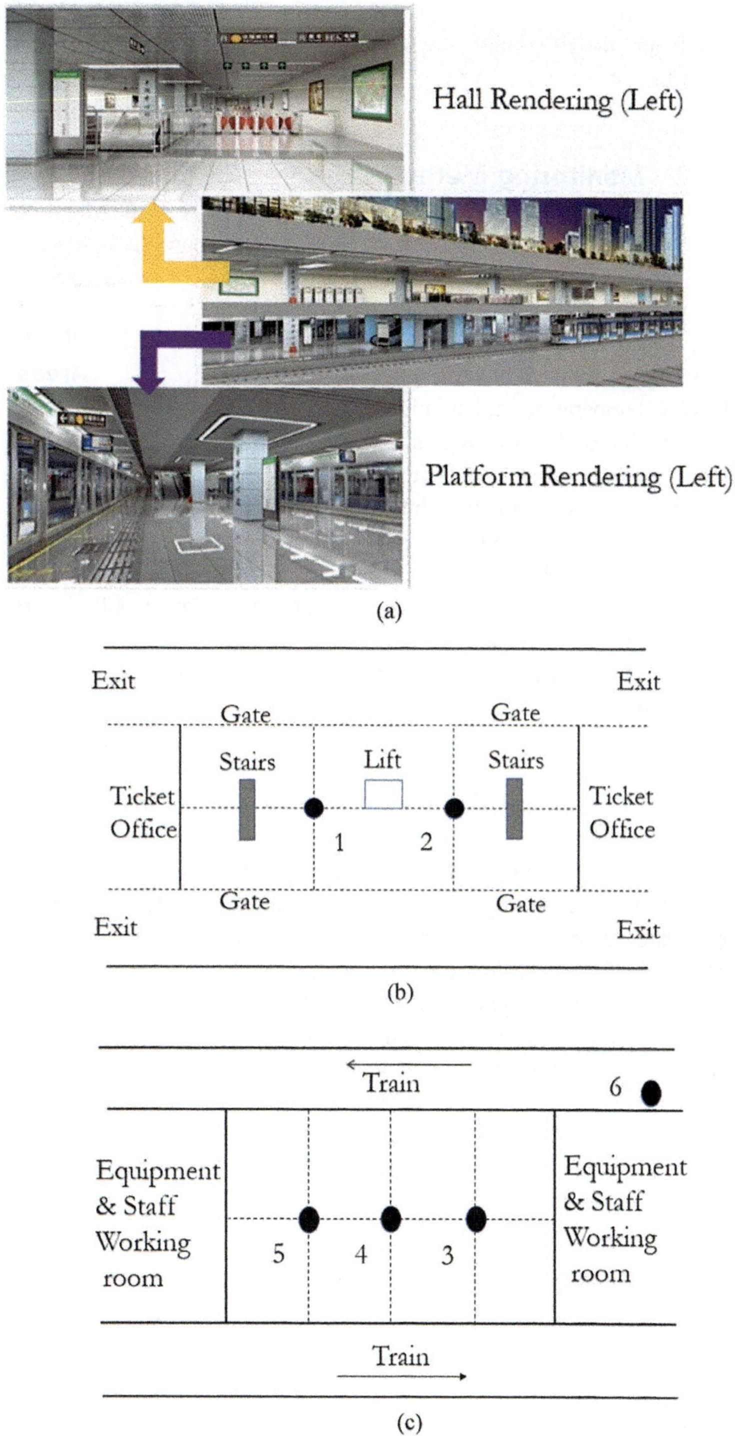

Fig. 13.2 Schematic diagram of subway structure (**a**); measuring points for **b** the hall and **c** platform

as train frequency, structure of the station, the depth, and the air condition system are the same, even the outside environment.

13.2 Monitoring Method and Equipment

The detailed parameters for the testing equipment are presented in Table 13.1. A portable Dust-trak II Aerosol monitor (Model 8532, TSI, USA) was used to monitor the concentration of PM10, PM2.5, temperature and humidity. Such equipment includes data-logging and light-scattering laser photometer for real-time aerosol mass readings. The data logging interval was set at 1 min. In order to identify the real fluctuation of PM10 and PM2.5 concentration in the subway, the data was measured for about continuous 2 weeks during the testing period. All the testing equipment have been calibrated before conducting the measurement.

13.3 Results and Discussion

According to standard of China, there are 6 levels in total for the atmosphere environment along with the average concentration value of the PM2.5 for 24 h. The first level is excellent with the concentration range of PM2.5 is 0–50 ug/m^3. The second level is 50–100 ug/m^3 and the light pollution (100–150 ug/m^3) is ranked as the level 3. Moderate pollution and heavy pollution are 150–200 ug/m^3 and 200–300 ug/m^3, respectively. Situations over 300 ug/m^3 are serious pollution, level 6. To compare the internal subway pollution against the external atmosphere environment, we chose the data that was different from outside pollution level, especially for the level 6, which is unusual but could cause serious problems to public health.

13.3.1 PM10 and PM2.5 for the Public Areas

The results about the average real-time concentration of PM10 and PM2.5 at different locations in the public areas at the Beigongda-Ximen station is presented as Table 13.2. Six representable days were picked up when the outside environment pollution ranked at different levels in Beijing. The concentration data for the outside environment was the average value during the measuring period. There were three measuring points at the platform and two at subway hall, while an average value of the measurements was calculated for comparison. All the temperature and humidity were also turned into average values due to a small variation.

It is observed from the Table 13.2, the pollution in the subway station varied in a wide range along with the outside environment, where the pollution level was from level 1 to level 6. In general, the concentrations of PM10 and PM2.5 were higher during the peak period than that at the non-peak period. The lowest PM concentration

Table 13.1 Instrument parameters of the testing equipment

Range	Sensor type	Time constant
0.001–400 mg/m^3	90° Light scattering	Adjustable 1–60 s
Flow accuracy	Zero stability	Resolution
±5% factory setpoint internal flow controlled	±0.002 mg/m3 24 h at 10 s time constant	±0.1% of reading of 0.001 mg/m^3, whichever is greater
Data logging	Log internal	Operational temp
45 days at 1 min samples	1 s–1 h	0–50 °C
Particle size range	Temperature coefficient	Storage temp
Approximately 0.1–10 μm	±0.001 mg/m^3/°C	−20 to 60 °C

Table 13.2 Measuring concentration and standard deviation of the Beigongda-Ximen station

Date	Location	PM2.5(ug/m^3)		PM10(ug/m^3)		Temp (°C)	Humid (%)
		Non-peak	Peak	Non-peak	Peak		
10.9 Sunday	Outside	18 ± 5	33 ± 4	36 ± 9	64 ± 8	18	36
	Hall	53 ± 8	65 ± 8	110 ± 17	148 ± 13	23	42
	Platform	61 ± 5	76 ± 12	128 ± 12	160 ± 8	23	42
10.10 Monday	Outside	92 ± 6	108 ± 8	127 ± 5	147 ± 16	20	40
	Hall	130 ± 14	146 ± 26	145 ± 12	165 ± 18	22	45
	Platform	165 ± 12	168 ± 23	167 ± 19	187 ± 15	23	30
10.11 Tuesday	Outside	216 ± 35	228 ± 34	439 ± 76	459 ± 26	25	51
	Hall	151 ± 14	179 ± 16	316 ± 31	356 ± 28	24	54
	Platform	161 ± 24	154 ± 25	339 ± 51	378 ± 45	26	50
10.12 Wednesday	Outside	80 ± 30	118 ± 38	102 ± 58	145 ± 18	20	37
	Hall	98 ± 18	138 ± 29	159 ± 45	156 ± 13	23	40
	Platform	126 ± 26	156 ± 34	175 ± 52	178 ± 18	24	40
10.13 Thursday	Outside	326 ± 3	345 ± 2	525 ± 82	536 ± 56	21	70
	Hall	263 ± 17	312 ± 16	462 ± 41	468 ± 41	27	54
	Platform	258 ± 20	298 ± 22	443 ± 42	446 ± 46	25	58
10.14 Friday	Outside	242 ± 34	265 ± 32	286 ± 26	294 ± 14	23	45
	Hall	196 ± 26	219 ± 17	276 ± 8	287 ± 25	24	34
	Platform	158 ± 20	168 ± 14	245 ± 14	264 ± 21	26	37
10.15 Saturday	Outside	186 ± 43	217 ± 23	276 ± 15	309 ± 36	17	37
	Hall	195 ± 34	224 ± 19	296 ± 24	298 ± 25	23	38
	Platform	183 ± 28	215 ± 17	265 ± 28	293 ± 35	25	40
10.16 Sunday	Outside	117 ± 37	169 ± 26	215 ± 38	269 ± 25	18	46
	Hall	146 ± 29	176 ± 15	178 ± 19	189 ± 14	24	42
	Platform	174 ± 32	198 ± 20	193 ± 26	224 ± 22	26	38

(continued)

Table 13.2 (continued)

Date	Location	PM2.5(ug/m^3)		PM10(ug/m^3)		Temp (°C)	Humid (%)
		Non-peak	Peak	Non-peak	Peak		
10.17 Monday	Outside	41 ± 8	53 ± 7	96 ± 17	101 ± 18	21	46
	Hall	48 ± 2	65 ± 3	96 ± 5	104 ± 8	23	41
	Platform	56 ± 3	78 ± 6	118 ± 8	138 ± 12	25	44
10.18 Tuesday	Outside	261 ± 21	380 ± 4	451 ± 51	520 ± 18	20	77
	Hall	235 ± 16	337 ± 14	384 ± 43	483 ± 28	23	65
	Platform	205 ± 17	319 ± 32	342 ± 36	403 ± 80	20	74
10.19 Wednesday	Outside	371 ± 12	398 ± 16	578 ± 29	690 ± 24	20	84
	Hall	317 ± 40	349 ± 29	561 ± 85	641 ± 73	23	73
	Platform	291 ± 80	250 ± 51	533 ± 181	425 ± 89	24	68
10.20 Thursday	Outside	87 ± 16	131 ± 25	138 ± 8	189 ± 17	16	55
	Hall	135 ± 16	163 ± 15	178 ± 27	216 ± 12	22	45
	Platform	165 ± 6	189 ± 14	194 ± 18	245 ± 8	24	48
10.21 Friday	Outside	43 ± 23	57 ± 34	68 ± 19	95 ± 4	11	70
	Hall	68 ± 16	72 ± 26	98 ± 15	110 ± 14	24	55
	Platform	93 ± 23	111 ± 7	156 ± 5	176 ± 3	23	48

values were observed on 17th October 2016 when the outside environment was excellent at level 1. The lowest average values of PM10 (PM2.5) were 96 ug/m^3 (48 ug/m^3) and 118 ug/m^3 (56 ug/m^3) respectively in the hall and the platform. On the other hand, the highest PM concentration values occurred on 19th October when the pollution of the outside environment was at the serious level (PM2.5 was 371 ug/m^3) in Beijing. The concentration of PM2.5 and PM10 were found at 349 and 641 ug/m^3 in the hall during the peak time; while they were 250 ug/m^3 (PM10) and 425 ug/m^3 (PM2.5) in the platform during the peak period. It is obvious from Table 13.2 that the PM concentrations inside the subway station were not always larger than that in the outside. For instance, the measurements of subway PM concentrations, on dates of 11th–15th, 18th, 19th Oct in 2016, were smaller than that in the outdoor environment. Such observation is different from the conclusion from some existing studies (Cheng and Yan 2011; Cheng et al. 2012; Chillrud et al. 2004, 2005; Gomez-Perales et al. 2004). In contrast, when the pollution of the outside environment was over 200 ug/m^3 (level 5), the result became opposite. In Fig. 13.3, the real-time

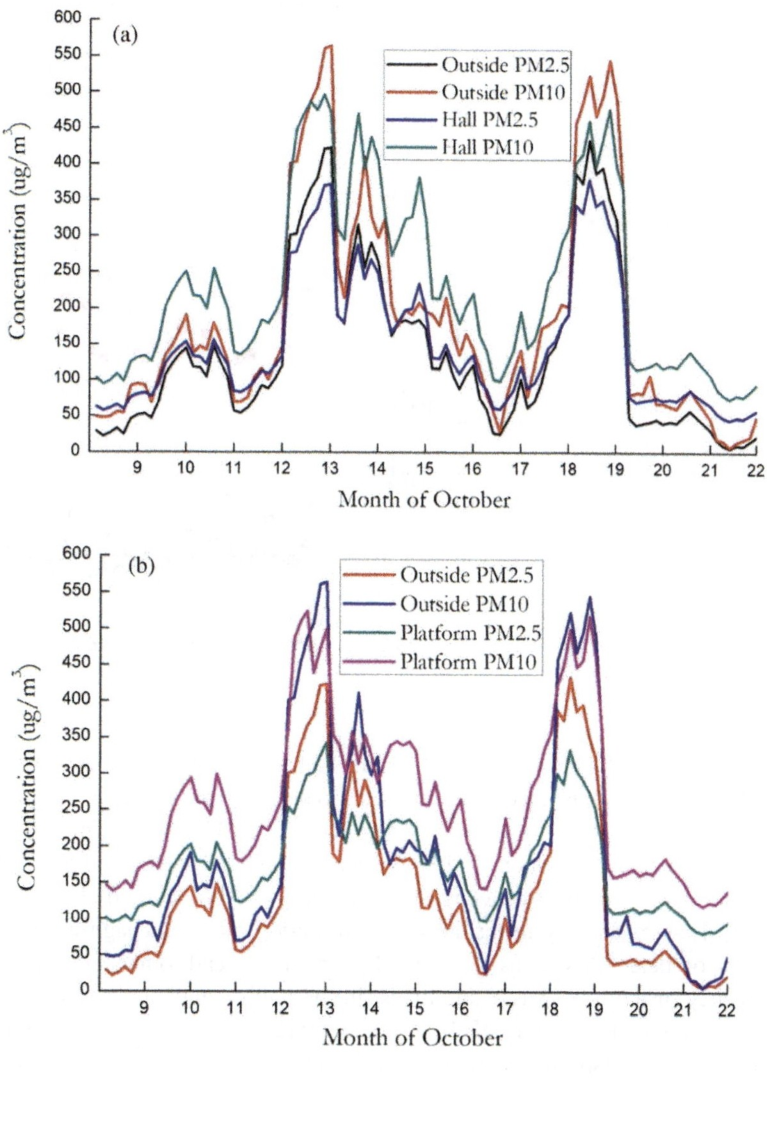

Fig. 13.3 The real-time concentration at hall (**a**) and platform (**b**)

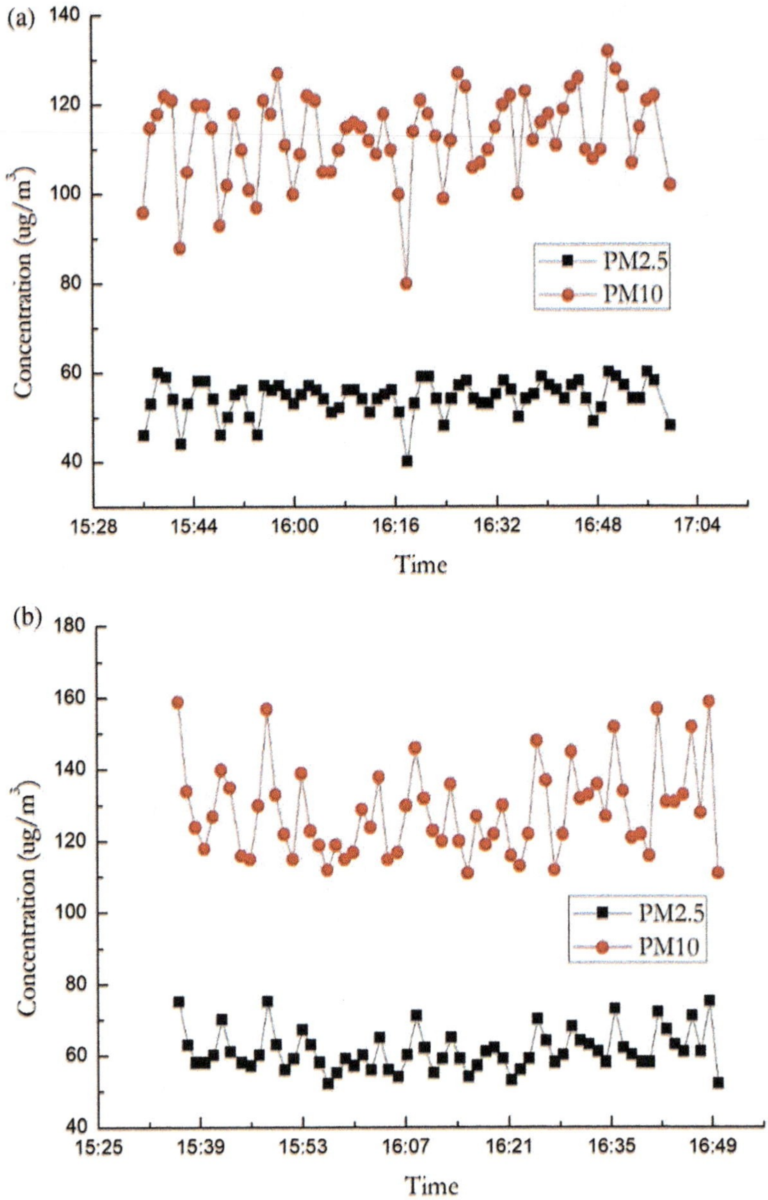

Fig. 13.4 The real-time concentration in ninth **a** for hall and **b** for platform

fluctuation of the PM10 and PM2.5 at the hall and platform were illustrated along with the outside PM values.

The detailed variation of PM concentrations against the train frequency are shown in Fig. 13.4a and b, by using the non-peak period data of date 9th Oct as an example. The frequency of the train was 7 min/times. In the hall area, the concentration of PM10 and PM2.5 ranges were respectively 40–61 ug/m^3 and 78–135 ug/m^3, as in Fig. 13.4a. In the platform, the concentration ranges of PM10 and PM2.5 were 112–159 ug/m^3 and 52–75 ug/m^3, respectively, as seen in Fig. 13.4b. The concentrations of PM10 and PM2.5 showed periodic changes affected significantly by the piston wind effect and other factors. There is about one-minute delay for PM concentration changing after the train arrived in platform, and the changing time in the hall was a little delayed at about two to

Table 13.3 Train exercise table during the measurement period

Train status	Entering	Parking	Starting	Leaving
Time	15:33	15:34	15:37	15:38
	15:45	15:46	15:49	15:50
	15:57	15:58	16:01	16:02
	16:09	16:10	16:13	16:14
	16:21	16:22	16:25	16:26
	16:33	16:34	16:37	16:38
	16:45	16:46	16:49	16:50

three minutes than that in the platform, owing to the construction structure of the subway station (Table 13.3).

13.3.2 PM10 and PM2.5 Concentrations in the Train and in the Employee's Working Areas

In addition to the hall and the platform, the interior environment within the subway train at different stations were measured in terms of the PM concentrations. The measuring points were located at the middle of two doors and the testing results are shown in Fig. 13.5. It was measured on the date when the outdoor concentration of PM2.5 was 265 ug/m^3. The equipment was set at 1 min/time and there were one to three measurements during train in the tunnel. In Fig. 13.5, the blue points represented the time that the train arrived the station and the train door opened. It was clear that the PM2.5 fluctuation was huge from 220–370 ug/m^3, which might be caused by the passenger flow and the air pressure changes between the platform and train when the train door opened.

The air quality in the employee's working areas is important for the subway staff. It is the first time that the concentration of PM10 and PM2.5 could be measured in the working areas as these areas are usually prohibited from entering. There were totally 8 kinds of rooms measured in Anhuaqiao station (line 8), as showed in Fig. 13.6. During the measurement, only lounge, working office and control room were equipped with the mechanical ventilation system to supply fresh air; while the meeting room was always closed. The measurement was conducted on 21th Oct 2016, when the outdoor environment was lightly polluted at level 3, with the average PM2.5 was 131 ug/m^3 and PM10 was 158 ug/m^3. In the ventilation system, the existed ventilation system cannot filter PM2.5 by using the coarse filter and medium efficiency filter. According to the testing results, the PM concentrations at all these rooms were higher than the outdoor data. The highest polluted areas were the equipment areas with facilities and poor ventilation, while the closed meeting room had the lowest PM pollutions.

13.3.3 The Ratio and Correlation of PM2.5 to PM10

The percentage and the correlation of PM2.5 that accounted for PM10 were different at different locations (Table 13.4). The average ratio of PM2.5 to PM10 outside the subway was 79.6% that was slightly higher than that in those locations inside subway including the train at about 68.7%. The working areas had the lowest ratio of PM2.5 to PM10 at 47.6%, while the ratio in the platform and the hall areas were almost 68.6% and 61.2%, respectively. Nevertheless, the ratio of the PM2.5 to PM10 in subway station was much lower than that before researches (Hosein et al. 2014; Huang et al. 2012; Kam et al. 2011a,b; Kamani et al. 2014; Karlsson et al. 2006).

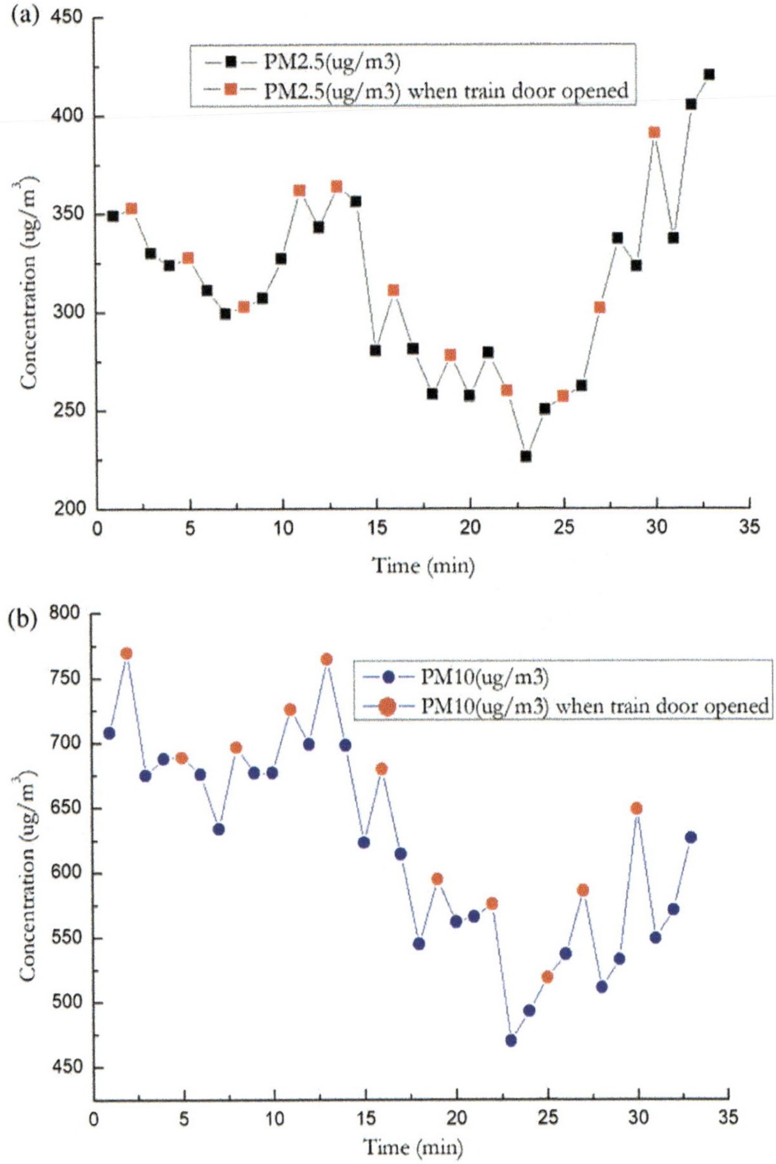

Fig. 13.5 PM concentration of train at different stations in line 14 **a** PM2.5; **b** PM10

The correlation analysis indicated that PM10 and PM2.5 were significantly correlated at Sig < 0.01, although the values were different at different locations. The highest coefficient (R^2) was 0.930 at the platform, while the lowest was 0.815 in the train. The results were similar to the researches made by Guo et al. (2014), Harrison et al. (1997), Hosein et al. (2014) and so on. The high coefficient indicated that the PM2.5 could predict from the concentration of PM10 as the two particles were highly correlated.

13.4 The Analysis of Influencing Factors

There are many factors that could affect the concentration of PM2.5 and PM10, which could be divided into four kinds: external factors,

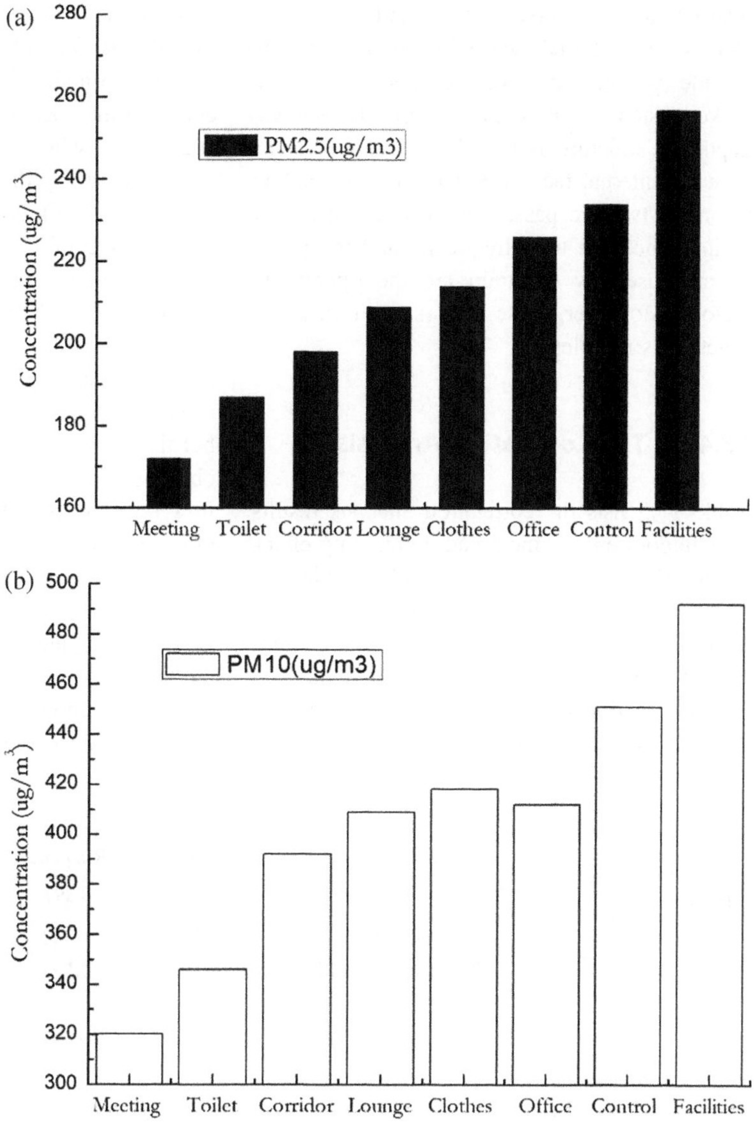

Fig. 13.6 Measuring results of working areas **a** PM2.5; **b** PM10

Table 13.4 The ratio (%) and correlation matrix of PM2.5 to PM10

PM2.5 to PM10	Outside	Hall	Platform	Train	Working office	Average
The ratio, %	79.6	61.2	68.6	68.7	47.6	65.14
R	0.921	0.911	0.956	0.903	0.92	0.9222
R Square	0.848	0.829	0.93	0.815	0.846	0.8536
Adjusted *R* square	0.845	0.827	0.93	0.813	0.843	0.8516
Sig.	0.000	0.000	0.000	0.000	0.000	0.000

internal factors, human factors and operational factors. The external factors include seasons, the outside weather and the outdoor traffic. The brake system, ventilation system, the subway depth, the structure of the subway and the service time are internal factors. For the human factors, the main two are passenger flow and the measuring time. The train frequency and the piston effect caused by the train are the operational factors. Moreover, these factors are mixed and function is turbulent.

13.4.1 The Correlation Analysis

Table 13.5 lists the correlation analysis results. The outdoor environment had dominating effects on the PM concentration in the subway platform ($R^2 = 0.897$). Combined with the general linear analysis, the linear regression equation was $Y = 75.370 + 0.877X$, which also indicated the high correlation of the outdoor environment to the PM2.5 concentrations in the subway station.

In Table 13.5, the passenger flow had very low correlation with the PM 2.5 concentration at $R^2 = 0.246$ and Sig = 0.135 > 0.05. It is different from the speculation made by the (Karlsson et al. 2005), who inferred that the passenger flow would cause significant influence on the PM2.5 concentration. Figure 13.7 characterizes the real-time change of PM2.5 concentration along the passenger flow, which shows the real character of PM10 and PM2.5 at the hall. The red points were the time when the passenger flow increased. After the passenger flow increasing, there was usually a slight fluctuation of the concentrations, but the average PM concentrations did not have obvious change. So we argue that the passenger flow could lead to fluctuation of the concentration, but does not affect the average PM2.5 concentration in subway, where the PM concentrations are mainly affected by the outdoor environment and the piston wind produced by the train.

For the temperature and humidity, the correlation was even lower at $R^2 = 0.138$ and Sig =

Table 13.5 The correlation analysis of different factors

Factors	R	R square	Adjusted R square	Sig.
Outside	0.947	0.897	0.871	0.00
Passenger flow	0.245	0.246	0.243	0.135
Temperature and humidity	0.135	0.138	0.134	0.159

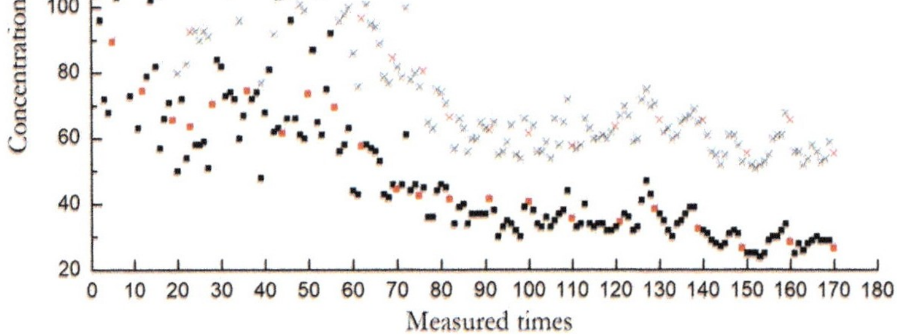

Fig. 13.7 Concentration fluctuation of PM2.5 at hall

0.159 > 0.05. The results from multivariate analysis, F = 1.265 and Sig = 0.295, also demonstrate the low correlation of the temperature and humidity on PM2.5 concentration.

13.5 Summary

In the Beigongda-Ximen station, the concentration of PM10 and PM2.5 changed a lot along with the outside pollution huge variety. At the hall and platform, the real-time PM10 and PM2.5 showed periodic changes. For the train, when the door opened, the concentration of the PM10 and PM2.5 would fluctuate significantly. In the working area, all rooms' pollution was higher than the outdoor environment; the highest pollution areas were equipment rooms, while the closed meeting room had the lowest PM concentrations. The average ratio of PM2.5 to PM10 outside the subway was 79.6%, slightly higher than that in the locations inside subway including the train at 68.7%. The working areas had the lowest ratio of PM2.5 to PM10 at 47.6%. The platform and hall for the public areas were 68.6% and 61.2%, respectively. Correlation analysis indicated that PM10 and PM2.5 were highly correlated. The outdoor environment and the service time of subway had significant effects on the concentration of PM2.5, while the passenger and temperature and humidity did not have obvious influence on it.

References

Cheng YH, Yan JW (2011) Comparisons of particulate matter, CO, and CO_2 levels in underground and ground-level stations in the Taipei mass rapid transit system. Atmos Environ 45(28):4882–4891

Cheng YH, Liu ZS, Yan JW (2012) Comparisons of PM10, PM2.5, particle number, and CO_2 levels inside metro trains traveling in underground tunnels and on elevated tracks. Aerosol Air Qual Res 12(5):879–891

Chillrud SN, Epstein D, Ross JM, Sax SN, Pederson D, Spengler JD, Kinney PL (2004) Elevated airborne exposures of teenagers to manganese, chromium, and iron from steel dust and New York City's subway system. Environ Sci Technol 38(3):732–737

Chillrud SN, Grass D, Ross JM, Coulibaly D, Slavkovich V, Epstein D, Sax SN, Pederson D, Johnson D, Spengler JD, Kinney PL, Simpson HJ, Brandt-Rauf P (2005) Steel dust in the New York City subway system as a source of manganese, chromium, and iron exposures for transit workers. J Urban Health-Bull NY Acad Med 82(1):33–42

Gomez-Perales JE, Covile RN, Nieuwenhuijsen MJ, Fernandez-Brem A, Gutierrez-Avedoy VJ, Paramo-Figueroa VH (2004) Commuters' exposure to PM2.5, CO, and benzene in public transport in the metropolitan area of Mexico City. Atmos Environ 38:1219–1229

Guo L, Hu YJ, Hu QQ, Lin J, Li CL, Chen JM, Li LN, Fu HB (2014) Characteristics and chemical compositions of particulate matter collected at the selected metro stations of Shanghai, China. Sci Total Environ 496:443–452

Harrison RM, Deacon AR, Jones MR (1997) Sources and processes affecting concentration of PM2.5 and PM10 particulate matter in Birmingham (U.K.). Atmos Environ 31(24):4103–4117

Hosein KM, Mahdi CYJ, Gholam HS (2014) Concentration and characterization of airborne particles in Tehran's subway system. Environ Sci Pollut Res 21:7319–7328

Huang J, Deng FR, Wu ShW, Guo XB (2012) Comparisons of personal exposure to PM2.5 and CO by different commuting modes in Beijing Chinag. Sci Total Environ 425:52–59

Kam W, Cheung K, Daher N, Sioutas C (2011a) Particulate matter (PM) concentrations in underground and ground-level rail systems of the Los Angeles metro. Atmos Environ 45(8):1506–1516

Kam W, Ning Z, Shafer MM, Schauer JJ, Sioutas C (2011b) Chemical characterization and redox potential of coarse and fine particulate matter (PM) in underground and ground-level rail systems of the Los Angeles metro. Environ Sci Technol 45(16):6769–6776

Kamani H, Hoseini M, Seyedsalehi M, Mahdavi Y, Jaafari J, Safari G (2014) Concentration and characterization of airborne particles in Tehran's subway system. Environ Sci Pollut Res 21(12):7319–7328

Karlsson L, Nilsson L, Moller L (2005) Subway particles are more genotoxic than street particles and induce oxidative stress in cultured human lung cells. Chem Res Toxicol 18:19–24

Karlsson L, Ljungman AG, Lindbom J, Moller L (2006) Comparison of genotoxic and inflammatory effects of particles generated by wood combustion, a road simulator and collected from street and subway. Toxicol Lett 165(3):203–211

Midander K, Elihn K, Wallen A, Belova L, Karlsson AKB, Wallinder IO (2012) Characterisation of nano- and micron-sized airborne and collected subway particles, a multi-analytical approach. Sci Total Environ 427:390–400

Moreno T, Perez N, Reche C, Martins V, de Miguel E, Capdevila M, Centelles S, Minguillon MC, Amato F, Alastuey A, Querol X, Gibbons W (2014) Subway platform air quality: assessing the influences of tunnel ventilation, train piston effect and station design. Atmos Environ 92:461–468

Part III

Sustainability in Communities and Cities

14 Genetic Algorithm for Transforming a Residential Building Cluster into Electricity Prosumers

Pei Huang, Marco Lovati, and Xingxing Zhang

Abstract

Smart grid is triggering the transformation of traditional electricity consumers into electricity prosumers. This chapter reports a case study of transforming an existing residential cluster in Sweden into electricity prosumers. The main energy concepts include (1) click-and-go photovoltaics (PV) panels for building integration, (2) centralized exhaust air heat pump, (3) thermal energy storage for storing excess PV electricity by using heat pump, and (4) PV electricity sharing within the building cluster for thermal/electrical demand (including electric vehicles load) on a direct-current micro grid. For the coupled PV-heat pump-thermal storage-electric vehicle system, a fitness function based on genetic algorithm is established to optimize the capacity and positions of PV modules at cluster level, with the purpose of maximizing the self-consumed electricity under a non-negative net present value during the economic lifetime. Different techno-economic key performance indicators, including the optimal PV capacity, self-sufficiency, self-consumption and levelized cost of electricity, are analysed under impacts of thermal storage integration, electric vehicle penetration and electricity sharing possibility. Results indicate that the coupled system can effectively improve the district-level PV electricity self-consumption rate to about 77% in the baseline case. The research results reveal how electric vehicle penetrations, thermal storage, and energy sharing affect PV system sizing/positions and the performance indicators, and thus help promote the PV deployment.

Keywords

Building cluster · Prosumer · PV optimization · Heat pump · Thermal storage · Electrical vehicle

14.1 Introduction

Buildings as electricity prosumers are growing in energy space as they not only produce energy from distributed energy resources, but also consume the generated energy locally, through heating, ventilation and air conditioning (HVAC) systems, appliances and electric vehicles (EV) etc. These have profound impacts on the smart grid value chain. It is also a harbinger of another transformation—the shift of "power", from being

P. Huang (✉) · M. Lovati · X. Zhang
Department of Energy and Community Buildings, Dalarna University, SE-79188 Falun, Sweden
e-mail: phn@du.se

M. Lovati
e-mail: mlov@du.se

X. Zhang
e-mail: xza@du.se

X. Zhang (ed.), *Data-driven Analytics for Sustainable Buildings and Cities*, Sustainable Development Goals Series, https://doi.org/10.1007/978-981-16-2778-1_14

concentrated in the hands of utilities as the sole owners/distributors of electricity, to electricity prosumers on a vastly distributed and decentralized basis (Christine 2013). With the intensive growth in photovoltaic (PV) panels, EVs, home batteries, distributed heat pumps (HP), thermal energy storage (TES) and direct-current (DC) grid, buildings offer great potentials for consumers and building owners to re-evaluate their energy practices (Parag and Sovacool 2016). As the electricity prosumers are increasing at urban or district scale, the building integrated or added PV installations are boosting with very large capacity in recent years, which bring many unknowns about the integration of smart gird infrastructure that need to be optimized (Parag and Sovacool 2016). To develop strategies for the future, policymakers and planners need knowledge of how many and where PV systems could be integrated effectively and efficiently into local energy infrastructure and markets.

Up to date, many researchers have devoted to the techno-economic optimization of PV at building level (Huang et al. 2018a). For instance, Gui et al. (2017) developed a genetic algorithm based optimization method to design the capacity, locations, tilt angles and azimuth of PV panels, with factors such as shapes and orientations of building exteriors and the surrounding obstacles considered. Their method is able to maximize the solar power output and thus reduces the capital investment per unit power output. Bingham et al. (2019) developed a non-sorting genetic algorithm based optimization method to design the envelop, PV systems, and battery storage of a residential building in Bahamas. Their study indicates that application of PV systems and battery storage can significantly reduce the cost and consumption of grid energy. Koskela et al. (2019) explored the optimal sizing of PV panels and batteries under different electricity pricing for an apartment building and detached houses in Finland. Their study shows that suitable electricity pricing can increase the profitability of applying PV panels and batteries. With the application of batteries and building load control technologies, O'Shaughnessy et al. (2018) analysed the improvements in PV energy self-consumption and net present value (NPV) using the renewable energy optimization (REopt) model. Oh et al. (2018) developed an integrated model (i-FEM) based on finite element method for estimating the techno-economic performance of the distributed solar generation system on building façades. Liu et al. (2019) investigated the design optimization of a PV-battery system combining heat pumps. Sensitivity analysis was also conducted by them under a range of PV capacities and battery prices to understand the impacts of heat pumps on PV-battery systems. Their study concludes that the use of heat pump can help increase the PV self-consumption and reduce the storage capacity. The abovementioned studies have considered solar resource and façade geometry, synergies with batteries and thermal envelope, electrical storage, electricity pricing, battery prices, and load controls. However, they didn't consider the synergies of energy sharing among neighbour buildings and the influence from local TES and EV penetration.

On district or above level, researchers have also been conducted regarding the techno-economic analysis of PV and its related systems. For instance, Heijde et al. (2019) developed a genetic algorithm based computation-efficient optimization tool, which uses the representative days to simplify the whole-year simulation, to determine the size of district solar energy systems and seasonal TES with the minimal operational costs. Shirazi et al. (2019) proposed an integrated techno-economic evaluation tool to identify the most appropriate PV installation façades in urban areas in Tehran of Iran. Their study shows that proper selection of the angles and building façades for installing PV panels can significantly increase the solar power production (e.g. 19%) and internal rate of return (e.g. 6%). Roberts et al. (2019) investigated the impacts of applying shared battery energy storage systems on the PV self-consumption and electricity bills of several apartment clusters in Australian. Their study indicates the shared batteries can effectively increase solar self-consumption and shave the peak demand of the building cluster. Zhang et al. (2012) investigated the energy and environment impacts of integrating PV power into electricity systems in Kensai of Japan, under various scenarios with different EV penetrations and heat pump capacities. It is found that EV and heat pump are helpful for keeping more PV power in the smart electricity systems. Notably, Rodríguez et al. (2018) proposed a PV-HP-thermal mass storage system for alleviating the energy poverty for a low-income housing district in Spain. In their study, the PV surplus electricity is used to power the heat pump to provide cooling/heating to improve thermal comfort of the occupants. However, the capacity of passive thermal storage (i.e. building thermal mass) is much smaller compared with active thermal storage, thus limiting the system flexibility and performance. Also, the energy sharing among different buildings and the EV penetrations is not considered. Regarding energy sharing among buildings, Shen and Sun (2016) compared the sizing of PV systems for a small building cluster with and without energy sharing enabled considering demand uncertainty. They study shows that enabling energy sharing can significantly reduce the required PV system capacity, since the surplus renewables of one building can compensate the renewable insufficiency of another. Huang et al. (2018b) investigated the operation of decentralized PV-battery system for a building cluster with energy sharing enabled. By comparing with the scenario that energy sharing is not allowed, their study indicates that energy sharing can significantly increase the cluster-level PV self-consumption and meanwhile reduce the electricity costs. These studies have conducted techno-economic analysis of district-scale PV from the aspects of the solar resources' maximization, seasonal TES, battery or heat pump integration, and power sharing integration. However, the integrated impacts of TES for excess PV electricity, EV as part of electrical load, and power sharing by DC grid are not fully studied on optimal design/operation of district/urban-level PV.

Regarding HPs and TES, many studies have been conducted for improving their performance and promoting their applications. For instance, Fischer and Madani (2017) conducted a comprehensive overview of applications for HPs in a smart grid. They conclude that HPs can be seen as core technology to connect the heating and electricity sectors. Nolting and Praktiknjo (2019) conducted validated simulations of realistic and flexible HP controls (i.e. time-of-use based controls and spot market price-based control), and assessed effects on energy efficiency and economic potentials compared to standard reference control algorithms. Sun et al. (2019) conducted a comprehensive analysis of fine-grained data collected from smart hybrid HPs (which perform smart switching between electricity and gas) and proposed a flexibility quantification framework to estimate the capability of HP demand shifting based on preheating. Their study results show that smart controls of hybrid HPs can deliver higher average COP values. By pairing HPs with TES, heat demand can be shifted to off peak periods or periods with surplus renewable electricity, and thus improved performance and increased flexibility can be achieved. Renaldi et al. (2017) developed a design and operational optimization model to assess the performance of HP-TES system. Their study results show that the integration of TES and time-of-use tariffs can reduce the operational cost of the HP systems and make the HP systems cost competitive with conventional systems. Psimopoulos et al. (2019) developed rule-based control algorithms for a HP system, which is integrated with TES and electrical storage, to minimize building energy usage and maximize self-consumption. Their developed method is able to reduce energy usage by 5–31% and the annual net cost by 3–26%. Baeten et al. (2017) proposed a multi-objective model predictive control strategy for a HP-TES system, which takes into account users' energy cost and environmental impacts. Their study shows that applying TES is effective in reducing the required peak capacity of HPs. But, when demand response is applied by using TES, the costs for consumer always increase compared to the case without demand response or TES. The abovementioned studies have systematically investigated the application of HPs and TES in improving the building performance. However, the integration of HPs and TES with EV load as well as the energy sharing among buildings, are rarely considered.

In particular, the EVs are increasing their influence in PV deployment and smart grid, which should be considered as part of the electrical demand in the optimal design of renewable-energy based building renovations. The Swedish government has set a goal that the 100% of the national energy used in vehicle fleets should be independent of fossil fuel by 2030 (Agency 2009). EVs, which can use the grid power and the potential renewable energy, are promising solutions to achieving these energy targets. Many studies have been conducted to investigate the EV energy usage patterns and estimate the EV load profiles. For instance, using an availability model which generates driving profiles by statistical analysis, Geth et al. (2010) analysed the impacts of both uncoordinated charging and coordinated charging on the load profiles in Belgium. Moreira et al. (2011) developed a discrete-state and discrete-time Markov chain model to simulate the EV motion and the energy usage. Shahidinejad et al. (2012) adopted a fuzzy-logic inference system to emulate the EV battery charging based on a large field-recorded driving database. By combining EV usage with synthetic activity generation of occupants' electricity-dependent activities, Grahn et al. (2013) used a Markov chain model to calculate the EV electricity consumption. They found that the EVs make up around 1/3 of the expected load during the peak hours and around 1/5 of the total daily electricity usage. Similarly, Munkhammar et al. (2015) modelled the EV charging states with a Bernoulli distribution and generated the EV charging patterns by a Markov chain model. Their study shows that large mismatch exists between the PV power production and EV power consumption. Munkhammar et al.'s (2013) study indicates that aggregating the multiple households' EV power usage and PV power generation will be more beneficial for increasing self-consumption of PV power than individual households. Fischer et al. (2019) proposed a stochastic bottom-up model to describe the EV usage, charging behaviour and the resulting electrical load profiles. Their study reveals that load peaks strongly depend on the deployed charging infrastructure and can easily increase by up to 3.6 times, and EVs will lead to an intensification and an approximately 45 min earlier start of peak load hours during evenings for working households in Germany. EVs will represent large electricity end-users in building sector and have large impact on PV system performance. However, the interactions of EVs with other energy systems, such as heat pump and TES, are rarely studied at cluster or district level when EVs contribute as part of building electricity load.

Besides, there are many existing studies that have presented the energy modelling at building cluster level, but most of them didn't fully optimize the dynamic synergies of PV generation, heat pump, EVs, TES integration, and building load sharing among neighbour buildings, as well as the dynamic interactions of local building energy systems. A research gap thus lies in the absence of detailed techno-economic optimization of the coupled PV-heat pump-TES-EV at cluster level (thus involving neighbour buildings as aggregated electricity prosumers) related to demand coverage of heating, domestic hot water (DHW), and other general appliances. Research questions are therefore raised up, for instance, how the optimal configuration of PV's NPV changes as the EV penetration increases, and how do the different key performance indicators (KPI) are affected. What share of the electric demand for buildings and mobility can be realistically and economically covered by PVs with present technologies? A global optimization of PV in each scenario is urgently desired.

Therefore, this study aims to optimize the capacity of installed PV panels at each building in a small residential district, by considering thermal/electricity loads, power sharing among neighbor buildings, heat pump, TES and EVs, in order to maximize the self-consumed electricity (SCE) when the system is profitable (i.e. positive NPV) during its economic lifetime (of 15 years). Due to factors such as surrounding shadings and orientation, installing PV modules in different positions can produce different amount of power, since the amount of solar radiation can be significantly different in different positions on building roof or facades. To maximize the PV power output, this study also optimized the positions of PV modules to be installed on the building roof or facades. The research results will be useful in testing the effective strategies of PV deployment as connecting with different EV penetrations and heat pump/TES capacities in Swedish residential district. In the future, the same method could be replicated on different types of building clusters for achieving improved KPIs.

The structure can be depicted as followings: Sect. 14.2 describes the overall renovation concepts for the building cluster; Sect. 14.3 clarifies the research methodology and key performance indicators considered; in Sect. 14.4, the modelling of the coupled system is presented. Section 14.5 summarizes the boundary conditions and input parameters for the simulation case. A series of optimization and the related sensitivity analysis are subsequently performed in Sect. 14.6. Section 14.7 finally presents discussion and outlook; while the brief conclusions are disclosed in Sect. 14.8.

14.2 Overall Energy Concepts for Building Cluster Renovation

14.2.1 Building Cluster Information

The studied building cluster is located in Sunnansjö, Ludvika, Dalarna region, Sweden. This demo site is a multifamily dwelling unit made of three buildings built in 1970/1973, as shown in Fig. 14.1. The cluster (three buildings) includes 48 apartments over three floors, and most of the apartment have one or two bedrooms. The total façade surface gross area of the complex is 2146 m^2, the total roof surface gross area is 1750 m^2, and the total heated area is 3861 m^2 (see Table 14.1). The energy consumption of the cluster is 165 kW h/(m^2 year), including operational electricity but not including electricity used in the flats for appliances and lighting. These buildings will be improved by a series of renovation plans including installation of PV, thermal energy storage, DC micro grid, EVs and heat pump systems.

14.2.2 Energy Concepts

With the purpose of improving the overall energy performance and reducing carbon emissions of the building cluster, the following interventions are being applied in renovating this building cluster. First, a centralized heat pump using exhaust air and ground as heat sources will be used for supplying the heating and hot water for all the three buildings. All the exhaust air in each building will be ducted to a heat exchanger unit, in which the waste heat will be recovered and then delivered to the centralized heat pump via a brine loop. A back-up pellet boiler is utilized to accommodate the peak heating needs. The PV can be installed on the roof and façades of the buildings. The PV energy is first used to power the electrical facilities in the buildings (e.g. fans, pumps, lighting, EV demands). After this part of electrical load is met, the remaining PV energy is considered as excess PV energy. A hot water storage is planned to store the excess PV energy in the form of heat, where the excess PV electricity power is transmitted to the heat pump to produce heating energy, and the produced heat is stored as the hot water. All electricity in the buildings, including that in the flats, as well as that supplied to the EV's is managed by one Energy Hub in each building, connected together via a DC micro grid. The DC sources (i.e. PV) and sinks (i.e. EVs and variable speed heat pump compressor) as well as batteries, if present, are connected directly to the DC micro grid. The overall energy concepts (the advanced techniques used in the building cluster renovation for improving the overall energy performance) of the renovation plans are presented in Fig. 14.2. In this subsection, the details of the major energy concepts are presented.

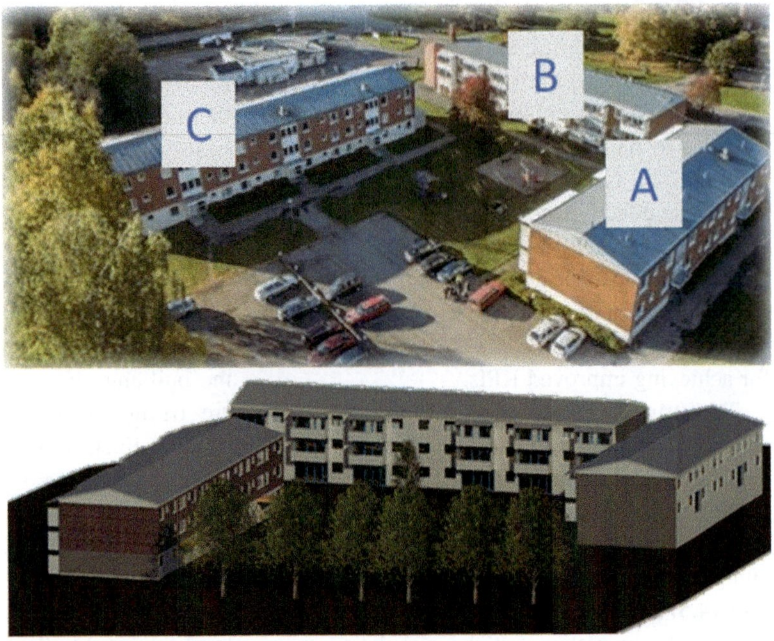

Fig. 14.1 Three buildings in the cluster for renovation in Ludivika, Sweden

Table 14.1 Area of each building used to assign a share of the private electricity demand

	Ground floor area (m^2)	Gross floor area (m^2)	Number of floors
A	381	1143	3
B	509	1527	3
C	552	1656	3
Total	1442	4326	

14.2.3 Click-And-Go PV Panels

Click-and-go is a concept that aims at making the installation/removal of the PV modules easier while improving the aesthetical quality. The mounting systems have been carefully designed so that they can be easily, conveniently, and flexibly installed/removed. The PV modules can be installed on either building roofs or façades. The roof mounting system has overlapping panels and has been certified by KIWA (IEC 61215) and meets building standard 'NEN 7250, Solar energy systems—integration in roofs and façades—building aspects' (NEN 2014). The mounting system for the façades is based on an equivalent click-and-go system, which can be easily mounted and demounted. In contrary to the pitched roof version with overlapping panels, the façade mounting system has flat mounted panels with a surrounding gap between the panels. Beside the easy mounting, the system has a high aesthetic value, as at the front side only the glass of the PV modules is visible and the mounting construction is invisible. Accumulation of dirt/stripes on glass is prevented simultaneously.

The PV panels are developed on basis of a portfolio of lightweight, aesthetic and click-and-go solutions. There are two lightweight PV module technologies developed in the EnergyMatching project: glass/glass and composite-based modules. The difference between them relies on the encapsulation material. This study uses the glass/glass-based modules. The manufacturing process of glass/glass PV modules is based on traditional manufacturing processes, adapted accordingly to allow the effective

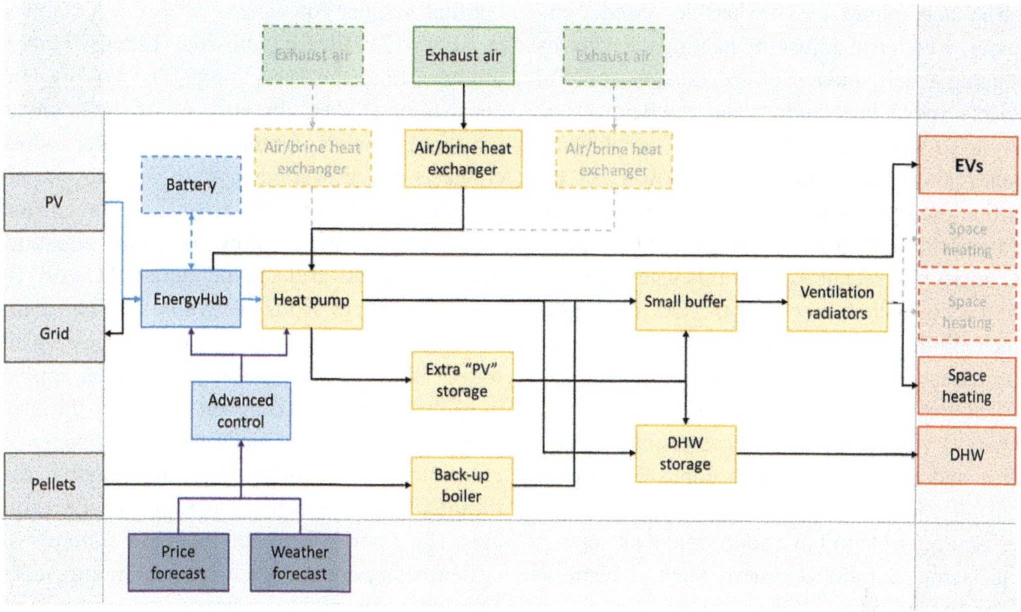

Fig. 14.2 Overall energy concepts for the building cluster

integration of PV cells, leading to a lightweight concept (range of weight between 10 and 15 kg/m^2).

14.2.4 Centralized Exhaust Air Heat Pump (EAHP) System

A centralized heating plant is proposed, which comprises a brine-water heat pump using exhaust air as heat source as well as a back-up boiler. The heat pump covers all the DHW load using a hot water storage tank. The heat pump also provides the space heating until its maximum capacity is reached. At peak heating loads, the back-up boiler is used to supply the remaining heating load. Each building has one heat recovery unit in the attic, to which the exhaust air in each room, including kitchens and bathrooms, is ducted by fans. The recovered heat from the ventilation air is then supplied to the heat pump as the heat source via a brine loop. An anti-freeze mixture is used in this loop so that the heat pump can operate with supply temperatures below zero. The amount that can be extracted from the exhaust air is dependent on the ventilation rate, which is generally fixed at the national requirements for hygienic air. In Sweden, the required air flow rate is normally 0.35 l/m^2 s, which results in roughly 0.5 ACH for normal ceiling heights (Concannon 2002). The design supply and return temperatures of water in ventilation radiators for space heating are 55 °C and 45 °C, respectively. The actual supply temperature is varied depending on the ambient temperature using a heating curve. The speed of the heat pump compressor (or pellet fed into the boiler) is controlled to maintain the desired supply temperature. When the required heating is below the minimum load of the heat pump, the heat pump goes into cycling mode to provide the desired load. The charging of the DHW storage is controlled using two temperature sensors in the storage. The control ensures that there is enough heat in the store to meet the expected load. The hot water circulation is supplied at about 55–57 °C with a return temperature around 52 °C, consistent with the Swedish requirement that the temperature at tapping points should be over 50 °C. During charging of the DHW, the heat pump is running at high power.

The heat pump has a variable speed compressor, which can adjust the heating capacity by changing the frequency of the compressor. For variable speed heat pumps, an inverter is typically used to convert DC power to variable frequency alternating current (AC) power. In the conventional heat pumps, this DC power is first created with a rectifier. With the application of DC micro grid technology, in this study the heat pump is connected directly to the DC micro grid, which eliminates the needs for extra rectifiers and thus reduces unnecessary convertion losses. At present, the available heat capacity is in the range of 6–60 kW. Up to eight such units can be cascaded and controlled by one controller, so the available capacity range can reach up to 480 kW. It is also possible to have additional heat sources in the same centralized system, such ambient air or the ground.

14.2.5 Micro Direct Current Grid—Energy Hub Concept

To use the DC power produced by PV panels, inverters are usually needed to convert the DC power into AC power that can be delivered by the conventional AC power distribution system. On the other hand, modern large loads, such as pumps, compressors, fans and EVs, are often operating with DC power with built-in AC-to-DC converters (Strunz and Abbasi 2014). The DC/AC converting at both the supply side and demand side not only causes dramatic electricity losses, but also reduces the system reliability due to increased complexity. To address these issues, the Energy Hub based on DC micro grid is recommended to replace the traditional AC distribution systems (Ayai et al. 2012).

Figure 14.3 presents the schematics of the Energy Hub DC micro grid. The Energy Hub closely integrates multiple energy systems of different energy carriers through convertors, energy distribution and storing components in an optimal manner for various energy use (Howell et al. 2017). Please note that Energy Hub is different from the Heat Pump. The Energy Hub consists of DC/AC inverters or AC/DC convertors and power optimization module, which connects the power generating and consuming facilities to form DC a micro grid. It converts and controls the energy flow in both directions between the DC grid and the facility AC grid. By using the Energy Hubs, the whole DC micro grid becomes an easy-to-use and energy-efficient power management system that can be applied in wide scales. The different nodes in the grid, such as PVs, EVs and loads, can exchange power efficiently and reliably by autonomous distributed control systems based on the DC voltage. The Energy Hub automatically controls all system components, such as heat pump, EVs, and other electrical appliance, to ensure optimized energy utilization within the DC grid.

The Energy Hub based DC micro grid is able to maximize the use of locally produced PV electricity by sharing with different end-users that have consumption at any given time. The system also supports peak load shifting to reduce power tariffs. For instance, load peaks of one user (e.g. EV charging) can be distributed to other users to reduce power drawn from the public AC grid. Meanwhile, by keeping as much of the electricity production and consumption on the DC grid, the converting losses can be significantly reduced (by up to 50%) and thus producing higher efficiency. The Energy Hub techniques used for constructing DC microgrid were developed by Ferroamp (2018). The operating voltage of the DC microgrid is 760 V. Loads that support a nominal DC voltage of 760 V can be powered directly from the DC grid. A minibus DC/DC converter has also been developed by Ferroamp to step down the 760 V DC grid voltage to the output voltage required by other DC loads (120–400 V), thus enabling more flexible applications. For details of the microgrid operation and control, please refer to Ferroamp (2019).

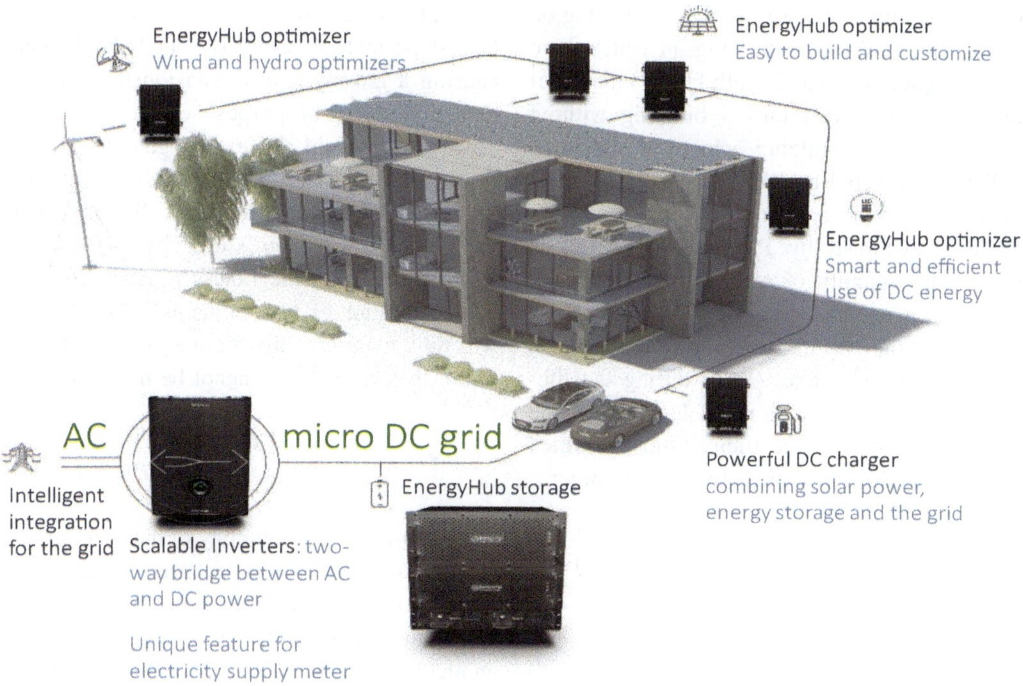

Fig. 14.3 Schematics of the Energy Hub micro DC grid (Ferroamp 2018)

14.2.6 Thermal Energy Storage for Excess PV Electricity by Heat Pump

As most of the buildings' energy is for thermal loads, this study uses hot water storage to store the excess PV power (after supplying other electricity loads) instead of electrical batteries. When there is excess PV power, the excess electricity will be used to power the heat pump for producing heat, which is stored in separate hot water tank. The stored heat can only be used for DHW purposes in the chosen system design. The algorithm varies the speed (and thus the heating capacity) of the heat pump in order to match the total electricity load with the PV supply. If there is a battery in the system, the utilization of thermal storage is prioritized. The battery is only used when the thermal storage is fully charged.

14.3 Proposed Method for Optimizing PV Capacity/Positions

14.3.1 Optimization Algorithm and Fitness Functions

The aim of the modelling is to understand which features the system should have to perform well over its lifetime, rather than to estimate the exact energy or economic output of a pre-designed system. For this reason, a set of features should be optimized according to a specific measure of performance. The features of the system are expressed in the optimization problem as parameters and the measure of performance is expressed as a fitness function based on genetic algorithm. The parameters are the quantity of PV capacity installed on each façade of the building and the quantity of electric storage. Focussing on the capacity of PV on a building, in which there are two façades (roof and south façade) the set of parameters [0,0] represents a building without PV, [1, 1] is a building where all the area available is covered by PV, [0,0.5] is a building where half of the area available on the south façade is occupied by PV while the roof is empty. The quantity of electric storage simply constitutes a parameter as it is. The aim of the fitness function used in this study is to maximize self-consumed electricity (*SCE*) during the lifetime of the system. Equation (14.1) expresses the lifetime cumulative electricity self-consumed weighted by electricity price for the consumer.

$$SCE = \sum_{t=0}^{N} c \cdot P_c(t, hoy) \quad (14.1)$$

where,

- SCE is the lifetime self-consumed electricity;
- t is the year of operation of the PV system where N is the planned lifetime of the system;
- c is the cumulative electricity produced by the PV system and consumed on site (contemporaneously or through electric storage);
- $P_c(t,hoy)$ is the "point in time" price of the electricity for the consumer depending on the year t and the hour of the year (*hoy*) (in case of day/night or summer/winter variations). This value was added to privilege, at the same level of self-sufficiency, the solutions that generate a higher economic output: this may seem a monetary consideration, but because of how the fitness function is designed, any profit would indeed constitute a sort of buffer that the algorithm trades in exchanges for an improvement in self-production until any profit is dissipated.

The fitness function in Eq. (14.1) presents a problem because it is monotonic relatively to the capacity, so it cannot decrease amid an increase in capacity and it will indeed increase as long as there is a gain in self-consumption. This feature of the fitness function would obviously cause the optimal set of parameters to converge on the largest possible capacity (i.e. [1,1,…,1]), hence causing a grossly over-dimensioned system and defeating the very purpose of an optimization process. To avoid the excess generated by the fitness function in Eq. (14.1) while still maximizing it and to avoid economically unprofitable solutions, the function is described as following Eq. (14.2), where the fitness function is described according to two domains. The algorithm will thus maximize the lifetime SCE at the condition that the system cannot be unprofitable.

$$Fitness function = \begin{cases} SCE & NPV \geq 0 \\ -SCE & NPV < 0 \end{cases} \quad (14.2)$$

In this way, the monotony of the function with respect to the capacity is used to guide the algorithm towards solutions that are not unprofitable, in fact once the NPV becomes negative, an increased capacity would increase the absolute value of the fitness function, rendering it more negative. In practice, the fitness function will maximize the self-sufficiency while guaranteeing that the system does not become a net cost during its lifetime.

The formula of NPV calculation is expressed by Eq. (14.3):

$$NPV = \sum_{t=0}^{N} \left(\frac{c \cdot P_c + s \cdot P_s - \omega_{PV} \cdot (CM_{PV,t} + CS)}{(1+i)^t} \right) - \omega_{PV} \cdot CI_{PV,0} - \omega_B \cdot CI_{B,0} \quad (14.3)$$

where,

- The quantity $c \cdot P_c$, represents the costs avoided for the electricity that was not purchased from the grid because of the self-sufficiency; P_c is the price for the electricity paid by the consumer: $c \cdot P_c$ is treated in the formula like an economic gain. An avoided cost and an earning are not different for any practical purpose in this optimization;
- Similarly $s \cdot P_s$ represents the revenues generated by the fraction of the electricity that is sold s multiplied by its price P_s: notice that the

price for the valorization of the electricity sold is usually lower than the price paid by the consumer (P_c).

- The letter ω is meant here to represent the capacity of a component: ω_{PV} is the capacity of the PV system in [kWp] while ω_B is the capacity of the electric storage (or battery) measured in kWh;
- Similarly, the capital letter C represents the costs and is found in three values: CM denotes the unitary costs for maintenance [€/kWp year]. CI stands for the unitary installation costs and is measured in [€/kWp] for PV and [€/kWh] for the electric storage. CS is the cost for the substitution of components and involves inverters and electric storage systems;
- The letter i in the denominator of the quantity in the annual sum indicates the discount rate applied for the investment as defined in Lovati et al. (2018).

The result of the optimization process strongly depends on the demand but, as Sect. 14.4.3 shows, the demand itself is influenced by the excess PV electricity (the excess PV power will be used by the heat pump), which is in turn influenced by the result of the optimization. The optimal PV system is generally characterized by some hours of over-production along the year: if this over-production is used to heat a thermal storage, the energy transfer to the local grid is reduced. The reduction of over-production would cause an increase in the optimal dimension of the PV system because it will effectively be an increase in electric demand, this causes a positive feedback loop (see Fig. 14.4) until convergence (i.e. when the average temperature of the extra storage is high enough to send electricity to the grid anyway).

14.3.2 Key Performance Indicators Considered

This study mainly considers four KPIs, self-consumption (SC), self-sufficiency (SS), expected levelized cost of electricity (LCOE), and

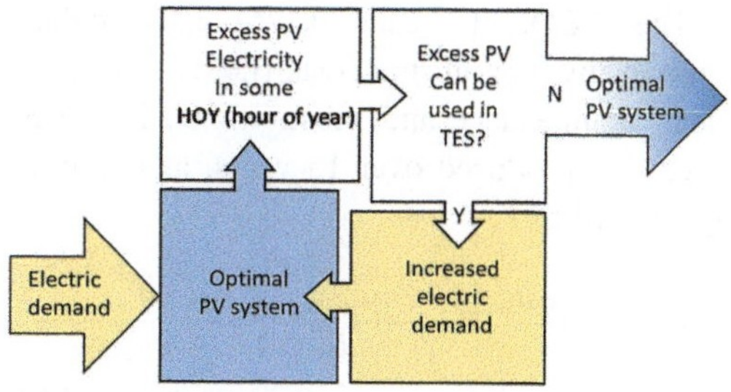

Fig. 14.4 Schematic representation of the feedback loop till convergence of the optimization process and the use of excess PV electricity in an extra thermal storage

expected self-consumed-LCOE ($LCOE_{self}$). These four KPIs are introduced in this section. The SC is the annual average of the rate (expressed as a %) at which the electricity produced by the PV system is consumed on-site. It is calculated by Eq. (14.4).

$$SC = \frac{E_{pv,onsite}}{E_{pv,onsite} + E_{pv,offsite}} \tag{14.4}$$

[kWh] is the aggregated PV power that is consumed on-site during one year period, and $E_{pv,offsite}$ [kWh] is the aggregated PV power that is consumed off-site (i.e. exported to the power grid). A larger SC indicates a better performance in terms of load matching.

The SS (expressed as a %) represents the annual average of the rate at which the electricity used by the building is provided by the PV systems. It is defined by Eq. (14.5).

$$SS = \frac{E_{pv,onsite}}{E_{d,whole}} = \frac{E_{d,pv}}{E_{d,pv} + E_{d,grid}} \tag{14.5}$$

where, $E_{d,pv}$ [kWh] is the aggregated electricity demand that is supplied by the PV system during one year period, and it is equal to $E_{pv,onsite}$ [kWh]. $E_{d,grid}$ [kWh] is the aggregated electricity demand that is supplied by the power grid. The sum of these two terms equals $E_{d,whole}$ [kWh] (i.e. whole electricity demand) of the building regardless of which source is providing it, a larger SS indicates a better performance as it refers to a building that is less reliant on the grid.

The LCOE [€ cent /kWh] is calculated through dividing all the costs (i.e. initial investment, maintenance and substitutions) by all the electricity produced over 15 years, as expressed by Eq. (14.6).

$$LCOE = \frac{Cost_{initial} + Cost_{maintenance} + Cost_{substitution}}{E_{pv,total}} \quad (14.6)$$

where $Cost_{initial}$ [€ cent], $Cost_{maintenance}$ [€ cent] and $Cost_{substitution}$ [€ cent] are the initial investment, maintenance cost and substitution costs, respectively, and $E_{pv,total}$ [kWh] is the aggregated PV power produced over the 15 years. The expected self-consumed LCOE [€ cent /kWh] refers to the LCOE for the electricity that is self-consumed, it is therefore obtained as the total costs of installation and maintenance divided by only the electricity self-consumed, as shown by Eq. (14.7),

$$LCOE_{self} = \frac{Cost_{initial} + Cost_{maintenance} + Cost_{substitution}}{E_{pv,onsite,total}} \quad (14.7)$$

where $E_{pv,onsite,total}$ [kWh] is the aggregated self-consumed PV power over the 15 years. Since the amount of self-consumed PV power is usually smaller than the total PV generations, the expected self-consumed LCOE is usually larger than the expected LCOE. In other words, this KPI considers the cost of electricity as if only the "onsite" share has been produced. This was made because the cost of production of a resource should refer to the share of this resource that is usable, and the over production from PV cannot be considered usable by the district. The excess PV electricity is in theory usable by someone else in the larger grid, but it cannot be guaranteed that all of the electricity will be utilized (especially in a future with a high penetration scenario for PV). In this sense this KPI can be interpreted as an extremely conservative value for the real cost of electricity by urban PV. Notice that if the electricity sold to the grid has value (i.e. $P_S \neq 0$) the $LCOE_{self}$ will be higher than the average price for the consumer (i.e. $LCOE_{self} > P_c$average), this is consequence that there are some revenues that can be used to purchase a larger system and so increase the SCE.

14.4 Modelling Approach for the Coupled Systems in Building Cluster

14.4.1 PV System Modelling and the Related Assumptions

The PV systems are modelled to ensure not only accurate operation, but also reducing the effort for collecting model inputs in the early design stage. In a complex context such as PV, the biggest losses are due to the partial shading of modules and arrays. In this chapter, the power profile of the PV system is estimated as proportional to the irradiation falling on the module, but corrected according to the cell temperature (Maturi et al. 2014) and a temperature coefficient as shown by Eq. (14.8) (Reich et al. 2012).

$$P_{PV,HOY} = PR \sum_{Mod=0}^{n} H_{Mod,HOY} \cdot \eta_{MOD} \cdot A_{MOD} \cdot c_T(Mod, HOY) \quad (14.8)$$

- $P_{PV,HOY}$ represents the power output of a PV system in a specific hour of the year (*HOY*);
- *PR* (performance ratio) is static performance ratio of 0.8 (Reich et al. 2012) that takes into account losses such as soiling or reflection. PR is defined as the ratio between the system yield (energy produced in time period over the nominal power) and a reference yield (the incident solar energy in time period *t* over the reference irradiance 1000 W/m^2);
- *Mod* is the latest module in the system characterized by its efficiency η_{MOD} and area A_{MOD};
- $H_{Mod,HOY}$ is the irradiation intensity [kWh/m^2] falling over a specific module *Mod* in a specific *HOY;*
- $c_T(Mod, HOY)$ is a temperature correction coefficient calculated as in 2019 and

dependant on the module temperature which is in turn determined by *MOD* and *HOY*. The temperature is found by the simplified relation $T_{Mod} = T_{ambient} + k \cdot H_{MOd,HOY}$ where $T_{ambient}$ [°C] is the ambient temperature retrieved from the weather file and k [m^2/kW] is the Ross coefficient described in Maturi et al. (2014) and Skoplaki and Palyvos (2009).

Aside from the modelling of the power production curve, the cost of the system is modelled as well: the initial costs are composed by total system costs (including modules, inverters, cables, structure installation and taxes), which are assumed to be directly proportional to the capacity installed following a linear relation. In reality, the unitary cost is probably bound to decrease for larger capacities, but due to the difficult estimation of the phenomena in an urban context (where large homogeneous production plants cannot be built), the price is assumed independent from the capacity. This approximation can be considered conservative as it will advantage smaller capacities during the optimization process. Likewise, the operational costs are also considered linearly correlated with the capacity of the system: these costs consist of maintenance costs and costs for the substitution of the inverter. The maintenance costs are an annual expense that should be paid proportionally to the capacity installed (expressed therefore in unit of €/kWp year), the inverter cost is an expense that comes once every 10 years and corresponds to the amount of 250 €/kWp. As for the calculation of the power, the exact number of inverters and the strings they serve is unknown, because of this the cost of inverters is included in the unitary price of the PV system at installation and amounts to 250 €/kWp in the years when it is substituted. The electric storage system is not interested by maintenance costs, but it generates initial and substitution costs every 10 years as well: these costs are proportional to the storage capacity, the initial cost of the electric storage is selected as an input while the substitution costs are averaged from the learning curve taken from three independent studies on the matter (Vartiainen et al. 2017; IRENA 2030; Leuthold 2014). It should be noted that given the current economics, the electric demand and the location, the electric storage suggested by the algorithm has an exceptionally low capacity.

14.4.2 Thermal and Electrical Loads of Buildings

TRNSYS 18 (Klein et al. 2017) is used to simulate the building and energy systems in a two stage process where an hourly space heating load file is generated by a building model in the first stage that is then used as input in the second stage to the HVAC system model. This in turn generates an hourly electricity load profile that is used as input by the PV optimisation tool (see Sect. 14.3.1).

In the first stage, all three buildings are modelled in one Type 56. A 3-D model built in the tool SketchUp is used to generate the geometry of the buildings. For building C, four zones are used, one for each of the floors plus one for the attic. For the other two buildings, where the ground floor has parts that contain flats and other parts that are not kept at normal room temperature, two zones are modelled for the ground floor in addition to the zones for the other floors and the attic, making five in total. The model is then converted into the non-geometric mode in order to have a faster simulation time, and material properties based on the real building were added. Ground coupling is done using a simplified approach with Type77 providing the ground temperature to which heat losses are calculated. The occupancy, electricity and DHW load profiles for the flats are derived by a stochastic model developed by Widén and Wäckelgård (2010) and the annual electricity usage has been calibrated to the measured data. The internal gains due to operational electricity are based on the measured values and assume 100% is converted to heat. The simulation model is then calibrated with the available measured data in order to achieve acceptable accuracy.

The HVAC system model is a simplified model of the proposed HVAC system, as shown by the TRNSYS model in Fig. 14.5. The space

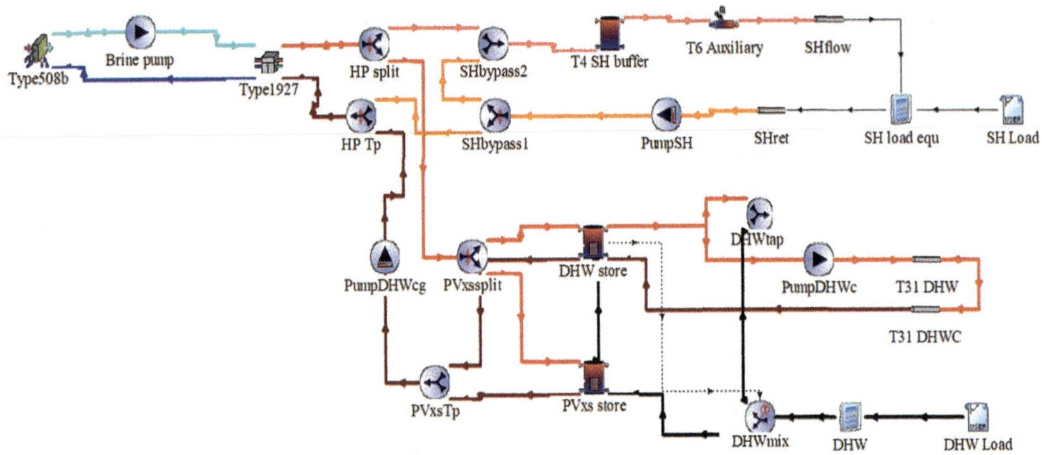

Fig. 14.5 HVAC system modelling in TRNSYS

heating and DHW loads are input as load files ("SH Load" and "DHW Load" respectively in Fig. 14.5), with one load file for the whole district. Similarly, only one exhaust air heat exchanger is modelled using Type508b as brine source to the heat pump, using the total ventilation flow rate for the district of 1460 l/s as input. The heat pump is modelled using Type1927, which uses a performance map with source and load inlet temperatures, source flow rate as well as compressor frequency as independent variables. A detailed performance map covering the range of operating conditions is provided by the manufacturer of the heat pump that is used in the system. The heat pump has a nominal heating capacity of 45 kW with a COP of 3.82 at B0W35 and 102 Hz compressor frequency. The space heat is controlled using a heating curve (a curve describing how supply water temperature should be set based on the outdoor air temperature) for a heating system with design temperatures of 55/45 °C at a design ambient temperature of −23 °C. "SH Load Equ" removes the heating rate specified in the load file "SH Load" from the space heating flow, while the two distribution pipes are for all the three buildings and based on the sizes, lengths and insulation standard in the buildings.

The heat pump frequency is controlled using a PID controller to supply the current flow temperature according to the space heating curve. If the maximum frequency does not give enough heat, a Type 6 auxiliary heater with 200 kW maximum heating rate adds heat in order to generate the required flow temperature. This auxiliary heater, representing the pellet boiler of the real systems, is thus only used for space heating, and supplies only as much heat as required to match the space heating load. A small buffer store of 370 L, modelled using Type 4, is located in series between the heat pump and the auxiliary heater.

The DHW of 2.5 m^3 and the extra store for excess PV in the form of heat ("PVxs store") are modelled with one Type 534 each, using five zones and connections at the top and bottom. The volume of the PVxs store is 3.5 m^3 for the case study. The U-value for the store heat losses is calibrated to give the measured heat loss from a typical DHW store at rated conditions. "DHW Load" and "DHW" provide the load mass flow rate at a time resolution of one minute together with the cold water temperature that varies with the seasons. The cold water is preheated in the extra hot water store ("PVxs store") in series with the DHW store. DHW circulation is connected to the DHW store and has pipes calibrated to give losses of 0.57 W/m^2 of living area, 2.2 kW in total for the three buildings. The DHW store is charged from the HP at full power using an on/off controller designed to maintain the store at the height of the sensor to between 51 and 56 °C.

During charging the three-way valves "HP split", "HP Tp", "SHbypass1" and "SHbypass2" are switched so as to send the flow from the heat pump to the stores and for the space heating loop to bypass the heat pump.

14.4.3 Storage of Excess PV Production as Thermal Energy in an Extra Thermal Energy Storage

The HVAC model also reads in the electricity use profiles for the building (flats and operational electricity) and the EVs as well as the production of the PV array, all with a time resolution of one hour, which is also the time resolution of the PV optimisation tool. The model then calculates the excess PV power available for running the heat pump, after use in the flats and for operational needs. If the heat pump is already running to supply space heat or to charge the DHW, no action is taken and the system operates as normal. If the heat pump is not in operation, the heat pump is turned on and the flow is switched to charge the PVxs store using the three-way valves "HP split", "HP Tp", "PVxssplit" and "PVxs Tp". The compressor frequency is then controlled so that its electricity load matches the available PV excess, using a function with excess PV power and condenser inlet temperature as independent variables. This is limited to the maximum compressor frequency and continues until the PVxs store is fully charged, 56 °C. The excess PV that is stored in the form of heat, is thus only used for supplying heat for DHW and can only be stored when there is no "normal" need for space heating or for charging the normal DHW store, and thus represents a conservative capability. The temperature in the store is varying from close to that of the mains cold water supply to 56 °C at the temperature sensor when fully charged. Thus, the preheating of hot water has to be achieved with a heat exchanger to avoid legionella problems, the preheated water being then heated to above 50 °C in the main DHW stores.

14.4.4 Electric Vehicle Load Generator

The EV load is generated by using the Grahn-Munkhammar model (Grahn et al. 2013). It simulates the EV home-charging based on standard settings of 0.2 kWh/km electricity use (including losses) and 24 kWh battery capacity available for trips, and a total distance driven per year of about 12,200 km as a Swedish average scenario. The used model considers the EV battery charging process is related to household activities (i.e. away, sleeping, etc.). For example, the EV owners usually charge the EVs after returning home from work, and thus charging process is usually activated in this period. The household activities are first computed by a discrete Markov-chain model. Then, based on the obtained household activities, the usage of EVs and the charging load profiles are calculated. The state of charge ($SOC_{i+1,j}$) of the jth EV battery in the (i + 1)th time interval is calculated by Eq. (14.9).

$$SOC_{i+1,j} = \begin{cases} SOC_{i,j} - \zeta(v, C_i^s) & \text{if consuming} \\ SOC_{i,j} + C^p \Delta t & \text{if charging} \\ SOC_{i,j} & \text{else} \end{cases} \tag{14.9}$$

When the EV is being used, the electricity consumption $\zeta(v, C_i^s)$ is calculated based on the EV velocity (v) and the season (represented by a seasonal coefficient C_i^s). When the EV is being charged, the *SOC* of the battery will increase at a constant charging rate of C^p(i.e. 2 kW used in this study). Δt is the time step for calculating the EV battery *SOC*. To prolong the service life of battery, full charging/discharging cycles should be avoided when using the battery, and thus a minimal *SOC* value should be considered. As depicted by Eq. (14.10) , the lower limit of *SOC* is determined by a fraction p_{dod}, which defines the minimal depth of discharge (DoD).

$$p_{dod} SOC_{max} \leq SOC_{i,j} \leq SOC_{max} \tag{14.10}$$

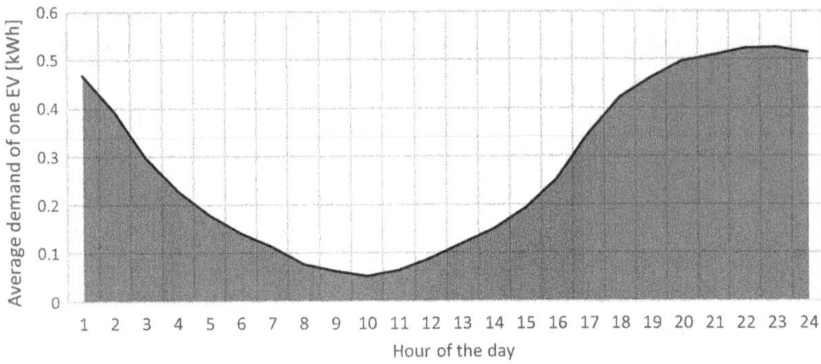

Fig. 14.6 Average EV charging load in one day

The EV load imposed on the building is then calculated based on the charging power of EV battery, as shown by Eq. (14.11).

$$P_{i,j} = \begin{cases} C^P & \text{if charging} \\ 0 & \text{else} \end{cases} \quad (14.11)$$

Figure 14.6 shows the hourly EV charging load in a typical day. The charging load is small during daytime and reaches the minimum during 9:00–11:00, while it reaches the maximum at night during 22:00–24:00. Note that in peak demand time (i.e. between 22:00 and 24:00) the EV demand is still lower than the charging power of the EV plug (i.e. 2 kW), this is due to the fact that the EV are not always charging contemporaneously.

14.5 Boundary Conditions, Input Parameters and Regulation for PV Electricity Sharing

The simulation tool requests a series of input parameters summarized in three categories in Fig. 14.7. The main input is a 3D model describing the building geometry and being used to calculate the irradiation matrix. This is expressed in W/m^2 for every hour of the year, and for each building unit surface. Every point represents a solar collector with a given area and is associated with an hourly irradiation. Hourly weather data for Borlänge regional airport, from the extended weather data set in TRNSYS, is used for all stages of the process: building simulation, PV optimisation tool and HVAC system simulation. The site of the weather station is roughly 45 km north-east of the building cluster in Sunnansjö, and has a similar climate. The average ambient temperature and annual global radiation are given in Table 14.2 together with key figures for the energy demands in the case study.

The PV module efficiency data is provided by LudvikaHem (i.e. the building owner) based on field measurements. The values of parameters related to economic analysis, including the electricity price, PV system price and electric storage price, are also provided by LudvikaHem. The price of the electricity for the consumer is assumed to be 0.16 €/kWh year round according to a long-term contract stipulated by the building owner. This arrangement is more profitable in terms of PV profitability compared to a monthly variable plan as, in the latter, the largest possible earnings are in periods when the radiation is unavailable. In a monthly variable plan the months where most earnings are possible for a PV system are only March and October. The electricity that is not contemporaneously self-consumed is assumed to be sent to the grid for 0.05 €/kWh. Nevertheless, it is assumed that the price paid by the energy provider for the excess PV electricity is going to decrease alongside the lifetime of the system or at best stay the same. For what concerns the EVs, the possibility to charge two EVs will be guaranteed immediately after the restoration project, nevertheless the long-term level of penetration of the EVs is unknown to this day. It is assumed that the

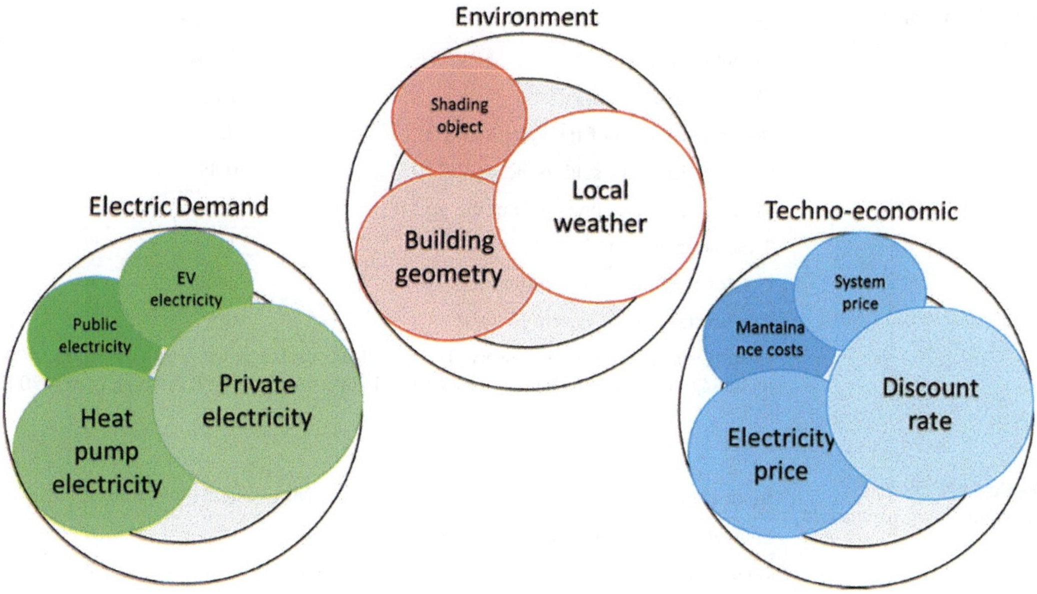

Fig. 14.7 Inputs for the optimization tool

Table 14.2 Key figures for the case study (all annual values) with no PV *system*

Quantity	Value
Global radiation [kWh/m^2]	971
Average ambient temperature [°C]	4.1
Space heating demand [MWh]	395
DHW demand [MWh]	80.5
DHW circulation demand [MWh]	17.7
Heat supplied by auxiliary heater [MWh]	214
Heat pump electricity [MWh]	78.2
Operational electricity [MWh]	42.5
Flat electricity [MWh]	89.0
EV demand (2 EVs) [MWh]	4.8

maximum penetration possible for the electric vehicle would be of one EV per family for a total of 48 EVs. Different optimization is therefore performed: with two EVs, with 25 EVs and with 48 EVs. Aside from these inputs, Tables 14.3 and 14.4 report the set of techno-economic parameters required for PV and other components.

Figure 14.8 represents the variation of electric demand due to the EVs on an annual basis and for the average day. The EV charging load is not much affected by the season in terms of cumulative demand, shown in Fig. 14.8a. This is because of two reasons: (1) The low operation temperature in winter will reduce the battery capacity, which can lead to reduced ranges. However, by increasing the charging frequency (i.e. how often the battery is charged) or the average depth of discharge, the overall cumulative charging loads (which can be considered approximately proportional to the product of ranges and charging frequency) are still likely to be stable. (2) For the EVs used in cold regions, large amount of electricity is needed for heating the interior of the car, leading to reduced

Table 14.3 Input parameters of PV system

Input name	Vvalue
Module efficiency	0.174
Mesh dimensions [m]	1.65 × 0.992
Performance ratio of the system at STC	0.8
Price of electricity sold to the grid[€]	0.05
Price of electricity bought from the grid [€]	0.16
Time horizon [years]	15
Cost of the finished PV system [€/kwp]	1420
Cost of the storage system [€/kwh]	670[a]

[a]Used Tesla powerwall: 1 Powerwall = 13.5 kWh usable power and costs 7.030,00 € (including taxes) + installation costs assumed 2000 € $\geq$ 9.030,00 € which is ca. 670 €/kWh

Table 14.4 Techno-economic input parameters

Input name	Min value	Max value
Annual maintenance costs [€/kwp year][a]	0	15
Linear annual growth of the electrical load	0	2
Linear annual efficiency losses	0.5	1
Annual discount rate	0	2
Linear annual growth of bought electricity	0	3
Linear annual growth of sold electricity	−1	0

[a]This cost does not include the substitution of inverters and batteries

available amount of battery-stored electricity for EV motion. However, in Sweden such amount of electricity is partly provided by the buildings (i.e. the heating process occurs in the car park before the EV usage), and the battery in the car does not need to supply heat to heat up the car before a journey. Since most of the electricity stored in the EV battery is still used for motion in winter seasons, the EV ranges will not be greatly reduced because of the increasing heating needs in Sweden. Please also note that even with assumptions that lead to a significant amount of heating by the car battery (greater winter EV charge demand), the results of the study would not change as this increased demand occurs only during the months with very little PV production, and when all PV can be used for other loads. In annual cumulative terms, each EV absorbs little over 1MWh so that the aggregated demand of 23 EVs requires an amount of energy that is almost equal to 30% of the baseload. In the hourly average load over the year, displayed in Fig. 14.8b, it is visible that the EVs are adding their demand mostly at night, and the additional load thins out during the daytime (especially during the late morning). In general, the annual behaviour of the EV demand can be considered advantageous for the PV installation, because the PV produces proportion proportionally more during the summer months when the rest of the load demand is the least. Nevertheless, the prevalence of the load at night risk to render the PV less useful unless electric storage is installed. On the other hand, an electric storage is extremely unlikely to be profitable as there is probably no over-production of PV electricity during the winter months (thus forcing the storage to have idle time and therefore reducing its profitability).

The construction of DC microgrid and the sharing of PV power among different buildings should obey the local regulation and policy and additionally needs approval from the grid operator. Until now, although many countries permit the feed-in of PV power to the power grid and clearly established regulations, renewable energy sharing among different electricity prosumers is

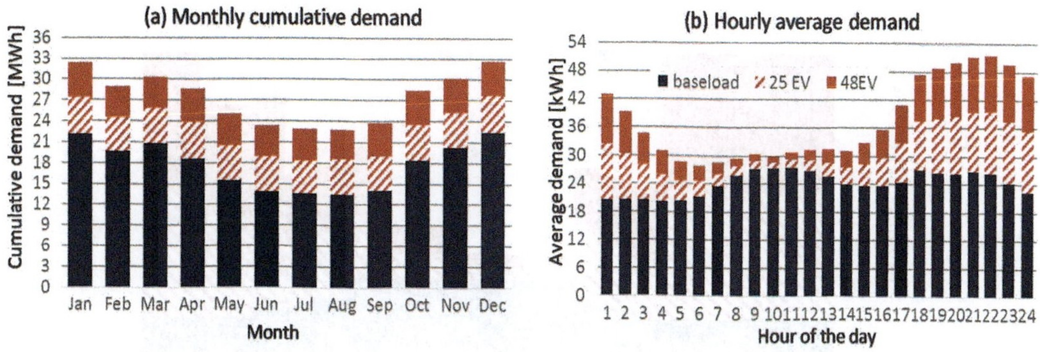

Fig. 14.8 Variation of building electric demand with EV demand considered **a** on annual basis and **b** in the average day (i.e. hourly average values over the year)

still not allowed and the related regulations are very unclear. In terms of regulatory condition in Sweden, DC micro grid has been deployed in building sectors for promoting the utilization of renewable energy in a few cases. When the local micro gird consists of DC lines that directly connect PV production plants, they are covered by the exemption under § 22 (a) of the IKN Regulation 2007:215 (MoIR 2013). While when the local micro grid consists of lines connecting buildings that are not equipped with PVs, the situation is a bit unclear for sharing the produced solar energy among buildings. A preliminary ruling from the Swedish Energy Markets Inspectorate is then needed by the grid owner for considering approving an exception from the grid concession. In this study, all the three buildings are planned with PVs, so an exemption is achieved according to the regulation, which means it is possible to share the produced PV electricity by micro grid among these three buildings. Unfortunately, in many other European countries, due to the concern of system reliability and safety, such micro grid application and energy trading among different small electricity prosumers are still not allowed.

To estimate the impact of the possibility of sharing electricity, the private consumption of the flats in the cluster is assigned differently to the three buildings. At first, the power demand is assigned proportionally to the area (obtaining ca. 21 MWh/m^{2}year), then three different occupancy schedules have been assigned to the three buildings. It is assumed that the building A is a retirement home and has therefore a similar occupancy to an hospital, while the other two buildings are regular households with most of the occupancy during the morning, evening and night. To estimate the occupancy profile of the three buildings, the schedules from Ahmed et al. (2017) are used, and the resulting occupancy profile of the whole cluster is shown in Fig. 14.9. By assigning shares of the total demand proportionally to the contemporary inhabitants, the demand profile in Fig. 14.10 is obtained. Building B and C have a slight peak before 8:00 and a sustained demand between 18:00 and 21:00. On the contrary, the A building features a strong and sustained demand between 10:00 and 16:00.

14.6 Optimization and Sensitivity Analysis Results

In this section, the optimal configuration of the PV system is shown in terms of capacity on the different roofs and façades in the building cluster, and different optimal PV configuration will be shown following an increasing order in terms of demand covered (see Fig. 14.11 and Table 14.5).

The first run of optimization (i.e. Scenario 1) is performed using as electric demand: public and private lights and appliances, electricity for the operation of the condominium devices, heat pump, heating of the DHW and two EVs. After

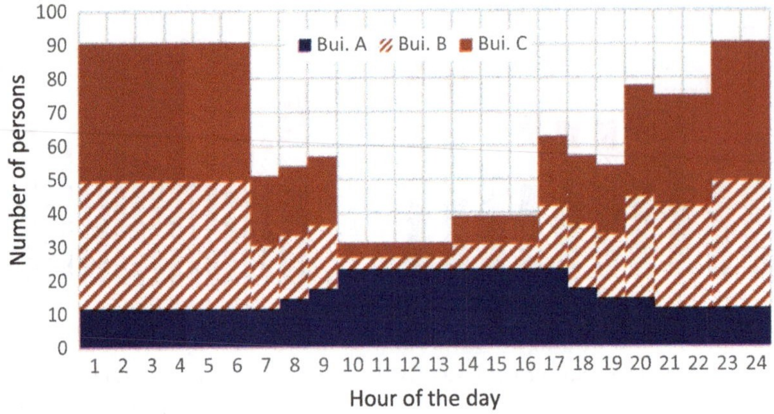

Fig. 14.9 Occupancy assigned to the three buildings in the cluster along 24 h (Building A is assumed in this simulation to be a retirement home)

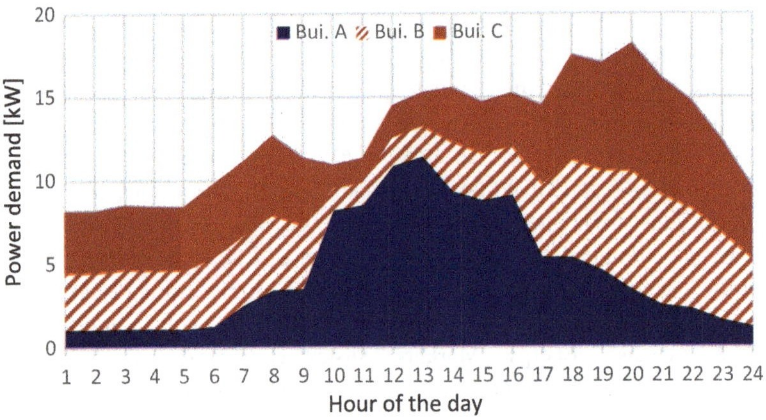

Fig. 14.10 Electric demand of January the first and relative quota assigned to the three buildings during the 24 h

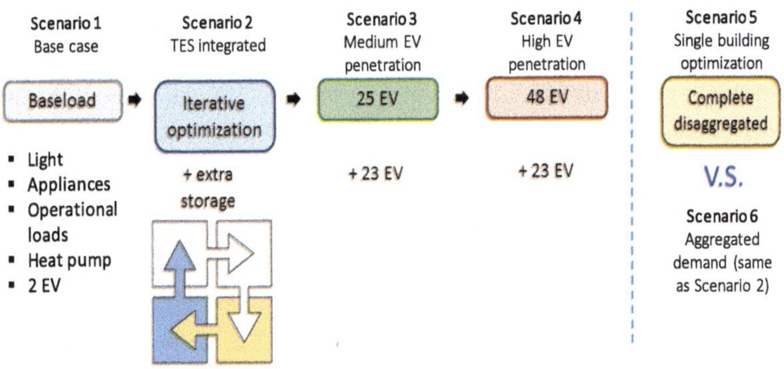

Fig. 14.11 Steps to produce the five optimal configurations analyzed in the study: the baseload consists of public and private lights and appliances, common operational loads, heat pump demand, domestic hot water and two EVs. The tone of color used in the chart is reported also in the 3D representation of the different configurations

Table 14.5 Configurations of systems and demands in each scenario

ID	Scenarios	Energy sharing	Thermal energy storage	EV number
1	Base case	Yes	No	2
2	TES integrated	Yes	Yes	2
3	Medium EV penetration	Yes	Yes	25
4	High EV penetration	Yes	Yes	48
5	Single building optimization	No	No	2
6	Aggregated demand (same as Scenario 2)	Yes	Yes	2

that, an iterative optimization (i.e. Scenario 2) is performed following the procedures described in the Sect. 14.3.1 in order to add an extra thermal storage for hot water. The iterative process is useful because it enables the control strategy of the heat pump to use the excess PV electricity when this is available. Once the demand is increased with the special additional thermal storage, progressively more EVs will be added (i.e. Scenarios 3 and 4) and the impact on the KPIs is assessed. All the optimization process explained in this section refers to an aggregated electrical demand for the whole cluster, this takes for granted that some form of electricity sharing exists among the three buildings. The last optimization (i.e. Scenario 5) explores the situation in which there is no possibility to exchange flat electricity among the buildings (except operational electricity). The positive impact of the energy sharing technology on the performance of the PV system is measured through the difference in optimal capacity and in performance KPIs.

14.6.1 Design Results of the Coupled System for the Base Load Scenario

The first run of optimization is the one that includes the smallest possible electric demand, in this scenario no extra thermal storage for the excess PV electricity is included and the number of EV is only the minimum of two in the whole cluster of buildings. Figure 14.12a shows the optimal configuration of PV modules over the roof and façade of the cluster. The screenshot on Fig. 14.12b shows in color-coded disks the annual cumulative irradiation over the different façades. The southern slope of the roof is clearly more irradiated than the rest of the surfaces made

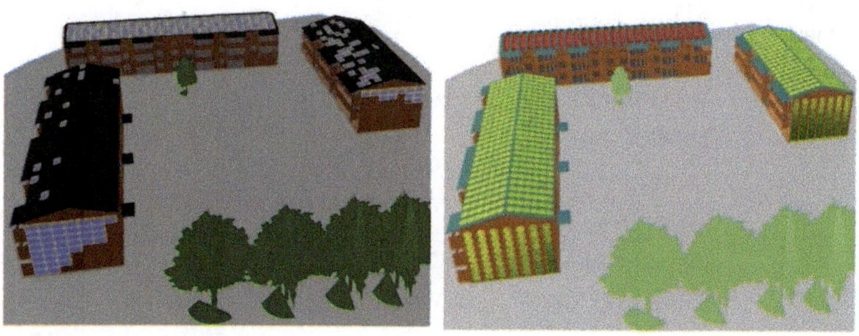

(a) (b)

Fig. 14.12 **a** The optimal configuration for the baseline case, most of the system is installed in the southern slope of the roof and on the southern façade; **b** Color-coded depiction of the annual cumulative irradiation, despite the higher area installed the façade is less irradiated overall available for a PV system averaging just below 1200 kWh/m^{2}year. The east and west slopes are a bit better irradiated than the south façades, despite this, a significant portion of the system is installed on the southern façade while much of the roof is still available. The southern façade turns out interesting for the optimization algorithm as it enjoys a more homogeneous irradiation throughout the year, during the winter season the solar angle is closer to the horizontal than the vertical and irradiates the southern façade more than the roof.

In Fig. 14.12a the modules on the east and west slopes of the roof have a dispersed pattern, because the placement of the modules is chosen randomly depending on areas with essentially equal radiation. Thus, the PV modules placed randomly by tool, which can be placed together in practice. In the case shown in Fig. 14.12b, the result of the tool is interpreted as follows: the southern slope is the most profitable surface and should be exploited as much as possible (ca. 45 kWp out of 65 kWp); the south façade should be used (ca. 6 kWp) even if large portions of the roof are still available, but the lower parts of the two façades should be avoided because of shading, while the east and west slopes of the roof should be used (ca. 14 kWp) and the exact position of the modules is not important from an optimization point of view. Table 14.6 shows the main KPIs reached by the optimal system for the baseload scenario.

Overall, the collection of KPIs can be considered satisfying, in fact it shows that it is possible to cover (contemporaneously to the production) ca. 20% of the electric demand of the cluster while retaining an excellent level of self-consumption of ca. 77%. Despite being at high latitude, the system reaches better results compared to other studies such as reference (Luthander et al. 2019). The reason for this might be the aggregation of the demand. A single-family house has a demand profile that is really hard to match for a PV system lacking storage (or with minor storage), this is due to the strong variability of the load that is characterized by an extremely low baseline and huge "spikes" or "peaks". The positive effect of the aggregation of the load is quantitatively discussed in Sect. 14.6.4.

It can be noticed that the optimal system often includes very small capacities of electric storage (see Tables 14.6, 14.7, 14.8 and 14.9), these capacities are trivial compared to the generation and loads, so cannot be making any meaningful contribution to the system. The result should be interpreted in general as the fact that the electric storage is unprofitable under these techno-economic conditions, nevertheless are reported for the sake of completeness. To avoid this aspect, it suffices to increase the capacity step for the optimization of the electric storage, in this study the capacity step used was 0.1 kWh.

Table 14.6 Main KPIs reached by the optimal system

KPI	Value
Installed capacity [kWp]	65.5
Installed storage capacity [kWh]	0.3
Installed area [m^2]	376.5
Capacity of electric storage [kWh]	0.3
System cost [€]	93,017
Expected self-consumed-LCOE [€ cent /kWh]	17.9
Expected LCOE [€ cent /kWh]	14.5
Self-consumption [%]	76.9
Self-sufficiency [%]	20.4
Annual cumulative production [kWh]	56,798
Annual cumulative balance production/consumption	0.3

Table 14.7 KPIs at the end of the three optimization processes

KPI	1st iteration	2nd iteration	3rd iteration
Capacity [kWp]	65.5	79.2	79.2
Battery Storage capacity [kWh]	0.3	0.0	0.4
Self-sufficiency [%]	20.4	24.8	25.1
Self-consumption [%]	76.9	78.3	79.4
LCOE self [€ cent/kWh]	17.9	17.8	17.7
LCOE [€ cent/kWh]	14.5	14.8	14.8
Residual demand [MWh]	170.8	160.1	159.2

Table 14.8 Various KPIs at the three levels of EV presence

KPI	2 EV	25 EV	48 EV
Installed capacity [kWp]	79.2	88.3	96.0
Installed storage capacity [kWh]	0.4	0.2	0.1
Expected self-consumed-LCOE [€ cent/kWh]	17.7	17.7	17.6
Self-consumption [%]	79.4	80.3	80.9
Self-sufficiency [%]	25.1	21.8	19.8
Annual cumulative demand [MWh]	213	274	330

Table 14.9. Selection of KPIs due to the impact of electricity sharing within the cluster

KPI	Building A	Building B	Building C	Disaggregated	Aggregated
Capacity [kWp]	23.1	24.5	14.5	62.1	79.2
Battery Storage capacity [kWh]	0.0	0.4	0.0	0.4	0.4
Self-sufficiency [%]	26.3	22.3	13.6	20.3	25.1
Self-consumption [%]	93.8	71.6	93.1	86.1	79.4
LCOE self [€ cent/kWh]	16.9	18.3	17.1	17.5	17.7
LCOE [€ cent/kWh]	16.2	13.9	16.1	15.4	14.8
Residual demand [MWh]	46.8	55.7	67.0	169.5	159.2

It should be noted that some external users might consume the electricity that is not self-consumed, this is out of the scope of the system. Furthermore, in a future where PV will become more pervasive, it is unrealistic to expect some demand by neighbouring clusters in over-production times. Compared to the present price of the electricity, the LCOE of the self-consumed fraction would be 12% higher (at almost 0.18 € against the present 0.16 €) while the whole electricity LCOE would be about 10% lower. Given the present costs and the economic modelling performed in this study, the tool suggests installing a very minor electrical energy storage quantity.

14.6.2 Impact of Thermal Storage Capacity on the PV Design and Overall Performance

The role of TES is to increase the self-consumption by enabling some part of the electricity to be used in a non-contemporaneous way. In Sect. 14.3.1, an iterative series of optimizations was described to be able to use the excess PV electricity. The ability of transforming the excess PV electricity in thermal energy generates an increase of the electric demand during the times of over-production. This encourages the next optimization process to install a higher capacity restoring the situation of overproduction in some hour of the year (HOY). The process is then repeated with the aim to exploit the over-production, nevertheless the average temperature of the storage is higher this time as the thermal need is limited, so the increase in electric demand is minor. The following optimization did not yield any increase of capacity and the process could be then considered converged. Table 14.7 shows the main KPI's for the first, second and third optimization. A volume of 3.5 m^3 was used in this case study for the PV excess thermal store.

The second iteration generates a large increase in the optimal PV capacity compared to the baseload scenario (+21%): this is not due to the increase in electric demand (the overall demand is basically unaltered as the increase during the central hours is compensated by a reduction in the evening), but due to the improved matching of the demand with PV production. The electric demand resulting from the use of excess PV electricity presents a bump during the central hours of the day. This feature makes it easier for the PV to match the electric demand and allows the optimization algorithm to install a larger capacity of PV. The two values for the LCOE shows how the iterations leave the LCOE almost unaltered. Nevertheless, it is possible to notice that the LCOE associated with the self-consumed quota is reducing while the overall one is increasing. This is not surprising since the self-consumption grows along the iterations boosting the cumulative electricity self-consumed and a larger system forces the algorithm to use position that are slightly less irradiated and therefore reduces the overall yield. The main result is the reduction in residual electric demand (i.e. the part that cannot be covered by the PV system + thermal storage) that achieves a −6.8% reduction over the whole process of roughly 11.6 MWh/year. Most of the reduction in demand is accomplished in the 2nd iteration when the PV capacity increases, but some reduction happens in the 3rd iteration and therefore is due solely to a better contemporaneity between production and consumption.

14.6.3 Impact of Electric Vehicle Variation on the PV Design and Overall Performance

PV optimization with the impact of variations of EV is shown in Fig. 14.13. The southern portion of the roof is the first one to be occupied by the PV system because it is the mostly irradiated part. With increasing presence of EVs, it is visible how the PV system grows in size. Despite having slightly higher irradiation compared to the façades, the east and west portions of the roof are not entirely utilized for the application of PV by the algorithm, the southern façades are used instead. The reason for this noticeable behaviour

(a) (b) (c)

Fig. 14.13 PV visualization with impact of variation of EV **a** two EV case; **b** 25 EV case; **c** 48 EV cases lies probably in it having a better performance during winter months, when the sun angles are closer to the horizontal and the electric demand is more prominent, the façade integration results therefore to be more profitable, thus prioritized by the algorithm.

Table 14.8 shows various KPIs at the three levels of EV presence. Despite a noticeable growth in the installed capacity, the larger growth of the demand forces the share of PV electricity to go down. There is a slight increase in self-consumption (not surprising considering that the whole system shifts towards larger load and larger capacity), thus a small reduction in the LCOE of the self-consumed electricity.

The results from this study are consistent with other similar analysis in Sweden. For instance, in Munkhammar et al. (2013) under different scenarios of PV capacities and EV penetrations, the self-sufficiency values vary within 20–30%, which is close to the values calculated in this study.

14.6.4 Impact of Electricity Sharing on the PV Design and System Performance

Using the optimization technique, the difference in performance between the optimal configuration for each building and the one for the whole cluster are addressed. The optimization for each building separately is the condition that would apply in the case in which there is no means of exchange of flat electricity among the different buildings within the cluster. In the case of the same heating demand and operational electrical load, Table 14.9 shows a selection of KPIs for two cases where the possibility to exchange flat electricity is extremely favourable for the PV market: (1) aggregated electricity sharing as a cluster, and (2) disaggregated cluster: no electricity sharing among neighbour buildings. In this example the aggregated cluster shows in fact an optimal installed capacity of a whopping 27.5% higher compared to the disaggregated one. This is because energy sharing makes the PV system more versatile (the whole cluster is more efficient at consuming the electricity produced on-site), hence improving its economic value. Because of the fitness function, where the self-sufficiency should be maximized, an increased value translates into a larger investment with the aim of increasing self-sufficiency, thus producing a larger overall PV capacity. This phenomena does not disagree with other studies such as reference (Shen and Sun 2016), where is shown that a smaller system can achieve the same level of performance of a larger one if sharing is taken into account. Also, in this case energy sharing would have allowed a smaller system to achieve the same performance of a larger one in a disaggregated scenario: but the aim is not to maintain the performance, but rather to out-perform it at the same price. In the disaggregated case, Building A has a better matching (better contemporaneity) because of the shape of the load matches the PV generations better (see Fig. 14.10). Building B has better yield thanks to the south slope and can afford to put some storage and highest capacity. Building C does not have a good yield nor contemporaneity. The higher capacity does generate more hours of over-production (self-consumption is reduced of ca. 7%), but the increase in capacity is more than enough to offset this effect causing an increase of the self-sufficiency of ca. 24%. These KPIs aside, also the LCOE has a benefit (in both ways it can be calculated) and the residual demand is reduced of about 6%.

Looking at the geometric patterns of installation (as shown in Fig. 14.14), it is visible that the south slope of the roof is not completely occupied by the PV system in the disaggregated case, this is an obvious source of inefficiency as that slope is the most irradiated part of the building and even if completely covered with PV does not cause significant over-production if applied to the whole cluster. In the disaggregated case, the possibility of installation on the southern slope is obviously limited by the lack of sufficient demand in the underlying building.

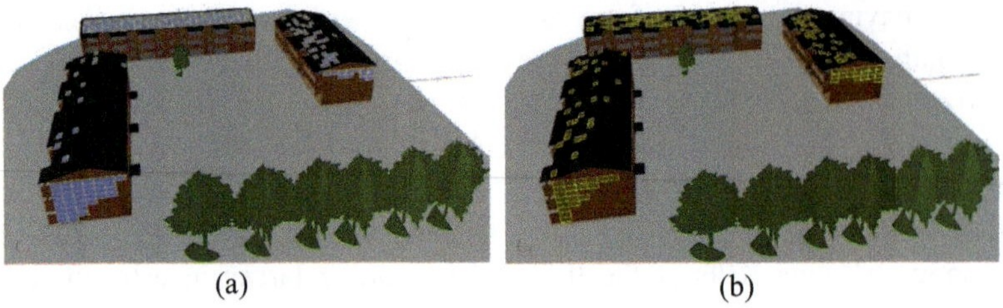

Fig. 14.14 PV visualization with impact of electricity sharing **a** Aggregated case; **b** Disaggregated case

14.7 Discussion and Outlook

For the demonstration building in Sweden, the most profitable surface for installing PV panels is the southern slope, since it is more irradiated ($\sim$1200 kWh/m^{2}year) compared with the other surfaces that are available for PV panel installation. The upper parts of the two south façades are the second most profitable locations for installing PV panels. This is because the south façades are irradiated more homogeneously throughout the year. The east and west slopes are the third most profitable surface for installing PV panels since they are not shaded and have relatively large solar irradiation.

The application of Energy Hub DC micro grid in buildings enables an easier, more convenient and more energy-efficient way for utilizing on-site produced renewable energy. This will help promote the deployment of renewable energy systems in the building sector. More importantly, the Energy Hub DC micro grid provides a platform for flexible energy sharing between different buildings. By enabling energy sharing among buildings, the buildings with surplus renewable energy generations can send their renewables to buildings with insufficient supply, thus achieving an improved match between the district-level renewable supply and electrical demand. The improved match helps boost the SC. Note that the significant increase in SC is contributed by the sharing mechanism at almost no extra cost, and thus should be possible to implement profitably in many cases, if national regulations allow this. Thus, the Energy Hub DC grid has huge potential to be applied in large scales for improving the SC of buildings. This is in line with Luthander et al. (2016), who investigated the impact of placement of meter and battery storage for a group of buildings and showed that centrally placed battery and metering for the buildings as a cluster resulted in increased self-consumption of the cluster compared to the case with buildings individually. It also showed that much less curtailment would be required if large amounts of PV were to be installed, and power input to the grid were to be limited. In this study, the considered three buildings all belong to residential buildings, which have similar occupancy schedules and load patterns. Such similarity limits the benefits from energy sharing, since the buildings may have surplus renewable generations or insufficient supply in the same period. When different types of buildings with different demand patterns are connected in one Energy Hub DC grid, more performance improvements in SCE are expected to be achieved. For instance, putting an office building and a residential building in one cluster, the potential renewable energy shortage of the office building during the daytime can be compensated by the surplus renewables from the residential building (Huang and Sun 2019).

Thermal energy storage is an efficient solution to improving the renewable energy self-consumption rate of buildings. The integration of thermal energy storage, together with appropriate control, will lead to increase in the optimal capacity of PV systems that maximizes the SC, as the charging of thermal energy storage is treated as extra electrical demand. By changing the shape of the electrical demand to match the renewable energy generations, an increased SC

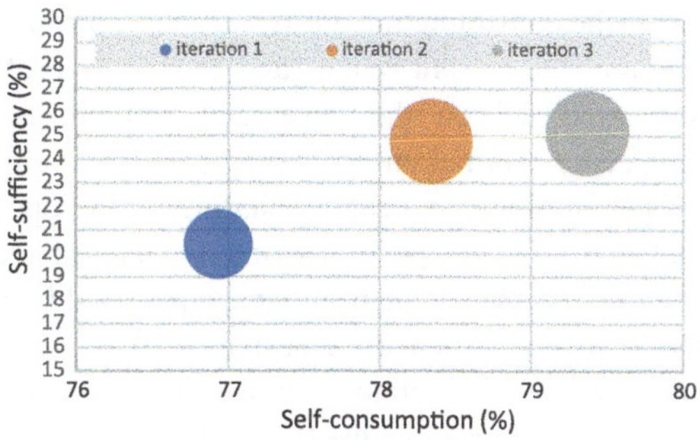

Fig. 14.15 Self-consumption and self-sufficiency scheme as proposed in Luthander (2018), the diameters of the bubbles represents the capacity of the PV system

can be achieved (i.e. the self-consumption increased from 77 to 79.4%) at the same time as the total installed capacity increased. Figure 14.15 presents the KPIs of the three iterations (refer to Sect. 14.6.2) using the scheme proposed in Luthander et al. (2019), as the authors points out the iteration 2 and 3 are aligned respect to the origin of the axes and the increase in self-production is therefore proportional to the increase in self-consumption.

It should be noted that this study is a single-objective optimization by the basic genetic algorithm, and previous research has shown that significant improvements can be achieved with more advanced algorithms (Psimopoulos et al. 2019), meaning that the results shown here are conservative compared to what can be achieved in practice. However, if thermal storage of excess PV in the form of heat is to be included in the PV optimisation tool, only basic algorithms will be feasible, together with a simplified model of the heat pump system. The energy sharing is likely to help reduce the capacity of TES to some extent. When energy sharing is enabled in the building cluster, the required optimal capacity of TES can be reduced, compared with the scenario in which energy sharing is not allowed. This is because the district-level electric demand can match the district-level renewable supply better as discussed previously and shown by Luthander (2018), and thus a smaller sized TES is needed to compensate the energy mismatch. The integration of TES does not affect LCOE too much, as both the renewable energy generations and the costs increase. Another potential application of the thermal energy storage is to conduct demand management in response to the varying electricity prices. By storing the grid power in low-electricity-price period in advance, economic savings can be achieved for the building cluster (Psimopoulos et al. 2019).

With increased EV penetration, the cost-optimal PV capacity will also increase due to the increased electrical demand. Increasing the number of EVs will lead to a slight increase in the SC, as the whole system shifts towards larger load and larger capacity. The EV profiles used have only a small seasonal variation. This is not always correct for the cold Swedish winter climate, when a significant amount of heating is required for the interior of the car, which increases the amount of electricity used for a given journey. How large this heating demand is depends on the duration of the journey and not its distance. The electricity demand of the buildings used includes electrical heating for cars in the car park before use, as is common practice in Sweden, so the battery in the car does not need to supply heat to heat up the car before a journey. For the location of the cluster, there is no traffic congestion for normal journeys and thus the energy use for motion will be much higher than that for heating, even in winter. Additionally, in the winter the electric load is far greater than the PV production, so any differences of EV load at this time of year would have very little impact on the results. In each hour of a day, the charging demand varies dramatically, reaching peaks at night and valleys during daytime. Such daily EV demand profiles have an opposite trend as the PV power generation, which reaches peak in the daytime and becomes zero at night. Thus, the increase of EV numbers will not promote a higher contemporary self-consumption rate for PV electricity. Similar to the application of Energy Hub DC micro grid, in the residential area of Sweden the particularly negative interaction between PV and EVs suggests to replicate the optimization process in other building clusters, instead of only residential buildings. The relation between PV and EVs could be more synergic in offices and commercial activities due to the better contemporaneity of production and demand, as EV's used for commuting are logically charged when people are at work. This study did not consider the scenario that EV batteries are allowed to charging the power grid and thus can help alleviate the grid stress in the peak-demand periods (Sun et al. 2018). In such scenario, EVs are used as mobile electrical energy storage which is charged in buildings with surplus renewable production and discharges in the buildings with insufficient renewable supplies (Barone et al. 2019). From this aspect, the deployment of EVs can help further increase the renewable energy self-consumption at the building cluster level.

According to 2030 Framework for climate and energy (Council 2014), the European commission bids to achieve a 32% share of renewable energy source (RES) by 2030. Increasing the capacity of renewable energy systems can help achieve this goal. However, the increased capacity will cause issues such as high investments and over-production. The energy concepts introduced in this study, such as energy sharing and TES integration, represent good solutions for the buildings in Sweden to achieving this '32% share of renewable energy source' target. They should also be easy to implement in many other countries on a technical level, but regulations would need to be revised in order to allow power sharing among buildings with or without PVs.

14.8 Conclusions

This chapter has presented a case study about transforming existing building cluster into electricity prosumers in Sweden. The core energy concepts, including click-and-go PV, centralized variable-speed heat pump, Energy Hub direct current microgrid that can share power between buildings and hot water thermal storage, have been introduced and applied for retrofitting an existing building cluster. An optimization method has been developed to design the capacity and positions of PV modules on each building, which aims at maximizing the self-consumed electricity under the constraint that the system has a positive lifetime net present value (and thus it is profitable). Based on the developed method, the impacts of thermal energy storage, electric vehicle penetrations, and energy sharing on the optimal capacity and positions of PV panels have been investigated. The results have revealed how those factors influence the design of PV systems and the system techno-economic performance, and thus help promote the PV deployment. More importantly, this study has demonstrated the feasibility for transferring the existing Swedish building cluster into smart electricity prosumers with higher self-consumption rates and energy efficiency and more intelligence, which offers good solutions for Sweden to achieving the '32% share of renewable energy source' target. The major findings are summarized as follows.

- The annual cumulative solar irradiation and homogeneity of irradiation are two significant factors affecting the PV power self-consumption, and thus they should be considered in the selection of locations for PV panel installation.
- The energy sharing can significantly improve the renewable energy self-consumption. The self-consumption could reach as high as 77% while maintaining a self-sufficiency above 20% in the baseline case, which is much higher than other studies at similar high latitude. This is because the aggregated electrical demand of multiple buildings eliminates the huge peaks featured by single building's demand, and thus can better match the PV power generations.
- The integration of thermal energy storage, together with suitable control for storing heat using PV excess production, will lead to increase in the optimal capacity of PV systems, as charging of thermal energy storage will increase electrical load. Due to an increased match between the electrical demand and power generation, the integration of thermal energy storage is beneficial for increasing renewable energy self-consumption, i.e. self-consumption increased from 77 to 79.4%. The

integration of thermal energy storage does not affect levelized cost of electricity too much, as both the power generation and the costs increase.

- The integration of electric vehicles will lead to increase in the optimal capacity of PV systems that maximizes the self-consumption, in this case study the self-consumption rate increased from 79.4 to 80.9% when EV number increased from 2 to 48. Meanwhile, due to the increased self-consumption, the levelized cost for the self-consumed electricity will be reduced slightly.
- Aggregating the building demand and supply by enable energy sharing will lead to increase in the optimal capacity of PV systems that maximizes the self-consumption rate, since energy sharing makes the PV system more versatile, and thus the whole cluster is more efficient at consuming the electricity produced on-site. The self-consumption will be reduced (i.e. 7.8% decrease), but this will be compensated by a dramatic increase in the self-sufficiency (i.e. 23.8% increase). The levelized cost of electricity is not affected by aggregating the building demand and supply.

In this study, the considered system is one centralized heat pump-thermal storage system for the three building. The energy sharing control is relatively easy as just one set of system needs to be controlled. When buildings have their own heating and storage systems, the district-level collaborative controls will become difficult. Future work is needed to develop advanced collaborative control strategies for building clusters, which can globally coordinate multiple systems and demands. One limitation of this study is that the thermal storage cost is not considered in the optimization. Future work will take into account storage costs for a more comprehensive optimization. Furthermore, the effect of the technologies analysed in this study could be as well investigated in terms of specific CO_2 emissions [kg CO_2-eq/MWh] (Huang et al. 2019).

References

Agency IE (2009) An integrated climate and energy policy framework: a sustainable energy and climate policy for the environment, competitiveness and long-term stability. Accessed on 20 May 2019. Available at https://www.government.se/content/1/c6/12/00/88/d353dca5.pdf

Ahmed K, Akhondzada A, Kurnitski J, Olesen B (2017) Occupancy schedules for energy simulation in new prEN16798-1 and ISO/FDIS 17772-1 standards. Sustain Cities Soc 35:134–144

Ayai N, Hisada T, Shibata T, Miyoshi H, Iwasaki T, Kitayama K-I (2012) DC micro grid system. SEI Tech Rev 75:132–136

Baeten B, Rogiers F, Helsen L (2017) Reduction of heat pump induced peak electricity use and required generation capacity through thermal energy storage and demand response. Appl Energy 195:184–195

Barone G, Buonomano A, Calise F, Forzano C, Palombo A (2019) Building to vehicle to building concept toward a novel zero energy paradigm: modelling and case studies. Renew Sustain Energy Rev 101:625–648

Bingham RD, Agelin-Chaab M, Rosen MA (2019) Whole building optimization of a residential home with PV and battery storage in The Bahamas. Renew Energ 132:1088–1103

Christine H (2013) Buildings transform into prosumers with the smart grid, smart grid library. Accessed on 20 Mar 2019. Available at http://www.smartgridlibrary.com/2013/05/20/buildings-transform-into-prosumers-with-the-smart-grid

Concannon P (2002) International energy agency energy conservation in buildings and community system programme. Technical note AIVC 57. Residential Ventilation. . Accessed on 20 May 2019. Available at https://www.aivc.org/sites/default/files/members_area/medias/pdf/Technotes/TN57%20Residential%20Ventilation.pdf

Council E (2014) 2030 climate & energy framework. Accessed on 23 Mar 2019. Available at https://ec.europa.eu/clima/policies/strategies/2030_en

Ferroamp (2018) The EnergyHub system. Accessed on 10 May 2019. Available at https://static.ferroamp.com/files/brochure/en/Ferroamp%20Brochure%20English%202018.pdf

Ferroamp (2019) EnergyHub system: optimized solar energy. Accessed on 13 Aug 2019. Available at https://ferroamp.com/energyhub-system/

Fischer D, Madani H (2017) On heat pumps in smart grids: a review. Renew Sustain Energy Rev 70:342–357

Fischer D, Harbrecht A, Surmann A, McKenna R (2019) Electric vehicles' impacts on residential electric local

profiles—a stochastic modelling approach considering socio-economic, behavioural and spatial factors. Appl Energy 233–234:644–658
Geth F, Willekens K, Clement K, Driesen J, De Breucker S (2010) Impact-analysis of the charging of plug-in hybrid vehicles on the production park in Belgium. In Melecon 2010–2010 15th IEEE Mediterranean electrotechnical conference: IEEE, pp 425–30
Grahn P, Munkhammar J, Widén J, Alvehag K, Söder L (2013) PHEV home-charging model based on residential activity patterns. Ieee T Power Syst 28:2507–2515
Gui N, Li J, Dong Y, Qiu Z, Jia Q, Gui W, Deconinck G (2017) BIM-based PV system optimization and deployment. Energy Build. 150:13–22
Howell S, Rezgui Y, Hippolyte J-L, Jayan B, Li H (2017) Towards the next generation of smart grids: Semantic and holonic multi-agent management of distributed energy resources. Renew Sustain Energy Rev 77:193–214
Huang P, Sun Y (2019) A clustering based grouping method of nearly zero energy buildings for performance improvements. Appl Energy 235:43–55
Huang P, Lovati M, Zhang X, Bales C, Hallbeck S, Becker A, Maturi L (2019) Transforming a residential building cluster into electricity prosumers in Sweden: optimal design of a coupled PV-heat pump-thermal storage-electric vehicle system. Appl Energy 255:113864
Huang P, Huang G, Sun Y (2018a) Uncertainty-based life-cycle analysis of near-zero energy buildings for performance improvements. Appl Energy 213:486–498
Huang P, Wu H, Huang G, Sun Y (2018b) A top-down control method of nZEBs for performance optimization at nZEB-cluster-level. Energy 159:891–904
IRENA (2017) Electricity storage and renewables: costs and markets to 2030. International Renewable Energy Agency, Abu Dhabi
Klein S, Beckman W, Mitchell J, Duffie J, Duffie N, Freeman T (2017) TRNSYS 18: a transient system simulation program. Solar Energy Laboratory
Koskela J, Rautiainen A, Järventausta P (2019) Using electrical energy storage in residential buildings—sizing of battery and photovoltaic panels based on electricity cost optimization. Appl Energ 239:1175–1189
Leuthold M (2014) Energy storage technologies battery storage for grid stabilization. In: IEA EGRD workshop on energy storage
Liu X, Zhang P, Pimm A, Feng D, Zheng M (2019) Optimal design and operation of PV-battery systems considering the interdependency of heat pumps. J Energy Storage 23:526–536
Lovati M, Salvalai G, Fratus G, Maturi L, Albatici R, Moser D (2019) New method for the early design of BIPV with electric storage: a case study in northern Italy. Sustain Cities Soc 48:101400
Lovati M, Salvalai G, Fratus G, Maturi L, Albatici R, Moser D (2019) New method for the early design of BIPV with electric storage: a case study in northern Italy. Sustain Cities Soc 48:101400
Luthander R (2018) Self-consumption of photovoltaic electricity in residential buildings. Doctoral thesis, comprehensive summary. Acta Universitatis Upsaliensis, Uppsala
Luthander R, Widén J, Munkhammar J, Lingfors DJE (2016) Self-consumption enhancement and peak shaving of residential photovoltaics using storage and curtailment. Energy 112:221–231
Luthander R, Nilsson AM, Widén J, Åberg M (2019) Graphical analysis of photovoltaic generation and load matching in buildings: a novel way of studying self-consumption and self-sufficiency. Appl Energy 250:748–759
Maturi L, Belluardo G, Moser D, Del Buono M (2014) BiPV system performance and efficiency drops: overview on PV module temperature conditions of different module types. Energy Procedia. 48:1311–1319
MoIR E (2013) Ordinance (2007: 215) on exemption from the requirement for network concession according to the Electricity Act (1997: 857). Accessed on 20 June 2019. Available at https://www.riksdagen.se/sv/dokument-lagar/dokument/svensk-forfattningssamling/forordning-2007215-om-undantag-fran-kravet-pa_sfs-2007-215
Moreira C, Lopes JP, Almeida PR, Seca L, Soares FJ (2011) A stochastic model to simulate electric vehicles motion and quantify the energy required from the grid
Munkhammar J, Grahn P, Widén J (2013) Quantifying self-consumption of on-site photovoltaic power generation in households with electric vehicle home charging. Sol Energy 97:208–216
Munkhammar J, Widén J, Rydén J (2015) On a probability distribution model combining household power consumption, electric vehicle home-charging and photovoltaic power production. Appl Energy 142:135–143
NEN (2014) NEN 7250, Solar energy systems—integration in roofs and façades—building aspects. Accessed on 20 May 2019. Available at https://standards.globalspec.com/std/10336983/nen-7250
Nolting L, Praktiknjo A (2019) Techno-economic analysis of flexible heat pump controls. Appl Energy 238:1417–1433
O'Shaughnessy E, Cutler D, Ardani K, Margolis R (2018) Solar plus: optimization of distributed solar PV through battery storage and dispatchable load in residential buildings. Appl Energ 213:11–21
Oh J, Koo C, Hong T, Cha SH (2018) An integrated model for estimating the techno-economic performance of the distributed solar generation system on building façades: focused on energy demand and supply. Appl Energ 228:1071–1090
Parag Y, Sovacool BK (2016) Electricity market design for the prosumer era. Nature Energy 1:16032

Psimopoulos E, Bee E, Widén J, Bales C (2019) Techno-economic analysis of control algorithms for an exhaust air heat pump system for detached houses coupled to a photovoltaic system
Reich NH, Mueller B, Armbruster A, Van Sark WG, Kiefer K, Reise C (2012) Performance ratio revisited: is PR> 90% realistic? In 26th EU PVSEC, Hamburg, Germany, vol 20, pp 717–726
Renaldi R, Kiprakis A, Friedrich D (2017) An optimisation framework for thermal energy storage integration in a residential heat pump heating system. Appl Energy 186:520–529
Roberts MB, Bruce A, MacGill I (2019) Impact of shared battery energy storage systems on photovoltaic self-consumption and electricity bills in apartment buildings. Appl Energ 245:78–95
Romero Rodríguez L, Sánchez Ramos J, Guerrero Delgado M, Molina Félix JL, Álvarez DS (2018) Mitigating energy poverty: Potential contributions of combining PV and building thermal mass storage in low-income households. Energy Convers Manage 173:65–80
Shahidinejad S, Filizadeh S, Bibeau E (2012) Profile of charging load on the grid due to plug-in vehicles. IEEE Trans Smart Grid 3:135–141
Shen L, Sun Y (2016) Performance comparisons of two system sizing approaches for net zero energy building clusters under uncertainties. Energy Build 127:10–21
Shirazi AM, Zomorodian ZS, Tahsildoost M (2019) Techno-economic BIPV evaluation method in urban areas. Renew Energ 143:1235–1246
Skoplaki E, Palyvos JA (2009) On the temperature dependence of photovoltaic module electrical performance: a review of efficiency/power correlations. Sol Energy 83:614–624
Strunz K, Abbasi E, Huu DN (2013) DC microgrid for wind and solar power integration. IEEE J Emerg Sel Top Power Electron 2(1):115–126
Sun Y, Yue H, Zhang J, Booth C (2018) Minimisation of residential energy cost considering energy storage system and EV with driving usage probabilities. IEEE Trans Sustain Energy 10(4):1752–1763.
Sun M, Djapic P, Aunedi M, Pudjianto D, Strbac G (2019) Benefits of smart control of hybrid heat pumps: an analysis of field trial data. Appl Energy 247:525–536
van der Heijde B, Vandermeulen A, Salenbien R, Helsen L (2019) Representative days selection for district energy system optimisation: a solar district heating system with seasonal storage. Appl Energ 248:79–94
Vartiainen E, Masson G, Breyer C, Moser D (2017) Improving the Competitiveness of Solar PV with Electricity Storage. In: 33rd European photovoltaic solar energy conference and exhibition, pp 2783–2789
Widén J, Wäckelgård E (2010) A high-resolution stochastic model of domestic activity patterns and electricity demand. Appl Energy 87:1880–1892
Zhang Q, Tezuka T, Ishihara KN, McLellan BC (2012) Integration of PV power into future low-carbon smart electricity systems with EV and HP in Kansai Area Japan. Renew Energy 44:99–108

15 Genetic Algorithm for a Coordinated Control to Improve Performance for a Building Cluster with Energy Storage, Electric Vehicles, and Energy Sharing

Pei Huang and Xingxing Zhang

Abstract

Existing studies have developed some advanced building side controls that enable renewable energy sharing and that aim to optimize building-cluster-level performance via regulating the energy storage charging/discharging. However, the flexible demand shifting ability of electric vehicles is rarely considered. For instance, the electric vehicle charging will usually start once they are plugged into charging stations. But, in such charging period the renewable generation may be insufficient to cover the EV charging load, leading to grid electricity imports. Consequently, the building-cluster-level performance is not optimized. Therefore, this study proposes a coordinated control of building prosumers for improving the cluster-level performance, by making use of energy sharing and storage capability of electricity batteries in both buildings and EVs. An EV charging/discharging model is first developed. Then, based on the predicted future 24 h electricity demand and renewable generation data, the coordinated control first considers the whole building cluster as one 'integrated' building and optimizes its operation as well as the EV charging/discharging using genetic algorithm. Next, the operation of individual buildings in the future 24 h is coordinated using nonlinear programming. For validation, the developed control has been tested on a real building cluster in Ludvika, Sweden. The study results show that the developed control can increase the cluster-level daily renewable self-consumption rate by 19% and meanwhile reduce the daily electricity bills by 36% compared with the conventional controls.

Keywords

PV · Electric vehicle · Energy sharing · Building cluster · Coordinated control

15.1 Introduction

Buildings represent large energy end-users worldwide (Zhang et al. 2020). In the E.U. and U.S, buildings currently consume over 40% of total primary energy usage (Cao et al. 2016). Renewable energy, which has much less carbon emissions and relatively lower costs compared with the conventional fossil fuel-based energy, offers a promising solution to meeting the large

P. Huang (✉) · X. Zhang
Department of Energy and Community Buildings, Dalarna University, 79188 Falun, Sweden
e-mail: phn@du.se

X. Zhang
e-mail: xza@du.se

X. Zhang (ed.), *Data-driven Analytics for Sustainable Buildings and Cities*, Sustainable Development Goals Series, https://doi.org/10.1007/978-981-16-2778-1_15

energy needs in the building sectors (Merrill et al. 2017). In this regard, distributed energy systems, such as PV panels, have gained popularity and are now widely installed in buildings (Zhang et al. 2016). For instance, the Swedish Energy Agency has set a target for 100% renewable electricity production by 2040, to which building integrated PV systems are planned to contribute 5–10% in electricity generation (Agency 2017). The integration of distributed energy systems has promoted the transformation of buildings' role from energy consumers to energy prosumers, i.e. energy consumers who produce energy for their own consumption using distributed energy technologies (Lovati et al. 2020). A popular type of energy prosumer is the zero energy buildings (ZEBs) (Huang et al. 2018a), which produce the same amount of energy as they consume. The transformation of buildings' role into energy prosumers also provides opportunities for collaborations among buildings to improve the overall cluster-level performances (Huang and Sun 2019a). When multiple building prosumers are involved in a building cluster, they can share their excessive renewables with others with insufficient generations (Fan et al. 2018). Such energy sharing can help improve the building-cluster-level renewable self-consumption rates and thus reduce the grid power usage (due to an increased share of renewable energy utilization). A study conducted by Luthander et al. (2016) shows that that even a simple energy sharing (i.e. aggregate electricity demand and supply) among 21 houses in Sweden can easily improve the PV power self-consumption by over 15%. When there is shared energy storage, the improvement in PV power self-consumption can reach 29%.

To achieve energy sharing among buildings, existing studies have developed a number of advanced controls. For example, Odonkor et al. proposed a control method of ZEBs using genetic algorithm and Pareto decision making based on an adaptive bi-level decision model (with a facilitator agent at cluster level and local systems at single NZEB level) (Odonkor and Lewis 2015). In such bi-level decision model, the individual building's systems (i.e. individual PV system and battery) are in the first level, and the centralized cooling system as well as an ice storage system are in the second level. Fan et al. proposed a collaborative demand response control of zero energy buildings for enhancing the building-cluster-level performances. In their method, the control of each building was conducted in sequence, and the optimization of one building's operation was based on the previously optimized buildings' operation (Fan et al. 2018). In each optimization, the daily hourly charging/discharging rates of the battery are set as variables to be optimized, and the economic cost and grid friendliness are set as the objective function. Prasad and Dusparic developed a Deep Reinforcement Learning based method for ZEB community (Prasad and Dusparic 2019). The ZEB community is modelled as a multiagent environment, where each agent represents a building. Every agent learns the optimal behaviour independently and is entirely responsible for making energy transactions on behalf of that building. The abovementioned controls optimize the building cluster performance in a bottom-up way, and they merely perform very limited collaborations among buildings.

With the purpose of maximizing the energy sharing within a building cluster, researchers have developed controls that directly use the building-cluster-level performances as the optimization targets. For instance, Gao and Sun (2016) developed a genetic algorithm based coordinated demand response control in which all the storage systems' charging rates were optimized simultaneously. Similarly, considering the problems caused by the independent microgrids operation, Zhang et al. (2018) proposed a coordinated control in which a cluster-level controller (i.e. an aggregator) was utilized to simultaneously manage local energy transactions among microgrids and energy exchanges with the grid. Both these two controls are easy to implement and effective in improving building-cluster-level performance. But with the increase of the number of buildings, such a straightforward coordination will face too many parameters to be optimized, causing excessive computation load. Such excessive computation loads make the straightforward method unfeasible to be applied in large NZEB clusters. To address the large computation, Huang et al. developed a top-down control for a cluster of building prosumers equipped with electrical energy storage system (Huang et al. 2018b). In their study, the optimal performances that can be achieved are first searched by advanced searching algorithm. Then the optimal performances at the top-level are divided into separate goals for each individual building at the bottom-level. Compared with the individual controls, their method can increase the daily load coverage by renewable energy by as much as 45%, reduce the daily peak energy exchanges with the power grid by as much as 80%, and meanwhile significantly reduce the daily operational costs. Similarly, in Huang and Sun 2019b a three-step demand response control algorithm is developed considering the dynamic pricing. Such control can flatten the electricity demand profiles via properly coordinating single buildings and thus maximize the benefits of both buildings and the power grid. Taking into account of the demand prediction uncertainty, in Huang and Sun 2019c a robust collaborative control is developed. Such control identifies the optimal operation strategy under the predicted uncertain ranges of demand, and thus it can maximize the renewable energy sharing robustly.

These existing controls can effectively improve the performances at building cluster level. However, electric vehicles (EV), which also play an important role in the building cluster scale energy systems, are usually considered as non-scheduled electrical loads (such as lighting) and their flexible demand shifting ability is rarely used (Taşcıkaraoğlu 2018; Huang et al. 2019). As a result, the flexible demand shifting ability of EVs are rarely considered together with the building control, leading to limited performance improvements at building cluster level (Barone et al. 2019; Dallinger et al. 2013). For instance, in practice the EV charging will start once they are plugged into charging stations. However, in such charging period the renewable generation may be insufficient to cover the EV charging load, leading to grid electricity imports. On the other hand, when there is surplus renewable generation, the EVs cannot be used as electricity storage if they have already been fully charged, leading to the surplus renewable energy exports. As a result, the overall building-cluster-level performance is not fully optimized.

The EV deployment is continuously increased and many governments have established policy or goals to promote the EV deployment. For instance, the French government set a target of 2 million EVs in 2020 (Merten et al. 2012). The Swedish government has set a goal that the vehicle fleets should be 100% independent of fossil fuel by 2030 (a large percentage should be achieved by EV deployment) (Xylia and Silveira 2017). The U.S. Federal government has enacted policies and legislations to promote the U.S. market for EVs, such as improvements of tax credits in current law, and competitive programs to encourage communities to invest in infrastructure supporting these vehicles (Agency 2010). The number of EVs on the road is projected to reach 18.7 million in 2030, up from slightly more than 1 million at the end of 2018 (Cooper and Schefter 2018). In the future, due to the large penetration, EVs will have large impacts on the grid power demands. Thus, it is necessary to make use of their potentials in demand regulation in the power grid and district energy systems.

By properly scheduling the EV charging loads, the batteries in EVs can be used as flexible energy storage to help regulate the electricity demands in the power grid. Existing studies have also developed some advanced controls for EVs at both individual level and aggregated level. At individual level, Islam et al. proposed a coordinated EV charging control based on a correlated probabilistic model of EV charging loads considering the stochastic charging behaviour (Shariful Islam et al. 2019). The charging control optimizes the power factors of PV and battery energy storage system to enhance the quality of service by minimizing the probability of voltage and current noncompliance. The application of the developed control on a three-phase IEEE 37-bus unbalanced distribution system using the real data of vehicles and solar PV shows it is effective in providing more quality of service. At aggregated level, Geth et al. developed a coordinated charging control for a number of EVs (Geth et al. 2010). In their control, a vehicle owner first indicates the point in time when the batteries should be fully charged. Then, the aggregator collects this information and calculates when each EV can start charging, based on two rules: *(i)* charging is most economically when the total demand (including the residential, industrial and EV consumption) is low, and *(ii)* the EVs can be charged during working hour in the working places. Case study shows that the coordinated charging can effectively decrease the peak load, as the coordination makes the charging load profile much flatter. Similarly, Usman et al. proposed an automated coordinated control of EV fleets, which can plan the charging strategy at the cheaper moments while keeping the vehicle charged enough to complete its scheduled trips (Usman et al. 2016). Their control uses a grid agent to grant tokens to the EVs in idle state based on the grid electricity prices. By shifting charging loads to low electricity price period (usually with low aggregated electricity demands in the power grid), this control can effectively increase the match between the available power and the consumed power. In (Kam and Sark 2015), three different smart EV charging control methods were proposed for increasing PV power usage in a microgrid: real-time charging based on PV power sufficiency; real-time charging based on PV power sufficiency and with vehicle-to-grid enabled (i.e. vehicle can discharge power back to grid); and linear programming based optimization (i.e. optimize all EVs' charging based on PV power production on a daily basis). Their study shows that coordinated EV controls can increase the PV power self-consumption by 13 ~ 38% and reduce the peak electricity demand by 27 ~ 67%. In this study, the optimization at the building side (e.g. battery charging/discharging and energy sharing) was not considered. Fachrizal and Munkhammar (Fachrizal and Munkhammar 2020) developed a centralized smart EV charging scheme for a residential building cluster, which optimizes the charging rates of all the EVs simultaneously considering the interaction of individual EVs and PV power production. This study innovatively considered EV smart charging in improving building-cluster-level performance. However, their study considers all the buildings in the cluster as one 'aggregated' building for simplicity, in which the operation of individual buildings is neglected. Such a simplification may not be practical. The abovementioned studies can effectively improve the economic performances of EV or EV fleets. However, these studies typically consider EVs as a separate role in the urban energy system and thus neglect their integration with the building controls. In the future scenario with increased number of building prosumers and EV penetration, EV control not integrated with building prosumers controls (i.e. buildings' energy sharing control) will limit the overall performance improvement potentials.

To sum up, existing studies have developed some advanced building side controls that enable renewable energy sharing and that aim to optimize building-cluster-level performance via regulating the energy storage's charging/discharging. However, the flexible demand shifting capability of EVs, which has been proven effective in enhancing building cluster-level performance [e.g. increase PV power self-consumption by over 10% (Fachrizal and Munkhammar 2020), reduce peak demand by 37% (Kara et al. 2015)], is not considered in the cluster-level controls. Therefore, this study proposes a coordinated control of building cluster with both energy sharing and the EV charging considered, with the purpose of improving the cluster-level performance by taking advantage of energy sharing and storage capability of electricity batteries in both buildings and EVs. An EV charging/discharging model is first developed, and then a coordinated control is developed for building cluster with the energy storage, EVs and energy sharing considered. Based on the predicted future 24 h electricity demand and renewable generation data, the coordinated control first considers the whole building cluster as one 'integrated' building and optimizes its operation as well as the EV charging/discharging using genetic algorithm. Then, the operation of individual buildings in the future 24 h is coordinated using nonlinear programming. For validation purpose, the developed control has been tested using the energy demand and supply data on a real building cluster in Ludvika, Sweden. The major contributions of this study to the subject have also been summarized and added to the introduction.

- A coordinated control method for building clusters has been developed for optimizing building-cluster-level performances.
- The developed control can regulate individual building's battery charging/discharging to maximize the renewable energy sharing and at the same time coordinate all the individual EVs' charging load in order to increase the building cluster renewable energy usage.
- The performances of the developed control have been compared with a conventional individual control (without energy sharing and smart EV charging) and a partial collaborative control (with energy sharing but without smart EV charging).
- The performance improvements have been analyzed in aspects of renewable energy self-consumption and economic saving.

The structure of this chapter is as followings. Section 15.2 describes the overall coordinated control for the building cluster. Section 15.3 presents the detailed building model and energy system models. In Sect. 15.4, the developed coordinated control is applied on a case building cluster. and its performance are compared with two existing scenarios. The brief conclusions are given in Sect. 15.5.

15.2 Coordinated Control to Improve Energy Performance for a Building Cluster

This section first introduces the energy sharing concept. Then, the detailed coordinated control, which takes account of the energy storage, EVs charging and renewable energy sharing among buildings, is introduced.

15.2.1 Energy Sharing

Energy sharing is an effective way to improve the overall performances at the building cluster level. In this study, the energy sharing is implemented by installing an energy sharing microgrid among the buildings (Huang et al. 2019; Ferroamp 2018), as depicted by Fig. 15.1. The renewable energy from Building A can be used to supply the electricity demands charge the EVs in Building B or C, or even be stored in the battery of Building B or C. Such renewable energy sharing can help increase the renewable self-utilization rates of the building cluster, and thus help improve both the economic and energy performances (Huang et al. 2018b).

15.2.2 A Coordinated Control to Improve Energy Performance for a Cluster of Building Energy Prosumers with Energy Storage, EVs, and Energy Sharing Considered

This section introduces the developed coordinated control. Figure 15.2 presents the flowchart of the developed method. The aim of the coordinated control is to coordinate the operation of energy storage (installed in each single building) and the EVs, to achieve the optimal cluster-level performances. The coordinated control consists of four steps. In Step 1, all the buildings in the building group are considered as a 'representative' building, and the electrical demand, renewable energy generation and load shifting capacity of the 'representative' building are predicted, i.e. its electrical demand/ renewable generation/demand shifting capacity equals the aggregated demand/ generation/

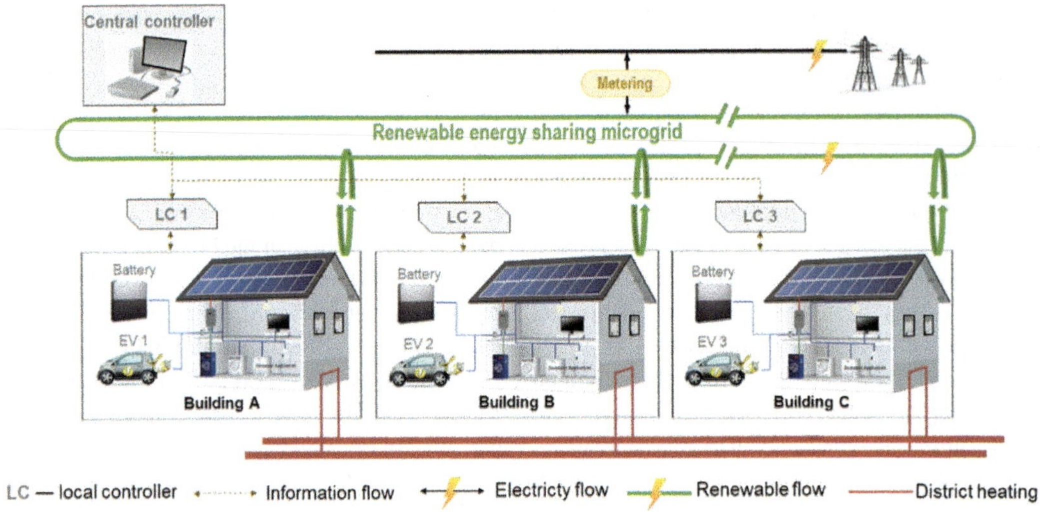

Fig. 15.1 Schematics of electricity energy sharing among buildings in a cluster

capacity of all buildings inside the cluster. In Step 2, the operation of the 'representative' building and the EV charging rates are optimized using genetic algorithm (GA). The performance of the 'representative' building, obtained by simultaneous optimization of the building and EV operation, is considered to be the best performances that the building group can achieve (Shen et al. 2016). In Step 3, the operation of each single building inside the building group is coordinated using non-linear programming (*NLP*) based on the 'representative' building's operation obtained from Step 2. In Step 4, the performances of the proposed coordinated control are compared with two existing controls, including a conventional individual control [Scenario 1 (Shen et al. 2016)], which does not enable renewable sharing and charge the EVs immediately after being parked, and an existing coordinated control [Scenario 2 (Gao and Sun 2016)], which enables full renewable energy sharing but also charges the EVs immediately after being parked. The details of each step are introduced below.

15.2.2.1 Step 1: Estimation of the 'Representative' Building's Demand and Storage

In this step, all the buildings inside the cluster are considered as a 'virtual' building. Its hourly electricity demand ($E^r_{d,i}$ (kW h)) equals the aggregated hourly electricity demand of each single building ($E^j_{d,i}$ (kW h)) (i indicates time with a unit of hour), its hourly renewable generation ($E^r_{s,i}$ (kW h)) equals the aggregated hourly renewable generation of each single building ($E^j_{s,i}$ (kW h)) and its load shifting capacity (CAP^r (kW h), i.e. battery capacity) is the aggregated load shifting capacity of each single building (CAP^j (kW h)).

$$E^r_{d,i} = \sum_{j=1}^{n} E^j_{d,i} \tag{15.1}$$

$$E^r_{s,i} = \sum_{j=1}^{n} E^j_{s,i} \tag{15.2}$$

$$CAP^r = \sum_{j=1}^{n} CAP^j \tag{15.3}$$

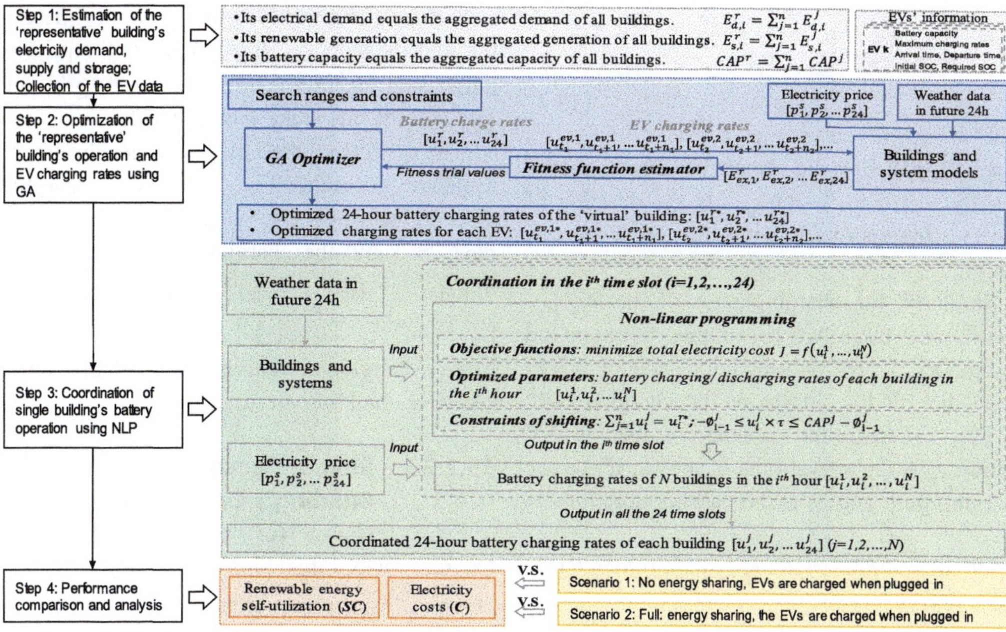

Fig. 15.2 Flowchart of the coordinated control to improve energy performance for a building cluster with energy storage, EVs, and energy sharing

15.2.2.2 Step 2: Optimization of the 'Representative' Building's Operation Using GA

The electricity demand and renewable generation of each individual building is calculated using the models presented in Sect. 15.3. The GA algorithm searches the optimal charging/discharging rates of both the battery (see Sect. 3.2 for the detailed models) and EVs that can minimize the electricity costs of the 'representative' building. For example, the EVs can be scheduled to be charged in periods with sufficient renewable generations while not charged in periods with insufficient generations. In the GA simulation, the inputs mainly include the battery charging/discharging rates (to be optimized), the EV charging rates (to be optimized), the EV parking periods, the future 24-h weather data, building parameters, and battery parameters. The EVs are different from the electrical battery, since they are not constantly connected into the buildings. This study uses four parameters to characterize an EV (e.g. the kth EV): arrival time to the charging port (t_k), parking periods in the charging port (n_k), initial state of charge ($SOC_{0,k}$), and the required state of charge when the car departs from the charging port ($SOC_{1,k}$). These parameters are considered known and will be used as inputs in the optimization.

In each generation of GA, trials of 24-h thermal storage hourly charging/discharging rates (i.e., $\left[u_1^v, u_2^v, \ldots u_{24}^v\right]$ kW) and charging rates of each EV (i.e. $[u_{t_k}^{ev,k}, u_{t_k+1}^{ev,k}, \ldots u_{t_k+n_k}^{ev,k}]$ kW) are generated by the GA optimizer. The representative building's hourly power demand ($E_{d,i}^r$ kW) and hourly renewable power generation ($E_{r,i}^r$ kW) in the future 24 h is then predicted using the building and system models (see models given in Sect. 15.3). The charging/discharging rates of the electrical battery should meet the following two constraints: (1) The battery charging amount could not exceed the remaining battery storage capacity. (2) The battery discharging amount could not exceed the stored electricity in the battery. These two constraints are expressed by Eq. (15.4) (Lu et al. 2015; Sun et al. 2018),

$$0 \leq \emptyset_0^r + (u_1^r + u_2^r + \ldots + u_i^r) \times \tau \leq CAP^r \quad \text{where} i = 1, 2, \ldots, 24 \tag{15.4}$$

where $\emptyset_0^v$ (kW h) is the amount of thermal energy initially stored in the tank, τ is the duration of battery charging/discharging (i.e., 1 h in this study).

Similarly, the charging rates of the *k*th EV should meet these two constraints, as expressed by Eq. (15.5). $SOC_{0,k}$ is the initial state of charge when the *k*th EV arrives at the charging port. CAP_k^{ev} (kW h) is the capacity of the *k*th EV battery. t_k is the arrival time of the *k*th EV at the charging port, and n_k is the parking duration.

$$0 \leq SOC_{0,k} \times CAP_k^{ev} + \left(u_{t_k}^{ev,k} + u_{t_k+1}^{ev,k} + u_{t_k+i}^{ev,k}\right) \times \tau \leq CAP_k^{ev} \quad \text{where } i = 1, 2, \ldots, n_k \tag{15.5}$$

In addition, the EV battery should be charged to a user-specified level ($SOC_{1,k}$) before they depart the charging port. This constraint is expressed by Eq. (15.6). When $SOC_{1,k}$ equals 1, it represents the EV users require the EV battery to be fully charged before they depart the charging port.

$$SOC_{0,k} \times CAP_k^{ev} + \left(u_{t_k}^{ev,k} + u_{t_k+1}^{ev,k} + \ldots + u_{t_k+n_k}^{ev,k}\right) \times \tau \geq SOC_{1,k} \times CAP_k^{ev} \tag{15.6}$$

This study considers the strategy to minimize daily electricity cost of the building group. Following this control goal, a fitness function is determined, as expressed by Eq. (15.7) (Salom et al. 2011).

$$J_{grid} = \min(Cost) \tag{15.7}$$

$$Cost = \sum_{i=1}^{24} E_{ex,i}^r \times \tau \times \chi_i, \begin{cases} \chi_i = \chi_{buy}, & if\ E_{ex,i}^r > 0 \\ \chi_i = \chi_{sell}, & if\ E_{ex,i}^r \leq 0 \end{cases} \tag{15.8}$$

where χ_i. (*kr/*(kW h)) is the electricity price in the ith time slot. χ_{buy} (*kr/*(kW h)) is the price of purchasing electricity from the power grid, and χ_{sell} (*kr/*(kW h)) is the feed-in-tariff.

The outputs of the GA search are the 'representative' building's battery charging/discharging rates ($[u_1^{r*}, u_2^{r*}, \ldots u_3^{r*}]$ kW) in the next 24 h and the charging rates of each individual EV ($[u_{t_1}^{ev,1*}, u_{t_k+1}^{ev,k*}, \ldots, u_{t_k+n_k}^{ev,k*}]$, $[u_{t_2}^{ev,2*}, u_{t_2+1}^{ev,2*}, \ldots u_{t_2+n_2}^{ev,2*}]$,... kW). The optimized battery charging/discharging rates of the 'representative' building are used in Step 3.

15.2.2.3 Step 3: Coordination of Single Building's Operation Using NLP

In this step, the single building's battery charging/discharging rates (i.e. u_i^j is the *j*th building in the *i*th hour) are coordinated using NLP based on the 'representative' building's operation (Zhao et al. 2015). The NLP is conducted in each hour and will be repeated 24 times for obtaining the building's daily operation. The fitness function of the NLP is expressed by Eqs (15.6) and (15.7), which aims at minimizing the electricity costs of the building group.

$$J_{NLP} = \min\left(Cost_{all,i}\right) \tag{15.9}$$

$$Cost_{all,i} = \sum_{j=1}^{n} \left(E_{d,i}^j \times \chi_i\right)^2 \tag{15.10}$$

In order to reduce the uneven allocation of the battery charging/discharging rates (otherwise only a few buildings take benefits from the demand response), the square of each building's operational cost is used in the fitness function. $E_{d,i}^j$ (kW h) is the energy demand of the *j*th building in the *i*th hour after applying the u_i^j (kW) amount of battery charging/discharging, which is calculated by the models presented in Sect. 15.3. χ_i (*HKD/*(kW h)) is the electricity price in the *i*th hour.

In the *i*th hour, the optimized parameters in the NLP are the hourly battery charging/discharging rates of all the buildings inside the building group, i.e., $[u_i^1, u_i^2, \ldots, u_i^N]$ (kW), where N indicates the number of buildings in the building group. The battery charging/discharging rates in each hour should follow the constraints below.

(i) The sum of battery charging/discharging rates of each building (u_i^j. (kW)) should equal the battery charging/discharging of the 'representative' building (u_i^{r*} (kW)) (obtained from Step 2).

$$\sum_{j=1}^{N} u_i^j = u_i^{r*} \tag{15.11}$$

(ii) For each single building, the electricity charging amount must be smaller than the remaining storage capacity of the battery, and the electricity discharging amount must be smaller than the amount of electricity stored in the battery. There are 2 N inequality constraints for N buildings.

$$-\emptyset_{i-1}^j \leq u_i^j \times \tau \leq CAP^j - \emptyset_{i-1}^j (j = 1, 2, \ldots N, \text{respectively}) \tag{15.12}$$

where τ is the charging duration (i.e., 1 h), CAP^j (kW h) is the battery capacity of the jth building, $\emptyset_{i-1}^j$ (kW h) is the electricity energy stored in the jth building's battery. $\emptyset_{i-1}^j$ (kW h) is calculated by Eq. (15.13).

$$\emptyset_{i-1}^j = \left(u_1^j + u_2^j + \ldots + u_{i-1}^j\right) \times \tau \tag{15.13}$$

In total, for a building group with N buildings, there are N unknown parameters to be solved, there is 1 equality constraints, and there are 2 N inequality constraints.

15.2.2.4 Step 4: Performance Comparison and Analysis

After obtaining the optimized operation of each single building, the performances of the proposed coordinated control are compared with two existing controls in aspects of renewable energy self-consumption improvements and economic cost savings. The two existing controls include a conventional individual control [Scenario *1* in Table 15.1 (Shen et al. 2016; Gao and Sun 2016)], which does not enable renewable sharing and charge the EVs immediately after connecting them, and an existing coordinated control [Scenario *2* in Table 15.1 (Huang et al. 2018b)], which enables full renewable energy sharing in the building cluster but charges the EVs immediately after connecting them. In both the two comparative studies, the EVs demand are first computed. Such load is added to the building electricity demand, which will then be used as inputs for battery charging/discharging controls. In Scenario 1 (i.e. an existing individual control) (Shen et al. 2016), GA was used for searching the optimal battery charging/discharging rates in each building, which is similar to the control optimization of the 'representative' building (see Step 2 in Fig. 15.2 without EV related variables). After obtaining the individual buildings' optimal operation, their electrical demands were aggregated for evaluating the building-cluster-level performances. In Scenario 2 (i.e. an existing coordinated control) (Gao and Sun 2016), the battery charging/discharging rates of all the three buildings are optimized simultaneously using GA, and the minimization of the building-cluster-level performance was used as the fitness function.

15.3 Buildings and System Modelling

This section introduces the building information and system modelling. Each building is installed with a renewable energy system (i.e., PV panels), an electricity storage system (i.e., battery), as well as an EV.

15.3.1 Building Modelling

This study considered a real building cluster located in Ludvika, Dalarna region, Sweden. This building cluster consists of three separate buildings, as shown in Fig. 15.3. The building

Table 15.1 Configuration of the three scenarios

Scenario	EV control?	Energy sharing?
1	Charged immediately when plugged in	No
2	Charged immediately when plugged in	Full sharing
3 (Developed control)	Charged at any time when parked	Full sharing

cluster (all the three buildings) includes 48 multifamily dwelling units over three floors, and most of the apartments have one or two bedrooms. The total façade surface gross area of the complex is 2146 m^2, the total roof surface gross area is 1750 m^2. These buildings will be improved by a series of renovation plans including installation of PV, battery storage, direct current (DC) micro grid, and EV charging station. It is assumed the heating is provided by district heating system. So, the PV panels will only need to provide power supply to the domestic electricity demand (e.g. lighting, TVs, dish wash). The schematics of the building cluster and the energy systems are shown in Fig. 15.1.

Until now, there are different models developed for modelling the electricity demand in residential buildings. For instance, Palacios-Garcia et al. (2018) developed a high-resolution model for calculating the electricity demand of heating and cooling appliances considering variables such as the number of residents, location, type of day (weekday or weekend) and date. In (Palacios-Garcia et al. 2015), a stochastic model for simulating lighting power consumption profiles in Spain was developed considering the number of household residents and differentiating between weekdays and weekends. In (Widén and Wäckelgård 2010), Widén developed a stochastic model for computing the occupancy and electricity load in Sweden. Since the occupancy schedules and lighting usage can vary significantly in different countries due to the culture and location difference, this study chooses Widén's model to calculate the electricity load profiles for the three individual buildings. Meanwhile, in order to achieve acceptable accuracy, the measured data about the annual electricity usage was used to calibrate the model.

15.3.2 Renewable Energy System Modeling

The power generation from the PV panel P_{PV} (kW) is calculated by Eq. (15.14) (Huang et al. 2018a; Klein et al. 2004),

$$P_{PV} = \tau \times I_{AM} \times I_T \times \eta \times CAP_{PV} \quad (15.14)$$

Fig. 15.3 Bird view of the case building cluster located in Ludvika, Sweden

where τ is the transmittance-absorptance product of the PV cover for solar radiation at a normal incidence angle, ranging from 0 to 1; I_{AM} is the combined incidence angle modifier for the PV cover material, ranging from 0 to 1; I_T (W/m^2) is the total amount of solar radiation incident on the PV collect surface; η is the overall efficiency of the PV array; CAP_{PV} (m^2) is the PV surface area.

15.3.3 Electrical Battery and EV Battery Modeling

This study used simplified electrical battery and EV battery models. The electricity stored in the battery is calculated using a simplified model, as expressed by Eqs. (15.4) and (15.5). It is estimated from the hourly charging rates (Sun et al. 2018). This study considers three EVs. Table 15.2 summarizes the capacity, maximum charging rates as well as the parking periods of each EV. EV 1, EV 2 and EV 3 are assumed to be charged in Building A, B and C, respectively. To consider the various EV usage, these three EVs are assumed to have different parking periods. EV 1 is assumed to be owned by a resident living in the building, and thus it is parked at night from 18:00 to 07:00 in the next day. EV 2 and EV 3 are assumed to be owned by some working staff in the building estate, and they are parked during daytime (i.e. one from 08:00 ~ 16:00 and the other from 09:00 ~ 17:00). The EV battery capacity and maximum charging rates are referred from the available EV models in the market in Ustun et al. (2013).

In all the three scenarios, the EVs are required to be fully charged before they leave the charging station. When the EVs arrive in a charging station in the home, a random SOC parameter (between 0 and 1) is assumed to represent the remaining storage in the EV battery.

15.4 Case Studies and Results Analysis

In the case studies, a typical summer week was selected to validate the developed coordinated controls. The weather data of Ludvika was used for modelling the local renewable generations. This section first presents the individual building's electricity demand and renewable generation information. Then, the detailed EV charging and battery charging results obtained from the two scenarios (see Step 4 in Sect. 15.2) and the developed control are compared and analyzed. Finally, the overall economic and energy performances are compared.

Table 15.3 summarizes the input parameters used in the case studies. According to the building dimension, 100 m^2, 200 m^2 and 300 m^2 roof areas are planned for installing PV panels in the three buildings, respectively. It was assumed each building is installed with an electrical battery with capacity of 20 kW•h and a maximum charging/discharging rates of 6 kW. The price of purchasing electricity from the power grid was set as 0.16 €/(kW h). Considering the negative impacts on the grid stability and safety, the feed-in-tariff was set as 0.05 €/(kW h), which is lower than price of electricity purchase (Huang et al. 2019). The price of electricity trading in the building cluster was set as 0.1 €/(kW h). Such price setting will provide incentives for energy sharing within the building cluster, i.e. the building owners can earn more by selling their excessive renewable energy to the building cluster than sell to the power grid, and vice versa.

Table 15.2 Capacity, charging limits and parking periods of the three different EVs, data obtained from Ustun et al. (2013)

ID	Battery capacity (kW h)	Maximum charging rates (kW)	Parking period
EV 1	22	4	18:00 ~ 07:00 (next day)
EV 2	27	5	08:00 ~ 16:00
EV 3	53	10	09:00 ~ 17:00

Table 15.3 Configuration of the PV and battery system and electricity prices

Input parameter	Value
Area of PV panel in building A (m^2)	100
Area of PV panel in building B (m^2)	200
Area of PV panel in building (m^2)	300
Battery capacity (kW h)	20
Battery maximum charging/discharging rates (kW)	6
Price of electricity sold to the grid (€/(kW h)) (Huang et al. 2019)	0.05
Price of electricity purchased from the grid (€/(kW h)) (Huang et al. 2019)	0.16
Price of electricity trading in the building cluster (€)	0.1

15.4.1 Building Electricity Demand, Renewable Generation and Electricity Mismatch

Figure 15.4 displays the hourly electricity demand, hourly PV generation, and the hourly electricity mismatch of the three buildings in the selected week. Note that the heating needs of the three buildings are assumed to be met by the district heating system. Thus, the electricity demand only includes the domestic electricity loads (i.e. lighting, washing machine, TV, etc.). The trends of PV power production of the three buildings are similar, since the solar irradiation is nearly the same for the three buildings which are located in the same location. As Building C has the largest roof area, more PV panels can be installed on its roof. Thus, it has the largest average PV production.

Power mismatch of each building is calculated as the deviation between the its hourly power demand and hourly renewable generation. A positive value of power mismatch indicates insufficient renewable generation (and thus grid power is needed), while a negative value of power mismatch indicates excessive renewable

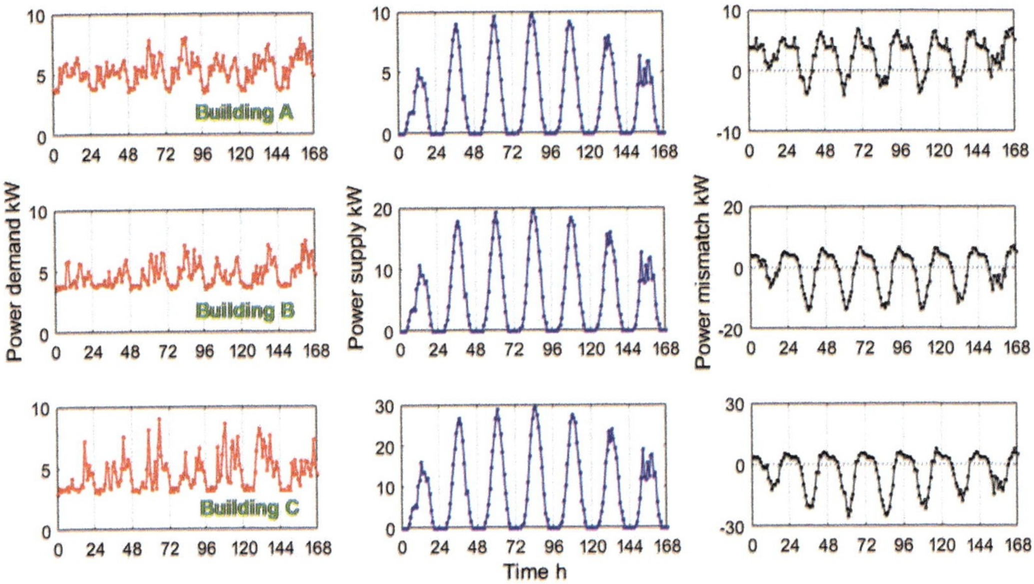

Fig. 15.4 The hourly power demand, renewable generation and power mismatch of three buildings in the selected summer week

generation (and thus selling electricity to the grid is needed). The diversity between the power mismatch provides good opportunities for the buildings to collaborative with each other in aspects of energy sharing. For instance, at noon (i.e. 11:00 ~ 16:00) in the first day, Building A has insufficient renewable generations (i.e. 7.6 kW h more demand), while Buildings B and C have excessive renewable generations (i.e. 24.7 kW h and 55.8 kW h more supply, respectively). Buildings B and C can share their surplus renewable generation with Building A to avoid grid power imports (for Building A) and power exports to the grid (for Buildings B and C), and thus help improve the overall performance at the building-cluster-level.

15.4.2 Detailed Battery Controls and Energy Flows

To have a close look at the charging of EVs and battery storage, as well as the energy flow in the system, the detailed operation in the first day of the selected week is presented and analyzed in this section. Note that the EV charging loads are exactly the same for the three scenarios. The initial SOCs when EVs arrive at the charging stations are the same for three scenarios. The initial SOCs upon arrival for the three EVs are 0.29, 0.61 and 0.62, respectively. All the EVs are required to be fully charged when they depart the charging stations, i.e. SOC equals 1. Figure 15.5 presents the State of Charge (SOC) of the three EVs' battery and the aggregated battery in the first day of the selected week. For Scenarios 1 and 2, since the EVs are charged at their maximum charging rates (i.e. 4 kW, 5 kW and 10 kW for the three EVs, respectively) immediately after being plugged into the charging ports, there is a stable increase in the SOCs for all the three EVs in the beginning of parking periods. In the developed control, the EVs are charged flexibly in the parking period. In some timeslots, they are charged at a high rate; while in some timeslots, they are charged at a low rate (or even zero). Despite the different charging patterns, all the EV batteries are fully charged (as specified in the case study, see Sect. 3.3) before they depart the charging ports in the three scenarios.

Regarding the battery storage usage, the aggregated battery has not been fully charged in Scenario 1, while it has been fully charged in Scenario 2 and the developed control. This is because in Scenario 1 the collaboration (i.e. renewable energy sharing) is not allowed among the buildings, while in Scenario 2 and the developed control, collaboration is enabled (see Fig. 15.6 for detailed energy sharing). The collaboration enables buildings to store their surplus renewables in other building's battery, thereby helping increase the overall battery utilization. Such increased battery utilization can help the building cluster keep more renewable energy onsite instead of exporting to the power grid, and thus contribute to increased renewable energy self-consumption rates.

Figure 15.6 depicts the electricity energy flow of the subsystems (i.e. electrical demands, renewable generation, EV demands and battery charging/discharging) in each building in the first day of the selected week for the three different scenarios. Figure 15.6 also shows the individual building's energy exchange with the building cluster (i.e. the other buildings) and with the power grid. A positive value of energy flow indicates energy demand, while a negative value indicates energy supply. For the power grid/building cluster, a positive energy flow indicates buildings export electricity to the power grid/building cluster, while a negative energy flow indicates buildings import electricity from the power grid/building cluster. The PV system produces electricity from 6:00 to 20:00. But for Buildings A and B, the amount of PV power production is less than the electricity demand in the early morning (6:00 ~ 8:00). As a result, there is grid power purchase in this period. While Building C has more power production in this period due to a larger PV system installed.

In Scenario 1, the energy sharing is not allowed among the buildings, so there is no

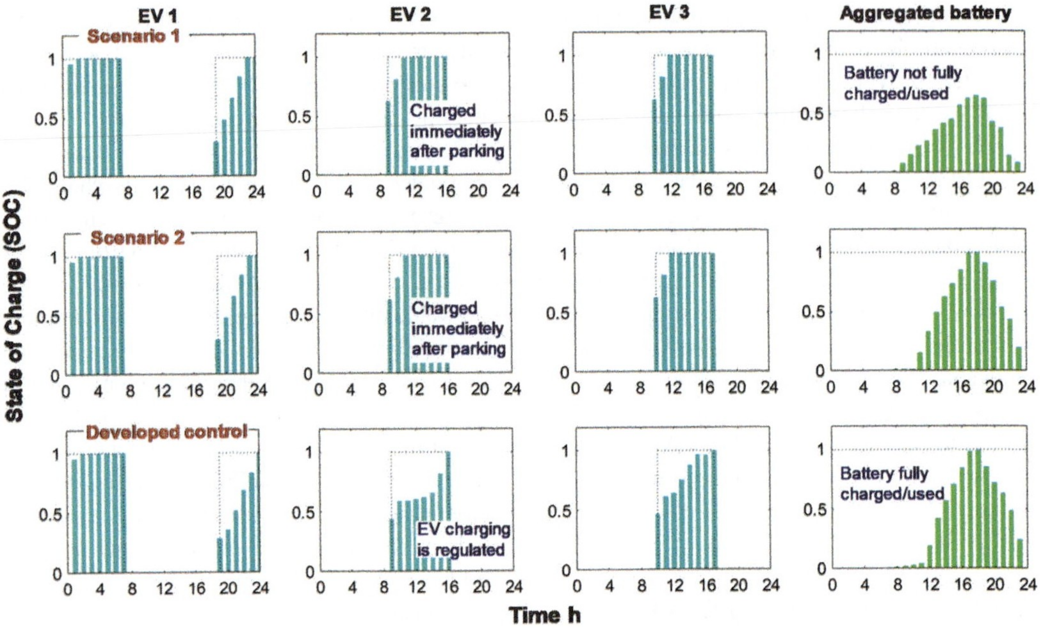

Fig. 15.5 State of Charge (SOC) of the three EVs and the aggregated battery in the first day of the selected week

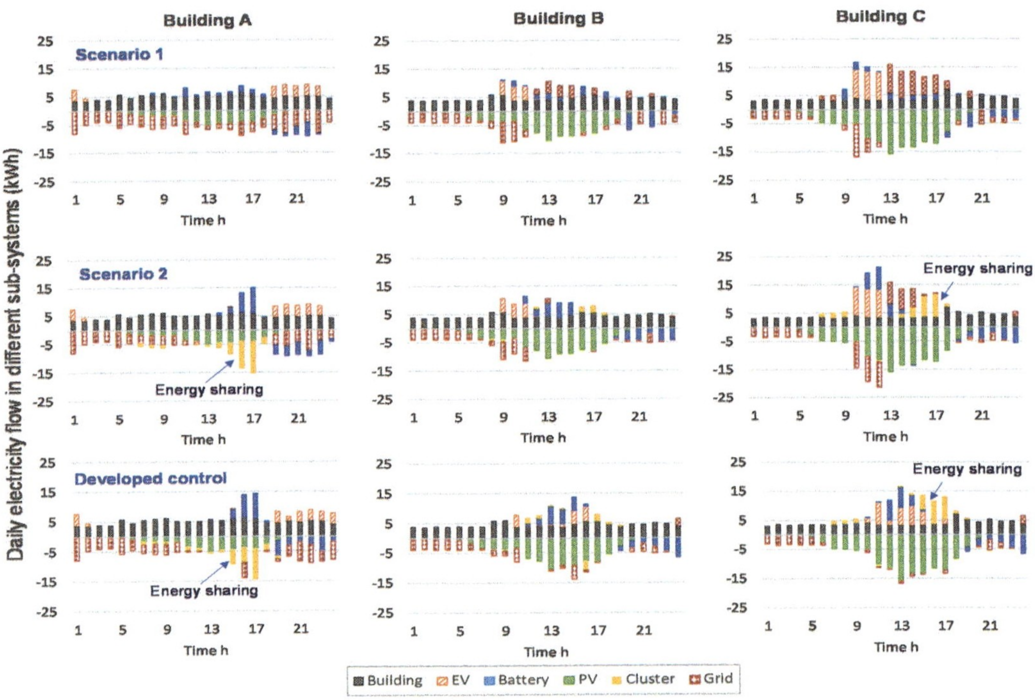

Fig. 15.6 Detailed electricity flow (of building, PV systems, battery and EV) in the individual building in each scenario in the first day of the selected week

energy exchange with the building cluster. Since the EVs are charged immediately after they are plugged into the charging stations, large EV loads can be observed for both Building B and C during 8:00 ~ 12:00 (one EV arrives at 9:00, and another EV arrives at 10:00). However, due to a lack of renewable generation in this period, a large amount of grid power is imported from the power grid (i.e. 19.4 kW h and 27.8 kW h electricity imports for Buildings B and C, respectively). At noon (i.e. 13:00 ~ 15:00), both Buildings B and C have surplus PV power production. However, due to a lack of energy sharing, such surplus PV power is exported to the power grid (i.e. 12.4 kW h and 27.8 kW•h electricity exports for Buildings B and C, respectively). Meanwhile, Building A purchased 5 kW h electricity from the power grid. In total, in this, Building A imported 102.7 kW•h electricity, Building B imported 55.6 kW•h electricity and exported 19.7 kW•h electricity, and Building C imported 45.6 kW•h electricity and exported 48.5 kW•h electricity.

In Scenario 2, energy sharing is enabled within the building cluster. At around 14:00, since the battery of Building C has already been fully charged, the surplus renewable energy from Building C is exported to Building A and stored in Building A's battery. Such energy sharing avoids the unnecessary renewable exports to the power grid, as compared with Scenario 1. In total, from 12:00 to 18:00, Building B shared 6.2 kW•h electricity with Building A, and Building C shared 22.3 kW•h electricity with Building A. Similar to Scenario 1, due to a lack of EV control, large EV charging demands can still be observed for both Buildings B and C during 8:00 ~ 12:00, leading to large grid power purchase (i.e. 17.8 kW h and 27 kW h electricity imports for Buildings B and C, respectively). In total, in this day, Building A imported 71.9 kW h electricity, Building B imported 41.5 kW h electricity and exported 1.5 kW h electricity, and Building C imported 43.2 kW h electricity and exported 23.4 kW h electricity.

In the developed control, the energy sharing is enabled among buildings and the EV charging is regulated. Similar to Scenario 2, the renewable energy sharing reduces the renewable energy exports to the power grid from Building C. In terms of EVs charging, due to the shifting of large EV charging loads to periods with more renewable energy generations in Buildings B and C, the developed control effectively reduces the grid power imports in the period 8:00 ~ 12:00 compared with Scenario 1 and 2. In total, in this day, Building A imported 74.5 kW h electricity, Building B imported 40.1 kW h electricity and exported 2.5 kW h electricity, and Building C imported 25.1 kW h electricity and exported 25.7 kW h electricity.

Figure 15.7 depicts the electricity energy flow of the building cluster (i.e. electricity demand), aggregated PV production, power grid, aggregated battery and three EVs in the first day of the selected week for the three different scenarios. The aggregated energy exchanges within the building cluster become zero in the aggregated level, since the amount of purchased electricity from the building cluster compensates with the amount of electricity sold to the building cluster.

In the period 9:00 ~ 12:00, for Scenario 1 and Scenario 2, large electricity demand occurs, as EV 2 and EV 3 are charged immediately after being plugged in. Unfortunately, the renewable energy generation is not sufficient in this period to meet the large demands. As a result, a large amount of grid electricity is purchased by the building cluster, i.e. 48.7 kW h and 52 kW h for Scenarios 1 and 2, respectively. In Scenario 3 (developed control), as EV 2 and EV 3 can be flexibly charged in any timeslot during the parking period, the controllers set relatively small EV charging rates in this period. Consequently, the amount of grid power purchase is significantly reduced in the developed control, i.e. 14.6 kW h. In the period 14:00 ~ 17:00, for Scenario 1, since there is no collaboration among buildings, only a small part of the surplus renewable energy is kept onsite, while a large part of the surplus renewables (i.e. 28.5 kW h) is exported to the power grid at a low price. In Scenario 2, contributed by the energy sharing within building cluster, more renewable energy can be stored in the battery. After the batteries in

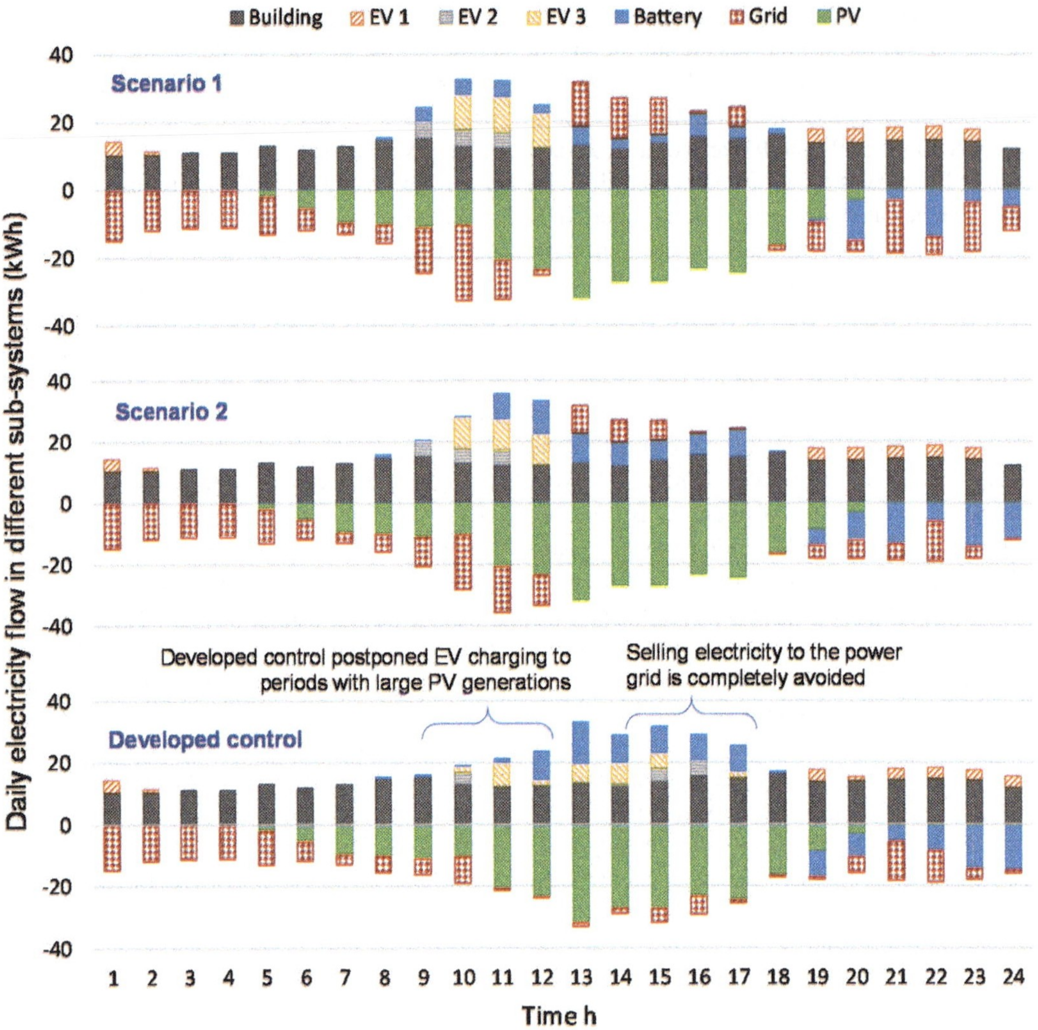

Fig. 15.7 Detailed energy flow (of building, PV systems, battery and three EVs) in the building cluster in each scenario in the first day of the selected week

the building cluster all being fully charged, only a small amount of surplus renewable energy (i.e. 14.1 kW h, which is only half of the amount of exported electricity in Scenario 1) is still exported to the power grid. Scenario 2 has better performance compared with Scenario 1. Since the batteries of EV 2 and EV 3 have already been fully charged in the period 9:00 ~ 12:00, there is no energy flow for them in the period 14:00 ~ 17:00. In the developed control, considering the large renewable energy production in this period, the controller shifts the charging load of EV 2 and EV 3 to this period. Part of the surplus renewable generation is stored in the building battery and part of the surplus renewables is used to supply the EV load. As a result, exporting renewable energy to the power grid is completely avoided. This can effectively improve the renewable energy self-consumption rate of the building cluster.

To sum up, in Scenario 1, the building cluster exported 41.3 kW h electricity to the grid and imported 177.0 kW h electricity from the grid. In Scenario 2, the building cluster exported 23.0 kW h electricity to the grid and imported 159 kW h electricity from the grid. Scenario 2 performs better than Scenario 1 (i.e. with reduced energy imports/exports) as energy sharing enables the building cluster to keep more renewable energy on-site. While using the developed control, the building cluster exported 0 kW h electricity to the grid and imported 135.6 kW h electricity from the grid. Scenario 3 performs even better than Scenario 2, as the controller shifts EV charging loads to periods with large renewable production and thus help keep more renewable energy used onsite in cases when the batteries in the buildings have been fully charged (see the aggregated battery SOC in Fig. 15.5).

15.4.3 Overall Economic and Energy Performance Comparison

This section compares the overall economic and energy performance of different controls. Table 15.4 summarizes the building-cluster-level daily electricity costs and renewable energy self-consumption rates. in different scenarios. Figure 15.8a compares the daily renewable energy self-consumption rates of the three scenarios in the selected week. The relative performances improvements of Scenario 2 and the developed control compared with Scenario 1 are also depicted. Compared with Scenario 1, Scenario 2 improved the renewable energy self-consumption by 6∼9%. This is because the collaboration enables buildings to share their surplus renewable energy with other buildings with insufficient supply and thus help reduce the electricity exports to the power grid (i.e. keep more renewable energy onsite). Compared with Scenario 2, the developed control further improves the renewable self-consumption rates by 3–11% (see Table 15.4). This is because the developed control makes use of the flexible charging ability of EVs. By shifting the EV charging load to periods with large renewable generation periods, more renewable energy can be used onsite, especially when the electrical battery storages are fully charged. In Day 1, the daily self-consumption rates are relatively higher than the other six days. This is because the amount of PV power production is relatively lower than other days (see Fig. 15.4 the PV power production profiles), and thus most of the PV power will be used to meet the electricity demand on-site.

Figure 15.8b compares the daily electricity costs of the three scenarios in the selected week. Due to increased renewable energy self-consumption rates and thus less grid power purchase, Scenario 2 achieves 11∼28% cost saving compared with Scenario 1, and the developed control achieves 7∼17% more cost saving compared with Scenario 2 (see Table 15.4). The relative improvements in economic performance is much larger than the relative improvements in daily self-consumption rates. This is because the building cluster purchase electricity from the power grid at a high price (i.e. 0.16 €/(kW h)) but sell electricity at a much lower price (i.e. 0.05 €/(kW h)). When the building cluster exports more renewables to the power grid (i.e. in Scenario 1), they will need to buy more electricity from the grid at a high price, as the aggregated daily electricity demand is fixed.

15.5 Conclusion

This study has proposed a coordinated control of building clusters for improving the cluster-level performance, with both energy sharing and EV charging considered. The developed coordinated control first uses a 'representative' building to represent the whole building cluster and optimizes its energy storage operation as well as the EV charging using genetic algorithm. The optimized performance of the building cluster is considered to be the optimal one that maximizes the energy sharing within the building cluster by coordinating individual building's operation. Then, non-linear programming is used to coordinate the operation of each individual building. For validation, the developed control has been tested using the energy demand and supply data on a real buildings cluster (with three EVs

Table 15.4 Building-cluster-level daily economic and energy performance

Day		1	2	3	4	5	6	7
Daily self-consumption	Scenario 1	83.7%	60.2%	60.9%	59.9%	65.1%	62.7%	75.2%
	Scenario 2	91.0%	64.2%	65.5%	63.4%	70.4%	67.3%	80.3%
	Proposed control	100.0%	66.2%	68.3%	65.5%	74.4%	73.2%	86.8%
Relative improvements	Scenario 2 versus 1	9%	7%	8%	6%	8%	7%	7%
	Developed versus 1	19%	10%	12%	9%	14%	17%	15%
Daily electricity costs (€)	Scenario 1	29.2	13.7	14.5	12.2	15.1	15.4	20.4
	Scenario 2	24.2	9.9	12.2	9.8	13.4	11.2	15.4
	Proposed control	21.7	9.0	10.8	8.6	12.4	8.5	13.0
Relative improvements	Scenario 2 versus 1	17%	27%	16%	20%	11%	28%	25%
	Developed versus 1	26%	34%	25%	29%	18%	45%	36%

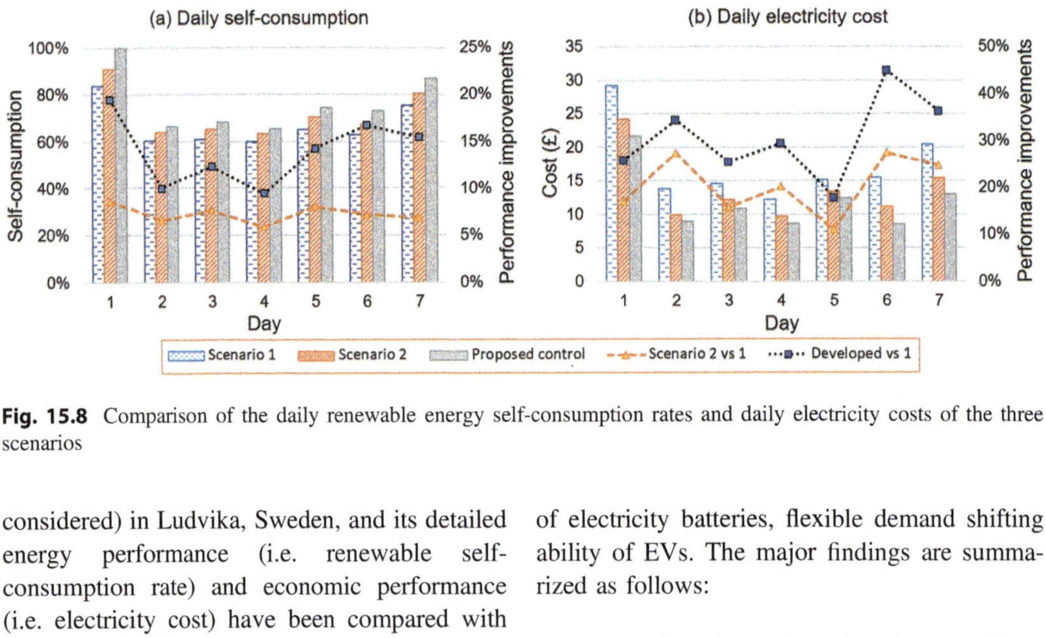

Fig. 15.8 Comparison of the daily renewable energy self-consumption rates and daily electricity costs of the three scenarios

considered) in Ludvika, Sweden, and its detailed energy performance (i.e. renewable self-consumption rate) and economic performance (i.e. electricity cost) have been compared with two scenarios (i.e. one does not enable energy sharing and one allows full energy sharing, both do not have EV charging controls). Case studies have shown that the developed coordinated control can effectively improve the renewable self-consumption rates and meanwhile reduce the electricity bills of the building cluster, by taking advantage of energy sharing, storage capability of electricity batteries, flexible demand shifting ability of EVs. The major findings are summarized as follows:

- The developed coordinated control provides a mechanism to coordinate each single building's operation and EV charging demands for improved building cluster performances.
- In aspect of renewable utilization, the coordinated control improved the daily self-consumption rates by as much as 19% compared with Scenario 1 (no EV control and no

energy sharing) and as much as 11% compared with Scenario 2 (no EV control but with energy sharing). This is because the developed control shifts the EV charging load to periods with large renewable generation periods, and thus more renewable energy are used onsite, especially when the electrical battery storages are fully charged.

- In aspect of economic costs, the coordinated control reduced the daily electricity costs by as much as 36% compared with Scenario 1 (no EV control and no energy sharing) and as much as 17% compared with Scenario 2 (no EV control but with energy sharing). This is because the developed control reduces the amount of high-price grid electricity imports.

This chapter concentrates on the development of control concept for the coupled PV-battery storage-EV systems in the case building cluster. So far, the renovation in this demo case is still under progress and the only pre-monitoring data before renovation has been collected. In the future with both PV and EV integrated, post-monitoring data will be collected and the experimental data from the demo site will be used to validate the simulation. In this study, the detailed driving patterns of EVs are not considered, and the SOC when they arrive the charging ports are determined by some random values. The mobility of humans is highly regular, and study shows that there is 93% potential predictability in user mobility (Song et al. 2010). Future work will take account of the predictive EV driving patterns in the optimization to achieve better performances. Meanwhile, the uncertainty in demand and renewable prediction is not considered in this study. Future work will try to develop more robust controls.

References

Agency USIE (2010) United States—Policies and legislation. Available at http://www.ieahev.org/by-country/united-states-policy-and-legislation/. Accessed on 25 Oct 2019

Agency SE (2017) Förslag till strategi för ökad användning av solel [Draft strategy for increased use of solar electricity]. Available at https://energimyndigheten.a-w2m.se/Home.mvc?ResourceId=5599. Accessed on 19 Dec 2019

Barone G, Buonomano A, Calise F, Forzano C, Palombo A (2019) Building to vehicle to building concept toward a novel zero energy paradigm: Modelling and case studies. Renew Sustain Energy Rev 101:625–648

Cao X, Dai X, Liu J (2016) Building energy-consumption status worldwide and the state-of-the-art technologies for zero-energy buildings during the past decade. Energy and Build 128:198–213

Cooper AI, Schefter KE (2018) Electric vehicle sales forecast and the charging infrastructure required through 203. Edison Electric Institute

Dallinger D, Gerda S, Wietschel M (2013) Integration of intermittent renewable power supply using grid-connected vehicles—A 2030 case study for California and Germany. Appl Energy 104:666–682

Fachrizal R, Munkhammar J (2020) Improved photovoltaic self-consumption in residential buildings with distributed and centralized smart charging of electric vehicles. Energies 13

Fan C, Huang G, Sun Y (2018) A collaborative control optimization of grid-connected net zero energy buildings for performance improvements at building group level. Energy 164:536–549

Ferroamp (2018) The energyhub system. Available at https://static.ferroamp.com/files/brochure/en/Ferroamp%20Brochure%20English%202018.pdf. Accessed on 10 May 2019

Gao D-C, Sun Y (2016) A GA-based coordinated demand response control for building group level peak demand limiting with benefits to grid power balance. Energy Build 110:31–40.

Geth F, Willekens K, Clement K, Driesen J, Breucker SD (2010) Impact-analysis of the charging of plug-in hybrid vehicles on the production park in Belgium. In: Melecon 2010—2010 15th IEEE mediterranean electrotechnical conference, pp 425–430

Huang P, Sun Y (2019a) A clustering based grouping method of nearly zero energy buildings for performance improvements. Appl Energy 235:43–55

Huang P, Sun Y (2019b) A collaborative demand control of nearly zero energy buildings in response to dynamic pricing for performance improvements at cluster level. Energy 174:911–921

Huang P, Sun Y (2019c) A robust control of nZEBs for performance optimization at cluster level under demand prediction uncertainty. Renew Energy 134:215–227

Huang P, Huang G, Sun Y (2018a) Uncertainty-based life-cycle analysis of near-zero energy buildings for performance improvements. Appl Energy 213:486–498

Huang P, Wu H, Huang G, Sun Y (2018b) A top-down control method of nZEBs for performance optimization at nZEB-cluster-level. Energy 159:891–904

Huang P, Lovati M, Zhang X, Bales C, Hallbeck S, Becker A et al (2019) Transforming a residential building cluster into electricity prosumers in Sweden: Optimal design of a coupled PV-heat pump-thermal storage-electric vehicle system. Appl Energy 255:113864

Kara EC, Macdonald JS, Black D, Bérges M, Hug G, Kiliccote S (2015) Estimating the benefits of electric vehicle smart charging at non-residential locations: a data-driven approach. Appl Energy 155:515–525

Klein S, Beckman W, Mitchell J, Duffie J, Duffie N, Freeman T et al (2004) TRNSYS 16–A TRaNsient system simulation program, user manual. University of Wisconsin-Madison, Solar Energy Laboratory Madison

Lovati M, Zhang X, Huang P, Olsmats C, Maturi LJB (2020) Optimal simulation of three peer to peer (P2P) business models for individual PV prosumers in a local electricity market using agent-based modelling. Buildings 10:138

Lu Y, Wang S, Sun Y, Yan C (2015) Optimal scheduling of buildings with energy generation and thermal energy storage under dynamic electricity pricing using mixed-integer nonlinear programming. Appl Energy 147:49–58

Luthander R, Widén J, Munkhammar J, Lingfors D (2016) Self-consumption enhancement and peak shaving of residential photovoltaics using storage and curtailment. Energy 112:221–231

Merrill L, Bridle R, Klimscheffskij M, Tommila P, Lontoh L, Sharma S et al (2017) Making the switch: From fossil fuel subsidies to sustainable energy. Nordic Council of Ministers

Merten J, Guillou H, Ha L, Quenard M, Wiss O, Barruel F (2012) Solar mobility: Two years of practical experience charging ten cars with solar energy. In proceedings of the 5th International conference on integration of renewable and distributed energy resources, Berlin,Germany (pp. 4–6)

Odonkor P, Lewis K (2015) Adaptive operation decisions in net zero building clusters. In *International Design Engineering Technical Conferences and Computersand Information in Engineering Conference* 57076: pp. V02AT03A029. American Society of Mechanical Engineers

Palacios-Garcia EJ, Chen A, Santiago I, Bellido-Outeiriño FJ, Flores-Arias JM, Moreno-Munoz A (2015) Stochastic model for lighting's electricity consumption in the residential sector. Impact of energy saving actions. Energy Build 89:245–259

Palacios-Garcia EJ, Moreno-Munoz A, Santiago I, Flores-Arias JM, Bellido-Outeirino FJ, Moreno-Garcia IM (2018) A stochastic modelling and simulation approach to heating and cooling electricity consumption in the residential sector. Energy 144:1080–1091

Prasad A, Dusparic I (2019) Multi-agent deep reinforcement learning for zero energy communities. In 2019 IEEE PES Innovative Smart Grid TechnologiesEurope (ISGT-Europe) (pp. 1-5). IEEE

Salom J, Widén J, Candanedo J, Sartori I, Voss K, Marszal A (2011) Understanding net zero energy buildings: evaluation of load matching and grid interaction indicators. In: Proceedings of building simulation, pp 2514–2521

Shariful Islam M, Mithulananthan N, Quoc HD (2019) Coordinated EV charging for correlated EV and grid loads and PV output using a novel, correlated, probabilistic model. Int J Electr Power Energy Syst 104:335–348

Shen L, Li Z, Sun Y (2016) Performance evaluation of conventional demand response at building-group-level under different electricity pricings. Energy Build 128:143–154

Song C, Qu Z, Blumm N, Barabasi A-L (2010) Limits of Predictability in human mobility. Science 327:1018–1021

Sun Y, Huang G, Xu X, Lai AC-K (2018) Building-group-level performance evaluations of net zero energy buildings with non-collaborative controls. Appl Energy 212:565–576

Taşcıkaraoğlu A (2018) Economic and operational benefits of energy storage sharing for a neighborhood of prosumers in a dynamic pricing environment. Sustain Cities Soc 38:219–229

Usman M, Knapen L, Yasar A-U-H, Vanrompay Y, Bellemans T, Janssens D et al (2016) A coordinated framework for optimized charging of EV fleet in smart grid. Procedia Comput Sci 94:332–339

Ustun TS, Zayegh A, Ozansoy C (2013) Electric vehicle potential in Australia Its Impact on smart grids. Ind Electron Mag IEEE 7:15–25

van der Kam M, van Sark W (2015) Smart charging of electric vehicles with photovoltaic power and vehicle-to-grid technology in a microgrid; a case study. Appl Energy 152:20–30

Widén J, Wäckelgård E (2010) A high-resolution stochastic model of domestic activity patterns and electricity demand. Appl Energy 87:1880–1892

Xylia M, Silveira SJEP (2017) On the road to fossil-free public transport: The case of Swedish bus fleets. Energy Policy 100:397–412

Zhang S, Huang P, Sun Y (2016) A multi-criterion renewable energy system design optimization for net zero energy buildings under uncertainties. Energy 94:654–665

Zhang B, Li Q, Wang L, Feng W (2018) Robust optimization for energy transactions in multi-microgrids under uncertainty. Appl Energy 217:346–360

Zhang S, Cheng Y, Olaide Oladokun M, Wu Y, Lin Z (2020) Improving predicted mean vote with inversely determined metabolic rate. Sustain Cities Soc 53:101870

Zhao Y, Lu Y, Yan C, Wang S (2015) MPC-based optimal scheduling of grid-connected low energy buildings with thermal energy storages. Energy Build 86:415–426

16 Genetic Algorithm and Mont Carlo Method for Global Sensitivity Analysis of Key Parameters Identification of Net Zero Energy Buildings Towards Power Grid Interaction Optimization

Yongjun Sun, Yelin Zhang, and Xingxing Zhang

Abstract

Utilizing renewable energy to meet the energy demand, net-zero energy building (NZEB) is considered a promising solution to the worsening energy and environmental problems. Due to the intermittent and unstable characteristics of renewable energy (e.g. solar energy), NZEB needs to frequently exchange energy with the power grid. Such frequent energy interactions can impose negative impacts on the grid in terms of power balance and voltage stability. Existing studies demonstrated that there exist many influential parameters to NZEB grid interaction. However, the impacts of influential parameters have not been systematically compared and the key parameters with critical impacts are still unknown. Without knowing the key parameters, researchers may mistakenly optimize non-critical parameters, thereby leading to limited performance improvements; or they have to take parameters more than necessary into consideration, thereby causing unnecessarily high computation loads. Therefore, this study proposes a novel method to identify the key parameters affecting NZEB grid interactions. In the method, global sensitivity analysis is adopted to quantitatively compare the impacts of 24 influential parameters in three major performance aspects including over/under voltage, grid dependence and energy loss. Meanwhile, Monte-Carlo method is used to simulate the parameter uncertainties. The identified key parameters have been verified through comparing their performance improvements and computation loads. Providing an effective way to identify key parameters out of numerous ones, the study results can substantially reduce the unnecessary considerations of non-critical parameters in design optimizations. Also, the identified key parameters can be used for improving NZEB grid interaction with limited computing power requirement.

Keywords

Net-zero energy building · Key parameter · Sensitivity analysis · Monte Carlo · Grid interaction

Y. Sun (✉) · Y. Zhang
Division of Building Science and Technology, City University of Hong Kong, Kowloon, Hong Kong, China
e-mail: yongjsun@cityu.edu.hk

Y. Zhang
e-mail: yelzhang2-c@my.cityu.edu.hk

X. Zhang
Department of Energy and Community Buildings, Dalarna University, 79188 Falun, Sweden
e-mail: xza@du.se

X. Zhang (ed.), *Data-driven Analytics for Sustainable Buildings and Cities*, Sustainable Development Goals Series, https://doi.org/10.1007/978-981-16-2778-1_16

16.1 Introduction

Net-zero energy buildings (NZEBs) are recognized as a promising solution to reducing the growing energy consumption and greenhouse gas emissions. A NZEB equips with renewable energy systems and generates as much energy as it consumes over a specific period (typically a year) (Huang et al. 2018a). The development of NZEBs has been legislated in many countries and organizations. For instance, the European Directive established the target that all new buildings should be NZEBs from 2021 (Huang et al. 2018b); the U.S. set the goal of net zero energy for all commercial buildings by 2050 (Doe 2008). Due to the intermittent and unstable characteristics of renewable energy (e.g. solar energy), NZEB needs to frequently exchange energy with the power grid. For instance, with insufficient/surplus renewable generation, a NZEB needs to import/export electrical energy from/to the power grid.

Such frequent energy interactions can impose negative effects on the grid in terms of power balance (Roos and Bolkesjø 2018) and voltage stability (Baetens et al. 2012). Regarding power balance, the frequent energy exports may result in an increased renewable penetration of the grid, thereby substantially increasing the technical difficulties and costs in maintaining the grid power balance (Brinkel et al. 2020; Holweger et al. 2020). It was reported that in the German 2030 power system, an expected increase of renewable penetration from 66 to 73% could result in the increase of the costly reserve from 226 to 1963 million euros (Roos and Bolkesjø 2018). Regarding voltage stability, such frequent energy interactions can escalate the overvoltage risk of the distribution lines. For instance, during the periods with surplus renewable generations, reverse power flow will occur at the low voltage feeder thereby raising up the feeder voltage and causeing the overvoltage problem. Baetens et al. (2012) found that in Belgium, due to the concern of overvoltage risk, up to 47% of solar power generations from the residential NZEBs were not allowed to be exported to the grid and thus wasted.

NZEB grid interaction is mainly determined by its power mismatch, i.e., the difference between renewable generation and energy demand. Thus, all the parameters affecting renewable generation and energy demands will be able to cause changes of the power mismatch and then can influence NZEB grid interaction. Meanwhile, since a storage system (such as a battery) can store the surplus renewable generation for later use, its buffering effect can reduce/eliminate the power mismatch and thus influence NZEB grid interaction. In this case, the parameters affecting the storage performance also need to be considered as NZEB grid interaction to be improved. Existing studies have demonstrated that there are various parameters affecting energy demand, renewable generation, and storage characteristics. All these parameters could eventually influence NZEB grid interaction but with impacts to different extents.

Regarding energy demand, existing studies have shown that different parameter optimizations can achieve substantial performance improvements. For instance, to improve energy and daylighting performance of a small office building, Fang and Cho (2019) optimized design parameters of building envelop, including window and skylight size and placement, and the study results showed that the energy consumption can be reduced by 20.2%. Adopting artificial neural network, Ilbeigi et al. (2020) optimized six design parameters to minimize the energy use of an office building in Iran. The study results indicated that such multi-parameter optimization can achieve up to 35% energy savings. Meanwhile, substantial energy savings can also be achieved by adopting novel advanced technologies to improve HVAC (heating, ventilation and air-conditioning) system COP (coefficient of performance). For instance, Wu et al. (2018) adopted energy recovery ventilator and ground source heat pump to improve the system COP from 2.42 to 3.38, which resulted in 14.7% annual energy savings.

Realizing various parameters may influence building energy demand to different extents, researchers also tried to identify the key parameters with critical impacts. For instance, Nasruddin et al. (2019) identified 10 key parameters for the energy use of an educational building in Indonesia, including temperature and relative humidity set-points, window to wall ratio (WWR), wall thickness and etc. In a comparative study, Yu et al. (2013) found that shading coefficient, WWR and wall U-value were the three most influential parameters to the energy use of high-rise residential buildings in the hot summer and cold winter region of China. The study results also indicated that roof U-value, building envelope solar absorptance had negligible impacts. For the energy use of apartment buildings in Turkey, Yildiz et al. (2012) found that U-value of wall and glazing, WWR and infiltration rate were the important influential parameters. It should be mentioned as different climates or building types considered, the identified key parameters could be different or inconsistent.

Regarding renewable generation, existing studies have demonstrated the existence of various influential parameters. The optimization of these parameters can improve the performance of load matching and renewable generation. For instance, Awad and Gül (2018) demonstrated that the load matching can be improved by optimizing the photovoltaic (PV) azimuth and tilt angle. In another study, Ma et al. (2014) optimized the size of a hybrid renewable system in a remote island and found that the load matching improvements, obtained from the optimized hybrid system, can saved more than 15% of the excess renewable energy. Similarly, Cao et al. (2013) indicated that up to 16.2% improvements can be achieved in load matching through optimizing the hub height and wind turbine (WT) number in a residential NZEB. To improve PV electrical efficiency, Alizadeh et al. (2018) proposed a pulsating heat pipe integrated PV system and demonstrated that the system can reduce the PV temperature by 16.1 °C, thereby increasing the energy generation by 18%. Through shape optimization and placement of deflector plates, Storti et al. (2019) enhanced wind energy harvesting by 30% in a vertical axis wind turbine (VAWT).

Regarding energy storage, many existing studies focused on battery size optimizations and inverter efficiency improvements. For instance, Sharma et al. (2019) optimized the battery size for a PV-powered NZEB to minimize the annual electricity bill with consideration of initial investment. Yu et al. (2016) investigated the impacts of battery size on grid independence for a net zero office building. They found that grid independence increased by 50% with a battery size of 130 kWh. The existing study also showed that the inverter energy loss can achieve 4.88% of the total renewable energy generation despite the inverter efficiency as high as over 90% (Ma et al. 2014).

Due to lack of systematic studies, among so many influential parameters, the key ones with significant impacts on NZEB grid interaction are still unknown. Although existing studies have identified the key parameters for building energy demand, such identification results are not applicable for NZEB grid interaction. The main reason is that, in comparison with building energy demand, NZEB grid interaction is influenced by much more parameters, and their influences are mixed and aggregated in a complex way. Without knowing the key parameters, researchers may mistakenly optimize non-critical parameters, thereby leading to limited performance improvements; or they have to take parameters more than necessary into consideration, thereby causing unnecessarily high computation loads.

Therefore, this chapter proposes a global sensitivity analysis-based method to identify the key parameters affecting NZEB grid interactions. Three indicators, namely over/under voltage, grid dependence and energy loss, are selected to assess the grid interaction performance. The study firstly identifies the key parameters using global sensitivity analysis. Meanwhile, Monte-Carlo simulation has been taken to consider the parameter uncertainties. Then, the reasons why the identified parameters have significant impacts are elaborated in detail. Next, the global sensitivity analysis results are verified by comparing the performance improvements and computation loads. Last, major findings are summarized in the conclusion part.

16.2 Methodology of Key Parameter Identification

Figure 16.1 shows the basic idea of the proposed method which includes two parts. The first part is to identify the key parameters and it includes three steps: input parameters collection, grid interaction performance assessment and global sensitivity analysis. In the first step, all the input parameters that may influence the grid interaction are collected and their ranges are identified with reference to existing studies. In the second step, the grid interaction performance is assessed using the Monte-Carlo simulation in which the parameter uncertainty samples are generated from the identified parameter ranges. The Monte-Carlo simulation, relying on repeated random sampling to obtain numerical results, is a commonly used method to assess the uncertainty impacts (Huang and Huang 2015). In the third step, based on the parameter uncertainty impacts obtained from the Monte-Carlo simulation, standardized regression coefficient (SRC) algorithm, a global sensitivity analysis method, is used to identify the key parameters.

The second part is to verify the sensitivity analysis results by comparing the performance improvements achieved from optimizing different parameter sets. In comparison with an unimportant parameter, a key parameter has

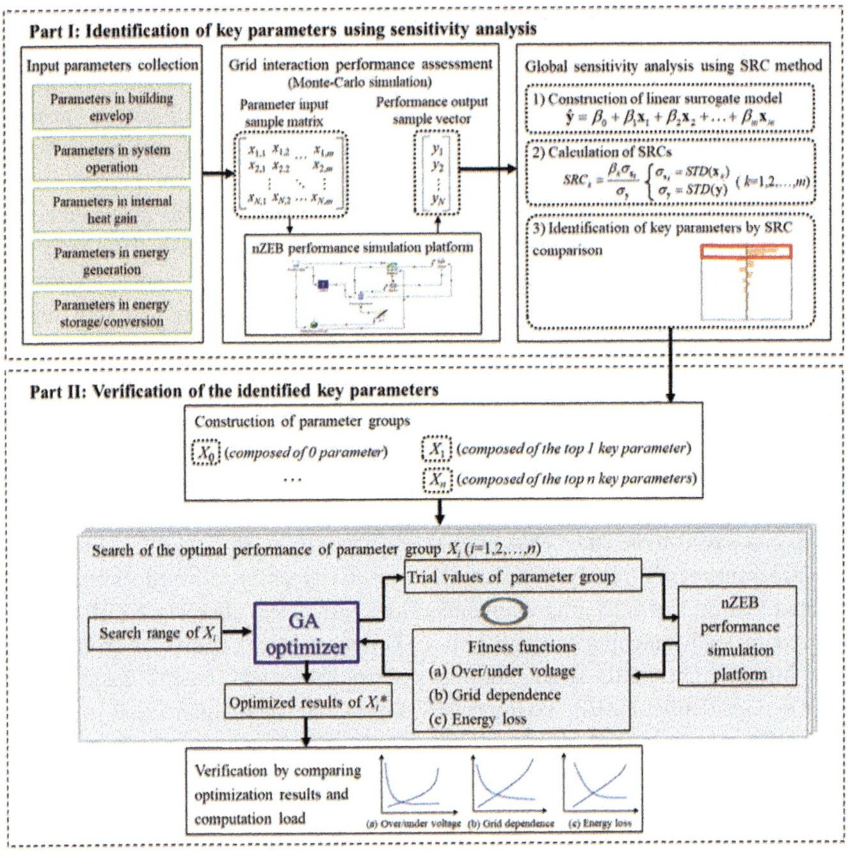

Fig. 16.1 Flowchart of the key parameter identification and verification method

more potentials to improve grid interaction performance under the same efforts. Thus, after separate optimizations, those parameters with significant performance improvements will be considered as key parameters. To verify the sensitivity analysis results, the following steps will be taken. First, based on the key parameters identified from sensitivity analysis in Part I, several parameter groups are constructed which consist of different numbers of key parameters. And then, the genetic algorithm (GA) optimizer is used to optimize the constructed parameter groups. Next, the performance improvements and computation loads of optimizations for different parameter groups are compared to verify the sensitivity analysis results.

16.2.1 Collection of Input Parameters

The grid interaction of NZEBs is caused by the mismatch between renewable generations and energy demands. Thus, the grid interaction performance is mainly influenced by energy demand, renewable generation, energy storage. Parameters that affect these processes should be considered as inputs. In this study, the main input parameters are classified into five categories, i.e. building envelope, system operation, internal heat gain, energy generation and energy storage/conversion. And 24 parameters in these categories are selected as input parameters of Monte-Carlo simulation and sensitivity analysis, as summarized in Table 16.1.

Table 16.1 Input parameters for sensitivity analysis (Li and Wang 2018; Sun 2015)

Category	Parameter	Base value	Range
Building envelop	Wall specific heat, kJ/(kg K)	1.4	0.8–2
	Wall U-value, W/(m^2 K)	0.8	0.2–1.5
	Wall solar absorptance	0.6	0.3–0.9
	Roof specific heat, kJ/(kg K)	0.93	0.45–1.4
	Roof U-value, W/(m^2 K)	0.8	0.2–1.5
	Roof solar absorptance	0.6	0.3–0.9
	Building height, m	3.5	3–4
	Internal shading factor	0.5	0–1
	Window to wall ratio	0.6	0.05–0.9
	Infiltration rate, ACH	0.75	0.5–1
System operation	System COP	3.5	3–4
	Ventilation rate, l/s/person	2.4	1.8–3
	Indoor temperature set-point, °C	24	22–26
	Relative humidity set-point, %	50	40–60
Internal heat gain	Occupant density, m^2/person	2	1.5–2.5
	Lighting demand, W/m^2	9	6–12
	Equipment demand, W/m^2	12	8–15
Energy generation	Generation capacity	Mean	1–1.2
	PV proportion	0.5	0.5–1
	PV electrical efficiency	0.14	0.08–0.2
	WT power loss	0.15	0–0.3
Energy storage/conversion	Battery capacity, kWh	Mean	(0–1) × $P_{con,daily}$ [a]
	Battery round-trip efficiency	0.8	0.75–0.85
	Inverter efficiency	0.95	0.93–0.97

[a] $P_{con,daily}$ is the daily energy consumption

In the design of a new building, designers select proper parameter values from a limited range suggested by design guidelines/handbooks. Considering the equivalent selection probability of each value, the study utilizes the uniform distribution to describe such uncertainty (Tian 2013). With reference to existing studies (Li and Wang 2018; Sun 2015), the ranges of the input parameter values are determined and presented in Table 16.1. Their base values are also given in Table 16.1 for the analysis of the identified key parameters. Note that in the retrofitting cases, other statistical distributions (e.g. normal distribution and triangular distribution), derived from actual measurements, can be used to quantify the parameter uncertainties (Huang et al. 2019a; Booth and Choudhary 2013).

16.2.2 Grid Interaction Performance Assessment Using Monte-Carlo Simulation

The Monte-Carlo simulation is a commonly used technique for uncertainty impact assessments. In this study, it is adopted to investigate the impacts of the parameter value uncertainties on the grid interaction performance. Based on the uniform uncertainty distribution given in Sect. 2.1, N parameter samples are generated using the random sampling approach (Huang and Huang 2015). For all the input parameters, an input sample matrix **x** is formed as below. The matrix consists of N sampled values of the m input parameters.

$$x = [x_1 \ x_2 \ldots x_m] = \begin{bmatrix} x_{1,1} & x_{1,2} & \cdots & x_{1,m} \\ x_{2,1} & x_{2,2} & \cdots & x_{2,m} \\ \cdots & \cdots & \cdots & \cdots \\ x_{N,1} & x_{N,2} & \cdots & x_{N,m} \end{bmatrix} \tag{16.1}$$

Next, the sample matrix **x** is imported into the NZEB model for estimating the uncertainty in the concerned performance indicator. The estimation process will repeat N times (corresponding to the N samples) for obtaining the performance vector **y**. In this study, three performance indicators are considered in terms of over/under voltage, grid dependence and energy loss. For each performance indicator, a performance vector **y** will be produced.

$$\mathbf{y} = \begin{bmatrix} y_1 \\ y_2 \\ \cdots \\ y_N \end{bmatrix} = \begin{bmatrix} f(x_{1,1}, x_{1,2}, \ldots, x_{1,m}) \\ f(x_{2,1}, x_{2,2}, \ldots, x_{2,m}) \\ \cdots \\ f(x_{N,1}, x_{N,2}, \ldots, x_{N,m}) \end{bmatrix} \tag{16.2}$$

16.2.2.1 Over/Under Voltage Performance Indicator

For the satisfactory operating of electrical and electronic devices, it is recommended to allow voltage variation at the prescribed limits, i.e. ±6% of rated voltage in Hong Kong (Hong Kong Electric 2019). Voltage variations have adverse effects on connected loads. For instance, overvoltage can cause insulation damage to electrical appliances, thereby leading to short circuits; while undervoltage causes equipment inefficient operation. The over/under voltage indicator (OV) (Mahmud et al. 2011) is the ratio of maximum voltage variation to the voltage variation limits, as shown in Eq. (16.3). For the safe and satisfactory operating of electrical devices, OV needs to be smaller than 100%.

$$OV = \max(OV_j) = \max\left(\left|\frac{\Delta U_j}{\Delta U_{limit}}\right|\right) \times 100\% \tag{16.3}$$

where OV_j is the over/under voltage in the jth hour; ΔU_{limit} is the absolute value of the allowed voltage variation in Hong Kong (6%); ΔU_j is the voltage variation (Mahmud et al. 2011) in the jth hour, which is calculated by Eq. (17.4),

$$\Delta U_j = \frac{P_{exchange,j} r + Q_{exchange,j} x}{U_0^2} \cdot l \tag{16.4}$$

where, U_0 is the nominal voltage (V); r and x are the resistance and reactance per kilometre respectively (Ω/km); l is the length of the feeder (km); $P_{exchange,j}$ (W) and $Q_{exchange,j}$ (Var) are the active and reactive power exchange between the NZEB and the power grid, respectively. $P_{exchange,j}$ and $Q_{exchange,j}$ are explained in details in Sect. 3.5.

16.2.2.2 Grid Dependence Performance Indicator

The grid dependence indicator (*GD*) (Salom et al. 2011) describes the degree of a NZEB's dependence on the power grid, which is evaluated as the ratio of time with power exchanges to the total counting time (see Eq. (16.5)). The value of *GD* is between 0 and 1. A larger *GD* means a heavier dependence of NZEB on the grid and thus a worse grid interaction performance.

$$GD = \left(1 - \frac{\Gamma_{P_{\text{exchange}}=0}}{\Gamma_{\text{tot}}}\right) \times 100\% \quad (16.5)$$

where, Γ is the time duration (h), and subscript $P_{exchange}$ is the power exchange between the NZEB and power grid (which is calculated by Eq. (16.18) in Sect. 3.5); Γ_{tot} is the total number of counting hours (8760 h).

16.2.2.3 Energy Loss Performance Indicator

The energy loss indicator (*EL*) (Ma et al. 2014) is calculated as the ratio of total energy loss $E_{total\text{-}loss}$ to the total energy generation of PV and WT. It mainly includes three parts, i.e. grid loss (EL_{grid}), battery loss (EL_{bat}) and inverter loss (EL_{inver}).

$$EL = EL_{grid} + EL_{bat} + EL_{inver} = \left(\frac{E_{gridloss}}{E_{totalgen}} + \frac{E_{batloss}}{E_{totalgen}} + \frac{E_{inverloss}}{E_{totalgen}}\right) \times 100\% \quad (16.6)$$

where, $E_{totalgen}$ is the total renewable energy harvested by PV and WT (kWh); $E_{batloss}$ is the total energy loss in the battery, which includes the charging/discharging loss and the self-discharge loss (kWh) (see Sect. 3.4); $E_{inverloss}$ is the total energy loss in the inverter, which is determined by the inverter efficiency and the energy passing through the inverter (kWh) (see Sect. 3.5); $E_{gridloss}$ is the total energy loss in the distribution network (kWh), which is calculated by Eq. (16.7) (Kashem et al. 2000; Jamil and Anees 2016),

$$E_{gridloss} = \sum_{j=1}^{8760} \frac{P^2_{exchange,j} + Q^2_{exchange,j}}{U_0^2} \cdot R \quad (16.7)$$

where, R is the total resistance of the feeder, $R = l \times r$ (Ω).

16.2.3 Global Sensitivity Analysis for Identifying Key Parameters

Based on the Monte-Carlo simulation results, global sensitivity analysis will be used to identify the key parameters with significant impacts on the performance indicators. Among different sensitivity analysis methods, the SRC method is selected in the study due to its simplicity and lower data requirement (Tian 2013; Hopfe 2009). In the SRC method, a linear multivariate model between the performance outputs (**y**) and the parameter inputs (**x**) is first constructed, as shown in Eq. (16.8).

$$\mathbf{y} = \beta_0 + \beta_1\mathbf{x}_1 + \beta_2\mathbf{x}_2 + \ldots + \beta_m\mathbf{x}_m \quad (16.8)$$

where, β_k (k = 1, 2, …, m) is the regression coefficient which can be calculated using least squares method (Domínguez-Muñoz et al. 2010).

The regression coefficient β_k will be further used to calculate the standardized regression coefficient SRC_k, as shown in Eq. (16.9). SRC_k directly reveals the importance of the parameter $\mathbf{x}_k$ in affecting the performance output. A higher absolute value of SRC_k represents a more significant impact of the input parameter. A positive SRC_k means that the performance output increases as the input parameter increases; while a negative SRC_k means that the performance output decreases as the input parameter increases.

$$SRC_k = \frac{\beta_k \sigma_{\mathbf{x}_k}}{\sigma_{\mathbf{y}}} \quad (16.9)$$

where, $\sigma_{\mathbf{x}_k}$ and $\sigma_{\mathbf{y}}$ are the standard deviation of input $\mathbf{x}_k$ and output $\mathbf{y}$, respectively.

16.2.4 Verification of the Key Identified Parameters

After the key parameter identification using the SRC method, the next step is to verify the identification results by comparing the performance improvements achieved from optimizing different parameter sets. The verification process consists of three steps: construction of parameter groups, search for the optimal parameters, and performance improvement comparison, as shown in Fig. 16.1.

In the first step, a parameter group X_i (i = 0, 1, 2, …, n) is composed of the top i key parameters (i.e. the parameters with the top i highest absolute SRC values). It should be noted that i = 0 represents a case in which no parameter is optimized. In the second step, genetic algorithm (GA) will be used to search for the optimal parameter values in X_i. In the search process, those parameters not included in X_i will be assigned random values based on their distributions to simulate the actual design scenario. During each search round, a group of trial values of X_i will be generated within the search space by the GA optimizer. Then, these trial values will be sent into the simulation platform for performance evaluation. The main performance considered is the grid interaction and thus the GA optimizer directly takes the three indicators as the fitness functions, see Eqs. (16.3), (16.5) and (16.6). In order to improve the comparison robustness, the optimization of each parameter group is repeated 10 times, and average values of the optimal performance and computation loads are used for comparison. In the last step, the average values of the optimal performance and computation loads are compared.

Through such comparisons, the parameters' significances will be obtained. If the optimization of a parameter leads to larger performance improvements than another one, this parameter will be considered as a more important parameter. On the other hand, if the optimization of one more parameter leads to limited/negligible performance improvements but with noticeable computation load increase, this parameter will be considered as a non-significant/non-key parameter.

16.3 Simulation Platform and System Modelling

With reference to existing NZEBs, a simulation platform is established using TRNSYS and MATLAB. Figure 16.2 shows the schematics of the platform and the energy system types selected from TRNSYS. It mainly includes a hybrid renewable system (i.e. PV and WT), an electrical battery, an inverter, several energy consuming systems (i.e. HVAC system, lighting system and other electrical equipment) and a power grid. In the platform, the renewable energy will firstly be used to meet the building energy demand, and then the surplus part will be stored in the battery. In case that the battery is fully charged, the surplus renewable energy will be exported to the power grid. In the following models, the parameter values will be assigned the base values of Table 16.1.

16.3.1 Building Model

Multi-zone model (i.e. type 56) from TRNSYS is used to simulate the NZEB. It is a one-floor office building with dimensions of 20 m (length) × 20 m (width) × 3.6 m (height). The single type window with a U-value of 5.68 W/m2K was selected from the window type library of TRNSYS. The values of other design parameters of the NZEB are summarized in Table 16.1. The total energy consumption includes energy used by the HVAC system, lighting system and other equipment. Since the building is located in a subtropical climate, heating is not needed. Cooling is provided by a central air-conditioning system to maintain the indoor temperature at a pre-defined set-point. Figure 16.3 presents the schedules of HVAC

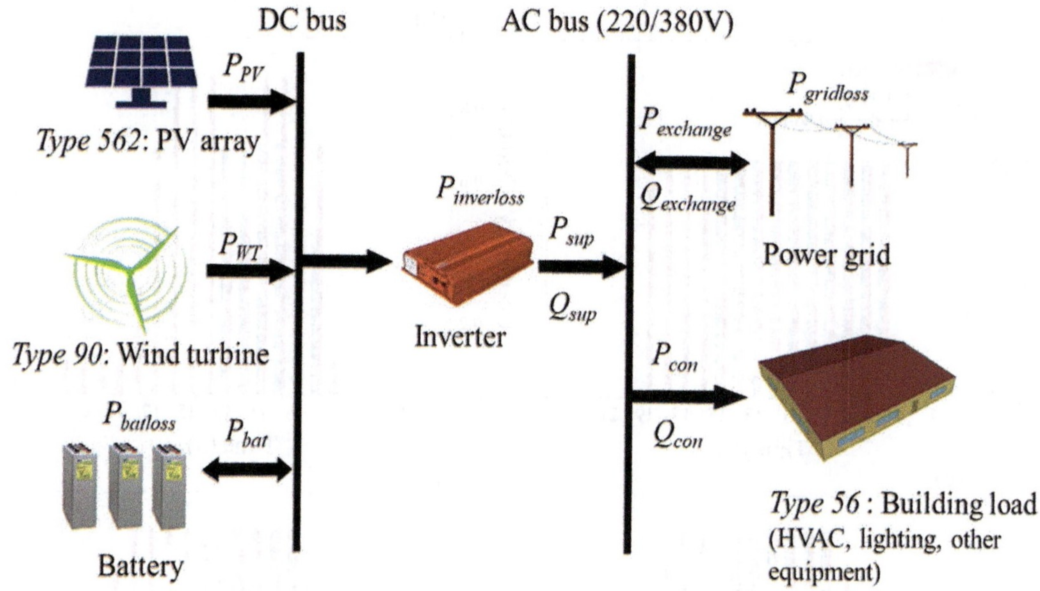

Fig. 16.2 Schematic diagram of the simulation platform

operation, occupancy, equipment, and lighting, with reference to Guideline on Performance-based Building Energy Code (2003). The energy consumptions of the lighting system and other equipment can be estimated based on their schedules and the rated power as shown in Table 16.1.

16.3.2 HVAC System

A typical configuration of HVAC system was selected and presented in Fig. 16.4. In the cooling water loop, a constant speed pump was used to supply cooling water to the chiller. A variable speed fan was used to deliver the ambient air to cool down the circulating water in the cooling tower. At the chilled water side, a constant speed pump was installed in the primary loop and a variable speed pump was adopted in the secondary loop. The speed of the secondary water pump was varied to deliver the needed amount of chilled water according to the varying cooling load. The constant speed fan-coil units were used to supply the cool air to different rooms (Huang and Huang 2015; Zhang et al. 2016). The sizes of the HVAC components were selected based on the peak cooling load of the building (Zhang et al. 2016).

In this HVAC system, the power consuming components include the chiller, two fans and three pumps. The power consumed by the chiller is calculated by Eq. (16.10) (Chai et al. 2019; Huang et al. 2019b).

$$P_{con,chiller} = FFLP \times \frac{Q_{cooling}}{COP} \tag{16.10}$$

where, $FFLP$ is the fraction of full load power; $Q_{cooling}$ is the actual cooling load (W); COP is the nominal coefficient of performance (COP) of the chiller, as shown in Table 16.1.

The power consumed by the variable and constant fans is estimated by the air flow rate $\dot{m}_{air}$ and the static pressure drop of the air flow ΔP_{air}, as shown in Eq. (16.11) (Chai et al. 2019; Huang et al. 2019b).

$$P_{con,fan} = \dot{m}_{air} \times \Delta P_{air} \tag{16.11}$$

The power consumed by the variable and constant speed pumps is calculated according to the water flow rate $\dot{m}_w$ and the water head ΔP_w,

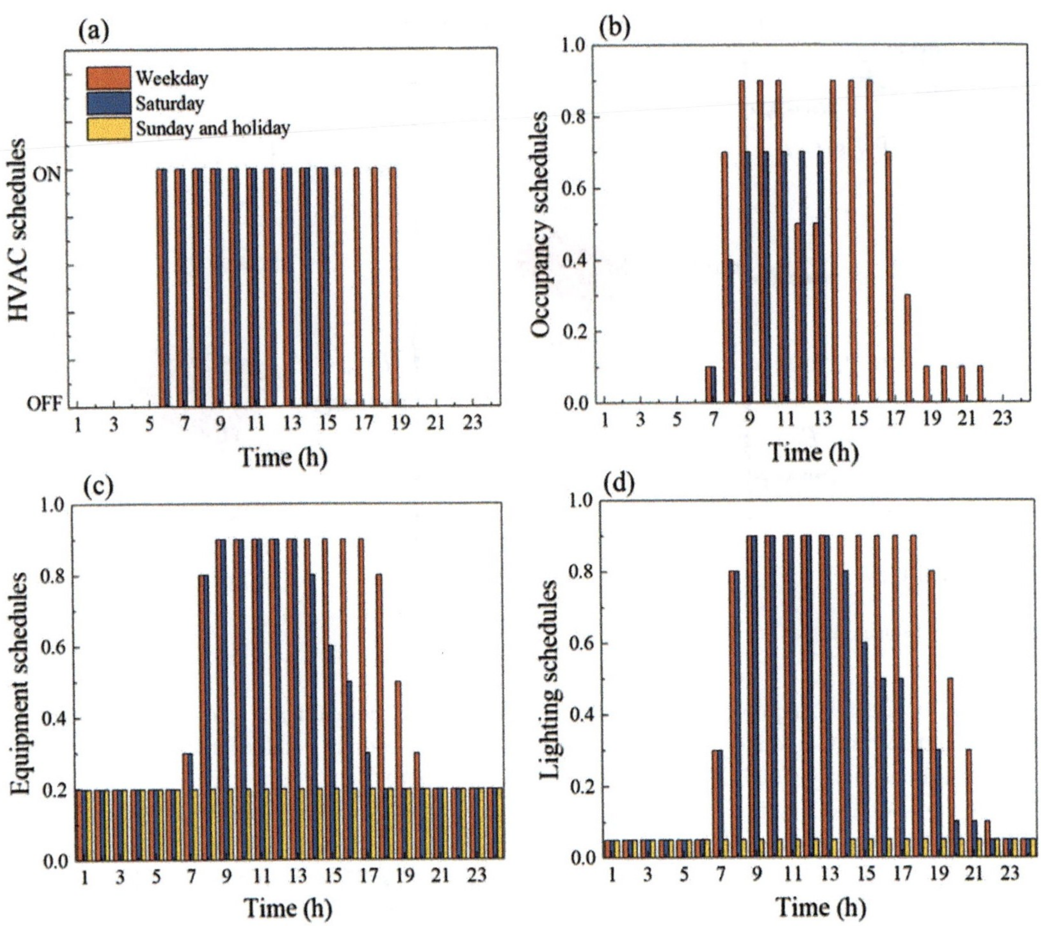

Fig. 16.3 Schedules of HVAC operation, occupancy, equipment, and lighting

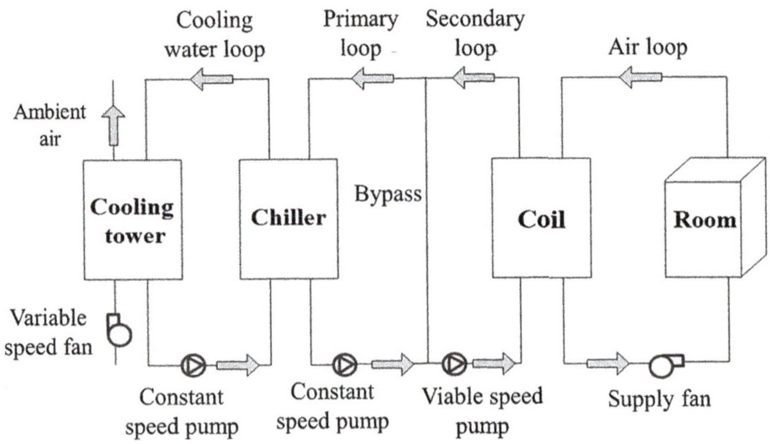

Fig. 16.4 Schematics of the HVAC system

as shown in Eq. (16.12) (Chai et al. 2019; Huang et al. 2019b).

$$P_{con,pump} = \dot{m}_w \times \Delta P_w \quad (16.12)$$

16.3.3 Renewable Energy System

Type 562 and type 90 from TRNSYS are selected to simulate the PV panel and WT respectively. The power outputs of PV (P_{PV}) and WT (P_{WT}) are calculated using Eqs. (16.13) and (16.14), respectively (Sun 2015; Zhang et al. 2016).

$$P_{PV} = (\tau\alpha)_n \times I_{AM} \times A_{PV} \times \eta \times I_T \quad (16.13)$$

where $(\tau\alpha)_n$ is the transmittance-absorptance product of the PV cover for solar radiation at a normal angle of incidence; I_{AM} is the combined incidence angle modifier for the PV cover material; I_T is the total amount of solar radiation on the PV collector surface (W/m^2); η is the overall efficiency of the PV array; and A_{PV} is the PV surface area (m^2).

$$P_{WT} = C_P \times \rho_{air} \times A_R \times v_a^3 \quad (16.14)$$

where, ρ_{air} is air density (kg/m^3); A_R is rotor area (m^2); v_a is wind velocity in the free stream (m/s); power coefficient C_p is a function of the axial induction factor.

The sizes of PV and WT were selected according to the PV proportion and generation capacity. PV proportion indicates the proportion of PV energy generation to the total energy generation. Generation capacity refers to the ratio of the annual renewable energy generation to the annual energy consumption. Due to the energy loss of the battery and inverter, the generation capacity could be slightly larger than 1. In this study, the PV proportion varies from 50 to 100%, and the generation capacity varies from 1 to 1.2.

16.3.4 Electrical Energy Storage System

The amount of electricity stored in the battery is calculated as the product of the battery capacity and the state of charge (*SOC*). It should be noted that the oversized batteries cause excessive initial costs and energy losses but with limited performance improvements (Guarino et al. 2015). In this case, following the existing study (Sun 2015), the upper limit of battery capacity was set as the average daily energy demand, as shown in Table 16.1.

The battery SOC is estimated using Eq. (16.15). In this equation, σ is the self-discharging rate taken as 0.02% (Agarwal and Kumar 2013); η_{bat} is the battery round-trip efficiency (see Table 16.1); $P_{bat,j}$ is the charging/discharging power of battery in the *j*th hour (W), which is calculated by the mismatch between power generation and power consumption. A negative value of $P_{bat,j}$ indicates battery charging and a positive value indicates battery discharging.

$$SOC_{j+1} = SOC_j(1-\sigma) - \frac{P_{bat,j}\eta_{bat}}{BC} \quad (16.15)$$

The calculation of the battery *SOC* is subject to the following constraints.

$$SOC_{\min} \le SOC_j \le SOC_{\max} \quad (16.16)$$

and

$$|P_{batj}| \le P_{bat,\max} \quad (16.17)$$

With reference to the existing studies (Ma et al. 2014), the allowable range of *SOC* is set to be 30–100% in the study, and the amount of maximum charging/discharging power $P_{bat,max}$ is set as 20% of the value of battery capacity (Yang and Lu 2007). If the battery is fully charged/discharged, the remaining electricity will be imported from/exported to the power grid.

16.3.5 Modelling of Grid Interaction

The energy interaction between the NZEB and the power grid includes active power exchange $P_{exchange}$ and reactive power exchange $Q_{exchange}$. They are calculated by Eqs. (16.18) and (16.19).

$$\begin{aligned} P_{exchange,j} &= P_{sup,j} - P_{con,j} \\ &= (P_{PV,j} + P_{WT,j} + P_{bat,j}) \times PF_{inver} \\ &\quad \times \eta_{inver} - P_{con,j} \end{aligned} \quad (16.18)$$

$$\begin{aligned} Q_{exchange,j} = Q_{sup,j} - Q_{con,j} &= P_{sup,j} \times \tan(\arccos PF_{inver}) \\ &\quad - P_{con,j} \times \tan(\arccos PF_{build}) \end{aligned} \quad (16.19)$$

where, PF_{inver} and PF_{build} are the power factor of inverter and energy consumers, respectively ($PF_{inver} = PF_{build} = 0.95$) (Tonkoski and Turcotte 2012; Shen et al. 2013).

Then, the obtained $P_{exchange}$ and $Q_{exchange}$ from Eqs. (16.18) and (16.19) are used as inputs for evaluating the grid interaction performance in terms of over/under voltage, grid dependence and energy loss. In these equations, the voltage at the middle-voltage/low-voltage (MV/LV) substation node (U_0) is set as 380 V. The length of feeder connecting the substation and the NZEB (l) is set as 400 m, and its impedance (r and x) is set to be 0.13 + j0.14 Ω/km following the designed guideline (Electrical and Mechanical Services Department 2015).

16.4 Case Studies and Analysis

This section presents the study results and associated analysis. In case studies, Hong Kong weather data (e.g. solar radiation and wind speed) in the typical year (https:energyplus.net/weather) were used. Figure 16.5 presents the daily wind velocity, solar radiation, ambient temperature and relative humidity in the typical year of Hong Kong. It can be observed that the maximum wind velocity and solar radiation can reach 9.7 m/s and 996.3 W/m^2, respectively. Figure 16.5b reveals that in the most time, the ambient temperature is above 25 °C and relative humidity is over 70%, representing a hot and humid climate. The simulation period was one year and the time interval was one hour. In the following parts, sensitivity analysis results were firstly presented. Then, detailed analysis was provided to elaborate the reasons why the identified parameters could have significant effects on the considered performance. Finally, the verification results were stated in terms of performance improvements and computation load variations when different parameter sets were optimized.

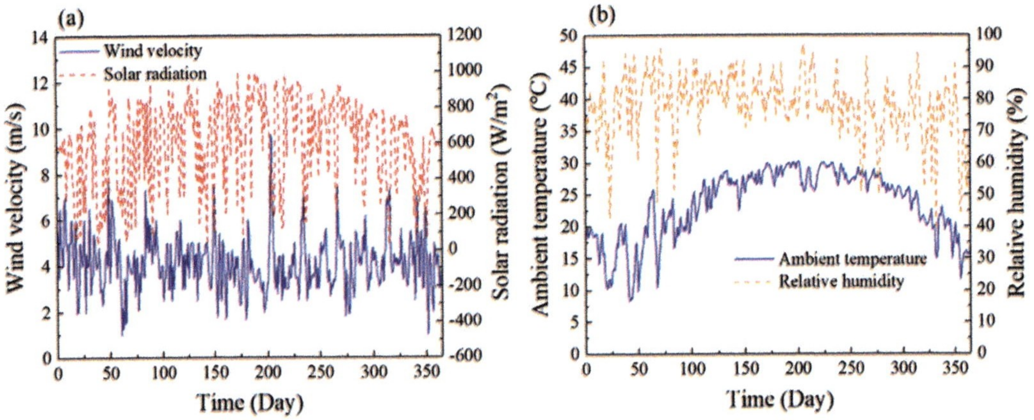

Fig. 16.5 Hong Kong weather data in the typical year (**a**) daily wind velocity and solar radiation; (**b**) daily ambient temperature and relative humidity

16.4.1 Key Parameter Identification Results via Global Sensitivity Analysis

Figure 16.6 shows the sensitivity analysis results of the three performance considered. The 24 input parameters were arranged in the order of SRC absolute values. A larger SRC absolute value indicates that the corresponding parameter plays a more important role in improving the concerned performance. The sign of the SRC represents whether the performance indicator has a positive or negative correlation with the parameter. For the over/under voltage, there were two identified key parameters, i.e. PV proportion and generation capacity. For the grid dependence, there were another two key parameters, i.e. battery capacity and PV proportion. For the energy loss, there were four key parameters, i.e. inverter efficiency, battery capacity, battery round-trip efficiency and PV proportion.

16.4.2 Analysis of the Identified Key Parameters for Over/Under Voltage

Figure 16.7 is used to elaborate the reason why the two identified key parameters had critical impacts on over/under voltage. Over/under voltage is mainly caused by the maximum/peak grid energy exchange. Figure 16.7a presents the daily peak power generations of WT and PV in the scenario of 50% PV proportion. It shows that the peak power generation of WT (appeared on the 204th day) was over 4 times of that of PV. Thus, in comparison with PV, WT was more likely to cause overvoltage problems due to its excessive peak power generation.

Figure 16.7b–d present variations of over/under voltage with PV proportion, generation capacity and COP respectively on the 204th and 205th day with severe overvoltage risks. As shown in Fig. 16.7b, when PV proportion

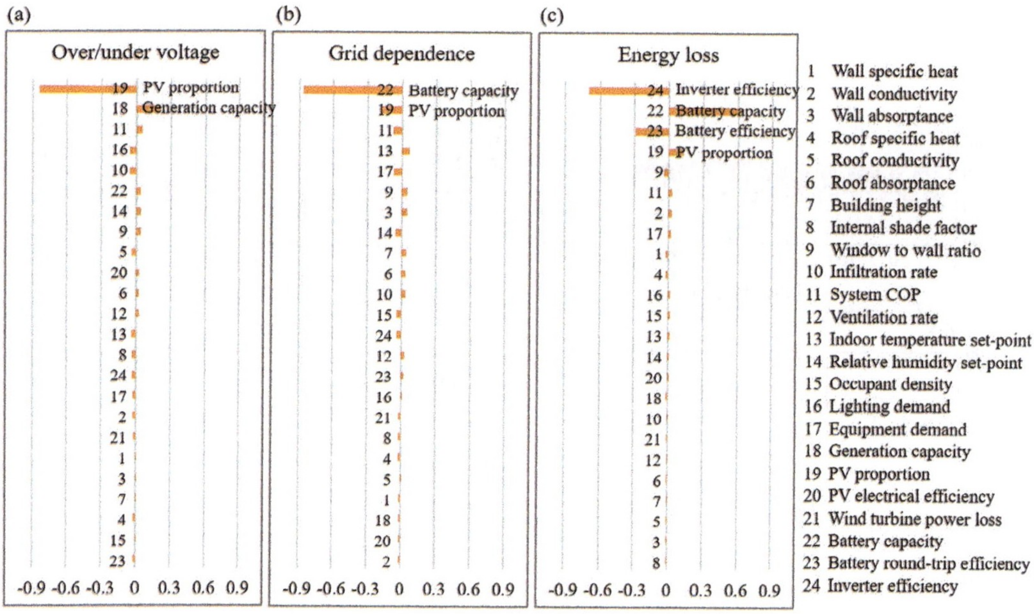

Fig. 16.6 SRCs of all input parameters for **a** over/under voltage; **b** grid dependence; **c** energy loss

increased from 50 to 100%, over/under voltage first rapidly reduced from 85.23 to 0% and then bounced back to 13.55%. Figure 16.7c shows that with the increase of the generation capacity from 1 to 1.2, overvoltage increased from 76.09 to 94.36%. The main reason was the change of the WT installation capacity. For instance, at a given PV proportion, the capacity increase directly lead to the increase of WT installation capacity, thereby substantially increasing the daily peak power generation and associated overvoltage risk, as shown in Fig. 16.7c. In contrast, overvoltage is almost unchanged as the system COP increased from 3 to 4, as shown in Fig. 16.7d. In other words, PV proportion and renewable generation capacity were of more significance in reducing overvoltage risk in comparison with system energy efficiency improvement.

16.4.3 Analysis of the Identified Key Parameters for Grid Dependence

Figures 16.8 and 16.9 are used to elaborate the reasons why the two identified key parameters played more important roles for grid dependence. Grid dependence is determined by power interaction between a NZEB and the power grid. The power interaction is jointly influenced by building renewable generation, energy demand and storage characteristics. Figure 17.8 presents the weekly power variations (including power

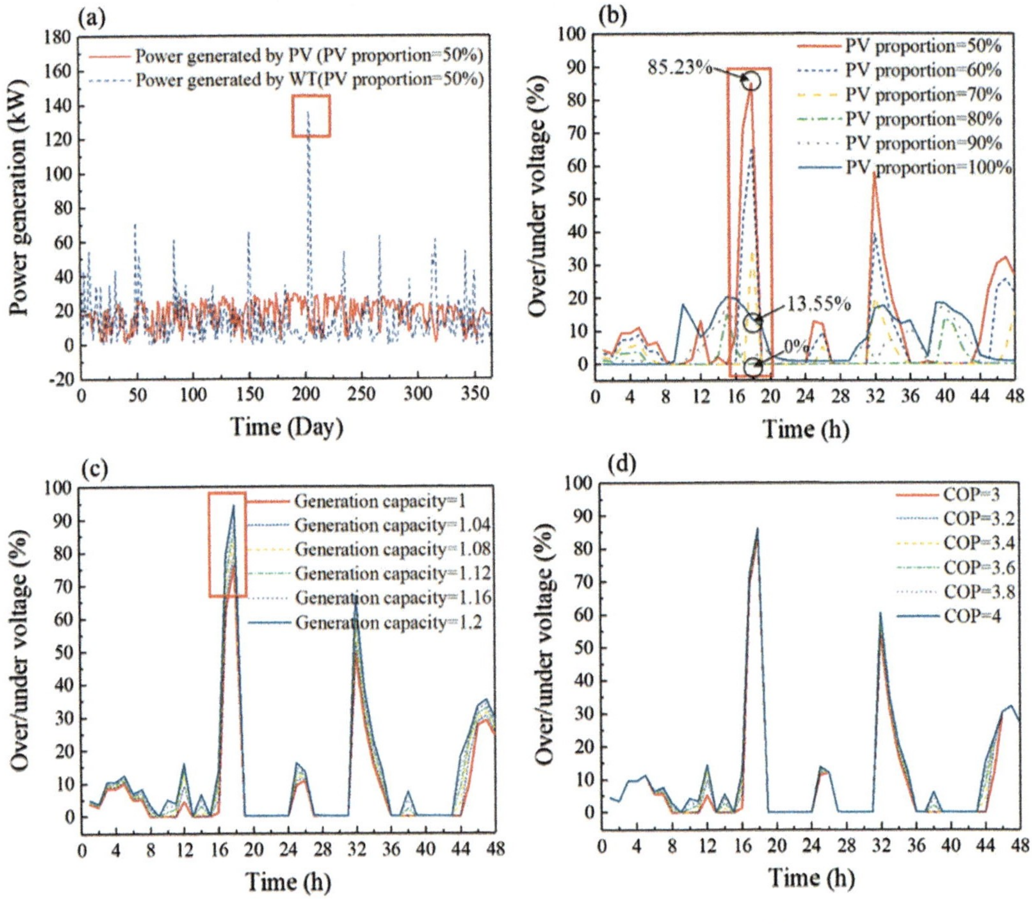

Fig. 16.7 **a** Power generation of PV and WT; variations of over/under voltage with **b** PV proportion, **c** generation capacity and **d** COP

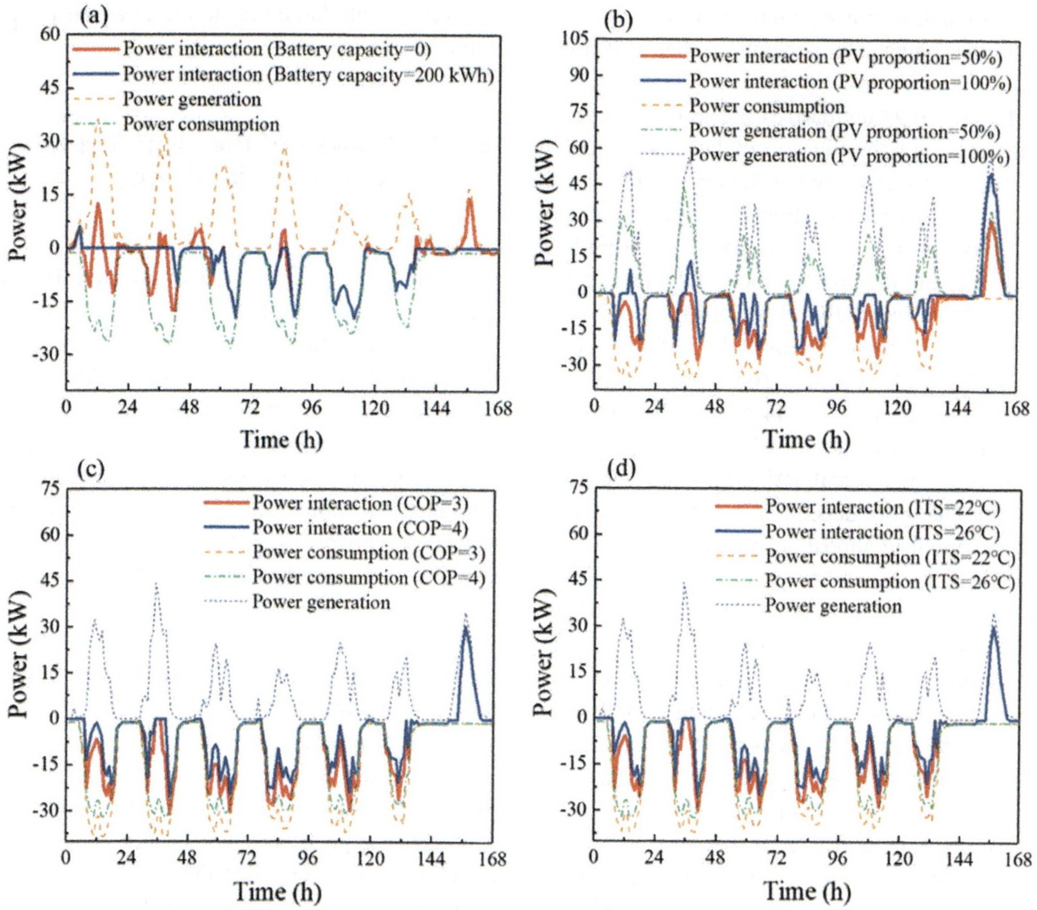

Fig. 16.8 Power variations with **a** battery capacity; **b** PV proportion; **c** COP; and **d** indoor temperature set-point

interaction between the NZEB and the grid, power consumption and renewable generation) as the four top-ranked key parameters considered. Figure 16.8a shows that with the battery capacity increased from 0 to 200 kWh, the grid independence period (i.e., the period requiring zero power interaction) largely increased from 0 to 80 h. This was because via storing the surplus renewable generation for later use, battery can effectively reduce or eliminate power mismatch.

Figure 16.8b shows the power variations with PV proportion. It was observed that PV proportion directly changed the renewable generation and thus caused the variation of the power interaction between the NZEB and the grid. As PV proportion increased from 50 to 100%, the grid independence period (i.e., the period requiring zero power interaction) increased from 23 to 46 h. Figure 16.8c and d show the power variations with COP and indoor temperature set-point (ITS) respectively. Although COP and ITS can largely change building power consumptions, the resulting variations of the power interactions and corresponding grid dependence were not substantially. For instance, as ITS increased from 22 to 26 °C, associated grid independence period only increased two hours, i.e. from 22 to 24 h. The main reason was that building power consumption had relatively stable trend and profile, and thus it had less impacts on the variation of power interaction in comparison with the renewable generation that had more dramatic variations.

Figure 16.9 summarizes the grid dependence variations with the four top-ranked parameters. It shows battery capacity has the largest impacts on the grid dependence. For instance, as battery capacity changed from 0 to 200 kWh (from Min to Max in Fig. 16.9), the grid dependence reduced substantially from 100 to 47.74%. As PV proportion increased from 50 to 100% (from Min to Max in Fig. 16.9), grid dependence first decreased from 63.44 to 56.05% and then slightly increased to 58.47%. It indicated that there was an optimal PV proportion for minimizing the grid dependence. The grid dependence remained almost unchanged with COP and indoor temperature set-point. Figure 16.9 also shows that with the increase of battery capacity, grid dependence decreased quickly at the beginning but slowly at the end. For instance, as the battery capacity increased from 0 to 40 kWh (+20% in Fig. 16.9), the grid dependence was largely reduced from 100 to 65.88%. Then, as the battery capacity kept increasing, the improvements in grid dependence became less effectively. The main reason was that a major part of the power interactions between the NZEB and the grid were of relatively small magnitudes, and thus they can be easily eliminated using a small sized battery. To eliminate the remaining power interactions with large magnitudes, batteries with dramatic size increases were required.

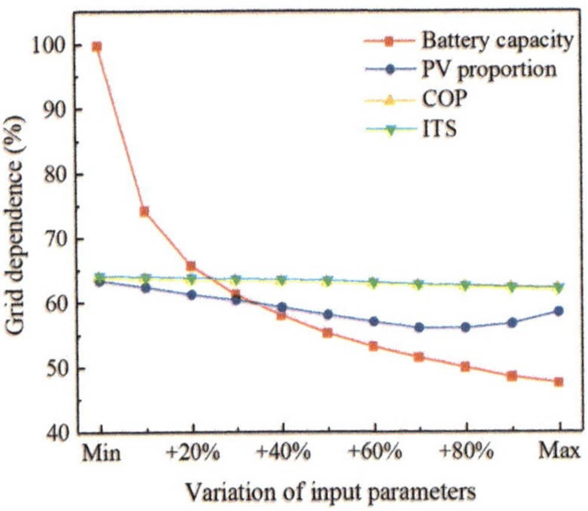

Fig. 16.9 Grid dependence variations with different parameters

16.4.4 Analysis of the Identified Key Parameters for Energy Loss

In the study, the energy losses in inverter, battery and grid feeder were considered. Figures 16.10 and 16.11 are used to elaborate the reasons why the four identified key parameters were important for energy loss. Figure 16.10 presents the variations of energy losses with the four top-ranked parameters. It shows that the inverter loss accounted for more than 50% of overall energy loss even with an over 90% high inverter efficiency. This was because all the renewable generations must be converted to alternative current (AC) via the inverter before their usage or deports to the grid. Due to the relatively low battery round-trip efficiency (e.g. 75–85%), the battery loss was also substantial. In total, the losses in battery and inverter accounted for about 90% of the total loss, and thus they had two most significant impacts on the energy loss. Figure 16.10 also reveals that the energy loss from the grid feeder was less than 10% of the total one. This was because the grid loss was mainly caused by the Joule effect of resistance in the power lines, and the line resistance was very small nowadays.

More importantly, Fig. 16.10 shows the energy loss variations with the four top-ranked parameters including inverter efficiency, battery capacity, battery round-trip efficiency and PV proportion. It can be observed that the inverter loss varied substantially with the inverter efficiency. As shown in Fig. 16.10a, when the inverter efficiency increased from 93 to 97%, the energy loss decreased from 10.00 to 6.11%. Figure 16.10b shows that with the increase of battery capacity from 0 to 200 kWh, the battery loss increased from 0 to 3.81%, and the total energy loss increases from 5.50 to 8.97%. The main reason was that a larger sized battery can provide more energy storage as surplus renewable generations existed, thereby leading to more

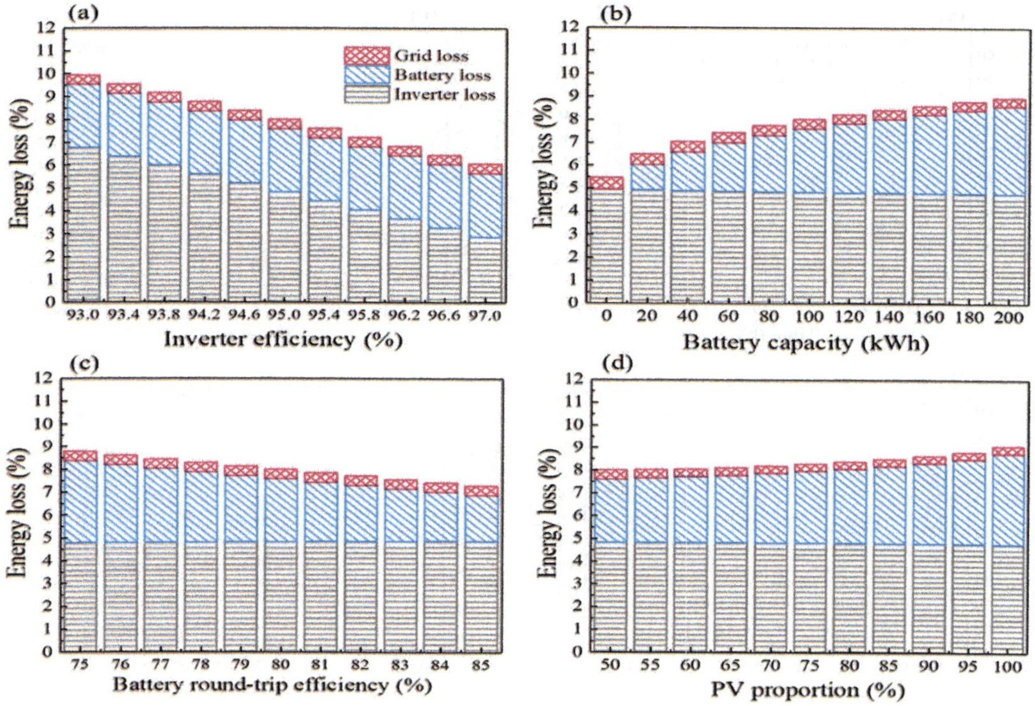

Fig. 16.10 Energy loss variations with **a** inverter efficiency; **b** battery capacity; **c** battery round-trip capacity; **d** PV proportion

battery energy losses in the charging and discharging processes. Figure 16.10c reveals that the total energy loss reduced from 8.83 to 7.34% with the increase of battery round-trip efficiency from 75 to 85%. Figure 16.10d shows that the battery loss increased gradually with the increase of PV proportion. The detailed reason was explained as below (in Fig. 16.11).

In Fig. 16.11a–c, a typical week was selected to compare the battery charge cycles at different PV proportions. When PV proportion was 50% (as shown in Fig. 16.11a), the power generation was less than the power consumption in most time of the week. In this case, very limited surplus renewable generation can be stored, and the battery SOC stayed at a minimum level for most of the week. Thus, only one charge cycle was performed in the whole week. In contrast, as PV proportion increased to 75 and 100%, the battery charge cycles increased to 3 and 5 respectively, as shown in Fig. 16.11b and c. As the PV proportion continuously increased, there exist more surplus renewable generations that can be stored for later use, thereby increasing the battery charge cycles. Figure 16.11d summarizes the annual battery cycle variations with PV proportion increase. It can be observed that the annual battery charge cycles increased from 151.5 to 221.7 with the PV proportion increase from 50 to 100%. Such an increase of the battery charge cycles was the reason why PV proportion increase can cause the increase of battery energy loss.

16.4.5 Verifications via Performance Improvement and Computation Load Comparisons

In the study, the verifications of the identified key parameters were performed by comparing performance improvements and computation loads. Figure 16.12 shows the performance

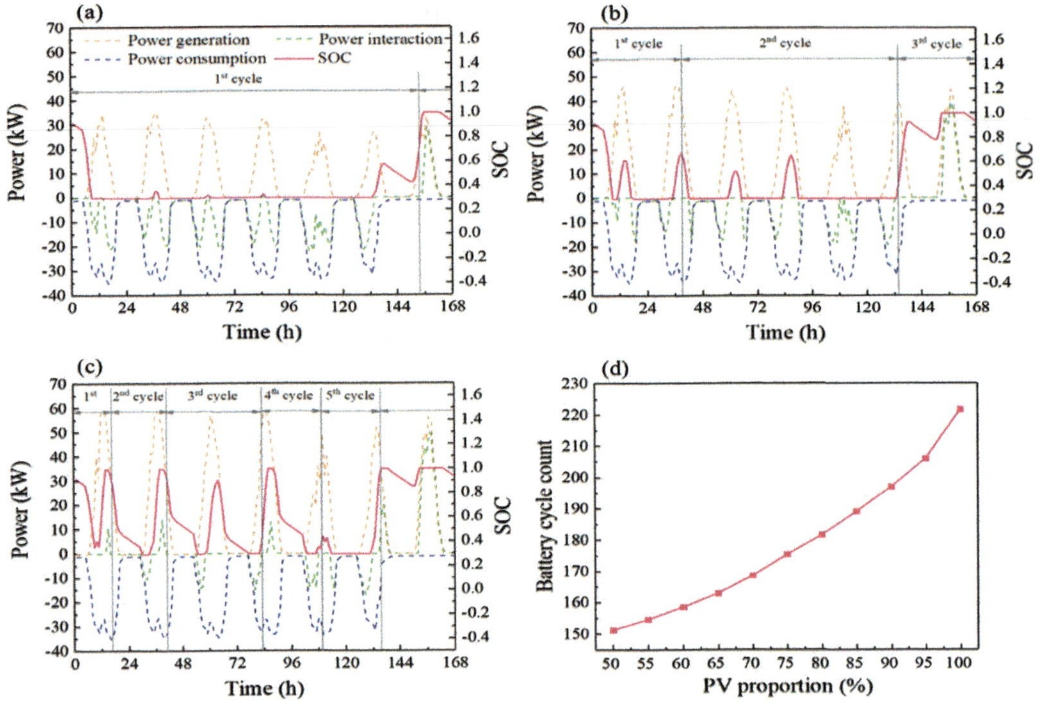

Fig. 16.11 Power, SOC and battery cycle variations under different PV proportion **a** PV proportion = 50%; **b** PV proportion = 75%; **c** PV proportion = 100%; **d** Annual battery cycle count

improvements and computation loads as different number of top-ranked parameters optimized. It should be mentioned that a smaller indicator value represents better performance. Figure 16.12a shows the variations of over/under voltage and associated computation load with different parameters optimized. The optimizations of the two key parameters (i.e. PV proportion and generation capacity) resulted in the substantial reductions of the overvoltage risk. In details, the optimization of PV proportion led to the largest overvoltage risk reduction, from 72.03 to 37.03%, and the optimization of the generation capacity further reduced the risk from 37.03 to 31.03%. In contrast, the other three optimization cases contributed little to the overvoltage risk reduction. Such different performance improvements verified that the two identified parameters (i.e., PV proportion and generation capacity) had critical impacts on NZEB overvoltage. On the other hand, with the more parameters optimized, associated computation loads increase rapidly from 0.92 to 2.94 h.

Figure 16.12b shows the variations of grid dependence and associated computation loads with different parameters optimized. As the battery capacity (the identified first-ranked parameter) was optimized, grid dependence substantially improved from 64.99 to 43.67%. The optimization of the identified second-ranked parameter (i.e. PV proportion) can achieve a further 11.74% improvement. In contrast, the optimizations of the non-key parameters including COP and ITS can only result in very limited performance improvements, i.e., 2.44% and 2.43%, respectively. Such different performance improvements verified that the two identified parameters (i.e., battery capacity and PV proportion) had critical impacts on NZEB grid dependence. Similar to the over/under voltage, the computation loads also rapidly increased from 0.86 to 3.13 h as more parameters were optimized.

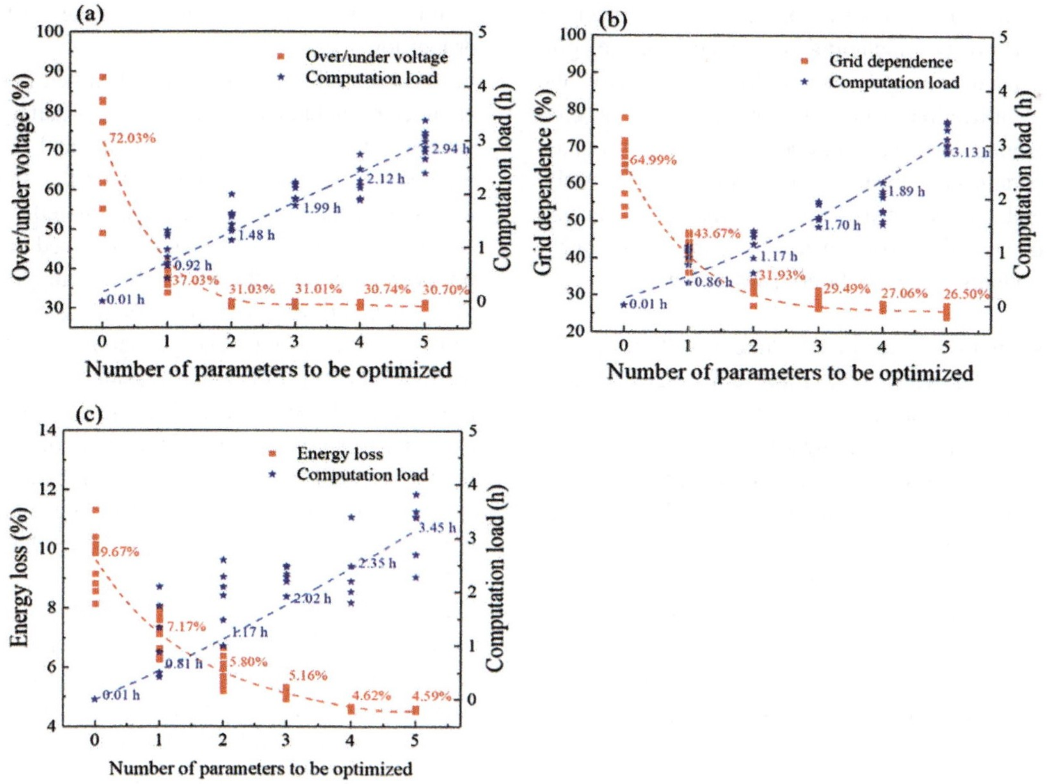

Fig. 16.12 Variations of different performance indicators and computation load with the number of optimized parameters **a** over/under voltage; **b** grid dependence; **c** energy loss

Figure 16.12c shows the variations of energy losses and associated computation loads with different parameters optimized. The optimizations of the four key parameters (including inverter efficiency, battery capacity, battery round-trip capacity, and PV proportion) achieved energy loss reductions of 2.5, 1.37, 0.64, and 0.54% respectively. In contrast, the optimization of the window to wall ratio (the fifth-ranked parameter) can only achieve negligible energy loss reduction, i.e. 0.03%. Such different performance improvements verified that the four identified parameters played more important roles in reducing the energy loss as compared with other parameters. It can also be observed that with the increase of the optimized parameters, associated computation loads increase rapidly from 0.81 to 3.45 h.

In summary, the optimizations of the identified key parameters can achieve substantial performance improvements, while the optimizations of the non-key parameters can only lead to negligible performance improvements but with rapid computation load increase. Thus, for computation-efficient optimization, the identified key parameters should be considered.

16.5 Summary

In this study, a novel method has been proposed to identify the key parameters affecting net-zero energy building (NZEB) grid interactions as parameter uncertainty considered. In the method, global sensitivity analysis has been adopted to systematically investigate the impacts of 24 influential parameters on three user-concerned performance including over/under voltage, grid dependence and energy loss. To verify the identification results, six parameter groups, containing different numbers of the top-ranked parameters, have been optimized separately, and then their performance improvements and computation loads were compared. The major findings of the study are stated as below.

- In the aspect of over/under voltage, photovoltaic (PV) proportion and renewable generation capacity are the two identified key parameters. In comparison with the renewable generations with excessively high peak power output, such as wind turbine (WT), PV energy generation is relatively more stable. Thus, an increase of PV proportion (or a reduction of WT proportion) can effectively reduce the peak renewable power generation, thereby substantially reducing the overvoltage risk. Meanwhile, with a given PV proportion, a reduction of the renewable generation capacity leads to a reduction of the WT installation capacity, which consequently reduces associated WT peak power generation and thus reduces the overvoltage risk.
- In the aspect of grid dependence, battery capacity and PV proportion are the two identified key parameters. In the study, as the battery capacity increased from the 0 kWh to the maximum one, the grid dependence rapidly reduced from 100 to 47.74%. The main reason was that battery can effectively reduce power interactions between the NZEB and the grid through storing the surplus renewable generation for later use. It must be mentioned that grid dependence did not improve in a simple linear way as the battery size increased (i.e., fast improvements at beginning but slow ones at end, as shown in Fig. 16.9). In this regard, battery capacity increase may no longer be effective to improve grid dependence after a certain extent. With relatively power generation, PV proportion increase can also improve the grid dependence, but the maximum improvement requires the optimization of the PV proportion.
- In the aspect of energy loss, the identified key parameters include inverter efficiency, battery capacity, battery round-trip efficiency and PV proportion. The study results showed that the losses in the battery and inverter accounted for about 90% of the total loss, which stressed the importance of their efficiencies in reducing the total energy losses. With an increased battery capacity, the stored energy amount will increase, thereby directly increasing the battery energy losses in the charging and discharging processes. In addition, a rise of PV proportion will cause an increase of battery charge cycles (as shown in Fig. 16.11), which consequently result in an increase of the battery energy loss.
- The study results have showed that the key parameters optimization can rapidly improve the considered performance, while the optimizations of other non-critical parameters can only slightly or even hardly improve the performance, as shown in Fig. 16.12. On the other hand, with more unnecessary/non-key parameters taken into considerations, the computation loads will quickly increase.

The study provided an effective solution to identifying the key influential parameters out of many, and thus unnecessary/non-critical parameters optimization can be substantially reduced. The identified key parameters can also be directly used to develop computation-efficient design optimization methods for improving NZEB grid interactions in the aspects of overvoltage, grid dependence, and energy loss.

References

Agarwal N, Kumar A (2013) Optimization of grid independent hybrid PV–diesel–battery system for power generation in remote villages of Uttar Pradesh, India. Energy Sustain Dev 17(3):210–219

Alizadeh H, Ghasempour R, Shafii MB, Ahmadi MH, Yan WM, Nazari MA (2018) Numerical simulation of PV cooling by using single turn pulsating heat pipe. Int J Heat Mass Transf 127:203–208

Awad H, Gül M (2018) Load-match-driven design of solar PV systems at high latitudes in the Northern hemisphere and its impact on the grid. Sol Energy 173:377–397

Baetens R, De CR, Van RJ, Verbruggen B, Driesen J, Helsen L et al (2012) Assessing electrical bottlenecks at feeder level for residential net zero-energy buildings by integrated system simulation. Appl Energy 96:74–83

Booth AT, Choudhary R (2013) Decision making under uncertainty in the retrofit analysis of the UK housing stock: implications for the Green Deal. Energy Build 64:292–308

Brinkel NBG, Gerritsma MK, AlSkaif TA, Lampropoulosa I, Voordenb AM, Fidderb HA et al (2020) Impact of rapid PV fluctuations on power quality in the low-voltage grid and mitigation strategies using electric vehicles. Int J Electr Power Energy Syst 118:105741

Cao S, Hasan A, Sirén K (2013) Analysis and solution for renewable energy load matching for a single-family house. Energy Build 65:398–411

Chai J, Huang P, Sun Y (2019) Investigations of climate change impacts on net-zero energy building lifecycle performance in typical Chinese climate regions. Energy 185:176–189

DOE U S (2008) Building technologies program, planned program activities for 2008–2012, Department of Energy, US

Domínguez-Muñoz F, Cejudo-López JM, Carrillo-Andrés A (2010) Uncertainty in peak cooling load calculations. Energy Build 42(7):1010–1018

Electrical and Mechanical Services Department (2015) Code of practice for the electricity (Wiring) regulations (2015 Edition). Electrical and Mechanical Services Department, Hong Kong SAR Government

Fang Y, Cho S (2019) Design optimization of building geometry and fenestration for daylighting and energy performance. Sol Energy 191:7–18

Guarino F, Cassarà P, Longo S, Cellura M, Ferro E (2015) Load match optimisation of a residential building case study: a cross-entropy based electricity storage sizing algorithm. Appl Energy 154:380–391

Guidelines on Performance-based building energy code (2003) Electrical and mechanical services Department, Hong Kong SAR Government

Holweger J, Bloch L, Ballif C, Wyrsch N (2020) Mitigating the impact of distributed PV in a low-voltage grid using electricity tariffs. Electr Power Syst Res 189:106763

Hong Kong Electric (2019) Supply rules. Hong Kong Electric Company Limited

Hopfe CJ (2009) Uncertainty and sensitivity analysis in building performance simulation for decision support and design optimization. Eindhoven University of Technology, Eindhoven, p 215

https://energyplus.net/weather

Huang P, Huang G, Wang Y (2015) HVAC system design under peak load prediction uncertainty using multiple-criterion decision making technique. Energy Build 91:26–36

Huang P, Wu H, Huang G, Sun Y (2018a) A top-down control method of nZEBs for performance optimization at nZEB-cluster-level. Energy 159:891–904

Huang P, Huang G, Sun Y (2018b) Uncertainty-based life-cycle analysis of near-zero energy buildings for performance improvements. Appl Energy 213:486–498

Huang P, Augenbroe G, Huang G, Sun Y (2019a) Investigation of maximum cooling loss in a piping network using Bayesian Markov Chain Monte Carlo method. J Build Perform Simul 12(2):117–132

Huang P, Fan C, Zhang X, Wang J (2019b) A hierarchical coordinated demand response control for buildings with improved performances at building group. Appl Energy 242:684–694

Ilbeigi M, Ghomeishi M, Dehghanbanadaki A (2020) Prediction and optimization of energy consumption in an office building using artificial neural network and a genetic algorithm. Sustain Cities Soc 102325

Jamil M, Anees AS (2016) Optimal sizing and location of SPV (solar photovoltaic) based MLDG (multiple location distributed generator) in distribution system for loss reduction, voltage profile improvement with economical benefits. Energy 103:231–239

Kashem MA, Ganapathy V, Jasmon GB, Buhari MI (2000) A novel method for loss minimization in distribution networks. DRPT2000. In: International conference on electric utility deregulation and restructuring and power technologies. Proceedings (Cat. No. 00EX382). IEEE, pp 251–256

Li H, Wang S, Cheung H (2018) Sensitivity analysis of design parameters and optimal design for zero/low energy buildings in subtropical regions. Appl Energy 228:1280–1291

Ma T, Yang H, Lu L (2014) A feasibility study of a stand-alone hybrid solar–wind–battery system for a remote island. Appl Energy 121:149–158

Mahmud MA, Hossain M, Pota HR (2011) Analysis of voltage rise effect on distribution network with distributed generation. IFAC Proc 44(1):14796–14801

Nasruddin S, Satrio P, Mahlia TMI, Giannetti N, Saito K (2019) Optimization of HVAC system energy consumption in a building using artificial neural network and multi-objective genetic algorithm. Sustain Energy Technol Assess 35:48–57

Roos A, Bolkesjø TF (2018) Value of demand flexibility on spot and reserve electricity markets in future power system with increased shares of variable renewable energy. Energy 144:207–217

Salom J, Widén J, Candanedo J, Sartori I, Voss K, Marszal A (2011) Understanding net zero energy buildings: evaluation of load matching and grid interaction indicators. Proc Build Simul 6:2514–2521

Sharma V, Haque MH, Aziz SM (2019) Energy cost minimization for net zero energy homes through optimal sizing of battery storage system. Renew Energy 141:278–286

Shen JM, Jou HL, Wu JC, Wu KD (2013) Five-level inverter for renewable power generation system. IEEE Trans Energy Convers 28(2):257–266

Storti BA, Dorella JJ, Roman ND, Peralta I, Albanesi AE (2019) Improving the efficiency of a Savonius wind turbine by designing a set of deflector plates with a metamodel-based optimization approach. Energy 186:115814

Sun Y (2015) Sensitivity analysis of macro-parameters in the system design of net zero energy building. Energy Build 86:464–477

Tian W (2013) A review of sensitivity analysis methods in building energy analysis. Renew Sustain Energy Rev 20:411–419

Tonkoski R, Turcotte D et al (2012) Impact of high PV penetration on voltage profiles in residential neighborhoods. IEEE Trans Sustain Energy 3(3):518–527

Wu W, Skye HM, Domanski PA (2018) Selecting HVAC systems to achieve comfortable and cost-effective residential net-zero energy buildings. Appl Energy 212:577–591

Yang H, Lu L, Zhou W (2007) A novel optimization sizing model for hybrid solar-wind power generation system. Solar Energy 81(1):76–84

Yildiz Y, Korkmaz K, Özbalta TG, Arsan ZD (2012) An approach for developing sensitive design parameter guidelines to reduce the energy requirements of low-rise apartment buildings. Appl Energy 93:337–347

Yu J, Tian L, Yang C, Xu X, Wang J (2013) Sensitivity analysis of energy performance for high-rise residential envelope in hot summer and cold winter zone of China. Energy Build 64:264–274

Yu ZJ, Chen J, Sun Y, Zhang G (2016) A GA-based system sizing method for net-zero energy buildings considering multi-criteria performance requirements under parameter uncertainties. Energy Build 129:524–534

Zhang S, Huang P, Sun Y (2016) A multi-criterion renewable energy system design optimization for net zero energy buildings under uncertainties. Energy 94:654–665

17 Local Energy Communities: Market Design Evaluation Using Agent Based Modelling

Marco Lovati, Pei Huang, and Xingxing Zhang

Abstract

Solar photovoltaic (PV) is becoming one of the most significant renewable sources for positive energy district (PED) in most countries, including Sweden. The lack of innovative business models and financing mechanisms are one of the main constraints for PV's deployment installed in local community. This chapter therefore analyses a set of peer-to-peer (P2P) business model for 48 individual building prosumers with PV installed in a Swedish community. It considers energy use behaviour, electricity/financial flows, ownerships, and trading rules in a local electricity market. Different local electricity markets are designed and studied using agent-based modelling technique, with different energy demands, cost–benefit schemes and financial hypotheses for an optimal evaluation. This chapter provides an early insight into a vast research space, i.e. the operation of an energy system through the constrained interaction of its constituting agents. The agents (48 households) show varying abilities in exploiting the common PV resource, as they achieve very heterogeneous self-sufficiency levels (from ca. 15 to 30%). The lack of demand side management suggests that social and lifestyle differences generate huge impacts on the ability to be self-sufficient with a shared, limited PV resource. Despite the differences in self-sufficiency, the sheer energy amount obtained from the shared PV correlates mainly with annual cumulative demand.

Keywords

Agent based modelling · Micro-grid · Self-sufficiency · Emergent systems · Prosumer · PV optimization

17.1 Introduction

Climate change is one of the main challenges that threaten the well being or the very existence of human society. This threat cannot be ignored because it can impact a wide range of natural ecosystems and socio-technical systems. In the last few decades numerous technologies have been discovered, or improved, that can dramatically reduce our greenhouse gas emissions: renewable or low carbon energy generation devices, energy storage systems, energy

M. Lovati (✉)
Department of Architecture, Aalto University, Otakaari 1, 02150 Espoo, Finland
e-mail: mlov@du.se

P. Huang · X. Zhang
Department of Energy and Community Buildings, Dalarna University, 79188 Falun, Sweden
e-mail: phn@du.se

X. Zhang
e-mail: xza@du.se

X. Zhang (ed.), *Data-driven Analytics for Sustainable Buildings and Cities*, Sustainable Development Goals Series, https://doi.org/10.1007/978-981-16-2778-1_17

efficiency, and carbon capture devices. The vast majority of countries and international institutions on the planet agree on the danger of climate change and on the need for action (Liu et al. 2020). In other words, since the political and social will to build a low carbon economy has been largely achieved, the focus in this chapter has been put chiefly on practical strategies and effective transition pathways. The subject is how to achieve a transition to a low carbon society in an economically beneficial way and without causing discontent.

17.1.1 How Change Can Happen

To transform the will for change in actual change, it is important to understand the causes and the mechanisms that activate change. In Giddens (1984) an important role in the evolution of technology is played by the interaction between socio-technical regimes, I.e. the existing dominating technology and the social structure it generated, and technological niches, I.e. newer, smaller and dynamic socio-technical entities that disturb the existing regime. In Geels and Schot (2007) the authors elaborate different transition pathways (I.e. transformation, reconfiguration, technological substitution, and de-alignment and re-alignment) elaborating upon previous work and criticisms. In particular (Geels 2002) is reported, which add new elements on the subject introducing the so called 'socio-technical landscape'. The socio-technical landscape is the sum of morals, beliefs, knowledge and ideas that can push the change in a socio-technical regime. In Suarez and Oliva (2005) different modifications of the socio-technical landscape are presented (I. e. regular, hyperturbolence, specific shock, disruptive, avalanche). Also (Scott 2013) speaks about the forces that drive a transition or the conservation of a socio-technical regime, which can therefore be seen as a socio-technical landscape. These forces are divided into three groups: regulative (e.g. laws and standards), normative (e.g. values and norms), and cognitive (such as beliefs and search heuristics). The study argues that the stronger of these forces is the cognitive one since is the most immersive and invisible for the actors under its influence. Other aspects that are fundamental in a transition are the selection pressure and the coordination of resources, these two are deeply interconnected according to Smith et al. (2005).

17.1.2 Micro-grids, Local Energy Communities, and Relative Research

Fortunately, the possibility to form energy communities, where energy can be locally shared, has been regulated at European level in the Clean Energy package presented by the European Commission (2020) and at Swedish level under § 22 (a) of the IKN Regulation 2007:215 (Riksdag 2020). This can be an opportunity for a new business model development within the energy sector, e.g. Peer-to-Peer (P2P) trading. In such business model, consumers and pro-sumers organize in energy communities, in which the excess production could be sold to other members (Parag and Sovacool 2016). The benefits are threefold as the pro-sumers could make an additional margin on their sale, consumers could buy electricity at a more advantageous price and the grid could be more stable and resilient. This can be a potential solution to promoting PV installation in a sustainable way, while reducing the reliance on subsidies.

In order to support new regulations, careful design and optimal modelling of P2P business models for PV penetration is necessary by analysing current state of affairs and proposing future ways of exchanging energy. Huijben and Verbong (2013) summarized three possible ownerships of PV systems, such as Customer-Owned (single ownership), Community Shares (multiple ownership) and Third Party ownership. Based on these possibilities, Lettner et al. (2018) further described three different system boundaries of a PV prosumer business concept (as illustrated in Fig. 17.1): Group (1) single direct use (one consumer directly uses the generated PV electricity on site), Group (2) local collective use of PV in one building (several

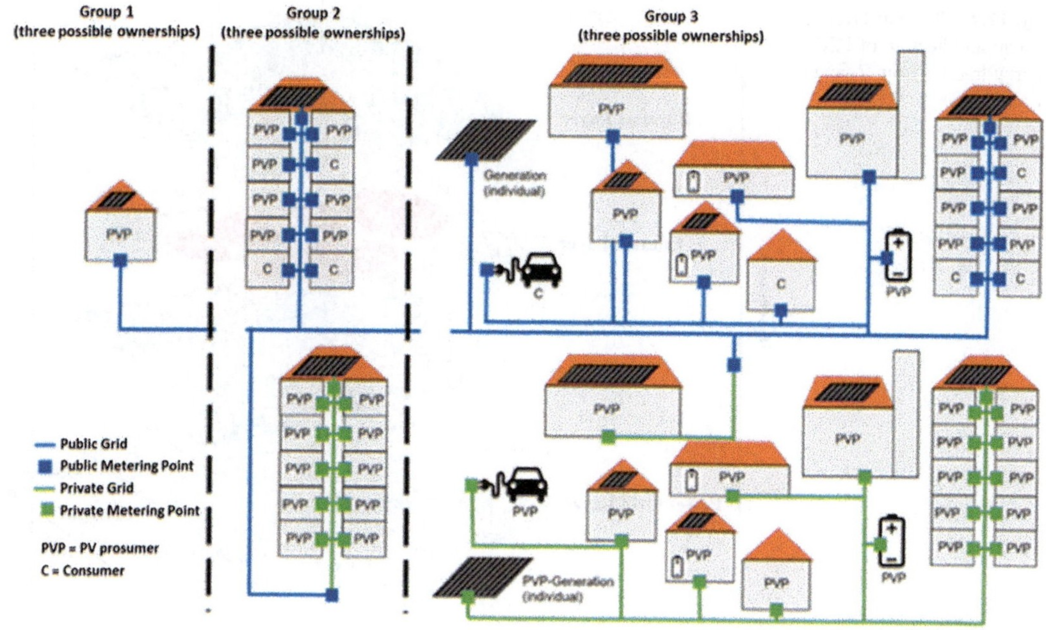

Fig. 17.1 Classification of integration concepts (Lettner et al. 2018)

consumers share the generated PV electricity with or without the public grid), and Group (3) district power model (PVs are installed in several buildings, where those prosumers directly consume locally generated PV power, and the PV electricity is further shared using public or private micro grid). It is possible to have different ownerships in each category of these boundary conditions, resulting in a large number of possibilities and uncertainties in the practical business operation. Learning and mapping (i.e. testing) a wide array of these possible designs and combinations are necessary. There are a few existing regulatory and modelling studies about the P2P PV-electricity trading. Community-owned PV system was surveyed as an innovative business model in Switzerland, where it can seemingly be a successful distribution channel for the further adoption of PV (Stauch and Vuichard 2019). Roberts et al., tested a range of financial scenarios in Australia, based on the P2P concept, to increase PV self-consumption and electricity self-efficiency by applying PVs to aggregated building loads (Roberts et al. 2019). Zhang et al. (2018) established a four-layer system architecture of P2P energy trading (as shown in Fig. 17.2, i.e. power grid layer, ICT layer, control layer and business layer), during which they focused on the bidding process on business layer using non-cooperative game theory in a microgrid with 10 peers. A price mechanism for the aggregated PV electricity exchange among peer buildings was also developed using either Lagrangian relaxation-based decentralized algorithm (Xu et al. 2017) or mixed integer linear programming (Nguyen et al. 2018). Jing et al. (2020) then applied the non-cooperative game theory to modelling the aggregated energy trading between residential and commercial buildings by considering fair energy pricing mechanism for both PV electricity and thermal energy simultaneously. Lüth et al. (2018) designed two local markets for decentralised storage (flexi user market—individually owned batteries) and centralised storage (pool hub market—commonly owned battery), based on a multi-period linear programming. It focused on the evaluation of two different ownerships of

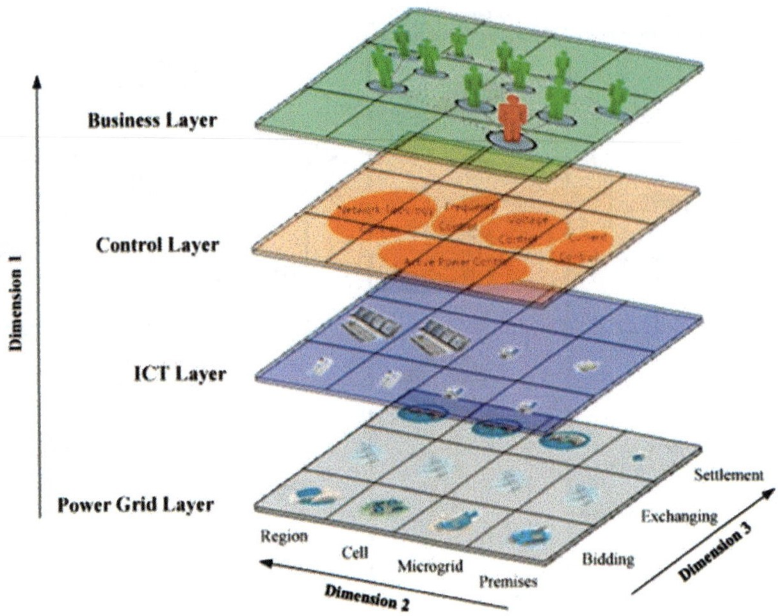

Fig. 17.2 The four layered system architecture of P2P energy trading from Zhang et al. (2018)

batteries and optimized P2P energy trading local markets. They indicated that the end users can save up to 31% electricity bills in the Flexi User Market and 24% in Pool Hub Market. Furthermore, two different ownership structures, namely the third-party owned structure and the user owned structure, were investigated in a P2P energy sharing network with PV and battery storage (Rodrigues et al. 2020). These existing studies almost cover all the four layers of a P2P network. The impact of other system and market components on the economic performance of PV P2P business models has been investigated, such as EV (Electric Vehicle) batteries (Tang et al. 2018), gas storage (Basnet and Zhong 2020), heat pump/hot water storage (Huang et al. 2019), advanced control (Thomas et al. 2019), energy cost optimization (Alam et al. 2019), bidding strategies for local free market (El-Baz et al. 2019), double auction market (Chen et al. 2019), local market designs (Sousa et al. 2019), integration of local electricity market into wholesale multi-market (Zepter et al. 2019), micro grid ICT architecture (Cornélusse et al. 2019) and grid operation (Almasalma et al. 2019) etc.

According to the above studies, a research gap is found in the lack of examination on full P2P energy trading process at the business layer in a local market for individual participant, which, in time sequence, consists of bidding, exchanging and settlement, under different local market conditions with various ownerships of PV systems and market rules. Bidding is often the first process when energy players (generators, consumers and pro-sumers) agree to trade energy with each other at a certain price for a specific amount of energy. Energy exchanging is the second process, during which energy is generated, transmitted and consumed. Settlement is the last process when bills and transactions are finally settled via settlement arrangements and payment (Zhang et al. 2018), which results in the final economic benefits. In cases of the physical network constraints, due to the varying energy demand and the intermittent generation of PVs, there are always mismatches between sellers and buyers. Such difference between electricity generation and demand are to be evaluated and charged/discharged during settlement stage.

17.1.3 Novelty and Aim

A number of studies have focused on the technical or economic aspects of the micro-grids and shared RES, but the endeavor has been tackled in a segmented way analyzing a narrow sample of possibilities among the vast search space of the business models. The existing studies have not yet fully test the effectiveness and compare the characteristics of various P2P business models, in case of heterogeneous peer (individual) energy supply/demand, and dynamic market rules for the full trading process on the business layer. There is a lack of a concise and efficient method yet to model.

Although the study in this chapter analyses only three different setups, it attempts to lay the groundwork for a systematic study of the subject. In other words, the results and the discussion presented in this chapter, although not conclusive by themselves, they are part of a well-defined search-space. This allows the outcomes to be interpreted from the perspective a larger systematic endeavor.

In summary, the elements of novelty of this chapter are described as the following:

1. The particular result of the study: to the knowledge of the authors, no study have linked the price of the electricity offered within a shared RES to both the risk of economic loss and the potentials for earning among the individual households within the shared micro-grid. Furthermore, the dominance of shear annual cumulative consumption over self-sufficiency in determining the earning potential in a shared RES is an unknown phenomenon. It deserves to be further analyzed (i.e. tested under different datasets) to be proven.
2. The examples of business models presented in the study are included in a well-defined search space map (see Fig. 17.3). This facilitates a systematic inquiry and offers a way to organize the results presented in the study of this chapter and in the follow-ups.

This chapter reports the results of a study of the P2P business model for 48 individual building prosumers with PV installed in a Swedish community. The aim is to discover 'latent opportunities' that were previously unknown and optimize the market design and its variables for the best benefit. It will have significant influence that integrates energy needs, supply and market rules. This chapter is expected to provide knowledge for policymakers to design a fair, effective and economical P2P energy framework. The research results will be useful to optimize PED's three functions (energy efficiency, energy production and flexibility) towards energy surplus and climate neutrality.

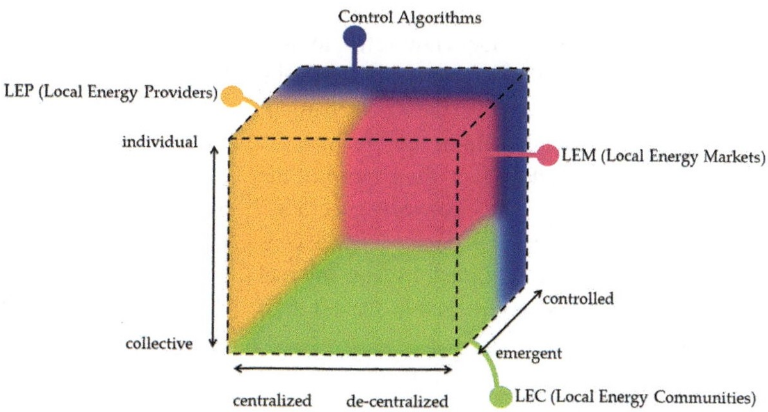

Fig. 17.3 District scale renewable energy systems behaviour map

17.2 Method

The definition of ownership structures from Huijben and Verbong (2013) distinguishes among customers, communities and third parties. In general, a similar distinction could be applied to the behaviour of the local grid instead to the ownership. In this way, the concept of ownership is not associated with the functioning of the grid and it is easier to describe hybrid forms (e.g. some share-holder of an energy provider, or more providers, which form a market although not prosumers etc.). Thinking about the behaviour of the shared system, a space can be defined according to three dimensions (see Fig. 17.3):

(1) The controlled versus emergent dimension describes how much there are rules or a controller that directs the exchanges, versus an emergent behaviour from the interactions between agents.
(2) The centralized versus de-centralized dimension describes how much the agents are equivalent among each other, versus the presence of few (potentially one) agents that concentrate some functions for a larger number of others.
(3) The individual versus collective dimension describes how much each agent controls and directs its own resources (i.e. PV, storage, demand-response resources etc..), versus having larger pools of agents who share some common resources.

The behaviour map does not refer to any specific levels (Zhang et al. 2018), although the last two (i.e. controls and business) are particularly affected from the volume of the map, in which they are located. In fact, the control of the energy and monetary flows between generation and demand points can be decided by a controller, which can be assigned by the internal rules of a community or emerged as the result of an auction.

17.2.1 Agent Based Modelling

Given the number and nature of the emergent behaviours in the behaviour map (i.e. Fig. 17.3), an agent based model (ABM) simulation was developed to get an insight on the energy and economic fluxes exchanged between the different actors in the local grid. Usually, every agent of the simulation represents one household in the local grid (i.e. a consumer or a pro-sumer), but producers are not excluded. Example of producers are energy providers. For instance, companies or investor interacts with the local grid without necessarily being served by it, or the parent grid, i.e. the larger grid in which the local grid is embedded. The local grid could be a micro-grid but also a secondary network, where the pro-sumers are allowed to have a certain level of control of the network.

In an ABM, each agent can interact with all the other agents by trading energy. Thus it can send energy in exchange for money or vice-versa. The movement of energy in the micro-grid is an emergent behaviour, which results from the interaction of a number of independent actors. This is opposed to a control algorithm, where the behaviour is set by a series of rules or conditions. Naturally, the freedom of the agents can be limited by the introduction of rules. For instance, a producer could be forced to prioritize the sale of renewable electricity to those consumers that have used the least of it in a given period. If the rules become tighter, the freedom of each individual agent is reduced. While if the rules are as tight as to completely limit any possibility of choice for the agents, the ABM degenerates into a control algorithm.

In the present study, the behaviour of the agents is extremely simplified: the consumers prioritize the purchase of electricity from the cheapest source available at any given time, on the other end the producers have the ability to set the price, and they do so according to the case as

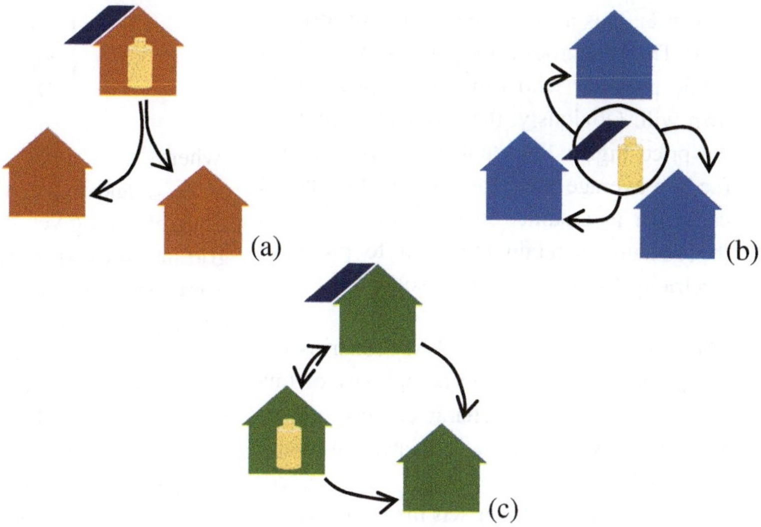

Fig. 17.4 Ownership structures organized in three main families: Local Energy Provider (LEP) (**a**), Local Energy Community (LEC) (**b**) and Local Energy Market (**c**)

explained in the following section (i.e. ownership structures and business models).

Figure 17.4 presents the possible ownership structures arranged in three main families, these are slightly different from those in Huijben and Verbong (2013) for the purpose of this study:

(1) Local Energy Provider (LEP) (a in Fig. 17.4): It occurs when a single agent owns the totality of the production or storage capacity of the entire local network and the other agents are strictly consumers. The owner of the plant can be either a producer or a prosumer.
(2) Local Energy Community (LEC) (b in Fig. 17.4): It is the case in which a communal plant is shared among all or a group of agents, the shares could be equally distributed or according to other principles such as energy used from the plant or the share of the initial investment.
(3) Local Energy Market (LEM) (c in Fig. 17.4): It is the most complex and free-form of all the structures, it is characterized by the presence of multiple producers, consumers and pro-sumers, in this arrangement the interaction between agents can reach significant complexity and the agents could achieve higher earnings by engaging in intelligent behaviours.

17.2.2 Ownership Structures and Business Models

In the case study examined (see following Sect. 17.2.4), a communal PV plant is shared among the different households in the building. This allows for all of the three basic ownership structures from Fig. 17.4 to be applied, because it is possible to create a LEM by having some household who own share of the large PV system. The ownership structure is intertwined with the business model and the rules of the market. In the following pages, the same communal PV plant is shared between the households in the local grid in three different market cases for LEC and LEP:

1. LEC gratis: in this arrangement, the electricity from the communal PV plant is given for free when available. All the households participate in the initial investment and in the Operation and Maintenance (O&M) costs of the plant according to equal shares.
2. LEC LCOE: in this arrangement, the electricity from the communal PV is given at production cost (i.e. without profit) and the revenues are divided among the shareholders. Although variable shares are possible, in this study, all the households are equal sharers in the LEC (i.e. initial investment and O&M costs, and revenues are shared equally).

3. LEP n%: This arrangement is a pure form of LEP. Thus the production plant is owned by a single provider who can set the price at its own will. Obviously, the provider cannot set the price higher than that of the parent grid (i.e. the average price for Swedish household consumer as assumed in the Sect. 17.2.4) as the consumers retain the right to purchase electricity from the cheapest source.

After these three cases, 6 LEM scenarios are analysed. Due to the inherent complexity of this ownership structure, the characteristics of this simulation are explained in a dedicated paragraph (i.e. Sect. 17.2.3).

In this study, the provider sets the price as half-way between the minimum of the local LCOE and the maximum of the consumer price from the parent grid. More precisely, the provider sets a price at a percentage n so that n = 0 is the LCOE, n = 100 is the price offered by the parent grid and n = 50 is half-way. This set-up is valid under the assumption that the LCOE of the system is lower than the price of the electricity for the consumer. Of course, if this assumption does not hold true, the provider will not be able to charge above market price and will thus operate at the minimum loss.

In all the arrangements, the consumer is programmed to buy electricity from the cheapest source. But by having a single source in the local grid, the choice is only between the local source and the parent grid. This implies that the price of electricity in the local grid must be at any time below the Swedish consumer price. If the local production is absent or insufficient (i.e. local consumption > local production), the demand shall be covered partially or totally by the parent grid. If the local production is not sufficient, in a given point in time, to cover entirely the demand, all the households will be served equally in terms of percentage of their demands as shown in the system of relations in (17.1).

$$\begin{cases} E_{local} = \eta \cdot D_{local} \\ E_{house} = \eta \cdot D_{house} \\ D_{local} = \sum D_{house} \end{cases} \tag{17.1}$$

where

$\mathbf{E_{local}}$ and $\mathbf{E_{house}}$ are the amount of electricity available in a given time for the aggregated local grid and for a specific household respectively. $\boldsymbol{\eta}$ is the self-sufficiency: a number between 0 and 1 that represents the share of the demand covered by locally produced electricity, note that is the same globally and for each household. $\mathbf{D_{local}}$ and $\mathbf{D_{house}}$ represent the aggregated demand and the demand of each single household respectively.

The equations in (17.1) imply that having a larger consumption when the local electricity production is scarce guarantees access to a larger amount of local energy, although equal in percentage. Another consequence of the relation in (17.1) involves the price of the electricity for each household: the price results from the weighted average (weighted on energy) of the prices from the different sources of electricity purchased. In the specific case of this study the price can be calculated with the relation (17.2):

$$P_{house} = P_{local} \cdot \eta + P_{parent} \cdot (1 - \eta) \tag{17.2}$$

where

$\mathbf{P_{house}}$, $\mathbf{P_{local}}$ and $\mathbf{P_{parent}}$ represent the electricity price for the individual household, the price for the energy produced locally and the price for the energy bought from the parent grid respectively. $\boldsymbol{\eta}$ is the self-sufficiency as defined for (17.1).

Considering that $\boldsymbol{\eta}$ is the same for every household in the local grid as shown in (17.1), the Eq. (17.2) implies that at any given time there is a unique price of the electricity within the local grid, which depends on the relation between the aggregated energy demand (D_{local}) and the aggregate energy production (E_{local}).

Thus, the price for the electricity is solely function of the Hour Of the Year (HOY) and is not function of any given household. This fact holds true also for the LEM case, in fact, in every time-step, the unique price in the micro-grid is equal to the average of the different prices of each available source. This average is weighted for the relative power of each source, thus, if a cheap source can satisfy a significant fraction of the demand, it will sensibly drive down the unique price. Of course, the ability of each single household to consume its own power, or at least to consume more power in cheap time-steps will affect the average price of electricity it pays (see Fig. 17.12).

To simplify, the agent based model can be described by a simple set of rules:

(1) Every household is represented by one independent agent in the simulation.
(2) Every agent has an energy balance in each HOY (Hour Of the Year). The energy balance is determined by its PV power (if it owns a PV system) minus its power demand in that particular HOY. If the balance is negative, the agent will be a net buyer in that HOY, otherwise it will be a seller. This rule implies that each agent can only sell electric power if it has already satisfied its own demand. Simply, each household can sell only excess PV production.
(3) Each seller can set the price for the power he has to export.
(4) If the electricity is offered by multiple sellers, the buying agent will buy preferentially by the cheapest source.
(5) If the aggregated demand of the district exceeds the offer of the cheapest source, the demand of each household is satisfied proportionally by the cheapest source. If, for example, the cheapest source covers 30% of the aggregated demand in that HOY, each household is provided 30% of its power demand by the cheapest source (see equations in 17.1).
(6) If the on-site renewable power exceeds the power demand in a certain HOY, the cheapest sources are consumed preferentially, while the more expensive ones risk to be in excess of the demand and sell part (or all) their power to the grid. Those who sell to the grid cannot set the price but are simply valued the price paid by the grid (which is always way lower than that of the local sellers).

17.2.3 The LEM (Local Energy Market)

The LEM, being a more loose aggregation of stakeholders, is open to higher complexity and is thus studied in more detail, in this Chap. 6 scenarios have been hypothesized to study different behaviours within a LEM.

17.2.3.1 Scenario 1

All residents agree to purchase the PV system, every household purchases an equal share of the total system and has thus the right to 1/48 of the power at any time (I.e. ca. 1.36 kW each). The price for the sale within the micro-grid is agreed for the long term as the summer grid price/1.2 (thus a static 1 SEK/kWh at the year 0), therefore whoever buys electricity from another household saves ca. 17% on the electricity cost in summer and 45% in winter.

17.2.3.2 Scenario 2

All residents agree to purchase the PV system, likewise scenario 1. The price for the sale within the micro-grid is agreed for the long term as 99% of the grid price, therefore whoever buys electricity from another household has almost no savings compared to the grid. In this case it is assumed that using local energy is perceived as a value in itself by the partecipants in the grid.

17.2.3.3 Scenario 3

Only 50% of the residents agree to purchase the PV system, every PV equipped household purchases an equal share of the total system and has thus the right to 1/24 of the power at any time (I.e. ca. 2.73 kW each). The price for the sale within the micro-grid is agreed for the long term as the summer grid price/1.2, likewise in scenario 1.

Table 17.1 PV capacities per household and prices in the 6 different scenarios

Scenario	PV capacity (kW/household)	Electricity price (at year 0) (SEK/kWh)
(1)	1.36	1
(2)	1.36	1.19 (summer), 1.78 (winter)
(3)	2.73 or 0	1
(4)	2.73 or 0	1.19 (summer), 1.78 (winter)
(5)	1.36	1 or 1.19 (summer), 1.78 (winter)
(6)	1.36	1 or dynamic

17.2.3.4 Scenario 4

Only 50% of the residents agree to purchase the PV system, every PV equipped household purchases an equal share of the total system likewise in scenario 3. The price for the sale within the micro-grid is agreed for the long term as 99% of the grid price, likewise in scenario 2.

17.2.3.5 Scenario 5

All residents agree to purchase the PV system, likewise scenario 1. The price for the sale within the micro-grid is left to the choice of the single household, 50% of the households decide to charge a high price (I.e. 90% of the grid, like case 2 and case 4), the others charge the summer price /1.2.

17.2.3.6 Scenario 6

All residents agree to purchase the PV system, likewise scenario 1. The price for the sale within the micro-grid is left to the choice of the single household, 50% of the households decide to adopt a dynamic price system based on their energy balance in every hour of the year. With this strategy the energy is sold at LCOE whenever the balance is more than double the average balance in that hour of the day. The other 50% charges the summer price /1.2 likewise scenario 1 and scenario 3 (Table 17.1).

17.2.4 Case Study Description

The agent based model is tested on a digital representation of a moderate size residential district (see Fig. 17.5) equipped with a shared PV system + DC micro-grid as described in Huang et al. (2019). The group of three buildings with three stories is located in Sunnansjö, Ludvika, Dalarna region, Sweden. The common PV system is formed by the arrays shown in Table 17.2. In total, there are 3 arrays on the roof and one on the southern façade (total 65.5 kWp).

The system capacity and the position of the arrays over the building resulted from an optimization process, presented in Huang et al. (2019), in order to maximize the self-sufficiency while maintaining a positive NPV over the lifetime. In this system, no electric storage was installed. The LCOE (Levelized Cost of Electricity) of the system was calculated to be about 0.83 SEK/kWh (0.077 €/kWh) under the following assumptions:

Fig. 17.5 Bird view of the small district in the case study (Huang et al. 2019)

Table 17.2 Characteristics of the shared PV system

Block	Facing	Tilt (Deg)	Capacity (kW_p)	Production (MWh)
B	South	18	28.4	22
C	East	18	15.9	10.4
A	West	18	15.9	10.3
A	South	90	5.3	3.4

- Local initial price of the turn-key system without taxation: 10,000 SEK/kWp (935 €/ kWp).
- Price of the inverter: 2500 SEK/kWp (234 €/ kWp) (changed 2 times over the lifetime). The number of changes was retrieved as the expected value assuming a lifetime of the inverter between 12 and 15 years.
- Planned lifetime of the system: 30 years.
- Maintenance costs for the system (substitutions, cleaning and inspection): 5109 SEK/year (477 €/year). This value is calculated as the expected value out of 100 stochastic simulations.
- Degradation of the performance of the system: ca. −1.15%/year.

The weather file and the production of the diverse arrays of PV have been calculated from PVGIS (Šúri et al. 2005). The load profile of the 48 households could not be published for privacy concerns. Thus, the study is presented using data generated by the LPG (Load Profile Generator) software (Pflugradt and Muntwyler 2017). Load Profile Generator is a tool that simulates the electric demand for residential light and appliances. The variability of the aggregated curve according to the number of households has been validated against a real low voltage grid consumption (Pflugradt et al. 2013). The electric demand is generated by simulating every household component as an agent. Its demand is determined by the power absorption and duration of use of devices among an available selection (see Fig. 17.6). These are chosen by the household components according to a set of activities and needs. The needs are modelled as counters that grow at each time-step: a high counter represents a need that is in urgent need of satisfaction. Different needs have different growth rates for each time-step, which means that some needs are to be satisfied more often than others.

The parent grid (i.e. the Swedish national grid) has been assumed to offer electricity for 1.8 SEK/kWh (0.17 €/kWh) from October to March and 1.2 SEK/kWh (0.11 €/kWh) from March to October. These prices have been assumed as a reasonable price for each single household at the annual cumulative level of consumption observed. According to (Eurostat, 2007–2019), the average price for household electricity in 2019 was 1.39 SEK/kWh

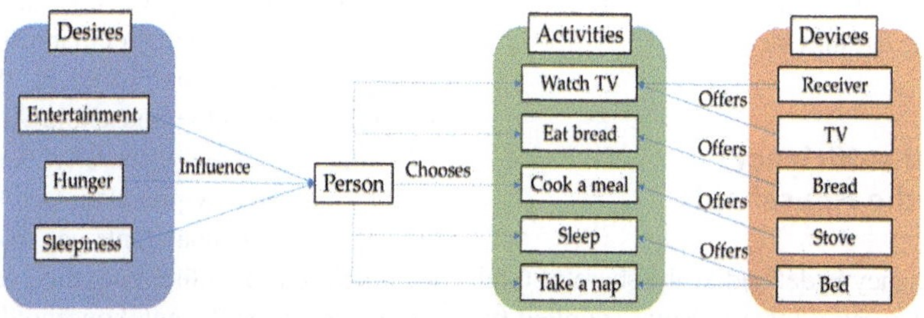

Fig. 17.6 Workflow diagram of the load electricity generation (Pflugradt and Muntwyler 2017)

(0.1297 €/kWh) for electricity transmission, system services, distribution and other necessary services. If VAT and levies are added, the average price would reach 2.2 SEK/kWh (0.2058 €/kWh) (Eurostat, 2007–2019). It is not clear what taxes can be avoided consuming locally produced electricity, but it is reasonable to believe that VAT can be avoided in both the LEC cases explored as the electricity is offered for free or at a price equal to production cost. Conversely, it is not possible to estimate how much of the base 1.39 SEK can be reduced thanks to the aggregation of the loads. The price of the electricity is not static but is projected to grow linearly over the next 30 years at a rate of +1%/year. This is under the assumption that the national grid will need liquidity to invest in the energy transition. Conversely, the revenues for the energy sold to the grid are set to be worth 0.3 SEK/kWh (0.028 €/kWh), but are assumed to shrink by 1.67%/year under the assumption that the increase in installation of PV will gradually discount the energy during sunny hours.

17.3 Results

The results section begins with a discussion about the self-sufficiency of the different households in the local network. It then proceeds with a techno-economic analysis of each arrangement to establish its features and its behaviour (i.e. distribution of risk and profit among stakeholders). Given that the local PV plant is unique, the movement of energy in the network is the same in all the arrangements, thus the self-sufficiency is a static figure throughout the arrangements.

17.3.1 Self-sufficiency of the Households

PV self-sufficiency is defined as the share of total demand in a household that is being supplied by locally generated electricity from PV system (Luthander et al. 2015). In this study, the system, as it is designed, allows to cover an estimated 20.2% of the annual cumulative demand of the district. This result is satisfactory for a system without any electric storage. For a reference, according to IEA 2020b the country, with the most electricity production from PV (i.e. Honduras), has an estimate PV self-sufficiency of 14.8% with the EU on average having 4.9%. It has been calculated in Lovati et al. (2019) and (Huang et al. 2019) that the economically optimal self-sufficiency of a conveniently aggregated system, even in absence of electric storage, is comfortably above any penetration level we see today (i.e. often above 20%). The economically optimal self-sufficiency sets a conservative limit of hosting capacity in an electrical system in a regime of self-sufficiency. The P50 (i.e. 50th percentile or median) household has a self-sufficiency of 18.5% as shown in Fig. 17.7a: this value is below the average value of the aggregated district because the slope of the increase is higher to the right of P50 (see Fig. 17.7a). The P50 (i.e. 50 percentile) household has a relatively low self-sufficiency because there is a positive correlation between annual cumulative demand and self-sufficiency (see discussion about Fig. 17.9). In general, the variability in self-sufficiency between the households in the micro-grid is high. The most self-sufficient household possesses in fact a value double of the lesser one (14.1 to 28.4%). This strong variability suggests that, even without any deliberate attempt for demand control, some households show habits, or a way of life, that can take out the most from the available PV energy.

Figure 17.7b and c show the share of the annual demand in different hours of the day or month of the year respectively: this is to say how much of the total annual demand is concentrated during a specific hour of every day or month along the year. In the household with the highest self-sufficiency, the electricity demand around 12:00 is particularly prevalent (see Fig. 17.7b). It indicates that its inhabitants use to cook at home for lunch. On the other end, the evening peak of the most self-sufficient household is way less prominent than in the lowest one. Looking at the prevalence throughout the months of the year (Fig. 17.7c), the difference is less marked

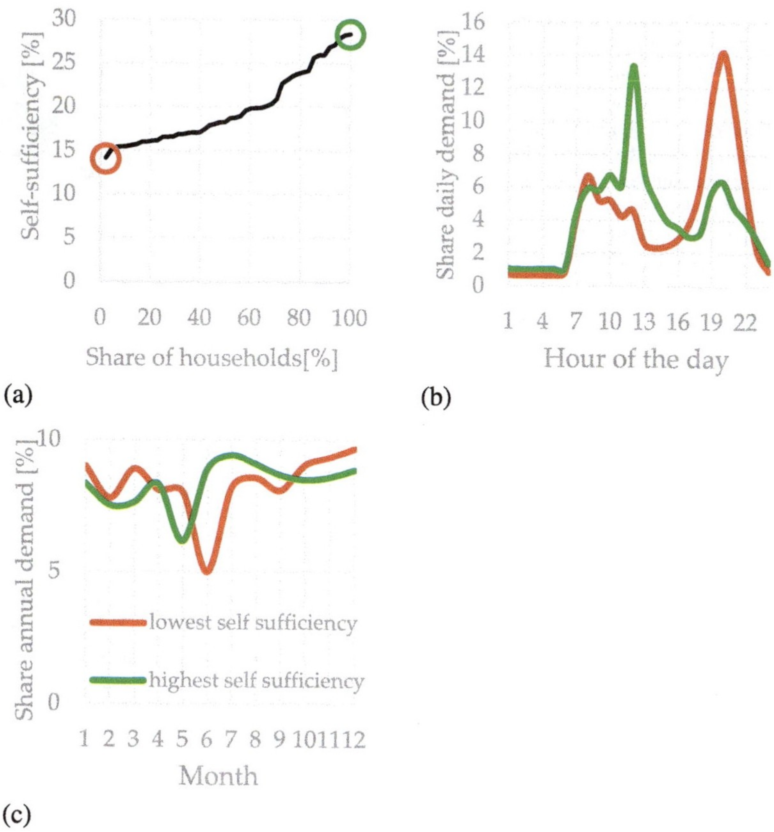

Fig. 17.7 Self-sufficiency of the apartments in the local grid. **a** is the distribution of self-sufficiencies across the 48 households, **b** shows the hourly average of the extreme households, **c** shows the monthly average consumption of the extreme households

compared to the daily average: both the households present a steep drop in sunny months which seems to indicate an absence due to summer holidays. The most self-sufficient household appears to have had an absence for holidays during May instead of June, as shown in Fig. 17.7c. This might be advantageous as it allows to use more PV electricity when the overall electricity demand of the district is lower and the radiation from the sun is higher. It should be noted that, in general, the best performing household presents a smaller dip in demand for the summer holidays, it is unknown whether it is due to a shorter holiday or at the presence of some household's components at home.

The examples shown in Fig. 17.7 highlight the two apartments that are extreme in terms of self-sufficiency. To infer more generalized information on the time of high consumption that favors high self-sufficiency (see Fig. 17.8) the following formula was used:

$$ISelfS_{ts} = \sum_{HH=1}^{48} \begin{cases} TP_{ts,HH} - TP_{ts,Tot} & \forall SelfS_{HH} \geq SelfS_{tot} \\ 0 & \forall SelfS_{HH} \leq SelfS_{tot} \end{cases} \quad (17.3)$$

where

$ISelfS_{ts}$ is the influence of high energy demand in a given time step (which could be an hour of the day or a month of the year). HH stands for HouseHold as the curve results from the sum of all the individual households. $TP_{ts,HH}$ and $TP_{ts,tot}$ are the typical power demand [W] of said time step(ts) for the nth household (HH) or the whole district (tot) respectively. The sum of all time-steps is then rescaled so that it is equal to 1.

In practice, the curve is influenced only by the households that have a self-sufficiency above average. It represent the influence (positive or negative) that the demand in each time-step has on the overall self-sufficiency. Unsurprisingly, Fig. 17.8a shows that a lower average demand in the evening and early morning hours is

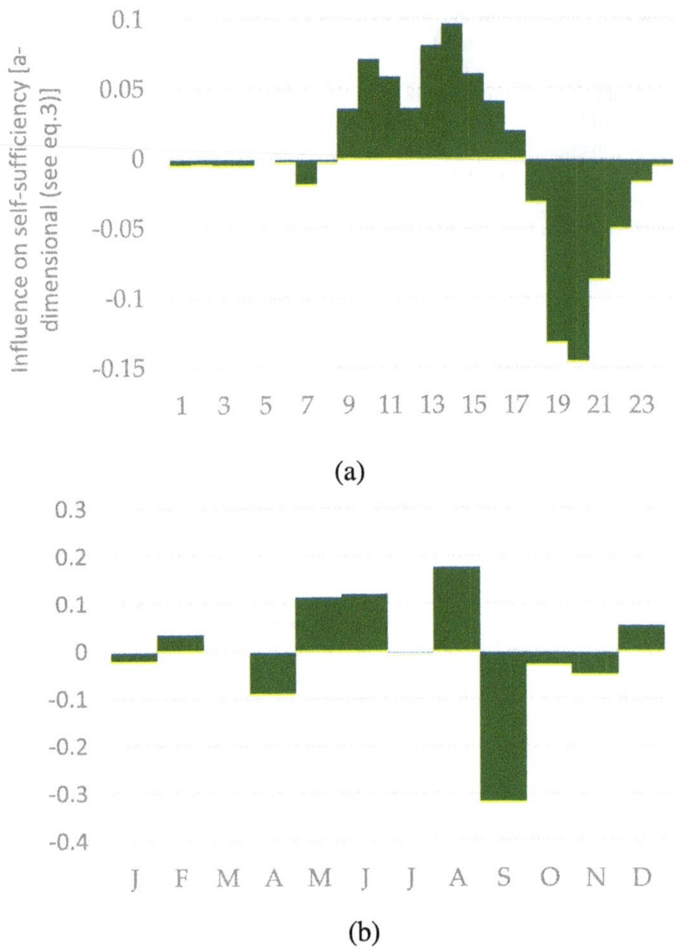

Fig. 17.8 Influence on self-sufficiency of high demand in **a** each hour of an average day; and **b** month of the year. The value is a-dimensional but it express the positive (or negative) influence of a high electric demand at a given time-step compared to all the others (see Eq. 17.3)

associated with high self-sufficiency. On the contrary, the central hours of the day are generally above average in highly self-sufficient households. It is interesting to notice how the electric demand at 12:00 is in general less beneficial for self-sufficiency than the hours around: this is somewhat counter intuitive, but it makes sense since at 12:00 the high general consumption due to lunch causes scarcity of renewable energy more often than in the hours immediately before or after. The signal on a monthly basis is not so easy to interpret. It appears to be beneficial to have above-average consumption in August and below-average in September: this is possibly due to a fraction of the households that went into holiday later in any given year. Given the sharp drop in irradiation of the month of September compared to July and August, it seems reasonable that going to holiday in September increases the self-sufficiency over the year.

17.3.2 Exploitation of the Common Renewable Resources: Sheer Cumulative Consumption Versus Self-sufficiency

Figure 17.9 shows the relation between the annual cumulative demand and the annual cumulative energy received from the shared PV system. These two variables are strongly correlated ($R > 0.9$), thus the quantity of energy consumed from the PV system can be assumed

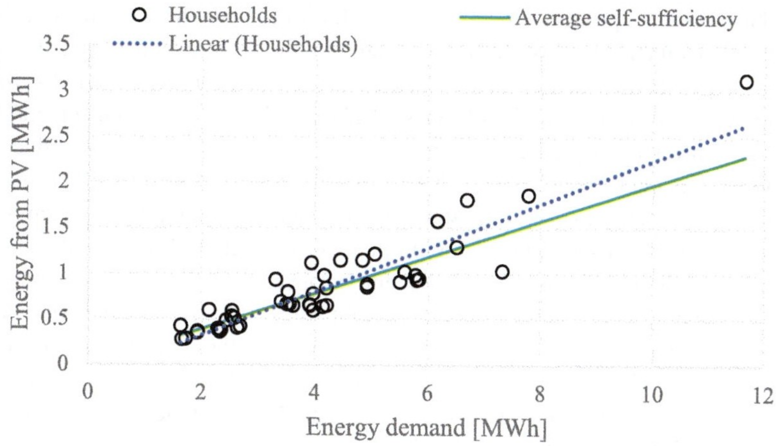

Fig. 17.9 Annual cumulative energy demand and annual cumulative energy used from the PV system for every household in the local grid

with good confidence from the annual cumulative demand alone (i.e. regardless of the self-sufficiency).

This aspect, although counter-intuitive, is a consequence of the highest variability in annual cumulative demand compared to the variability in self-sufficiency: if in fact the highest self-sufficiency is two times the lowest one, the highest cumulative demand is almost 5 times the lowest one (excluding the highest value as an outlier, otherwise is more than 7 times). The strong prominence in variability of cumulative demand compared to self-sufficiency reduces the variation in self-sufficiency as a mere noise compared to the other variable (as visible in Fig. 17.9). Furthermore, as self-sufficiency is a share of the demand, it does not have much importance in absolute terms when applied to households with low cumulative demand. This fact represents somewhat a hindrance as it implies that increasing overall consumption works better than improving self-sufficiency to seize larger quantities of scarce local renewable resources. Nevertheless, it is not clear what power has an individual household to change its cumulative energy demand. Further investigation on the aspects that influence the cumulative energy demand (e.g. number of people in the household, cooking habits, holiday habits etc..) is needed to assess whether it is something that the inhabitants can change. If each household has significant power on the cumulative energy consumption, it is reasonable to fear a sharp increase in the overall consumption after the installation of the communal PV system. It should be acknowledged that the lack of data with respect to other households might focus the attention of the inhabitants on their own energy demand advising them to increase the self-sufficiency. Another interesting aspect, shown in Fig. 17.8, is that the linear interpolation of the household data points has a steeper slope than the average self-sufficiency of the 48 households. This means that the household with the highest annual cumulative consumption also has, on average, a highest self-sufficiency. The highest slope of the interpolation implies that at low consumption the self-sufficiency of a household tend to be lower than average, while at higher consumption tends to be higher. A correlation analysis between annual cumulative consumption and self-sufficiency found a positive, albeit weak, correlation ($R \approx 0.2$). Although it is weak and thus uncertain, the correlation suggests that highly consuming households might have more contemporaneity with the production from PV. This might be due to larger households having some members who stay at home during daytime, or to electric consumption by people who spend daytime at home being larger overall.

17.3.3 LEC Gratis

In this arrangement, the households in the district are shareholders of the system. So they can use the electricity produced by the system for free when available. In this study, the shares of the PV system are equal. Each household will therefore have to pay 13,646 SEK (1275 €) of initial investment plus ca. 342 SEK/year (32 €/year) for maintenance and substitution of the inverter. Different ownership structures are possible, but the business model should be modified to avoid loopholes in the risk–benefit balance. For instance, equal shares could be distributed to a sub-group of the households (i.e. there are consumers who do not hold shares). In this case, a price of the electricity for non-owners should be established (see section LEP n%).

Figure 17.10 shows the difference in price between the energy offered by the parent grid and the energy available within the local system. The chart shows monthly values, which refer to the average cost of the electricity that month in the grid. We know from the section "Ownership structures and business models" that at any given time the price of the electricity is unique within the micro-grid and depends from the relationship between production of PV and demand (see Eqs. 17.1 and 17.2). The bars in Fig. 17.10 are the average of all the electricity prices of the respective month weighted by the aggregated electric consumption in that month. Obviously, since the energy, not met by the local production, is bought from the parent-grid, the external price has an influence on the internal one. In simpler terms, the internal price of the electric energy in one month, because the Eq. (17.2) with Plocal = 0, is proportional to the residual demand. Notice that, due to the higher external price, the drop in cost of electricity during the months of March (month 3) is similar to that in April (month 4) despite a lower self-sufficiency.

Even if the price of the electricity is the same within the micro grid at any given point in time, the average price paid by each household varies according to the time patterns of consumption. A household will enjoy a lower average price when they consumed a large share of its annual consumption at times when the electricity was free (or at least cheaper). This is to say that a higher self-sufficiency will lower the average price. However, in terms of gross economic benefit (i.e. the sum that can be saved), it is not the average price that matter, but the cumulative energy received for free. In this sense, the conclusion from Fig. 17.9 is troublesome as the earnings are not due to the ability to obtain a higher self-sufficiency, but simply to the sheer cumulative consumption. In Fig. 17.11, the households in the micro-grid are divided in 3 groups of 16 elements each according to their annual cumulative consumption. As in Fig. 17.9, the correlation of the KPI (Key Performance Indicator) with annual cumulative consumption is evident. In fact, the lifetime economic balance is determined solely by the savings, thus by the sheer quantity of energy that is received by each household. From Fig. 17.11a it is visible how being in the upper third of the cumulative consumption charts guarantees substantial earnings (IRR: internal rate of return from 1.9 to 6%), in case of the initial investment of about 13,646 SEK (1275 €/household). Conversely, the low-consumption households are doomed to economic losses, which means they are unable to recover the investment itself.

If the relation between annual cumulative consumption and lifetime earnings would become known by the households in the local grid, there is a risk that there would be a considerable increase of the cumulative demand after the installation of the communal system. This fact, although potentially reducing the risk for those investing in the system (especially in a LEP case), would counteract the purpose of reducing consumption of electricity from the grid.

17.3.4 Lec Lcoe

If the energy is sold at production cost (LCOE), instead of being given for free, the difference in lifetime balance from the different households are greatly reduced, but they persist. In this case, the advantage associated with the use of energy from the system is influenced by the stake of ownership of the system. In general, it can be noted that the lifetime earnings (i.e. Figure 17.11a and b) follow a linear transformation from the extreme inequality (as in Fig. 17.10a),

Fig. 17.10 Monthly difference in price between the energy offered by the parent grid and the average paid by the shareholders in a LEC gratis arrangement

to a situation of complete equality of earnings (if a LEC grid-price is hypothesized), where no benefit is obtained by the use of on-site electricity. In the hypothesis, a benefit for self-consumed electricity would spur increased self-sufficiency. A balance should be found between risk for the low consumption households and reward for the consumption of local renewable energy.

17.3.5 LEP N%

In this arrangement, the PV system is owned by a single provider who has the right to set the price. Obviously, since the parent grid has the ability to supply 100% of the demand of the district, the owner cannot set the price higher than the electric grid lest being completely out-bid (e.g. no household would use the owner's energy). In this study, the provider sets the price as half-way between the minimum of the local LCOE and the maximum of the consumer price from the parent grid. More precisely, the provider sets a price at a percentage n so that n = 0 is the LCOE, n = 100 is the price offered by the parent grid and n = 50 is exactly half-way in between.

Table 17.3 shows how the annual revenues, the balance over the lifetime and the real IRR change according to the price at which the electricity is sold.

Notice how with n = 0% (i.e. the electricity sold at production cost of 0.83 SEK/kWh), the balance and thus the IRR result are negative. This is due to the fact that the self-consumption of the system is not 100% (it is in fact ca.85%). In other words, not all the energy produced by the PV system is consumed by the households in the local grid. Therefore, part of the production is sold to the grid below LCOE and results in a moderate loss over the lifetime. The existence of this loss justifies the use of a LCOE adjusted for self-consumption as described in Huang et al. (2019). This loss also explains why, under LEC LCOE arrangement, some households experience economic losses over the lifetime when the electricity by the communal system is given at price of cost (see Fig. 17.11b). When the electricity is sold at LCOE, the IRR of the PV system is negative, thus holding its shares leads to a loss unless the benefit for cheaper energy outweighs the costs.

Applying an n = 9.43% does not result in any loss or gain over the lifetime of the system. It can be argued that no investor would like to take any risk to have an expected NPV (Net Present Value) of 0 at the end of the lifetime with a discount rate of 0. Nevertheless, there are potential business models for large homeowners such as general contractors or municipalities who could substitute part of the roof and façade cladding with BIPV thus avoiding the cost of an

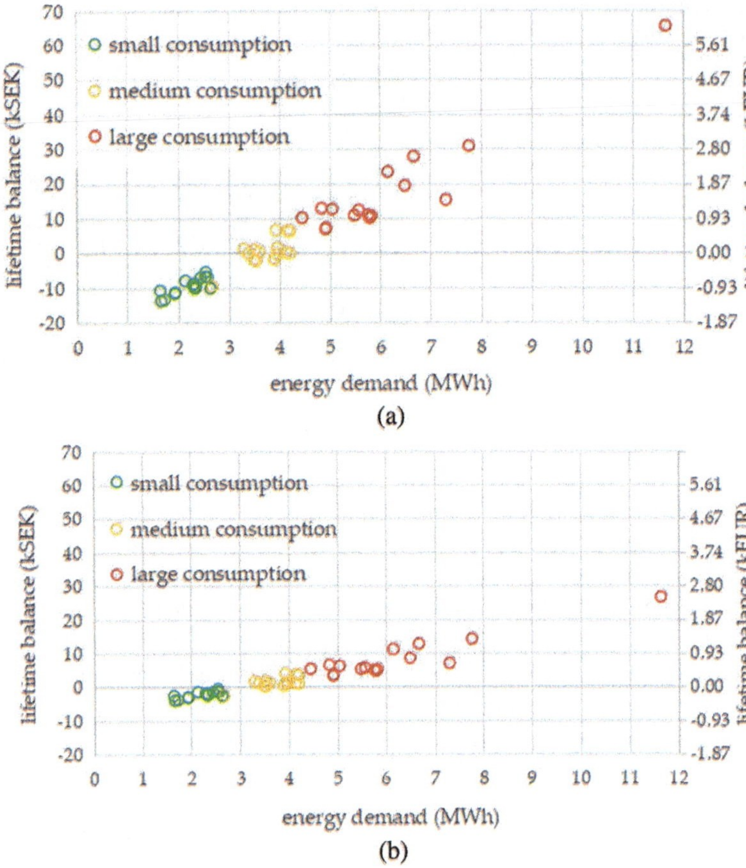

Fig. 17.11 Cumulative balance over the lifetime of the system against the annual energy demand. The households have been divided in 3 groups, each of 16 specimens, according to their cumulative consumption. **a** LEC Gratis, **b** LEC LCOE

alternative material. Furthermore, this price tag is extremely interesting as price of sale from LEC. It in fact presents the advantage of expected lifetime economic balance in positive ground for each household.

A good business opportunity is finally offered by the n = 100%. This price, while suggesting a real IRR around 3% for the LEP, offers the occupants the opportunity to largely increase their share of renewable energy use without having to pay any upfront cost. In this case, the households have no economic benefit in installing the PV, but they have no risk nor upfront investment and could receive information about their own self-sufficiency by the provider, e.g. with a monthly email.

17.3.6 LEM

Figure 17.12 describes the average hourly price for scenario 1 during the different hours of the day within the district. The grey bands represent the variability between different households of the district. If a bar is longer, it means that some households have an average price that is significantly lower than others in that hour of the day. The red ticks represent the average price in that hour for the whole district. It is immediately visible that between 7 P.M. and 4 A.M. the price is almost stable at one point five SEK. The price is stable because there is no electric storage installed in the micro grid and therefore, when a photovoltaic system is not producing, the price is

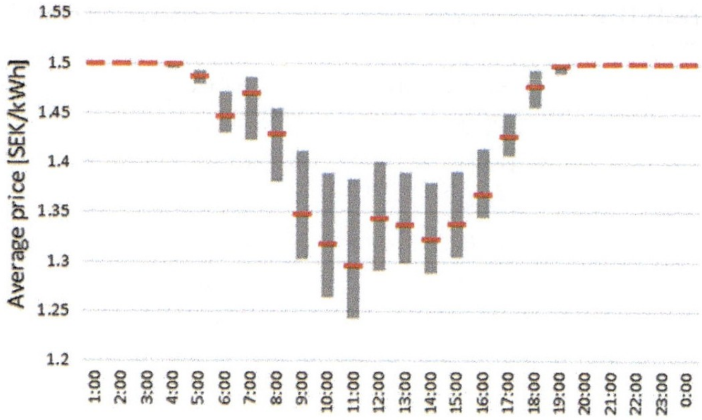

Fig. 17.12 Average price for each hour of the day and its variability within the district (LEM, Scenario 1)

that of the power provided by the electric grid. Considering that the night-time consumption of the district remains stable throughout the year, And that the the electricity is sold at 1.8 SEK/kWh in winter and 1.2 SEK/kWh in summer, the night price paid by the whole district equals almost exactly the average of the two (I.e. 1.5 SEK/kWh). During daytime the price is lower, there are two reasons for this phenomenon. The first reason is that each household owns a share of the local PV system, and therefore all the electricity produced by their own share already belongs to them, and is thus free for themselves. The second reason is that they can purchase electricity from their peers, which price is consistently lower than that of the grid. For these reasons, those households that are able to use larger share of their own electricity, or at least of the electricity from their peers, can enjoy a lower price for the electricity. It is visible, how indeed expected, that the price of electricity is generally lower during the central hours of the day. Furthermore, it can be observed that times of the day of comparatively higher price, correspond with moments of high electric demand. For example, it can be seen how the price is comparatively higher at seven and eight am (when people prepare and consume breakfast) and at noon (when many prepare and consume lunch). This is due to the fact that, despite a large photovoltaic production in that hour, the outlier high demand forces the whole district to supply part of its demand from the electric grid.

Figure 17.13 shows the relationship between savings and revenues for each single households within the micro grid. All the sixth scenarios described previously are shown in the chart. In general, the savings are obtained either by using the electricity produced by one's own system, therefore saving 100% of the price from the grid, or else by using the electricity from a peer household at a discounted price. On the other end, the revenues are obtained either by selling electricity to the grid, or else by selling electricity to another household, the latter providing a much higher price. In each of the charts, every household is displayed as a circle, its position on the x axis represents its annual revenues, while its position on the y axis represents its annual savings. The color of the circle line represents its belonging to a different category, according to the number of people living in the household. In general, these charts should be studied in their relative difference between each other, rather than in their absolute values. In fact, the absolute value of revenues and savings are not interesting when the cost of initial investment and those for the maintenance of the system are not taken into account. To see the lifetime techno-economic performance of the system the internal rates of return should be investigated. Observing the colour in all the cases it is visible that the smaller

Table 17.3 Annual revenues, lifetime balance and internal rate of return (real) of the investment by different prices set by the owner

N (%)	Revenues (SEK)	Balance (SEK)	Balance (€)	IRR (%)
0	34′553	−94′058	−8′790	−0.5
9.43	37′689	0	0	0.0
25	42′864	155′247	14′509	0.7
50	51′174	404′553	37′809	1.6
75	59′484	653′859	61′108	2.3
100	67′794	903′165	84′408	2.9

households, I.e. those which have a smaller electric demand tend to show lower savings and higher revenues. This phenomenon is quite unsurprising because In the example considered every household purchases an equal capacity to all the others. Therefore, the small households will have a PV capacity that is larger relative to their demand, and will, thus, export and sell a larger fraction of the electricity that they produce.

There is a noticeable difference between the scenario 1 and the scenario 2. In the scenario 1 every household agreed to sell their electricity at a significantly lower price compared to what they agree in the scenario 2 (see Table 17.1). Because of these, the scenario 1 seems to be particularly advantageous for the largest consumers. This is due to the fact that the largest consumers have high savings but low revenues, and therefore, the sale of electricity at a lower price from others leads to higher saving for them, while it does not impact their revenues significantly. On the contrary, the smallest consumers, that have in general lower savings and higher revenues, will benefit from higher price of the electricity. In this case, they will be able to increase greatly their revenues without changing too much their savings. In fact, being smaller, most of their savings comes from their own PV system, which is already over-dimensioned compared to their size. Going back to the largest consumers, a large fraction of their savings implies purchasing electricity from the smaller peers. Another noticeable aspect is that in scenario 2, all the points are almost linearly correlated. This is due to the high price of the electricity sold, which is almost the same of the cost of electricity from the grid. The smaller remaining differences can be explained by the correlation between the private electric demand and the demand of the whole district. If an household over-produce electricity when the whole district is in over-production, Its performances will be slightly lower because it will often sell to the grid. In the opposite case, when an household overproduces at times when there is need from other households, then its performance will be slightly higher.

Scenario 3 and 4 reflect scenario 1 and 2 in terms of price, but they have the peculiar aspect that only half of the households choose to purchase a PV system. This state of affair would be very important for any practical application of the micro-grid. This is because, in practice, it is really difficult to convince 100% of the tenants in a multi-family dwelling to participate, and especially to invest money, in a PV system. In a realistic setting it is expected that part of the population is unwilling to invest in the system, nevertheless, in the simulation is assumed that they decided to participate in the micro-grid as simple consumers (i.e. those who do not own any part of the system). The assumption is pretty safe because being a simple consumer only requires to always purchase the electricity from the cheapest source. In this way, to participate as a simple consumer does not have any initial cost, but it might have a benefit during the lifetime of the system. If scenario 1 is compared to scenario 3, it is visible that the points in the latter overwhelmingly outperform those in the former both in terms of revenues and in terms of savings. It is tolerably intuitive that, if only half of the households own a PV system, their revenues will

Fig. 17.13 Savings versus revenues for each household in the microgrid. Both axes are in [SEK/year]

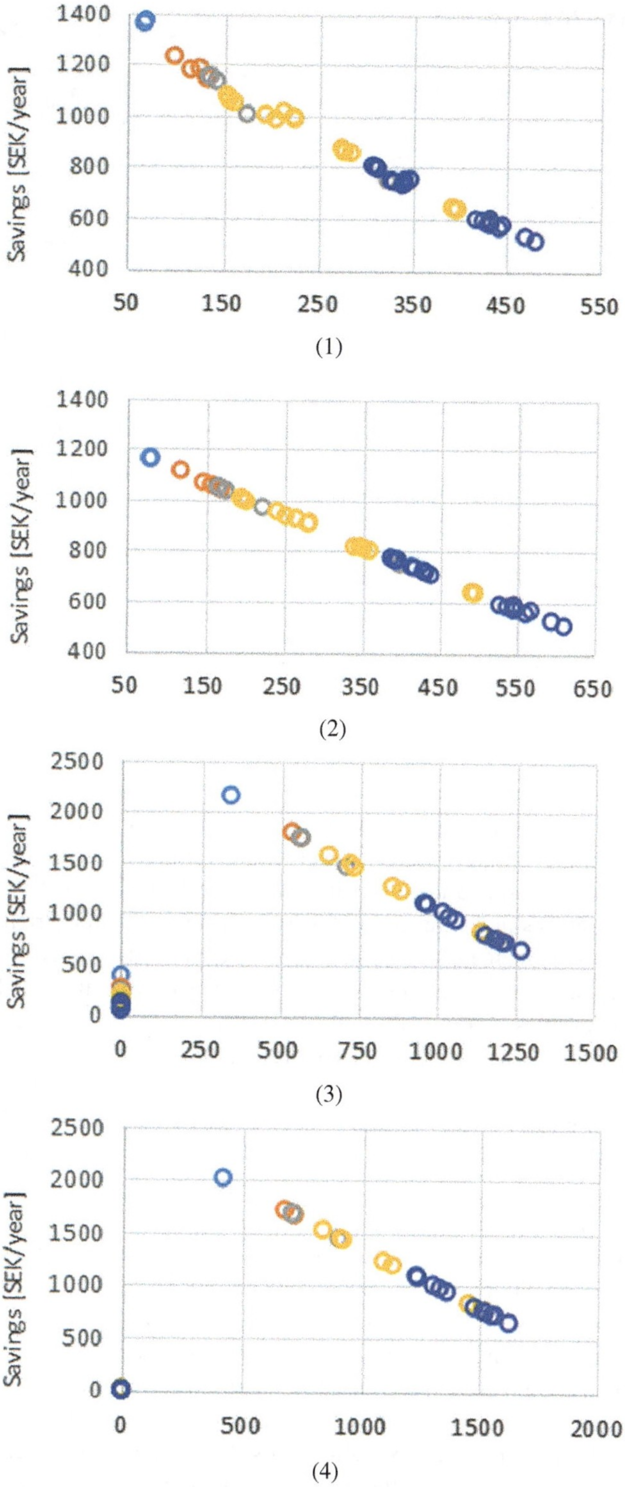

Fig. 17.13 (continued)

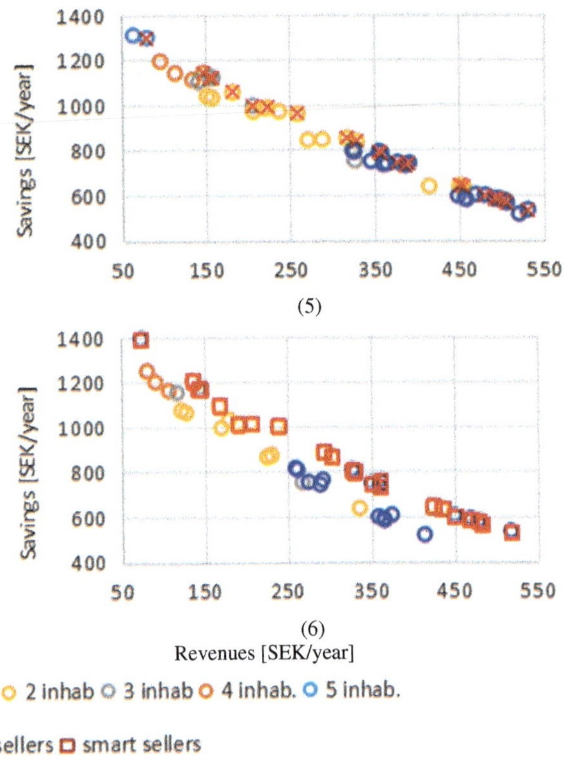

increase. In fact, all the households who do not own PV can only buy the electricity from those who own it, and therefore, the whole local market moves in the direction of a "seller's market". Also the increase in savings is readily explained, In fact, since the optimal capacity is unchanged in every scenario, every PV owner has at his disposal a larger capacity. This fact implies that there is more available electricity for self-consumption in every HOY, even at times of a relatively high private electric demand. This spare over-capacity favours an increase in self-sufficiency. Looking at the bottom left corner of the chart for scenario 3, it can be seen how there is a benefit in terms of savings also for those who do not own a PV system. Of course, these savings are minor compared to those of the other households, and this is due to two specific reasons. The first one is that, by lacking a PV system of their own, these households do not have their own electricity for free, and therefore can only purchase electricity from their peers. The second reason is that every household, before to sell electricity, satisfies his own demand, and therefore those in the micro-grid who do not own a PV system can only benefit from the left-over electricity from the others. In other words the household without PV can only purchase electricity when they happen to be in need of power at times when others are in over-production. Scenario 4, like the scenario 2, presents a linear correlation between the revenues and the savings of each household. Like scenario 2 over scenario 1, also scenario 4 presents a relatively higher revenues and lower savings compared to scenario 3, thus favouring the smallest consumers. Furthermore, like scenario 3, it presents a sharp contrast between the PV owners and the other households. It should be noted, though, that this time there is absolutely no benefit in participating in the micro-grid, or at least the benefits are so tiny that cannot be seen by the naked eye. There is nevertheless a benefit for those household, it is the possibility to increase the share of renewable on-site electricity in their energy consumption. It can be expected that, given the absence of initial investment, most consumer would be willing to increase their renewable energy share. It is presented to them the possibility to save the planet with a cost-less, thus effortless, action.

In the first four scenarios the price for the sale of PV power And the number of PV owners has been changed. Nevertheless, in all these cases, every PV owner agreed to maintain the same price as everybody else. In scenario 5 the hypothesis is made that a half of the PV owners prefer to sell at the lower price (I.e. the price of the whole group of PV owners in scenario 2 and 4). Meanwhile the other half of the owners prefers to sell at the same price of scenario 1 and 3. Observing the revenues and the savings in this arrangement, it can be noted that, in general, the sellers who decide to sell for a lower price (I.e. shown as 'cheap sellers'), enjoy a higher revenues compared to the others. First of all, let's consider that savings are not affected by the price for the sale of one's own electricity, but rather by the price for available energy from the other households. For every saving level in the chart, the lowest selling households, which are marked as 'cheap sellers', appear to be on the right side compared to the others, this means that they have managed to obtain higher revenues. There are two factors at play when measuring the revenues in this type of market (I.e. where there are two different prices groups). The first factor is the sheer revenues per KWh sold, this it acts by lowering the revenues for the so called 'cheap sellers'. The second factor regards the ability to effectively sell your electricity within the micro grid at all (i.e. the capacity of not be in over supply, Thus forced to sell most of your power to the grid). If a different price was chosen, the result might have been different, but in this case the increased revenues deriving from an higher price scheme are not enough to offset the increased instances in which the electricity cannot be sold due to high price and low demand. Also in scenario 6, the last one, the group of households was divided into 2 sub groups. As in the previous scenario, some households where selling their electricity at a lower price compared to others. This time, though, the expensive sellers were given the ability to change the price according to a behaviour of their own. The mechanism used to change the price was set according to the simple principle explained in the description of scenario 6. In practice, these households, identified in the chart as 'smart sellers', will sell their power at the LCOE of the system, which is lower than the static price of the cheap sellers, whenever they have an outlier high energy balance. In other words, when a smart seller has an outlier, low power consumption or an outlier high power production from PV, it will sell its electric power at the lowest possible price. This strategy is extremely simple and is prone to numerous fallacies. In fact, if for example a particular household is on holiday during an unpopular period (I.e. when nobody else is on holiday), its power demand would be unusually low, thus resulting in an outlier high balance. This could cause it to sell at LCOE in a time in which the electricity is indeed in high demand throughout the district. In this example, the household would be selling at the lowest possible price in a time in which the maximum price would still manage to sell to the peers. Conversely, if an household will experience an outlier high demand for its own reasons, it might find itself selling its available power dearly, while there might be plenty of energy available for everyone. In this case it will be forced to sell most its power to the electric grid. Despite the simplicity of this strategy and its obvious flaws, it is visible from the chart (Fig. 17.13 (6)) that such a simple behaviour is good enough for outsmarting the cheap sellers in the competition for the sale of electric power. Given any savings level, the smart sellers undeniably manage to obtain higher revenues. This is a very important result because it shows that it is possible to create an effective strategy without knowing the consumption of the other agents in the micro grid, thus, avoiding privacy issues.

17.4 Discussion and Outlook

17.4.1 Social and Cultural Differences Among Households Have a Huge Impact on Self-sufficiency

In the local grid, if the renewable energy is not enough to cover the electric demand during a specific hour, the aggregated self-sufficiency is assigned to each household regardless of its demand (see Eqs. 17.1 and 17.2). A large difference in terms of self-sufficiency has been observed within the 48 households, with the individual self-sufficiencies spanning from ca. 14% to more than 28% (see Fig. 17.7a). Considering the absence of active strategies to increase the self-sufficiency in the cluster, such large differences can be attributed only to socio-cultural factors and spontaneous lifestyle choices. From Fig. 17.7b it appears that the most self-sufficient household has on average the peak of energy consumption at noon (possibly due to home cooking), while the least self-sufficient one has usually its peak consumption at 8 P.M. Differences are visible also over the different months of the year but their effect is not as clear as in the hours of the day. The large differences observed in self-sufficiency, having no active engagement or use of demand-shifting technologies, invites a deeper analysis and understanding of the existing electric demand and the factors which affect self-sufficiency.

17.4.2 High Cumulative Energy Demand is More Effective Than High Self-sufficiency in Exploiting the Shared Renewable Resource

Despite the large variation in self-sufficiency, it has been observed that the sheer amount of energy used from the system is mainly determined by the annual cumulative demand (see Fig. 17.9). This phenomenon, albeit counter-intuitive, is due to the fact that the variability of cumulative demand far outweighs the variability in self-sufficiency (the largest being 5 or even 7 times the smallest one). In other words, the fraction self-consumed is not significant when applied to a group of households whose entire demand is hardly significant compared to others. This fact is problematic because the energy savings (i.e. the main earning mechanism of the investment in some market designs) come from the amount of PV energy consumed, and not from the self-sufficiency reached. The relation between annual cumulative consumption and cumulative energy from PV is transposed in the relation between energy consumption and lifetime balance (see Fig. 17.13). The balance in a LEC gratis arrangement (Fig. 17.11a) is almost completely determined by the cumulative consumption, with the self-sufficiency being reduced to a noise in the linear relation. Moreover, if the households are divided in 3 groups according to their cumulative consumption, the biggest consumers all have positive balance and the smallest consumers all have a negative one. This aspect suggests that, if the communal PV system is installed under a LEC gratis arrangement, the shareholders might increase their electric demand in a bid to outdo each other's energy consumption. This behaviour would possibly defeat the purpose of installing on-site renewables in the first place. It should be also considered that, due to privacy laws and standard practice, each individual household is likely only aware of its own electric demand and self-sufficiency. This lack of data might drive each household to work on improving self-sufficiency instead of annual cumulative demand. It should also be remembered that the earnings are savings, thus increasing the cumulative demand would anyway lead to an increase in the energy bill. In this sense, the increased exploitation of the common electricity through increased cumulative demand would happen only if increased consumption is perceived as a value, for example through the purchase or increased use of energy hungry appliances for cooking or DIY (Do It Yourself) purposes. How easy or difficult it is to change self-sufficiency compared to cumulative demand should also be considered to assess the likelihood of one outcome over the other. For example, cumulative demand might be strongly constrained by working schedule or number of household members. These aspects reiterate the need for a deeper study on the aspect of demand that influence self-sufficiency. From the perspective of the investment in PV, both the changes in behaviour envisioned would increase self-consumption, hence earning potential.

17.4.3 Different Selling Prices Generates Various Business Opportunities

Assuming that the shared PV system is owned by a single entity in a LEP (Local Energy Provider) arrangement, this entity enjoys freedom in setting the price for the sale of electricity. This freedom is nevertheless constrained by the LCOE of the PV system and by the price offered by the parent grid. If the LEP sells electricity at a higher price than the parent-grid it will have no purchaser among the households. This happens because the grid has the capacity to satisfy 100% of the demand of the whole district at any time. For this reason, a coefficient "n" has been devised so that: n = 0 is the LCOE of the local system and n = 100 is the sale of energy at the exact same price as from the parent grid. It has been shown that at n = 0, despite selling at production cost, the lifetime balance is < 0. This is due to the self-consumption being below 100% (i.e. ca 85%), hence ca. 15% of the energy produced being sold at spot price (i.e. 0.3 to 0.15 SEK/kWh or 3 to 1.5 € cent/kWh). This loss also explains why in the LEC LCOE arrangement some households still have a negative lifetime balance, as demonstrated in Fig. 17.11b. Another interesting selling price is the one obtained with n = 9.43% because this is the price at which no profit nor loss is made from the LEP. This price tag, albeit unattractive as an investment for a third-party PV owner, presents an interesting mean for building owners to substitute other claddings on their properties. Using this selling price offers in fact a building material that, contrary to every other, does not cost anything over its lifetime. If applied as common price in a LEC it allows all the household to have a positive lifetime economic balance, yet to have individual differences in earnings. It should be said that this price was determined at the end of a previous run when the overall self-consumption was already known. In a real case, to obtain such an equilibrium, the price should be updated at any point in time according to the evolution of self-consumption and energy prices. Selling energy at the price of the parent grid (n = 100) could be an interesting investment as it guarantees the LEP with a real IRR of around 3%, it provides no economic benefits for the household consumers but it gives them the ability to boost their reliance on renewable without any upfront cost nor risk. Furthermore, the possibility for the households to buy voluntarily sized shares of the LEP could kick start a set of tantalizing business opportunities.

17.5 Conclusions

In the study, a newly developed agent based model was tested on a shared PV system serving a small district comprising 48 apartments in a local community. Different ownership structures were explored. The LEC arrangement was studied both with the electricity given for free to all the equal shareholders or given at a price (in the study the LCOE). For the LEP, because the free offering would make no sense, an array of different prices was tried (see Table 17.3).

17.5.1 Key Findings

The main findings of the study are reported as follows and interpreted in the corresponding paragraphs in the discussion section:

- Social and cultural differences among households have a huge impact on self-sufficiency: the households were simulated without introducing any demand-response measure or smart control. Yet, some households achieved a self-sufficiency of almost 30% using the common PV system while others stopped short of 15%. also in the LEM case, where a more accurate study of the energy demand were made, a similar result was obtained.

- High cumulative energy demand is more effective than high self-sufficiency in exploiting the shared renewable resource: despite the large differences observed in self-sufficiency among households, the quantity of energy received from the shared system has been determined almost completely by the annual cumulative demand rather than by self-sufficiency.
- Different selling prices generates various business opportunities: different value of n%, as defined in the Sect. 17.2.2, generate advantage and interesting features for diverse stakeholders. For instance, a very low n% (i.e. <10%) generates a strong drive for the shareholders to self-consume as much PV energy as possible, but it contains a risk for the least consuming ones. Higher n% (i.e. from ca. 10 to 100%) are interesting for building owners and BIPV solutions and, amid increasing n%, become more and more interesting for third party energy providers.

17.5.2 Follow-up Studies

The present study shows a plain set-up and a narrow set of possibilities, but it sets the stage for a broader class of studies. In principle, some of the simplifying assumptions employed in this study should be removed in favor of a higher realism and a more complex modelling, nevertheless models that are too complex for the level of uncertainty and for the input data available should be avoided.

For instance, it is tempting to change the present model for the prices from the parent grid (i.e. static seasonal price + long term linear trends for sold and bought electricity) into a spot-price + distribution costs. However, while the change reflects reality better, the long term (multi-annual) modelling of the spot-price would be a daunting task and affected by huge uncertainty. Because of this reason, it might pay off to just maintain a simplified model for the prices (i.e. 2 seasonal prices for purchase and sale + time of day variation), but to perform a stochastic simulation with variability in the time-evolution of the prices. In other words, any further complexity addition should only be determined by the use case of the model. And, for this model, the use case is the market design to finance and maintain a fair and remunerative local electric energy system.

On the other end, there are several low hanging fruits that can be easily harvested: for example, while in this study the price was always set by either a unique actor (be it a community or a provider) or by two groups of actors (such as in scenario 5 and 6 of the LEM). It would be interesting to explore the effect of different prosumer setting each an arbitrary price and explore their interaction. In this sense, one more step could be to endow the agents with some level of intelligence, beyond the simple behaviour of the smart sellers, and let them adjust the price reacting to the environment to maximize potential economic gains.

In the present study, there are devices and loads that have not been investigated, such as EVs and electric storages, in the local grid. These features, given a simplified enough model, are extremely easy to be implemented and can constitute a game-changer in the effectiveness of a business model.

Another interesting and potentially prolific research direction would be the study of the demand itself. Given the large variation of self-sufficiency found among the different agents participating in the micro-grid, it is possible to find correlation with socio-economic and life-style parameters such as median age, work-home schedules, number of members in an household etc. This does not constitute information in itself, but it can lead to different results according to the different shared renewable systems. In other words, each social-mix might demand a different system (capacity of PV, capacity of electric storage).

Regarding the demand, it is of paramount importance to consider how often a house remains vacant due to change or death of the owner. These aspects should be investigated in terms of impact over each business model, but also in terms of risk-mitigating effect of larger local grids. It shall not be forgotten that lower risk can allow lower IRR for the investment, thus unlock wider market niches. The vacancy of the households is as well affected by socio-economic parameters and median age of the households, these aspects likely present spatial variability in different parts of the city and the world.

References

Alam MR, St-Hilaire M, Kunz T (2019) Peer-to-peer energy trading among smart homes. Appl Energy 238:1434–1443

Almasalma H, Claeys S, Deconinck G (2019) Peer-to-peer-based integrated grid voltage support function for smart photovoltaic inverters. Appl Energy 239:1037–1048

Basnet A, Zhong J (2020) Integrating gas energy storage system in a peer-to-peer community energy market for enhanced operation.Int J Electr Power Energy Syst 118:105789

Chen K, Lin J, Song Y (2019) Trading strategy optimization for a prosumer in continuous double auction-based peer-to-peer market: A prediction-integration model. Appl Energy 242:1121–1133

Cornélusse B, Savelli I, Paoletti S, Giannitrapani A, Vicino A (2019) A community microgrid architecture with an internal local market. Appl Energy 242:547–560

E. Commission (2020) Clear energy package [Online]. Available: https://ec.europa.eu/energy/topics/energy-strategy/clean-energy-all-europeans_en. Accessed 03 June 2020

El-Baz W, Tzscheutschler P, Wagner U (2019) Integration of energy markets in microgrids: a double-sided auction with device-oriented bidding strategies. Appl Energy 241:625–639

Geels FW (2002) Technological transitions as evolutionary reconfiguration processes: a multi-level perspective and a case-study. Res Policy 31(8–9):1257–1274

Geels FW, Schot J (2007) Typology of sociotechnical transition pathways. Res Policy 36(3):399–417

Giddens A (1984) The constitution of society: Outline of the theory of structuration. Univ of California Press

Huang P, Lovati M, Zhang X, Bales C, Hallbeck S, Becker A, Bergqvist H, Hedberg J, Maturi L (2019) Transforming a residential building cluster into electricity prosumers in Sweden: Optimal design of a coupled PV-heat pump-thermal storage-electric vehicle system. Appl Energy 255:113864

Huijben JCCM, Verbong GPJ (2013) Breakthrough without subsidies? PV business model experiments in the Netherlands. Energy Policy 56:362–370

IEA (2020a) IEA EBC annex 83, [Online]. Available: https://annex83.iea-ebc.org/. Accessed 2 June 2020

IEA (2020b) Snapshot of global PV markets

Jing R, Xie MN, Wang FX, Chen LX (2020)Fair P2P energy trading between residential and commercial multi-energy systems enabling integrated demand-side management. Appl Energy 262:114551

Lettner G, Auer H, Fleischhacker A, Schwabeneder D, Dallinger B, Moisl F (2018) Existing and future PV prosumer concepts

Liu W, McKibbin WJ, Morris AC, Wilcoxen PJ (2020) Global economic and environmental outcomes of the Paris Agreement. Energy Econ 90:104838

Lovati M, Salvalai G, Fratus G, Maturi L, Albatici R, Moser D (2019)New method for the early design of BIPV with electric storage: a case study in northern Italy. Sustain Cities Soc 48:101400

Lüth A, Zepter JM, del Granado PC, Egging R (2018) Local electricity market designs for peer-to-peer trading: the role of battery flexibility. Appl Energy 229:1233–1243

Luthander R, Widén J, Nilsson D, Palm J (2015) Photovoltaic self-consumption in buildings: a review. Appl Energy 142:80–94

Nguyen S, Peng W, Sokolowski P, Alahakoon D, Yu X (2018) Optimizing rooftop photovoltaic distributed generation with battery storage for peer-to-peer energy trading. Appl Energy 228:2567–2580

Parag Y, Sovacool BK (2016) Electricity market design for the prosumer era. Nat Energy 1:1–6

Pflugradt N, Muntwyler U (2017) Synthesizing residential load profiles using behaviour simulation. Energy Procedia 122:655–660

Pflugradt N, Teuscher J, Platzer B, Schufft W (2013) Analysing low-voltage grids using a behaviour based load profile generator. In: International conference on renewable energies and power quality

Riksdag S (2020) Regulation (2007: 215) on exemptions from the requirement for a network concession pursuant to the Electricity Act (1997: 857).[Online]. Available: https://www.riksdagen.se/sv/dokument-lagar/dokument/svensk-forfattningssamling/forordning-2007215-om-undantag-fran-kravet-pa_sfs-2007-215. Accessed 3 June 2020

Roberts MB, Bruce A, MacGill I (2019) A comparison of arrangements for increasing self-consumption and maximising the value of distributed photovoltaics on apartment buildings. Sol Energy 193:372–386

Rodrigues DL, Ye X, Xia X, Zhu B (2020)Battery energy storage sizing optimisation for different ownership structures in a peer-to-peer energy sharing community. Appl Energy 262:114498

Scott WR (2013) Institutions and organizations: ideas, interests, and identities. Sage publications

Smith A, Stirling A, Berkhout F (2005) The governance of sustainable socio-technical transitions. Res Policy 34(10):1491–1510

Sousa T, Soares T, Pinson P, Moret F, Baroche T, Sorin E (2019) Peer-to-peer and community-based markets: a comprehensive review. Renew Sustain Energy Rev 104:367–378

Stauch A, Vuichard P (2019) Community solar as an innovative business model for building-integrated photovoltaics: an experimental analysis with Swiss electricity consumers. Energy Build 204:109526

Suarez FF, Oliva R (2005) Environmental change and organizational transformation. Ind Corp Chang 14 (6):1017–1041

Šúri M, Huld TA, Dunlop ED (2005) PV-GIS: a web-based solar radiation database for the calculation of PV potential in Europe. Int J Sustain Energ 24(2):55–67

Tang Y, Zhang Q, Mclellan B, Li H (2018) Study on the impacts of sharing business models on economic performance of distributed PV-battery systems. Energy 161:544–558

Thomas L, Zhou Y, Long C, Wu J, Jenkins N (2019) A general form of smart contract for decentralized energy systems management. Nat Energy 4:140–149

Xu Z, Hu G, Spanos CJ (2017) Coordinated optimization of multiple buildings with a fair price mechanism for energy exchange. Energy Build 151:132–145

Zepter JM, Lüth A, del Granado PC, Egging R (2019) Prosumer integration in wholesale electricity markets: synergies of peer-to-peer trade and residential storage. Energy Build 184:163–176

Zhang C, Wu J, Zhou Y, Cheng M, Long C (2018) Peer-to-peer energy trading in a microgrid. Appl Energy 220:1–12

District Household Electricity Consumption Pattern Analysis Based on Auto-Encoder Algorithm

18

Yuan Jin, Da Yan, Xingxing Zhang, Mengjie Han, Xuyuan Kang, Jingjing An, and Hongsan Sun

Abstract

The energy shortage is one key issue for sustainable development, a potential solution of which is the integration with the renewable energy resources. However, the temporal sequential characteristic of renewable resources is different from traditional power grid. For the entire power grid, it is essential to match the energy generation side with the energy consumption side, so the load characteristic at the energy use side is crucial for renewable power integration. Better understanding of energy consumption pattern in buildings contributes to matching different source of energy generation. Under the background of integration of traditional and renewable energy, this research focuses on analysis of different household electricity consumption patterns in an urban scale. The original data is from measurement of daily energy consumption with smart meter in households. To avoid the dimension explosion phenomenon, the auto-encoder algorithm is introduced during the clustering analysis of daily electricity use data, which plays the role of principal component analysis. The clustering based on auto-encoder gives a clear insight into the urban electricity use patterns in household. During the data analysis, several feature variables are proposed, which include peak value, valley value and average value. The distinction analysis is also conducted to evaluate the analysis performance. The chapter takes households in Nanjing city, China as a case study, to conduct the clustering analysis on electricity consumption of residential buildings. The analysis results can be further applied, such as during the capacity design of district energy storage.

Y. Jin · D. Yan (✉) · X. Kang · H. Sun
School of Architecture, Tsinghua University, Beijing 100084, China
e-mail: yanda@tsinghua.edu.cn

Y. Jin
e-mail: jin-y17@mails.tsinghua.edu.cn

X. Kang
e-mail: kangxy18@mails.tsinghua.edu.cn

H. Sun
e-mail: shs@mail.tsinghua.edu.cn

X. Zhang · M. Han
Department of Energy and Community Buildings, Dalarna University, 79188 Falun, Sweden
e-mail: xza@du.se

M. Han
e-mail: mea@du.se

J. An
Beijing Key Lab of HVAC, School of Environment and Energy Engineering, Beijing University of Civil Engineering and Architecture, Beijing 100044, China
e-mail: anjingjing@bucea.edu.cn

X. Zhang (ed.), *Data-driven Analytics for Sustainable Buildings and Cities*,
Sustainable Development Goals Series, https://doi.org/10.1007/978-981-16-2778-1_18

Keywords

Household · Electricity consumption · Auto-encoder

18.1 Introduction

Residential energy consumption accounts for around 20–30% electricity consumption in different countries and is of great potential for energy conservation (Teeraratkul et al. 2018). In 2016, the household electricity consumption takes up 27.2% proportion of the world total electricity consumption (IEA Electricity 2018). With the development of smart grid and sustainable energy integration, the combination of renewable energy becomes more promising to save energy and to face the challenge of energy shortage. During the integration of the renewable energy source with the traditional electricity grid, because of the profile discrepancy, it is necessary to make the energy generation and end use side comply with each other. Based on the developing technology of energy consumption measurement, such as smart meter, it is easier to analyse the electricity use with different temporal resolutions (Kong 2018; Ponocko et al. 2017). Since then, a lot of researches on the electricity profile and demand response have arisen, and the new smart grids also bring many challenges to the demand response research (Godina 2018; Nan et al. 2018; Setlhaolo et al. 2014; Vázquez-Canteli and Nagy 2019). The electricity use data with different temporal resolution possesses different potentials for research and engineering application. The hourly data contributes to the load profile study in one day and instructs the strategy of demand response. While the daily data helps with the fuel supplies, maintenance operations (Stoll 1989) as well as energy storage design.

Many researchers have conducted analysis based on the daily electricity use data. Valor et al. (2001) conducted research on relationship between daily electricity consumption and air temperature, with the objective to simulate the response of energy consumption to weather conditions. The research showed that the energy consumption was strongly related to the outdoor air temperature, which indicated the influence of temperature on the energy consumption of heating or cooling. The analysis divided the data from the weekdays and the weekends (including holidays). Compared with the weekends and holidays, the electricity use in weekdays presented stronger and clearer correlation. Beccali (2004) proposed a combined approach based on unsupervised and supervised neural networks to forecast and predict the electric energy demand, correlating the electricity use with the weather data, including temperature, relative humidity and global solar radiation. Besides the regression and prediction work, many researchers focus on the clustering analysis. To instruct the building management for forecasting and detection of abnormal energy use, Li et al. (2010) made classification of the energy consumption in buildings with outlier detection methods. The features of the energy consumption were firstly extracted, used for the abnormal outlier detection as well as the classification work. Wen et al. (2019) proposed a shape-based clustering method to recognize the electricity use patterns in residential buildings. There are also many clustering methods about the electricity use data, such as Dirichlet process mixture model, k-means method, fuzzy clustering and decision trees (Granell et al. 2015; Ryu 2016; Qi et al. 2017). Zhou et al. (2017) also used an improved fuzzy c-mean method to cluster monthly electricity use data. Different algorithms meet the requirement for data with different scale and amount. In the process of clustering analysis, k-means clustering method shows advantage than other algorithms (Wen et al. 2019). However, it should be noticed that the k-means method is based on the distance analysis among different observations. Once the dimension of the observation reaches to a large scale, the "dimension explosion" shall happen, which makes Euclidean distance meaningless and the clustering analysis may not work. In this chapter, the data contains the daily electricity use data for a whole year, with 365 values, and the dimension of the data is relatively high. Therefore, to deal with the clustering analysis of such high dimensional data, the dimension reduction should be introduced.

In this research, an auto-encoder based method is proposed to solve the high-dimension problem during the clustering analysis of electricity use data. As for the data pre-processing, the k-nearest neighbour analysis is used to conduct the outlier analysis. One case study with daily electricity data for one year is conducted to illustrate the proposed clustering analysis method. Based on the clustering analysis, the typical daily and monthly electricity pattern in the urban scale are obtained. Also, the peak and valley values of the typical electricity pattern are analysed. The primary results will help with the capacity design of district energy power station. Based on this research, further the model for district electricity use will be established, which will be used for analysis of power integration with renewable resource.

18.2 Methodology

This chapter focuses on the clustering analysis of electricity use data, which possesses high-dimensional characteristic. The data is firstly processed with data cleaning, dimension reduction and outlier analysis. Here, as for the dimension reduction, the proposed improved method of auto-encoder algorithm to extract principle component with much lower dimension. Then the electricity data is clustered according to the principle components. Finally, the evaluation and illustration of each cluster is performed. The clustering results are further used for energy management and demand response analysis. The overall methodology illustration is as shown in Fig. 18.1.

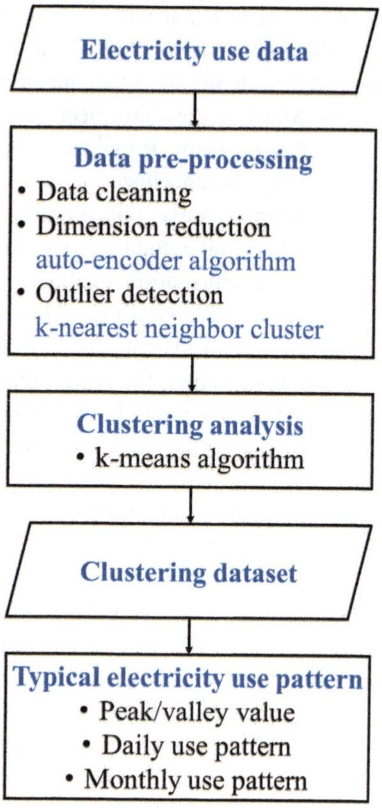

Fig. 18.1 Methodology illustration of auto-encoder based clustering analysis

18.2.1 Data Pre-processing

The data pre-processing is divided into three parts, (1) data cleaning, (2) dimension reduction and (3) outlier detection.

Firstly, data cleaning is required. Because of the errors of measurement and recording, the electricity use data may not be complete. In this chapter, if the data is less than the 90% length of the full observation, this observation will be deleted from the whole dataset. As for the remaining observations, the blank electricity use will be filled with the average daily energy consumption in this month to represent the average electricity use.

Secondly, this research focus on daily electricity consumption data for a certain time, which is considered as high-dimensional. The traditional clustering methods are distance-based or density-based, among which the k-means algorithm (Hartigan and Wong 1979) presents superiority of clustering performance. However, the high-dimensional data may bring about the so-called "curse of dimensionality" to clustering analysis, which means that with the increase of observation's dimension, the Euclidean distance may be invalid to depict the relationship among observations. Many studies have been conducted to deal with this problem. Principal component analysis and shape-based clustering method are proposed by Wen to solve this problem (Wen et al. 2019). The similar methods are listed as wavelet packet decomposition method, discrete Fourier transform, piecewise linear regression, etc. (Khan, et al. 2014; Torriti 2014; Eichinger et al. 2015). In this chapter, the cluster method is improved by the introduction of auto-encoder (Li 2017), which plays the role of dimension reduction.

The auto-encoder is an unsupervised neural network, including three layer-structures, the input layer, output layer and the hidden layer(s). The particularity is that the input layer and the output layer are exactly the same during the model training (Fig. 18.2). Through well training process, normally with backward propagation (Rummelhart 1986), the model can well represent nearly all the information from input layer to output layer. In this way, the useful information is extracted and preserved in the neurons of hidden layers, the information in hidden layers is considered to be adequate for the representation of the almost all the details of input layers. The structure from input layer to hidden layer(s) is called the encoding process, and the structure from hidden layer(s) to output layer is called the decoding process. If the dimension of the hidden layer is less than the dimension of the input observation, then the high-dimensional data is reduced to low-dimensional data in hidden layer.

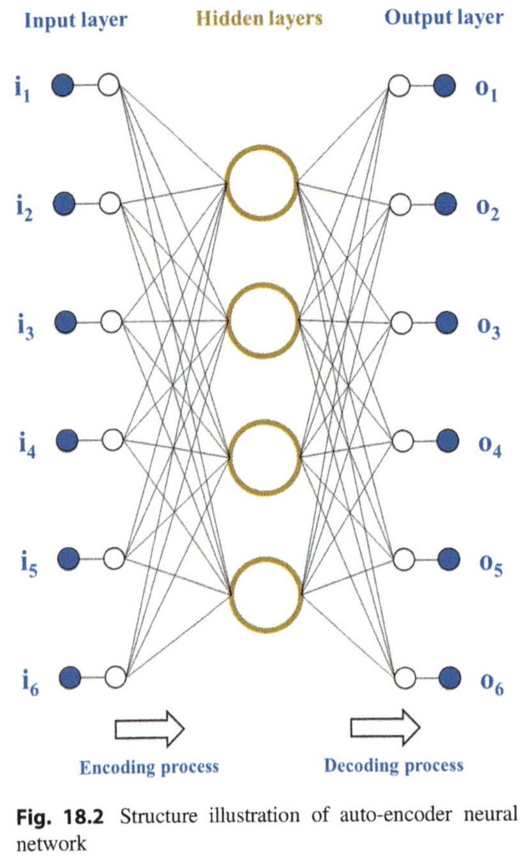

Fig. 18.2 Structure illustration of auto-encoder neural network

According to the dimension scale of the original data, the stacked auto-encoder is introduced when the dimension of the data is rather high. The whole neural network is symmetrical, with stacked auto-encoder networks shown in Fig. 18.2, which indicates multi-time dimension reduction during the model training. Hence, the stacked auto-encoder network reduces the dimension of the data step by step, and the middle of hidden layer is the results of the dimension reduction, as the core hidden layer. The detailed stacked auto-encoder network is further illustrated in Sect. 18.3.

Thirdly, due to the abnormal electricity use may occur no matter related to the malfunction of the smart meter or the consumption boundary confusion, outlier detection is necessary during the data pre-processing. Here, the consumption boundary confusion means that the region classified as residential however the building used as commercial or office functions, and this may occur in the case study. In this chapter, the households amount is also at a large scale, so an efficient outlier detection for raw data with large amount is introduced (Liu 2017).

The outlier detection, also called anomaly detection, is used to detect the abnormal or deviant observations in a dataset. Mainly there are two procedures of the outlier detection, outlier ranking and determination. Normally, an index or score is defined to represent the abnormal and anomalous level of one observation. In this chapter, the outlier detection combines k-nearest neighbours and local outlier factors to conduct the outlier detection (Breunig et al. 2000). Considering the dataset with large amount observations, the k-nearest neighbours will firstly conduct to help with the rough category and reduce the computation cost (Cover and Hart 1967). Then in the scope of their k-nearest neighbours, the index of local outlier factor (LOF) is estimated to represent the "outlier level" of each observation. The "outlier level" depends on how isolated the object is compared with the surrounding neighbours (Breunig et al. 2000). It can be understood as the average of the ratio of the local reachability density of a sample and those of its k-nearest neighbours. The inlier observation's LOF tends to be close to 1, while outliers tend to have a larger LOF value. By this means, the outlier observations of the electricity dataset are recognized and removed.

18.2.2 Clustering Analysis

After dimension reduction processing, the results of lower dimension are used to make clustering analysis. In this chapter, k-means algorithm is recommended. To decide the proper cluster amount, the Calinski-Harabaz index is introduced, under the circumstance that the true cluster label cannot be obtained. The higher index value indicates that the analysis possesses better defined clusters.

18.2.3 Typical Electricity Use Pattern

After the clustering analysis of the dataset, the results for principle components also the original electricity use pattern can be illustrated and further analysed with different temporal resolution. For each cluster, the average daily electricity use data of the households in this cluster can be calculated and analysed. As for the daily electricity consumption profile, the feature indicators like peak value, valley value and average value are analysed. For the monthly electricity consumption data, the peak value and valley value can also be analysed. From the analysis, the relationship between electricity consumption and the season and temperature can be inferred.

18.3 Case Study

In this chapter, the residential electricity use data from Nanjing city, Jiangsu province, in China is used as the case study. The buildings include apartment buildings and detached buildings. The daily consumption data is measured by smart meter from January 1, 2014 to December 31, 2014, lasting for one year in total. The data is recorded in the temporal resolution of 1 h. This dataset is used to depict the clustering analysis based on auto-encoder algorithm, which shows potential of daily electricity consumption data analysis to district power storage capacity design and integration with renewable energy.

18.3.1 Data Pre-processing

For the first step of data pre-processing, the valid days of electricity consumption data are analysed and used for data cleaning. The household with amount of valid days less than 90% will be deleted from the dataset. Figure 18.3 shows the histogram of the statistical results of the amount of the valid days for all the households in the dataset. After the data cleaning, there are 17,031 valid households in the dataset. As for the vacant data, the monthly average electricity consumption will be filled into the blanks.

The auto-encoder network is established based on package called "Scikit-learn" in python (Pedregosa et al. 2011), with the symmetrical structure and same output layer with input layer, however. The stacked auto-encoder is trained through the neural network training, and the detailed parameter settings and network structure is as shown in Table 18.1. The middle hidden-layer compromise 3 neurons, which is the reduced dimension of the electricity use data.

The outlier detection is conducted to avoid the abnormal data due to malfunction of the measurement equipment. The k-nearest neighbours are introduced, and the k is set to be 1,000.

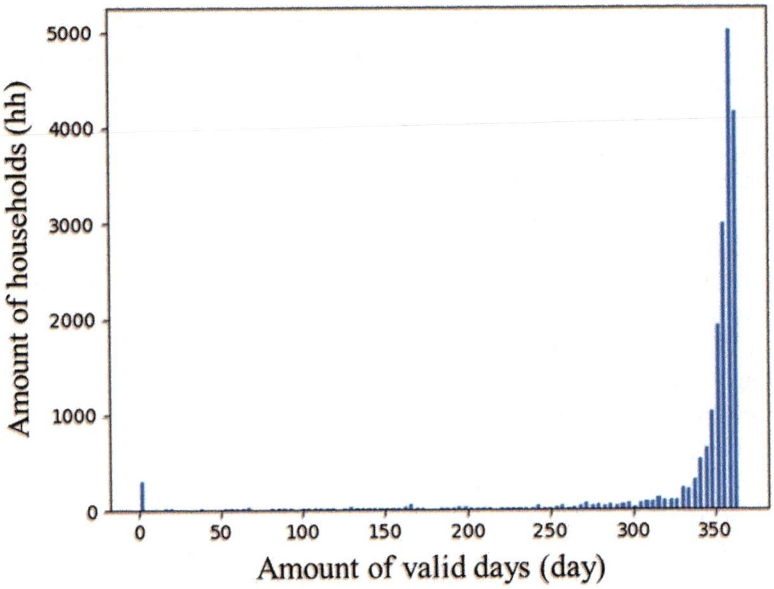

Fig. 18.3 Histogram of amount of valid days of all the households in the original dataset

Table 18.1 Parameter settings and structure of auto-encoder network

Solver	α	Activation function	Network size of hidden layers
Kingma and Ba (2015)	0.001	f(x) = max(0, x)	(180, 60, 12, 3, 12, 60, 180)

The local outlier factor (LOF) for each household is computed according to the 1000 nearest neighbours fold data. The larger LOF value refers to more anomaly. The histogram of LOF values is illustrated in Fig. 18.4. There are 5.9% households with LOF larger than 1.5, and this figure shows the results of LOF under 1.5. After the outlier detection, the inlier households of the electricity consumption data are 14,478 households.

18.3.2 Clustering Analysis

With the training of auto-encoder network, the output layer is close to the input layer, which has already been standardized, and the 3 neurons is considered as the dimension-reduced factors with almost all the key feature of the daily electricity consumption data lasting for one year. The information of middle hidden-layer is named as core layer for easy understanding. The clustering analysis is performed on the core data. Figure 18.5 shows the results for the clustering of the core data. The meaning of each value in the core data is not specific, however, the information from the core layer is adequate to represent the characteristic of the input data. According to the cluster labels for each household, the average daily electricity consumption data is also analysed, which is shown in Fig. 18.6. The proportion and values of each cluster centre are shown in Table 18.2.

18.3.3 Typical Electricity Use Pattern

Based on the cluster analysis results, the feature indicators of electricity consumption in each cluster are analysed. Figure 18.6 has already illustrated the average electricity use pattern for each cluster. For further application of

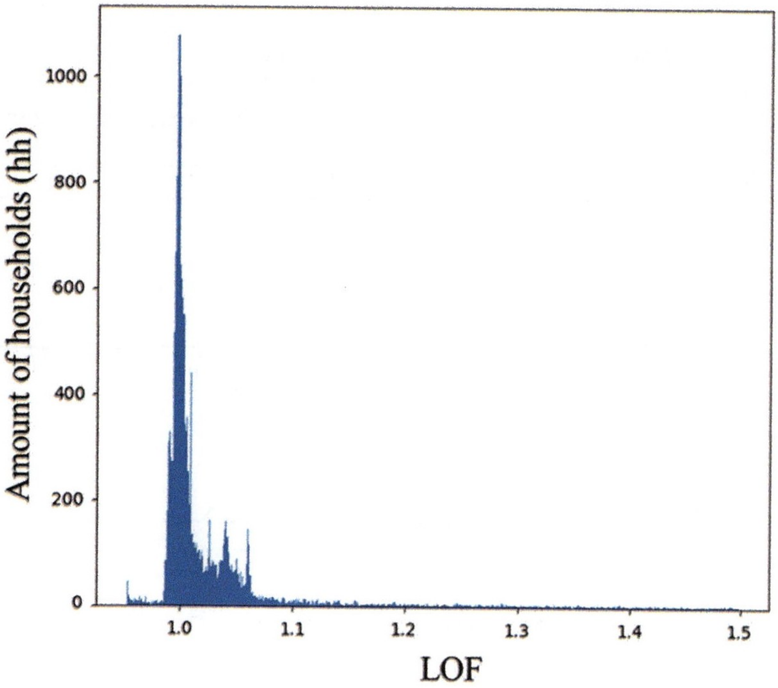

Fig. 18.4 Histogram of local outlier factors during the outlier detection process

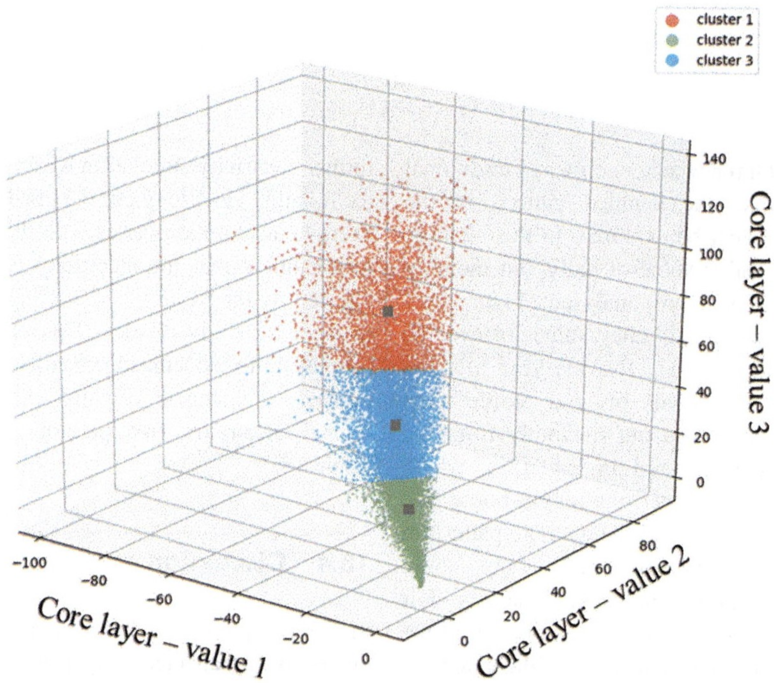

Fig. 18.5 Clustering results of core data

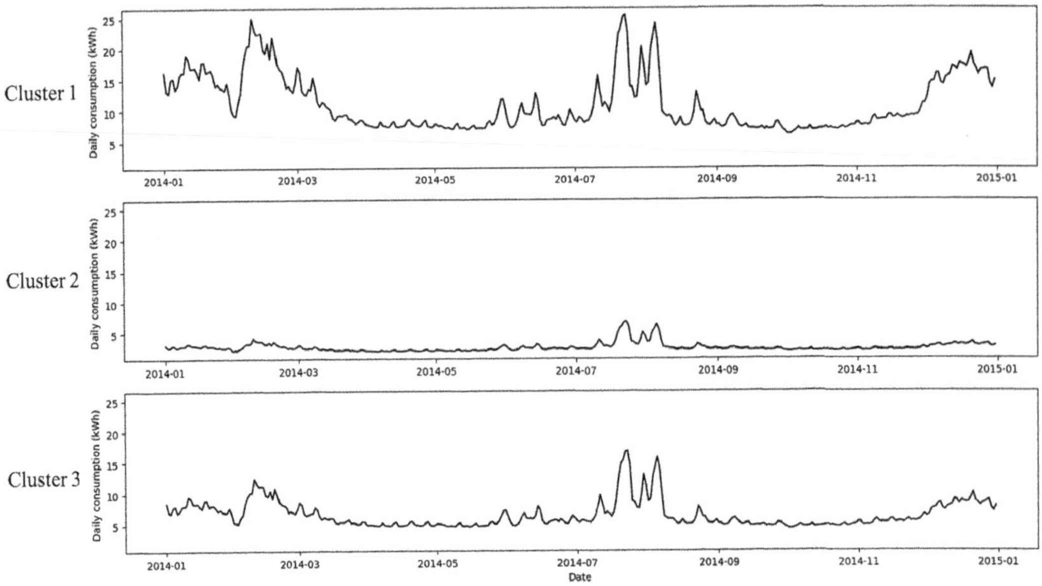

Fig. 18.6 Clustering results of average daily electricity consumption data

Table 18.2 Proportion of each cluster of electricity consumption data

Cluster	Proportion (%)	Core layer value		
		1	2	3
1	21.9	−34.6	34.9	87.1
2	36.6	−8.0	6.2	19.2
3	41.5	−19.9	17.0	50.0

integration with renewable energy and traditional energy resource, the detailed pattern of the electricity use shall be explored. In this chapter, the peak and valley value of daily and monthly electricity use data are analysed. Figure 18.7 illustrates the peak and valley values of average daily electricity use data for one year of three clusters. Ten households of load profile from each cluster are picked out randomly from the raw dataset and depicted in Fig. 18.8, which shows difference between the single household electricity use pattern and the average pattern in one district. The randomness and variety should be considered during the future analysis. For example, the histogram of peak values of all the households in each cluster is illustrated in Fig. 18.9. The peak value of single household is higher than the peak value of average electricity use, which should be carefully considered during further electricity simulation work. In Fig. 18.10, the monthly electricity use data is illustrated. The energy consumption varies with different season, which infers that the electricity use is strongly affected by the heating and cooling consumption in the case study district. This characteristic is promising to be considered for electricity use model establishment. Finally, the summary of typical electricity use pattern are shown in Table 18.3.

18.4 Discussion

In this chapter, the clustering analysis is conducted on urban electricity use data. To depict the relationship among the different clusters, one of analysis of variance, called Duncan's test (Bernhardson 1975; Harter 1960) is performed to

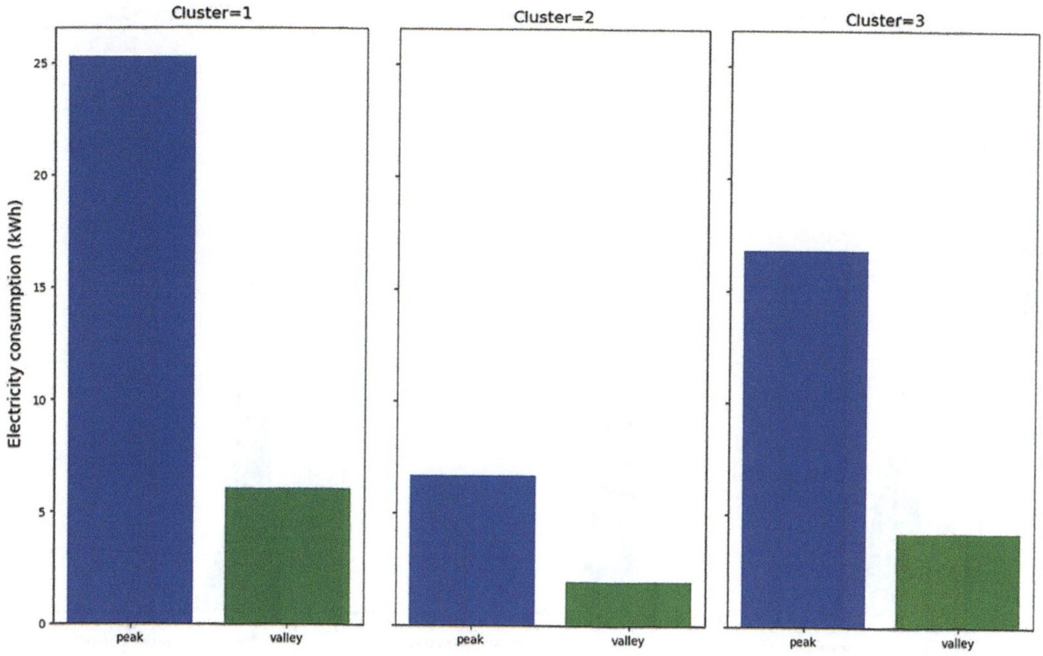

Fig. 18.7 Peak value and valley value of average electricity use data for each cluster

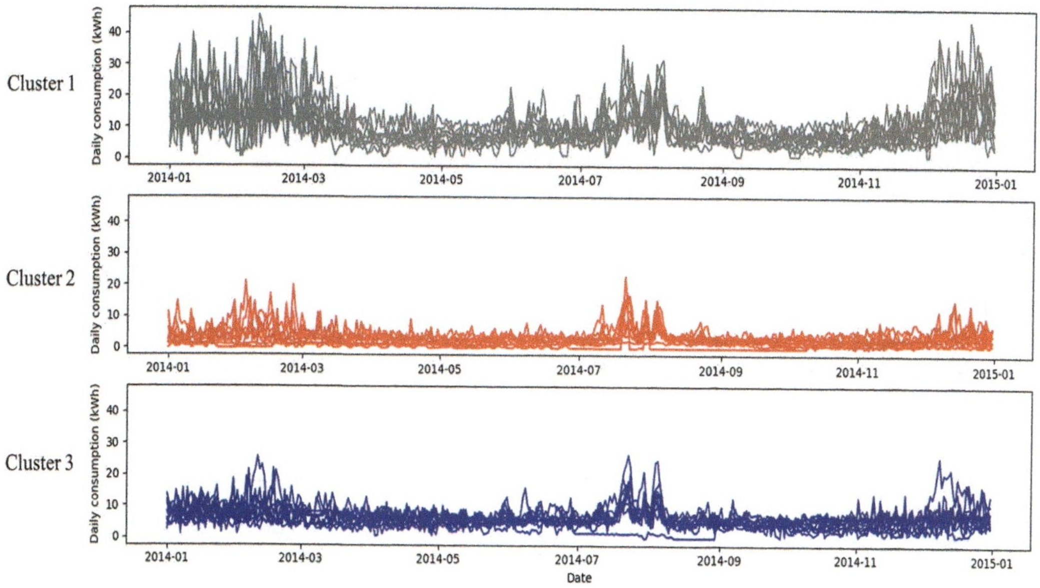

Fig. 18.8 Electricity use data of random ten households for each cluster

the cluster results to analyse the correlation among each cluster centre, which shows the significance of different cluster centres and relationship among them. Normally, the variables are listed in the sequence of the average value. Through Duncan's test, the significance and relationship among different clusters can be analysed, normally according to the average and

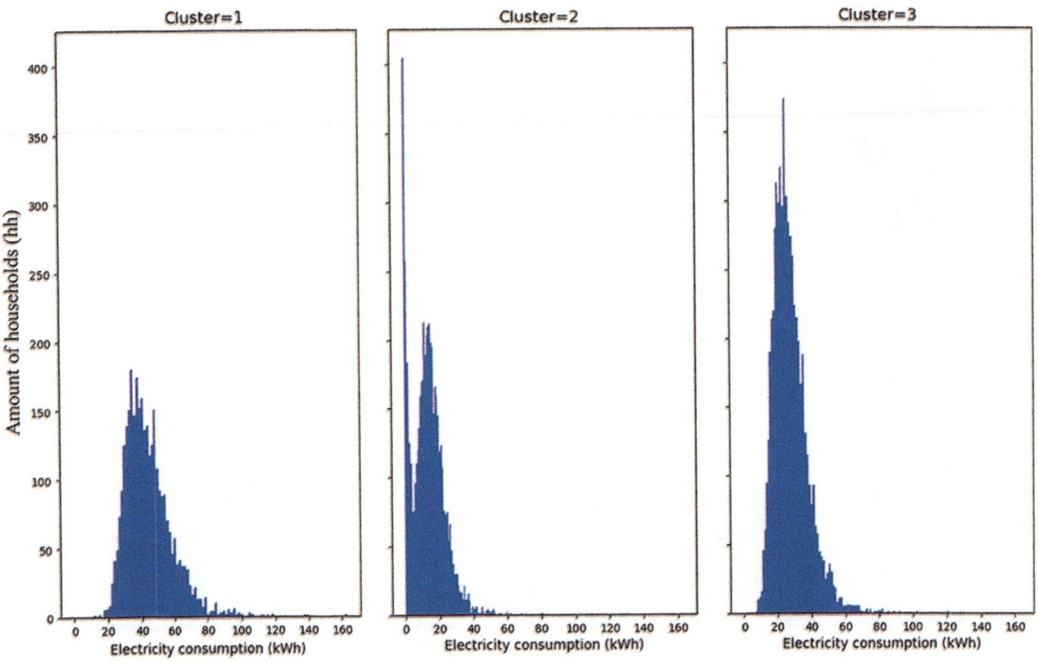

Fig. 18.9 Histogram of peak value of households in each cluster

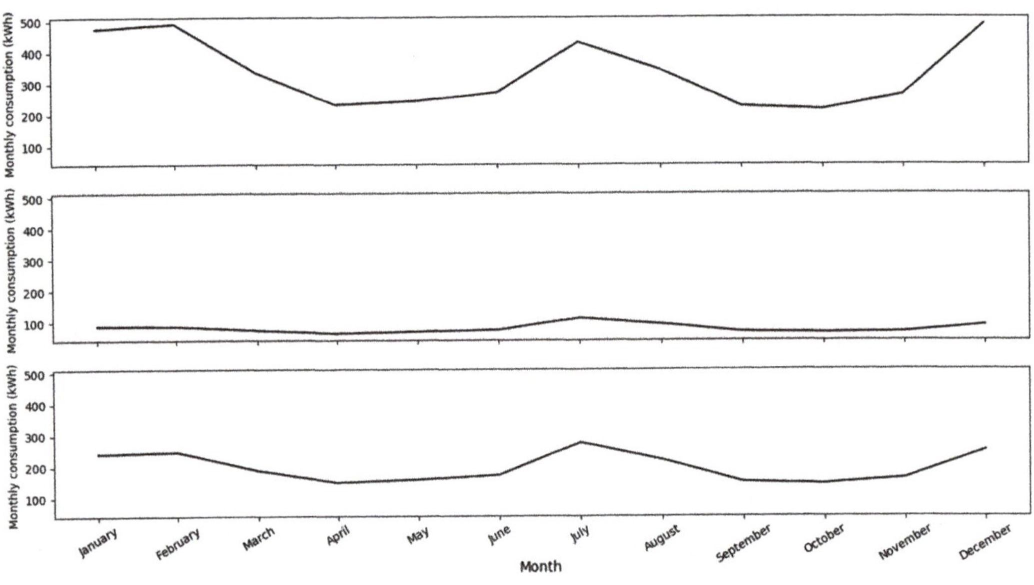

Fig. 18.10 Average monthly electricity use for each cluster

variance values. Table 18.4 shows the results for Duncan's test. The three category under the 0.05 α value shows significant distinction, which verifies the performance of clustering analysis.

This chapter is promising to provide a novel insight of the electricity consumption data analysis. From the temporal resolution of month, the seasonal characteristic can also be analysed.

Table 18.3 Summary of typical electricity use pattern for each cluster

Cluster	Proportion (%)	Average daily electricity consumption (kWh)	Daily peak value (kWh)	Daily valley value (kWh)	Monthly peak value (kWh)	Monthly valley value (kWh)
1	22	10.9	25.3	6.1	488.3	213.0
2	37	2.6	6.7	1.9	110.3	62.1
3	41	6.4	16.8	4.2	274.0	142.7

Table 18.4 Table of Duncan's test results

Cluster	N	$\alpha = 0.05$ category		
		1	2	3
2	365	2.55		
3	365		6.41	
1	365			10.92
Significance		1.000	1.000	1.000

Because the clustering is conducted through daily use data, the seasonal features may depict more relation between the daily use and the seasonal characteristic. More correlation work between the electricity use and weather condition can be studied, and detailed energy use model in residential buildings in urban scale is going be established.

Furthermore, there still exists challenging work to improve. The outlier detection is limited when the outlier data accounts for a larger proportion, so the outlier recognition should be improved. Meanwhile, the cluster analysis should also be combined with more specific and detailed information including the household information and weather condition, so that the analysis will be more solid.

18.5 Summary

This chapter proposes a novel clustering analysis method based on auto-encoder algorithm, to deal with the analysis of daily electricity consumption data. Firstly, the data is pre-processed using a novel method of outlier detection based on k nearest neighbours and local outlier factor. To avoid the "curse of dimensionality" during the clustering analysis, the auto-encoder algorithm is combined with the clustering analysis, which plays a role as dimension reduction. By this means, the electricity data can be analysed and clustered in a more refined temporal resolution, which includes more detailed information. The results of this chapter are promising for further integration between the residential grid and renewable energy resources, such as the power storage capacity design and evaluation of the solar panels for residential buildings.

References

Beccali M et al (2004) Forecasting daily urban electric load profiles using artificial neural networks. Energy Convers Manage 45(18–19):2879–2900

Bernhardson CS (1975) 375: Type I error rates when multiple comparison procedures follow a significant F test of ANOVA. Biometrics 31(1):229–232

Breunig MM et al (2000) LOF: identifying density-based local outliers. In: ACM sigmod record. ACM

Cover TM, Hart PE (1967) Nearest neighbor pattern classification. IEEE Trans Inf Theory 13(1):21–27

Eichinger F et al (2015) A time-series compression technique and its application to the smart grid. VLDB J—Int J Very Large Data Bases 24(2):193–218

Godina R et al (2018) Model predictive control home energy management and optimization strategy with demand response. Appl Sci-Basel 8(3):19

Granell R, Axon CJ, Wallom DCH (2015) Clustering disaggregated load profiles using a Dirichlet process mixture model. Energy Convers Manage 92:507–516

Harter HL (1960) Critical values for Duncan's new multiple range test. In: Aerospace Research Labs Wright-Patterson AFB OH

Hartigan JA, Wong MA (1979) Algorithm AS 136: A K-means clustering algorithm. J R Stat Soc Ser C (Appl Stat) 28(1):100–108

IEA, Electricity (2018)

Khan J et al (2014) PMU data analysis in smart grid using WPD. In: IECON 2014–40th annual conference of the IEEE Industrial Electronics Society. IEEE, New York

Kingma D, Ba L (2015) Adam: a method for stochastic optimization

Kong WC et al (2018) An extensible approach for non-intrusive load disaggregation with smart meter data. IEEE Trans Smart Grid 9(4):3362–3372

Li C et al (2017) Building energy consumption prediction: An extreme deep learning approach. Energies 10 (10):1525

Liu H et al (2017) Efficient outlier detection for high-dimensional data. IEEE Trans Syst Man Cybern: Syst 99:1–11

Li XL, Bowers CP, Schnier T (2010) Classification of energy consumption in buildings with outlier detection. IEEE Trans Industr Electron 57(11):3639–3644

Nan SB, Zhou M, Li GY (2018) Optimal residential community demand response scheduling in smart grid. Appl Energy 210:1280–1289

Pedregosa F et al (2011) Scikit-learn: machine learning in Python. J Mach Learn Res 12:2825–2830

Ponocko J, Milanovic JV (2017) Application of data analytics for advanced demand profiling of residential load using smart meter data. In: 2017 IEEE Manchester Powertech

Qi Y et al (2017) Load pattern recognition method based on fuzzy clustering and decision tree. In: 2017 IEEE conference on energy internet and energy system integration (EI2). IEEE, New York

Rummelhart DE (1986) Learning internal representations by error propagation. Parallel Distrib Process: I. Found 318–362

Ryu S et al (2016) Customer load pattern analysis using clustering techniques. KEPCO J Electr Power Energy 2(1):61–69

Setlhaolo D, Xia XH, Zhang JF (2014) Optimal scheduling of household appliances for demand response. Electr Power Syst Res 116:24–28

Stoll H (1989) Least-cost electric utility planning. Wiley, New York

Teeraratkul T, O'Neill D, Lall S (2018) Shape-based approach to household electric load curve clustering and prediction. IEEE Trans Smart Grid 9(5):5196–5206

Torriti J (2014) A review of time use models of residential electricity demand. Renew Sustain Energy Rev 37:265–272

Valor E, Meneu V, Caselles V (2001) Daily air temperature and electricity load in Spain. J Appl Meteorol 40 (8):1413–1421

Vázquez-Canteli JR, Nagy Z (2019) Reinforcement learning for demand response: a review of algorithms and modeling techniques. Appl Energy 235:1072–1089

Wen LL, Zhou KL, Yang SL (2019) A shape-based clustering method for pattern recognition of residential electricity consumption. J Clean Prod 212:475–488

Zhou K, Yang S, Shao Z (2017) Household monthly electricity consumption pattern mining: a fuzzy clustering-based model and a case study. J Clean Prod 141:900–908

Digital Mapping of Spatial Energy Use for Buildings in City

19

Samer Quintana, Pei Huang, Mengjie Han, and Xingxing Zhang

Abstract

Urban energy mapping plays a crucial role in benchmarking the energy performance of buildings for many stakeholders. This study examined a set of buildings in the city of Borlänge, Sweden, owned by the municipality. The aim was to present a digital spatial mapping of both electricity use and district heating demand. A toolkit for top-down data processing and analysis was considered based on the energy performance database of municipality-owned buildings. The data were initially cleaned and transformed using the Feature Manipulation Engine tool (FME) and then it was geocoded using a python script with an application program interface (API) for OpenStreetMap. The dataset consists of 221 and 89 geocoded addresses for, respectively, electricity and district heating monthly consumption for the year 2018. The electricity use and heating demand in the building samples were about 24.06 kWh/m^2 and 190.99 kWh/m^2 respectively, for which great potential for saving heating energy was observed. The digital mapping revealed a spatial representation of identifiable hotspots for electricity uses in high-occupancy/density areas and for district heating needs in districts with buildings mostly constructed before 1980. These results will provide a comprehensive means of understanding the existing energy distributions to stakeholders and energy advisors. They also facilitate strategy geared towards future energy planning in the city, such as energy benchmarking policies.

Keywords

Energy mapping · GIS · Geocoding · UBEM

S. Quintana (✉) · P. Huang · X. Zhang
Department of Energy and Community Buildings, Dalarna University, 79188 Falun, Sweden
e-mail: ssq@du.se; samer.quintana@angstrom.uu.se

P. Huang
e-mail: phn@du.se

X. Zhang
e-mail: xza@du.se

S. Quintana
Department of Engineering Sciences, Uppsala University, Uppsala, Sweden

M. Han
Micro Data Analysis, Dalarna University, 79188 Falun, Sweden
e-mail: mea@du.se

19.1 Introduction

Buildings represent large energy end-users worldwide. In the E.U. and U.S, buildings currently consume over 40% of total primary energy usage (Huang et al. 2020). With sights set in the

X. Zhang (ed.), *Data-driven Analytics for Sustainable Buildings and Cities*,
Sustainable Development Goals Series, https://doi.org/10.1007/978-981-16-2778-1_19

new paradigm shift regarding energy production, efficiency and climate change, Sweden will implement strategies to reach national targets of energy efficiency in the building sector by 2050. According to these targets energy use per square metre should decrease by 20% by 2020 and 50% by 2050, in comparison with use in 1995. This is a national target for energy efficiency in the housing sector (Ministry of Sustainable Development Sweden 2006). In 2010, over 50% of the world's population were living in urban areas. By 2050, this number is expected to reach 75% (UN-Habitat 2009). Urban development and the expansion of cities, through the modification of land uses (from natural to artificial) modify the local energy budget and wind patterns. Such a transformation has significantly changed the microenvironment and the related energy usage in urban cities (Torabi Moghadam et al. 2019). The mapping of urban building energy plays a crucial role in understanding the multitude of agents that take part in the energy performance of buildings, which will set up the benchmarks in different districts for various stakeholders.

In the context of sustainable cities, spatial visualization is a very effective approach that can help decision-makers in the urban planning process to create future energy transition strategies and implement energy efficiency and renewable energy technologies. Geographic Information System (GIS) techniques can be used for visualizing the energy demand or production in buildings, from urban to regional, or even to a national scale. Some of these visualization techniques are: the thematic 2D map (Mhalas et al. 2013); the 'hit maps' (i.e., aggregated data in 3D charts) (Murugesan et al. 2015); the 3D city models with semantic objects (Gröger and Plümer 2012). There are many studies using GIS techniques to visualize the energy data in building stocks. For instance, Mattinen et al. developed a method for estimating and visualizing the energy use and greenhouse gas emissions from a residential building stock located in Kaukajärvi district, Finland (Mattinen et al. 2014). Using such visualization model, they also analysed the impacts of behavioural and technical changes on the energy performance in the building stock. Finney et al. made a comprehensive mapping of heat sources and sinks in Sheffield City, the UK (Finney 2013). Based on the heat source mapping, they linked these smaller systems to create a combined-heat-and-power based urban-scale network of energy generation and delivery. Huang et al. used GIS technique to obtain the roof area in Kowloon district in Hong Kong. Using the obtained roof area, they evaluated the solar power potential of the whole district by installing rooftop PV panels. Based on the mapped solar power potentials, they developed an optimal design method to sit the public charging stations (Huang et al. 2019). Similarly, Ramachandra and Shruthi used the GIS technique to map the wind energy resources of Karnataka state, India. Based on the wind power mapping, they analysed their variability considering spatial and seasonal aspects (Ramachandra and Shruthi 2005).

In Swan and Ugursal's study, the modelling approaches for energy consumption in a number of buildings were classified into bottom-up or top-down approaches (Swan and Ugursal 2009). The bottom-up approach is more appropriate when there is a need for evaluating the energy consumption based on a high detailed level of data and the ability to model technological systems (Kavgic et al. 2010). Bottom-up models can be divided into two types: deterministic (or engineering) and statistical. The statistical methods search for correlations, utilizing a sample of information in energy bills as a source of data for energy modelling and analysing the link between energy consumption and a range of different variables (e.g. building shape, age, and occupant behaviour) (Nouvel et al. 2015). They can also consider socioeconomic effects in the equations. They calculate reliable consumption based on the available information on the current status of buildings. However, due to their strong dependency on available historical consumption data, these methods are restricted to predict the impact of new technology options and energy saving potential after applying refurbishment measures (Torabi Moghadam et al. 2018). The deterministic methods are detailed models which are based on thermodynamic relationships and heat transfer calculations (Bruse and Fleer 1998). The main advantage of an engineering-based method is the ability of predicting energy saving potentials for buildings if some renovation measures are to be implemented (Mauree et al. 2017). These modelling approaches require a large amount of information about the building structures and parametric input for estimating the energy usage of a set of reference buildings of the stock based on a numerical model. Additionally, the evaluation of urban planning scenarios is computationally extensive, and the availability of construction and geometrical data needed as input for the models is very scarce. The top-down approach treat the entire residential sector as one energy sink. The top-down methods are suitable for a large-scale analysis and not for the identification of the possible improvements at the building at urban and local levels (United Nations 2015). Compared with the bottom-up approaches, the top-down methods are relatively easy to develop based on the limited information provided by macroeconomic indicators such as price and income, technology development pace, and climate. As summarized by Swan and Ugursal, the top-down approach has advantages including long-term forecasting in the absence of any discontinuity, inclusion of macroeconomic and socioeconomic effects, simple input information required and encompasses trends (Swan and Ugursal 2009).

Although there are existing studies in mapping energy uses in different cities, spatial energy analysis in local municipality are necessary as they will be different in various city and culture contexts. Specific consideration should be paid to the differences between cities when it aims to optimize the integration of urban energy systems operated in buildings, and promote renovation and renewable energy systems. This is because cities differ from each other at the local, national and international levels in the perspectives of geography, socio-economy, culture, infrastructure, and information platform. The types of cities and districts will determine the kind of users and needs, and consequently the nature (qualitative and quantitative) of the policy/regulation schemes and the calibration/adjustment of the energy infrastructures. The citizen's behaviours and needs/preferences of energy may be different from each other in different cities, which can lead to a great difference in the energy demand. Within the same framework of transforming to sustainable and liveable city, different areas must not only adopt standardized approaches, but also consider the specificities at the local level. Dedicated research into local city and district is therefore of paramount importance to ensure the proper mix between international/national scenarios and local measures.

The urban energy mapping and analysis for Borlänge city have not yet been done. Therefore, this study aims to cover the research gap by studying a set of buildings owned by the municipality of Borlänge, Sweden. The initial step of the study is to give a spatial mapping of both electricity use and district heating demand. A top-down approach was considered based on the energy consumption data of the municipality-owned buildings. It was expected that this study would be able to provide insights that allow the understanding of the existing local energy distributions. It also facilitates strategy geared towards future energy planning in this city.

This chapter is structured as follows: methods illustrates the data source and the methodology used to process the data; in results and discussion, both statistical and spatial analysis are presented; a conclusion is further depicted after.

19.2 Methods

19.2.1 Data Sources

Acquiring the necessary data to create an urban model can be a difficult endeavour. New general data protection regulation laws (GDPR) by the European Parliament regulates how the data must be acquired, handled and stored while protecting the privacy of the individuals (European Parliament 2016). Energy consumption data is sensitive information that fall into the new regulation, greatly complicating the data acquisition. Depending on the data resolution, storing the information can be complicated, may be not kept for large periods of time or stored in obsoletes systems making difficult to be used.

The primary source of data used for this model is provided by Tunabyggen, a municipality owned company that constructs, manages and rents a set of buildings in the Borlänge municipality. The data is provided in PDF format, containing a total number of 375 pages of monthly data for electricity demand, district heating and hot water flow rate for the year 2018. The geographical information is obtained from the official Swedish surveying institution, Lantmäteriet, specifically, the vector data for the property information and LiDAR data for the Borlänge municipality. Other social statistics and specific data such as building year of construction, percentage of occupation, demographics and typologies are acquired from *hitta.se*, which is a Swedish search engine that offers telephone directory, addresses and maps. To complete and validate the model, it is necessary some extra information that was obtained by visual inspection such as the number of floors, area and shape of the roofs. The flowchart, Fig. 19.1, further describes the processes, databases and validation operations.

19.2.2 Data Extraction

The first step in the process is to extract the information from the data source provided. The PDF archaic data structure format must be transformed into a common format that can be used by other applications. In order to extract the data a custom python script is written to parse out the information. Then, the data is further inspected for missing data and error correction. From the 375 pages in PDF format, a total of 262 addresses and 463 entry points of monthly data for electricity (kWh), district heating (MWh) and flow rate (m^3) for the year 2018 were extracted.

19.2.3 Geocoding

The extracted addresses from the data source are further expanded to include the city and the country information. Then, it is run through a python script, using an application program interface (API) for OpenStreetMap, Fig. 19.2, gives the script that uses pandas, geopy libraries. In parallel, another script was used to connect to Google Maps API geocoding services. Two outputs from each geocoding service are obtained with the longitude and latitudes of the addresses. The output format for the coordinate system is the standard LL-WGS84. The location for a total of 222 out of the 262 entry points were found on the first iteration.

19.2.3.1 Geocoding Validation

The results are plotted and further inspected for validation. During this process, the locations are geocoded and manually centred in the property area, as displayed in Fig. 19.3. The green dots are the geocoded locations and the brown dots are the manually centred locations. The output becomes to 238 out of the 262 total addresses, leaving a total of 24 addresses and 31 entry points that are not able to be geocoded due to unspecific naming until manual visual inspection and analysis of the context is performed. The final result generates a total of 250 geocoded addresses and 12 unclarified ones.

19.2.3.2 Area Merger Code, Area Validation

Next parameters are extracted from the Swedish survey database Lantmäteriet. The building property vector information is provided in a shapefile (.shp) format, a digital vector storage format for storing geometric location and associated attribute information.

Using the Feature Manipulation Engine (FME) tool, shown in Fig. 19.4, it is possible to extract and calculate the areas for the geocoded addresses points. This information is compared to the visual inspected area, in order to analyse its accuracy. The extra information stored in the shapefile is incorporated to the dataset. This information includes a building description, coordinates in the Swedish reference system SWEREF-99-TM and a unique object identity.

Fig. 19.1 Flowchart for data processing, extraction, geocoding and validation

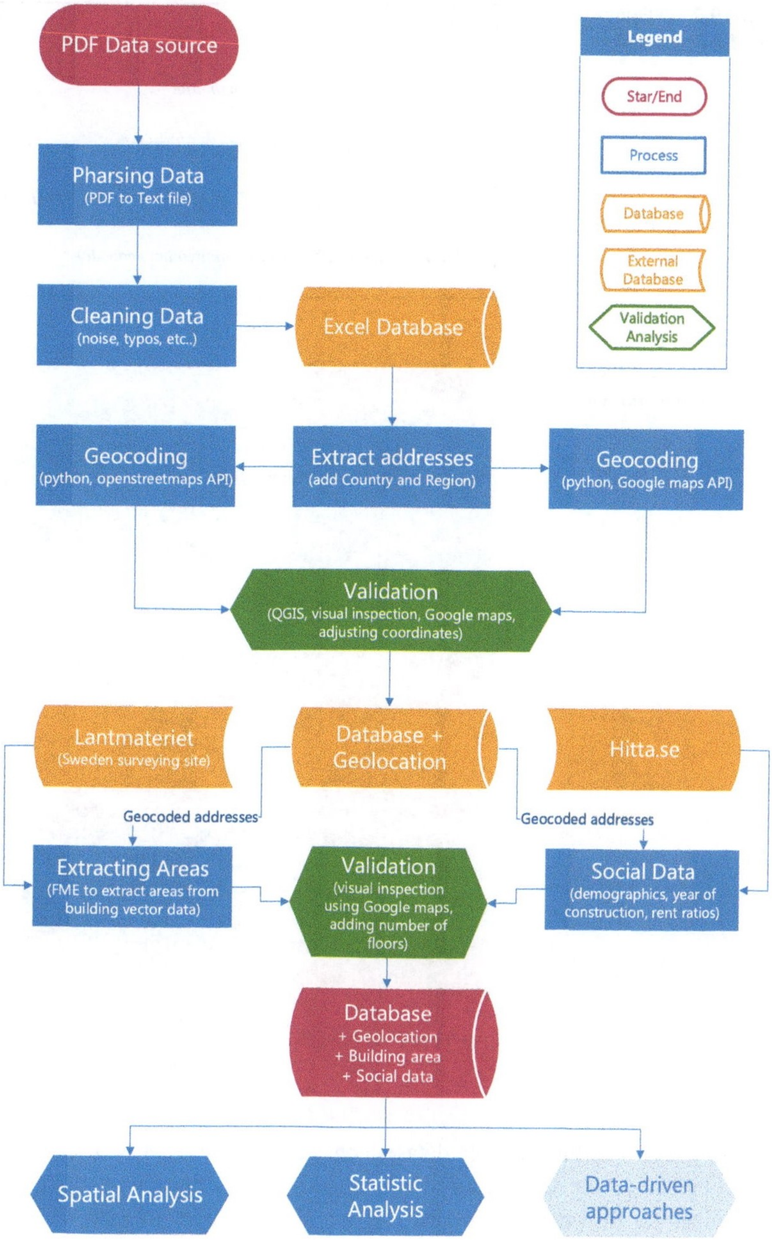

19.2.3.3 Data Processing

All the different sources of information are finally combined together and inspected for errors or inconsistencies. The total building area is calculated using the number of floors and the buildings vector surface areas. Finally, the results for the energy consumption, electricity and district heating in kWh/m^2 for the year 2018 were obtained.

From the 250 addresses that were geocoded, 28 addresses were excluded from the analysed dataset due to missing, erroneous or abnormal

Fig. 19.2 Python script for OSM API geocoder

```
df=pandas.read_excel ("addresses.xlsx")

from geopy.geocoders import Nominatim

nom=Nominatim(user_agent="my-application")

df["address"]=df["address"]+", "+df["City"]+", "+df["State"]+", "+df["Country"]

df["Coodinates"]=df["address"].apply(nom.geocode)
```

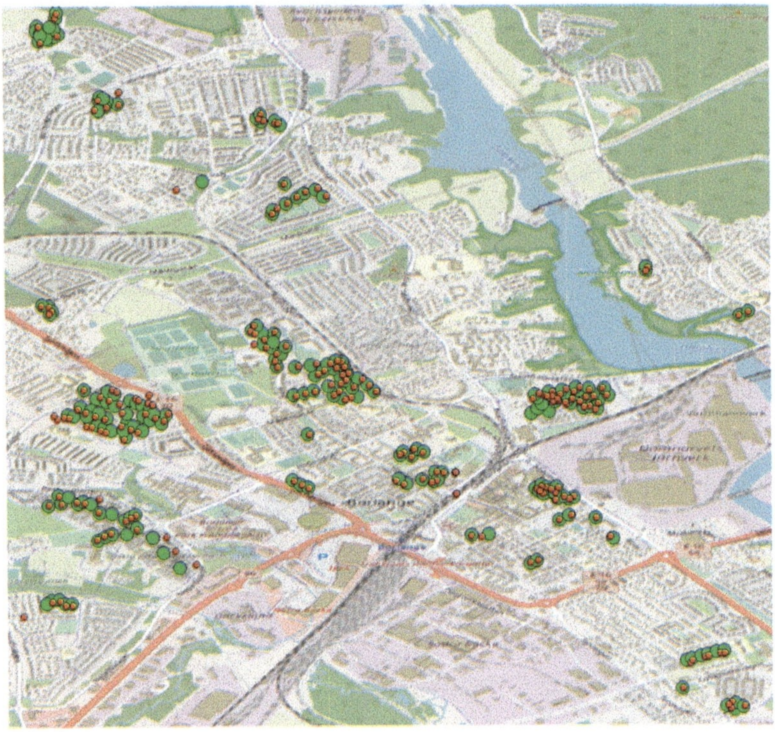

Fig. 19.3 Geocoded data will adjusted coordinates

information. The final sample dataset consists of 221 buildings for the electricity data and 89 buildings for district heating data.

19.3 Results and Discussion

19.3.1 Statistic Data Analysis

In the considered building samples, all of the buildings are residential buildings and the related facility buildings (such as laundries, storage, etc.). The energy use is normalized by dividing the energy use by the heated floor area. The definition of the heated or living floor area has a large impact on the magnitude of the area-specific energy requirement. In Sweden, the heated floor area is defined as the floor area that is heated more than 10 °C. As a result, in this study, we assume the heated floor area is averagely 87% of the total external floor area for the analysis (Mata and Kalagasidis 2009). In addition, electricity demand is further normalized by considering the occupancy ratio of each building. For the heating demand there is no need to consider the occupation ratio, as it is common in

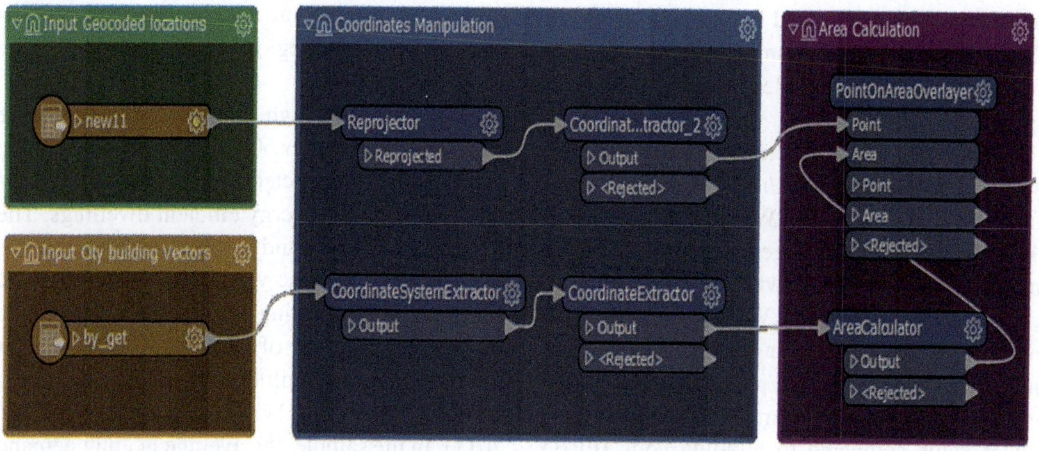

Fig. 19.4 Processes in FME area merger workflow

Sweden for the heating systems to stay on, even if there is no occupancy in the building.

The annual electricity demand for lighting and appliances in the building samples are illustrated in Fig. 19.5.

The mean electricity demand of 221 building samples was 24.06 kWh/m^2, with a total range from minimum 0.02 kWh/m^2 to maximum 189.89 kWh/m^2. Comparing to the average electricity demand of 30–36 kWh/m^2 in Swedish context (Mata et al. 2013), the average electricity demand of the building samples is reasonably low. The median electricity demand was 12.72 kWh/m^2, which means that 50% of the building samples demand less electricity than this value. Furthermore, over 76% of the building samples achieves lower electricity use than 30 kWh/m^2.

The Swedish Housing Agency's building rules (Boverket 2011) stipulate requirements for the energy performance of buildings depending on their use, end-use heating system and climate zones. The energy performance (heating demand) requirements are given as the specific energy use, comprising the purchased energy for space heating, domestic hot water and electricity for fans and pumps but excluding electricity for

Fig. 19.5 Annual electricity demand for building samples

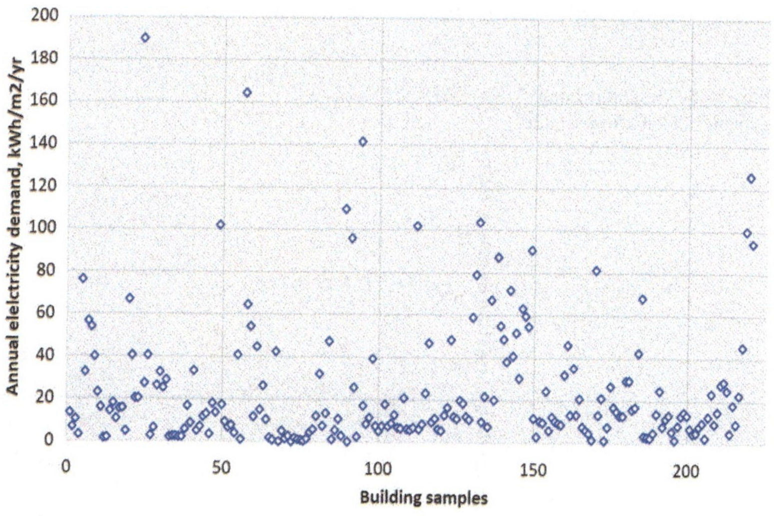

household appliances and lighting (Dodoo and Gustavsson 2014). The annual heating demand for the building samples are displayed in Fig. 19.6.

The mean heating demand of 89 building samples was 190.99 kWh/m^2, with a total range from minimum 0.41 kWh/m^2 to maximum 492.52 kWh/m^2. Borlänge city belongs to climate zone II in Sweden, where the new building code requires up to 110 kWh/m^2 energy use for non-electric heated buildings (with district heating) annually. In addition, passive houses criteria even have higher requirements with up to 35% lower value compared to building code (FEBY 2012). Thus, the average heating demand in the building samples was much higher than either the building code or the passive house standard, about twice the requirement stipulated by building code, and three times of the requirement of the passive house standard. The median heating demand was 145.43 kWh/m^2, which means that 50% of the building samples demand less heating than this value. Approximately 25% of the building samples achieves lower heating demand than 110 kWh/m^2. The difference between the different counties is clear. In Gävleborg, it was found that the average heating demand was about 185 kWh/m^2 in 2010. When across the whole Sweden, the average annual energy use for heating in single- or two-dwelling was reported at about 158 kWh/m^2 per year in 2014 (Swedish Energy Agency 2015). So, the heating use in Borlänge city stays at a high level when compared to the closed regions and the average figure over the country.

However, this high energy demand can be understood since over 60% of the buildings in the sample were constructed before 1980, and therefore it may not be energy efficient dwellings. The annual heating demand average varies considerably depending on the year of construction of the building. For buildings built after 1980, the heating demand was of about 97–98 kWh/m^2 in 2004, while those built before 1980 used heating from 120 to 133 kWh/m^2 per year (Pallardó 2011). In the sample, the average heating demand for buildings constructed before 1980 was about 187.98 kWh/m^2 per year, where these buildings account for 90,651 m^2 of the heated floor area. So, there is great potential (about 4532–5439 MWh/year) for these buildings built before 1980 to improve their energy performance through renovations such as, increasing the thermal insulation of the walls/roofs, upgrading windows and heating radiators for example.

19.3.2 Spatial Data Analysis

A digital mapping method was applied in this study to compile and format the energy data into a virtual image and thus to produce a general map of energy use in Borlänge city based on building samples, which offers appropriate

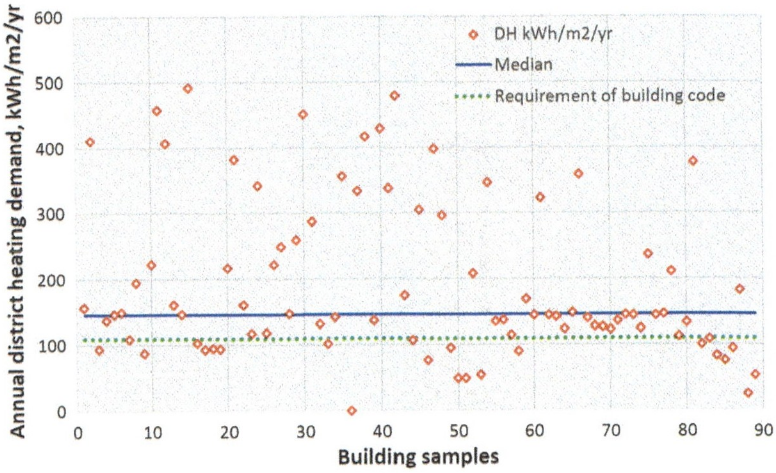

Fig. 19.6 Annual heating demand for building samples

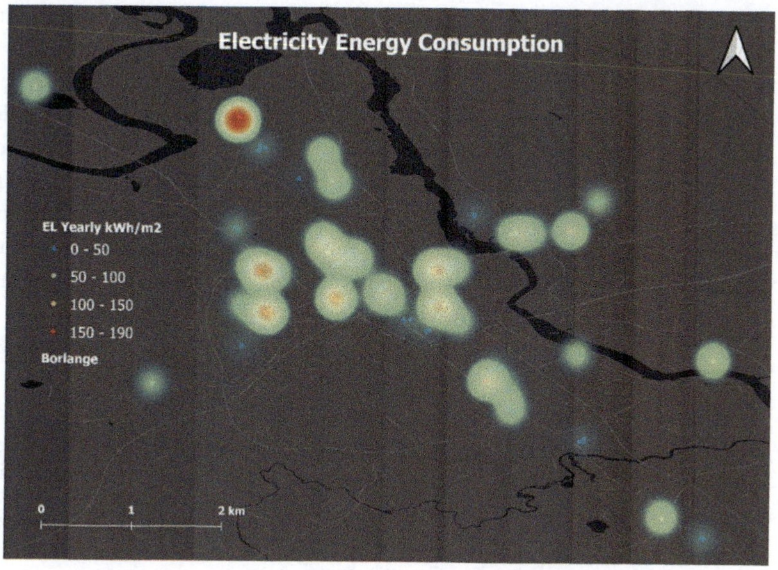

Fig. 19.7 Digital mapping of electricity use in Borlänge city based on building samples

representations of the dedicated areas and districts. Using a Geographic Information System (GIS) tool—QGIS, it was able to visualize the sample energy data on the spatial map of Borlänge. Using the yearly electricity and heating demand in the unit of kWh/m^2 as the weight factor, longitude and latitude of the addresses, two digital maps were generated as shown in Figs. 19.7 and 19.8, respectively for electricity use and heating demand.

These digital maps provide an interactive and scalable way of visualizing the energy use across the city, which is used to spot abnormalities or faulty energy data points. These maps also illustrate a spatial idea of identifiable hotspots for electricity uses in high-occupancy/dense areas. For district heating demands it shows hotspots with buildings mostly constructed before 1980. For instance, some of the hotspots can be easily identified as several student's accommodation

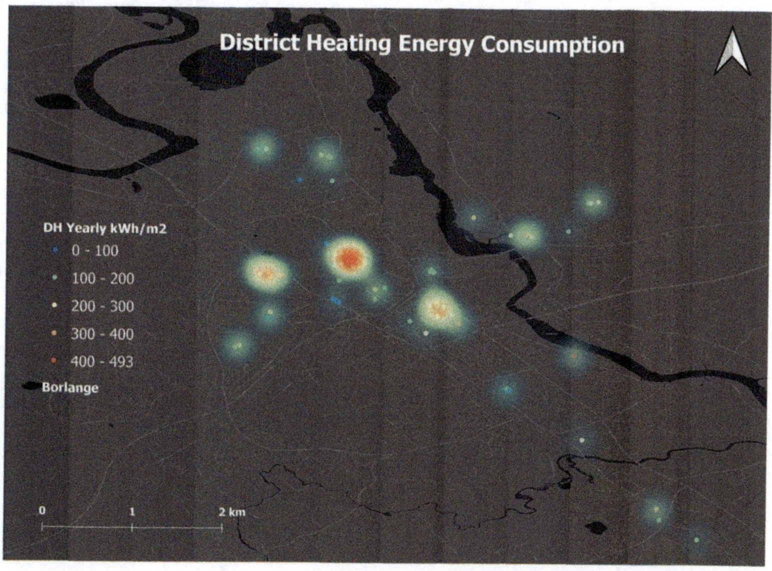

Fig. 19.8 Digital mapping of heating demand in Borlänge city based on building samples

areas in the northwest quadrant. These highly dense buildings have high electricity consumption since the occupants remain indoor for the most learning and living activities, but at the same time these buildings have relatively low heating needs as the buildings are well maintained and insulated. It is observed from these two maps that electricity use is mainly relied on the occupancy density, where higher population per floor area usually results in higher electricity use. On the other hand, district heating demand is dependent on the building itself, where poorly-insulated building leads to higher heating need. As a result, electricity use and heating demand do not always appear in the same district/area since they are influenced by different parameters. This offers clear insights for planning of urban energy infrastructure and distributions, as well as the potential contributions from local renewable energy source (RES) systems. For instance, more electricity distribution or RES power generation is necessary for high-dense residential areas, while higher heating should be distributed to those areas with buildings mostly constructed before 1980.

19.4 Conclusion

A dedicated spatial analysis of both electricity use and district heating demand in a Swedish local-city context was completed, through a toolkit for top-down digital mapping. The average electricity demand in Borlänge building samples was 24.06 kWh/m^2, which was reasonably lower than the average value in Sweden. The mean value of heating of the building samples was 190.99 kWh/m^2, which was much higher than either the building code or the passive house standard. The heating use in Borlänge city stays at a high level when compared to the closed regions and the average figure over the country. In particular, there are great potentials (about 4532–5439 MWh/year) for the buildings built before 1980 to improve the energy performance.

The digital maps provided a spatial representation of the identifiable hotspots for electricity uses in high-occupancy/density areas and for district heating needs in districts with buildings mostly constructed before 1980. Electricity use and heating demand do not always appear in the same district/area since they are influenced by different parameters. The overall result offers clear insights for the planning of urban energy infrastructure and distributions, as well as a potential contribution from local RES implementation.

References

Boverket (2011) Boverket's mandatory provisions and general recommendations, BBR, BFS 2011:6 with amendments up to BFS 2018:4, p 154

Bruse M, Fleer H (1998) Simulating surface–plant–air interactions inside urban environments with a three dimensional numerical model. Environ Model Softw 13(3):373–384. https://doi.org/10.1016/S1364-8152(98)00042-5

Dodoo A, Gustavsson L (2014) Effect of energy efficiency requirements for residential buildings in Sweden on lifecycle primary energy use. Energy Proc 61:1183–1186. https://doi.org/10.1016/j.egypro.2014.11.1049

European Parliament (2016) Regulation (EU) 2016/ 679 of the European Parliament and of the council—of 27 April 2016—on the protection of natural persons with regard to the processing of personal data and on the free movement of such data, and repealing Directive 95/46/EC (General Data Protection Regulation), p 88

FEBY (2012) Specification for zero energy, passive and low-energy houses

Finney KN et al (2013) Modelling and mapping sustainable heating for cities. Appl Therm Eng 53(2):246–255. https://doi.org/10.1016/j.applthermaleng.2012.04.009

Gröger G, Plümer L (2012) CityGML—interoperable semantic 3D city models. ISPRS J Photogramm Remote Sens 71:12–33. https://doi.org/10.1016/j.isprsjprs.2012.04.004

Huang P, Ma Z, Xiao L, Sun Y (2019) Geographic information system-assisted optimal design of renewable powered electric vehicle charging stations in high-density cities. Appl Energy 255:113855. https://doi.org/10.1016/j.apenergy.2019.113855

Huang P, Lovati M, Zhang X, Bales C (2020) A coordinated control to improve performance for a building cluster with energy storage, electric vehicles, and energy sharing considered. Appl Energy 268:114983. https://doi.org/10.1016/j.apenergy.2020.114983

Kavgic M, Mavrogianni A, Mumovic D, Summerfield A, Stevanovic Z, Djurovic-Petrovic M (2010) A review of bottom-up building stock models for energy consumption in the residential sector. Build Environ

45(7):1683–1697. https://doi.org/10.1016/j.buildenv.2010.01.021
Mata É, Kalagasidis AS (2009) Calculation of energy use in the Swedish housing, p 52
Mata É, Sasic Kalagasidis A, Johnsson F (2013) Energy usage and technical potential for energy saving measures in the Swedish residential building stock. Energy Policy 55:404–414. https://doi.org/10.1016/j.enpol.2012.12.023
Mattinen MK, Heljo J, Vihola J, Kurvinen A, Lehtoranta S, Nissinen A (2014) Modeling and visualization of residential sector energy consumption and greenhouse gas emissions. J Clean Prod 81:70–80. https://doi.org/10.1016/j.jclepro.2014.05.054
Mauree D, Coccolo S, Kaempf J, Scartezzini J-L (2017) Multi-scale modelling to evaluate building energy consumption at the neighbourhood scale. PLoS ONE 12(9):e0183437. https://doi.org/10.1371/journal.pone.0183437
Mhalas A, Kassem M, Crosbie T, Dawood N (2013) A visual energy performance assessment and decision support tool for dwellings. Visual Eng 1. https://doi.org/10.1186/2213-7459-1-7
Ministry of Sustainable Development Sweden (2006) Swedish national roadmap for implementing the European Environmental Technologies Action Plan (ETAP). Regeringskansliet. Available: www.sweden.gov.se
Murugesan LK, Hoda R, Salcic Z (2015) Design criteria for visualization of energy consumption: a systematic literature review. Sustain Cities Soc 18:1–12. https://doi.org/10.1016/j.scs.2015.04.009
Nouvel R, Mastrucci A, Leopold U, Baume O, Coors V, Eicker U (2015) Combining GIS-based statistical and engineering urban heat consumption models: towards a new framework for multi-scale policy support. Energy Build 107:204–212. https://doi.org/10.1016/j.enbuild.2015.08.021
Pallardó GG (2011) Energy consumption in tertiary buildings in Sweden case-study: M-building at Lund University—LTH. Lund University
Ramachandra TV, Shruthi BV (2005) Wind energy potential mapping in Karnataka, India, using GIS. Energy Convers Manage 46(9):1561–1578. https://doi.org/10.1016/j.enconman.2004.07.009
Swan LG, Ugursal VI (2009) Modeling of end-use energy consumption in the residential sector: a review of modeling techniques. Renew Sustain Energy Rev 13(8):1819–1835. https://doi.org/10.1016/j.rser.2008.09.033
Swedish Energy Agency (2020) New regional energy statistics for single- or two-dwelling buildings. http://www.energimyndigheten.se/en/news/2011/new-regional-energy-statistics-for-single–or-two-dwelling-buildings/. Accessed 10 May 2020
Torabi Moghadam S, Toniolo J, Mutani G, Lombardi P (2018) A GIS-statistical approach for assessing built environment energy use at urban scale. Sustain Cities Soc 37:70–84. https://doi.org/10.1016/j.scs.2017.10.002
Torabi Moghadam S, Coccolo S, Mutani G, Lombardi P, Scartezzini J-L, Mauree D (2019) A new clustering and visualization method to evaluate urban heat energy planning scenarios. Cities 88:9–36. https://doi.org/10.1016/j.cities.2018.12.007
United Nations (2015) Global initiative for resources efficient cities. United Nations Environment Programme. Available: http://unep.org/
UN-Habitat (2009) Global report on human settlements 2009: planning sustainable cities. Earthscan: for UN-Habitat, Nairobi

Machine Learning and Artificial Intelligence for Digital Twin to Accelerate Sustainability in Positive Energy Districts

20

Jingchun Shen, Puneet Kumar Saini, and Xingxing Zhang

Abstract

Positive Energy Districts (PED) require integration of different systems and infrastructures for the optimal interactions among buildings, stakeholders, mobility, energy systems and ICT systems. Digital twin is a coupled approach for new forms of modelling and analysis based on big data and machine learning/artificial intelligence, which combines capacities of virtual model, data management, analytics, simulation, system controls, visualization and information sharing. Digital twin is regarded as a potential solution to optimize PEDs. This chapter presents a comprehensive review about digital twins for PED from aspects of concepts, working principles, tools/platform and applications, in order to address the issues of both 'how digital PED twin is made' and 'how digital PED twin optimizes liveability'. Further challenges and opportunities are brought forward for discussion. The outcome of the review is expected to provide useful information for optimizing the liveability of the urban environment in line with social, economic and environmental sustainability.

Keywords

Positive energy districts · Digital twin

20.1 Introduction

According to European Strategic Energy Technology (SET) Plan Action 3.2, positive energy districts (PED) are the essential part of comprehensive approaches towards sustainable urbanization including technology, spatial, regulatory, financial, legal, social and economic perspectives (Europe 2019). Urban development is moving from building solutions to PEDs in order to achieve EU's energy and climate targets. It has planned more than 100 PEDs by 2025 in EU, which will require coupled and dynamic interaction among buildings, users and energy systems, mobility and ICT infrastructures (SET-Plan action 3.2 2018). PEDs are defined as energy-efficient and energy-flexible urban areas with surplus renewable energy production and net zero greenhouse gas emissions. Active information exchange and analysis will be necessary that enable balancing and optimization, peak shaving,

J. Shen (✉) · Puneet KumarSaini · X. Zhang
Department of Energy and Community Buildings, Dalarna University, 79188 Falun, Sweden
e-mail: jih@du.se

Puneet KumarSaini
e-mail: pks@du.se

X. Zhang
e-mail: xza@du.se

X. Zhang (ed.), *Data-driven Analytics for Sustainable Buildings and Cities*,
Sustainable Development Goals Series, https://doi.org/10.1007/978-981-16-2778-1_20

load shifting, demand response and reduced curtailment of renewables, and district-level self-consumption of energy.

The integration of digital methods can be a solution to the challenges in PEDs. Buildings and districts can be designed to be more vibrant, efficient and resilient if they are modelled, analyzed and tested before they are built. Digital twin refers to the creation of digital model/platform by monitoring, modelling and optimizing the PEDs as a complex multi-physics system based on real-time big data sets (Woods and Freas 2019). Digital twin integrates internet of things (IoT), artificial intelligence (AI), machine learning and analytics, to create living digital simulation model that update and change as needed. A digital twin model continuously learns and updates itself from multiple sources to represent its near real-time status.

Digital twin of PEDs enables a revolutionary way to accelerate sustainability of the society, in terms of energy transition, circular economic and climate change. In a digital twin platform, sensors will be set up to collect all kind of information, such as occupancy (mobility), temperature, moisture, energy consumption, renewable production, CO_2 concentration, costs, waste, carbon footprint, etc., creating the 'brain of PEDs'. With such big data sets, digital twin model can be used to assess energy demand/supply, indoor air quality, thermal comfort, CO_2 emissions, expenses for operating and maintenance, building renovation/replacement needs (including recycle of waste construction material), carbon emissions and payback periods of energy saving measures over lifetime. This therefore optimizes PED's three functions in energy efficiency, energy production and flexibility, towards energy surplus and climate neutrality.

Although there are several existing projects and reports about digital twin for PEDs, it significantly lacks of a systematic review and summary about the current R&D status in this area, in order to identity current working limits and future research directions. This chapter therefore presents a comprehensive review about digital twins for PED from aspects of concepts, working principles, tools/platform and applications, in order to address the issues of both "how digital PED twin is made" and "how digital PED twin optimizes livability". Further challenges and opportunities are discussed.

20.2 Concept and Working Principle for PEDs

A digital PED twin usually consists of four important components: a virtual model of PED, sensor network integration, data analytics and stakeholder layer. Figure 20.1 displays a schematic of how digital PED twin is made. The virtual model is a visualization process for a PED that can derive from 3D models extracted from BIM or the custom 3D models of PED. The information and data within a digital PED twin can collected and transfer by various sensor networks to create a real-time monitoring, which often include weather conditions (temperature, solar irradiation etc.), material (new or wasted), cost, carbon emissions and footprint, facility/system status, indoor air quality (IAQ), inhabitant behaviour, electric vehicle (EV) mobility, energy demand and local RES supply. These data sets will be further analysed by machine learning or AI approaches, and exchange actions/decisions with different stakeholders for operations. In this sense, the stakeholders may refer to public institution/government, property owner/manager, inhabitant, urban planner, engineer, financial company, utilities, service provider, and so on. This dynamic interaction allows for real-time analytics, informed decision-making, resource efficiency, and comfort enhancement (Khajavi et al. 2019).

Ideally, in a digital PED twin, the real-time data is collected and transferred to data analytic centre, where all kinds of data are analysed along with their complex systems for either prediction or optimization for multi objectives of liveability, such as maintaining IAQ level, improving energy self-consumption, maximizing economic benefits, and minimizing carbon emissions, etc. The predicted or optimized information is further, if

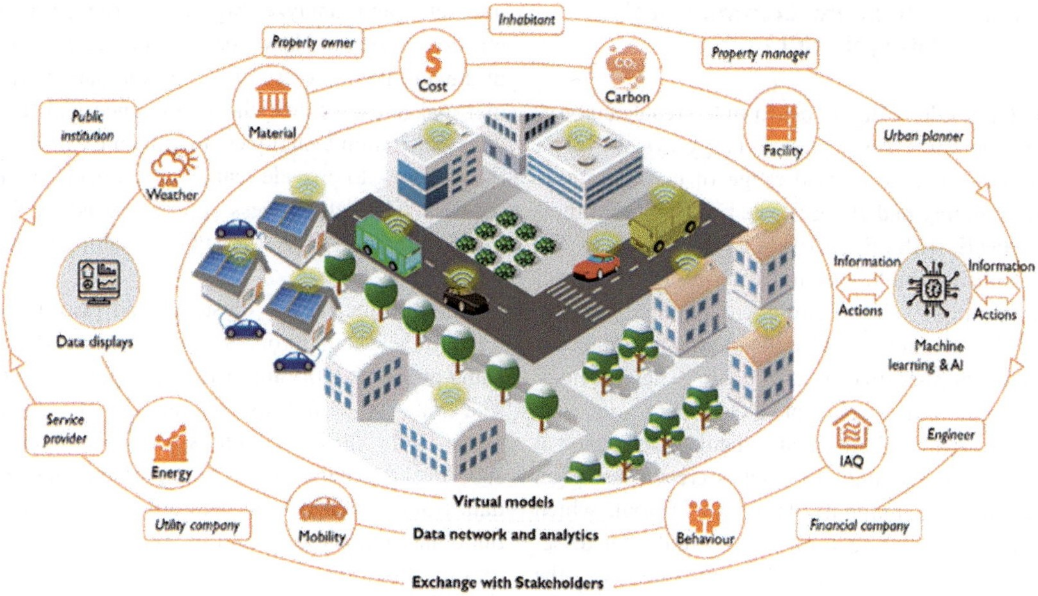

Fig. 20.1 Schematic design of key components in a digital PED twin

necessary, sent to stakeholders for decisions, or returned to the individual systems in reality for regular operations. Such digital twin increases system resilience by considering interdependent systems and optimizing the decision/operation of the future.

The considerable benefits of a digital PED twin can be anticipated. The digital twin environment will facilitate interaction and collaboration between all stakeholders involved in a PED's life cycle, by enabling integrated data-information-knowledge-decision sharing capabilities. Such activity will further increase public and individual awareness. Since data and feedback are real-time, it can increase energy systems' flexibility and robustness during operation. A PED-centred digital twin Environment can aggregate all data through the whole life cycle of a PED from design and construction (renovation), to operational and demolishment phases, hence improving the livability by more resource-efficient, economic and environmental decision taking (Alonso et al. 2019).

20.3 Digital Twin Platform and Tools

A digital twin is a combination of several modules such as computer model, physical model, communication services, and data analytics. These modules work in synchronization to monitor, learn, and optimize the complete system operation. A digital twin can capture the relationships that define such modules and allows new levels of analysis for complex environments. However, the implementation of the digital twin concept may require new processes, methods, and novel platforms to interact with each of these modules (Qin et al. 2017). Similar to the diversification in the 3D modelling techniques, there does not exist a common digital twin platform. The reason is that solution to provide digital twin does not lie in technology, but on the methodology and processes used to provide these solutions (Theiet 2020). Therefore, this section intends to provide an overview of various available tools, which are used, for digital twin in the built environment and PED context.

20.3.1 Intelligent Communities' Lifecycle (ICL)

ICL provides a digital platform to create, analyse, and optimize the complex energy systems. The tool can assess a broad range of configurations for building and energy systems throughout their lifecycle (IESVE 2020). ICL makes use of the established building simulation tools to interact with the broader environment required for DT implementation. This results in an interconnected platform that makes use of IoT data, sensor readings, OpenData, or any other source to investigate the built environment performance at any level. The digital model is created using an intelligent community design (iCD) tool, which utilizes open street maps to build the three-dimension model and BIM interoperability of the case building. The input from iCD is used for dynamic simulation performance of the system across the entire building lifecycle. This is enabled by the virtual environment (VE) platform to carry daylight, energy performance, and life cycle assessment of the system. Furthermore, the data from the real case building is obtained and analysed to identify operational issues, risk mitigation, and understanding system interactions using an energy management information platform. The utilization of the tool is commercially demonstrated with over 100 projects all over the World (IESVE 2020).

20.3.2 BuildingMinds

The tool is developed by BuildingMinds GmBh, Germany, in collaboration with Microsoft, USA. Similar to ICL, BuildingMinds (BuildingMinds 2020) provides a common platform to obtain, integrate, and analyze big data from various physical systems. However, the unique features of the tool lie in the use of a common data model that makes use of artificial intelligence and data democratization techniques, to efficiently process the databank to provide real-time feedback to the services. The data process approach used by the tool is shown in Fig. 20.2 (Policy Lab and Spatial Services 2019).

Initially, the data is clustered and prioritized based on a specific process, and further validated using available tools and performance indicators. The data is stored as "common data' with specific attributes and entities for further analysis. This also enables users to import and analyze data from existing digital and analog sources to build an interoperable real-time representation of existing building entities.

20.3.3 Predix

Predix is a digital twin platform developed by General electrical, USA with a focused development of cost-effective and predictive maintenance strategies for industrial components such as turbines, and motors (Ge.com 2020). It has been reported that a major share of manufacturers suffers from excessive costs related to materials, labour, packaging, and shipping. The reasons identified to these expenses include lack of component performance prediction, which might result in un-scheduled asset downtime, maintenance, and late shipments (Macfarlane 2020). However, predictive analytics enabled by digital twin is used as a vital tool to help reduce all of these issues. The dynamical linkage of each digital twin element with construction, engineering, and operational data can help to

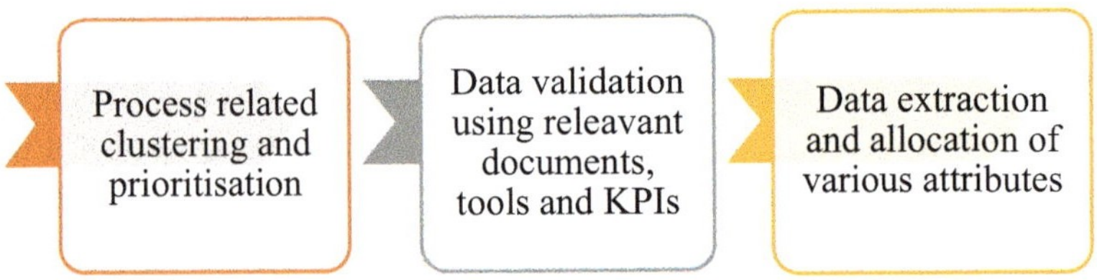

Fig. 20.2 Data process approach used by BuildingMinds

streamline the element controls. Predix claims to be the world's first industrial internet platform, to manage digital twin and associated analytical models to connect data, intelligence, and people. The platform consists of an asset performance management kit that designed to optimize asset operations. The kit also includes HMI and supervisory control and data acquisition system to implement intelligent control applications.

20.3.4 Ecodomus

Ecodomus provides a constellation of four submodules required to create a DT of the asset, as shown in the Fig. 20.3. The software addresses the full lifetime approach of the building from the design phase to the decommissioning phase to achieve short- and long-term efficiency gains (Ecodomus.com 2020). The conjunction of building management software and BIM model with facility operational tools helps to understand and get critical insights on the operational systems of the building. However, the geographical information for the asset such as topography and site-wide information is obtained using Ariel equipment, such as drones to scan the field, and used as input to create a realistic digital representation of the asset. The tools have been commercially demonstrated for managing a digital twin of a wastewater treatment plant in Hamilton, New Zealand.

20.3.5 Other Platforms/Tools

There exist a few other platforms, which can support digital twin. However, most of them have a focus on industrial applications. Examples for such tools are Akselos, iTwins, and Seebo (Warner 2018). Tools such as feature manipulation engine also provide a data integration platform with applications in multiple sectors such as transportation, commercial, and utilities. The tool can be used to automate the data integration workflow and distribution to save time and effort (Warner 2018). Moreover, there are several existing EU-funded Horizon 2020 projects looking at digital twin to address various sectors, e.g. food-water-energy nexus, and industrial O&M. One such project is CRUNCH using urban living labs that are scattered throughout the EU and beyond. The digital twin of these living labs will try to mimic the flows that occur in a food-water-energy nexus (CRUNCH 2018). Project Dasher also provides a digital twin kiosk to explore the various possibilities of using digital twin for building applications to add value in building operations and support better decision making (Dasher360.com 2020). The rest possible

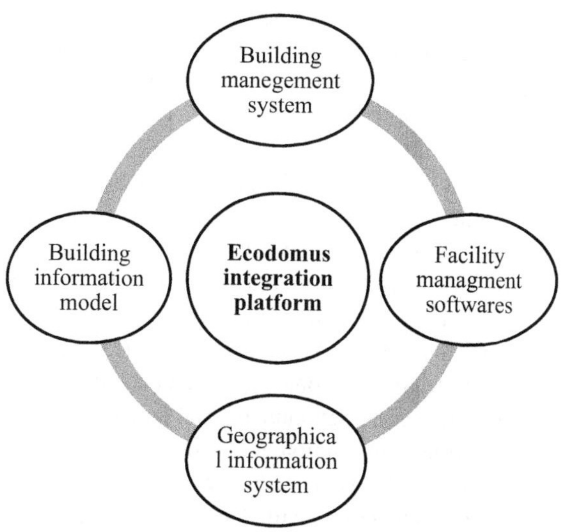

Fig. 20.3 Submodules of Ecodomus integration platform

platforms or tools are mainly developed based on existing GIS models of cities/districts (i.e. 3DEXPERIENCity platform), BIM models of buildings and transportation, data collection networks (i.e. TerriaJS), as well as data communication and visualization platforms.

Although the full implementation of digital twin might happen in near future, however, several platforms and tools are offering easily adaptable starter elements, development, of which will result in much more connected models for higher overall benefit to the built environment, industry and PEDs (Boje et al. 2020). However, there are few concerns, such as security for aggregated resources, validation, and authentication of the data, which needs to be addressed by these platforms (Qin et al. 2017).

20.4 Applications of Digital PED Twin

Until year 2020, there are several successful digital PED twin developments (Research and Markets 2020). In Europe, a digital twin project in Helsinki, Findland, has been developed on CityGML, which is a semantic, expandable information Open Geospatial Consortium model that can describe objects (Heiskanen 2019). Renne city in France establishs a digital 3D model for various urban studies (such as for urban mediation with citizens), and for urban development purposes (such sunshine simulation, noise modelling, tree shadow impact on buildings) (Poppe 2016). Rotterdam city in Netherland applies a digital twin for managing the city's infrastructure assets (Research and Markets 2020). In North America, Pasadena California in United States develops a useful supervisory tool for the city's public sector players. Meanwhile, Portland Oregon in United States plans to construct a digital transportation activated by residents' cellular data (Fischer 2019). Waterfront Toronto in Canada stands on digital twin technology launch a public advocate of waterfront revitalisation, and urban innovation organisation Sidewalk Labs, which aims to combine urban design with the latest in digital technology to address the challenges that cities face, such as housing affordability, transportation and energy use (Doyle 2019). In the Middle East, a project in Dubai focuses on the user experience by digital twin. Jaipur city in India underpins a digital twin project on urban planning and supervision (Research and Markets 2020). Yingtan city in China develops its first 5G Digital Twin. In Shanghai in China, immersive digital twins in railway engineering establish new practices to deliver sewage treatment plant (Parrott and Warshaw 2017). The detail description of specific digital twin projects can be found in Table 20.1.

From the demonstrated pilot projects, it can be concluded that the digital twin concept usually consists of three distinct parts: (1) the physical asset, from community to city; (2) the logical constructed digital/virtual product, the associated virtual three-dimensional digital replica; and (3) communications in between contained by specific applications. The communications usually take place on certain types of platform. The most popular digital twin city solution suppliers are Alphabet, Autodesk & Esri, Bentley, Cityzenith, Dassault systems, Engie Ineo/Siradel, Microsoft, NTT Data Corporation, Siemens and IES VE (Research and Markets 2020; Institute for Manufacturing 2020).

Meanwhile their connections between the physical items and the digital/virtual replica is continued data flows that streams from the physical product to the digital/virtual product, as well as information that generate from the digital/virtual platform to the physical environment. In a summary, the primary functions collected from mentioned projects are: Prediction, Simulation, Monitoring, Lifecycle, Sensing, Optimisation, Internet of Things, Artificial Intelligence, Building Information Modelling, knowledge processing with data sets and web-based data integration (Boje et al. 2020).

From the aspect of digital twin evolution, it has evolved from the monitoring platforms, the intelligent semantic platforms and the agent driven socio-technical platforms. The whole evolution represents a continuous growth in terms of both lifecycle integration and supply chain

Table 20.1 Summary of existing application examples

Project description	Platform and objectives
2019, New South Wales state, Australia (Policy Lab and Spatial Services 2019) It included digital visualisations of the local government areas that comprise the Western Sydney City Deal and Greater Parramatta to the Olympic Peninsula. It achieved the integration of live transport feeds as well as infrastructure building models that developed with the CSIRO's Data61 digital research network	Platform/tool: *TerriaJS* It generally aims to upgrade the existing state's spatial data from 2D to real-time 3D and 4D, following the recommendation of State Infrastructure Strategy in 2018. The platform will further engage with local, state and federal government agencies and industry bodies offering benefits at national, state and local government levels; from disaster management through to bus schedules for city's future needs
2017, Dundalk Institute Of Technology, Ireland (IESVE 2017) It created a 90 acre virtual campus energy model, which covering all campus buildings, streetlights, transformers, the 850 kW wind turbine and the 125 kW electricity storage battery. By analyzing the results on ICL, the direct improvements were: (1) Wind Turbine meets 52.8% of demand; (2) Electricity Storage meets only 2.2% of demand; (3) 10.5% Transformer Losses, (4) 45% of demand is met by the grid	Platform/tool: *ICL* It overarches a Net Zero Energy Campus. The virtual campus model was to: (1) Analyse all annual energy sources and demand for the campus; (2) Simulate energy demand data; (3) Validate pre-existing renewable investments and calculate return of investment (ROI) on improvement options. Finally, some suggestions were given towards its net-zero energy goal by disposing of the Electrical Storage unit, assessing renewable energy networks and investing in other energy conservation methods
2019, Virtual Singapore (Qin et al. 2017; Systèmes 2019) The virtual country project is a $73 million, continuous live digital replica urban laboratory that created by the Singapore National Research Foundation. It consists of 3D semantic model and the associated platform which incorporates real-time dynamics, as components of buildings and infrastructures, as well as information about demographics, climate or traffic. The project not only targets to translate urban relevant design tasks into data-rich driven totally, but also further turn into a virtual test bed at urban level, which ensures very large-scale simulations, like wind, noise, traffic simulation to support the city's R&D work. This project is the most advanced digital twin to date	Platform/tool: *3DEXPERIENCity platform* Because large cities are usually characterised by a dynamism in several symbiotic entities that conceals highly complex social structures and services that normally makes urban planning a complicated systematic work. The introduction of 3D virtual representation of real country offers a common and holistic way for designing better urban centres. With the help pf big data, cloud computing and virtual reality in the city-state of Singapore, it has been already beneficial for urban planners, citizens and business partners with: (1) optimizing a better accessibility solution in a specific area without any construction work; (2) estimating emergency situations and establishing the most suitable evacuation protocols; (3) providing real-time monitoring of the Project
2019, Boston 3D Model, USA (Patrick 2018) Evolving from the previous craved wooden downtown model in the 1980s, Boston Planning and Development Agency now prompted to an in-house digital twin. The general interests are to spatially understand environmental impacts using cutting-edge technology with data-driven workflows. The upgraded model has capabilities of geographic information system software and the expertise coming from Autodesk & Esri professional services	Platform/tool: *GIS-based 3D city model* The digital twin model allows to capture the entire city and determine real-world impacts to make timely decisions. It developed a GIS-based 3D twin that includes both quantitative and qualitative analysis workflows. In the coming future, designers are able to use metrics and a standardized process and procedure to evaluate projects, including planning and development, flood modelling, shadow studies, and line-of-sight evaluation
2018, University of Cambridge and the west Cambridge shire in UK (Institute for Manufacturing 2020) A digital twin has been applied to infrastructure system management in, University of Cambridge and the west Cambridgeshire in UK. The whole project has been	Platform/tool: *integrated platform with sub models/systems from Bentley, GeoSLAM, Topcon, and Redbite* Within the structured 3 work packages as: (1) data development; (2) data integration and (3) applications, the overall goals of this activity are to:

(continued)

Table 20.1 (continued)

Project description	Platform and objectives
carried out in partnership with along three interconnected work packages: • Bentley Systems for 3D BIM modelling building; • GeoSLAM for detailed context capture scan, • Topcon for a low-level-detailed 3D geometry and photogrammetry of the West Cambridge Site using drone and vehicle-based scanning and camera devices • Redbite for asset management solution. Redbite will further develop APIs that can be used to integrate the asset data with the 3D BIM model through Bentley's AssetWise operational analytics solution	(1) Demonstrate the impact of digital modelling and analysis of infrastructure performance and use on organizational productivity; (2) Provide the foundation for integrating city-scale data to optimize city services such as power, waste, transport and understand the impact on wider social and economic outcomes; (3) Establish a 'research capability platform' for researchers to understand and address the major challenges in implementing digital technologies at scale; (4) Foster a research community interested in developing novel applications to improve the management and use of infrastructure systems

integration (Boje et al. 2020). The associated data consumers and stakeholders are resident, supply chain, operator, engineer, manager, designer, planner and government/local authority.

20.4.1 The First Tier Generation

It works close to an enhanced version of BIM on construction sites to data. The limitations lie in information requirements for subsequent lifecycle stages and extensibility in associated complex computations. Typically, the evolved maturity elements at this stage are (Savian 2020):

- reality as-built data set capture (e.g. point cloud, drones, photogrammetry, or drawings/sketches);
- spatial information connected to 3D model;
- connect model to more static data (e.g. documents, drawings, asset management systems);

20.4.2 The Second Tier Generation

It moves a major progress forward with intelligent semantic platforms, providing a primary knowledge base development. But there are inadequate actuation capabilities in dealing with complex information interactions.

Typically, the evolved maturity elements at this stage are added with (Savian 2020):

- enrich with dynamic one-directional data flow (e.g. from Internet of Things, embedded sensors);
- establish two-way data integration and interaction (human-to-machine and machine-to-machine).

20.4.3 The Latest Generation

It has advanced knowledge leverages with the use of AI-enabled agents. Relying on the previous intelligent semantic platform, it elaborates AI technologies, such as machine learning, deep learning, data mining and analysis capabilities to construct a self-reliant, self-updatable and self-learning digital twin projects. Typically, the evolved maturity elements at this stage can be finalized with aspirational autonomous operations and maintenance (Savian 2020).

20.5 Challenge and Opportunity

20.5.1 Data Analysis and Semantic Interoperability

It is observed that most of the existing studies and applications emphasize the creation of a digital PED twin, rather than how to optimize it for operation and maintenance. Most studies have completed excellent virtual models of PED and integrated large-scale sensor network, but it still lacks of knowledge and skill in data analysis and interoperable interaction with different stakeholders.

The ability of a digital PED twin is to capture the complex-and-dynamic relationships of different components in PEDs, which allows new levels of analysis of complex environments. However, it now lacks of the studies to run analysis of real-time operations and different future scenarios, which aim to explore their impacts across the PED systems for new insights that enhance the ability to take more holistic approaches to building/PED design, energy strategies, and transportation planning etc. For instance, how inhabitants change their mobility behaviour in response to the increase of EV numbers; what is the impact of distributed PV installation on local network and storage systems, as well as local electricity market via different business models; what is the impact of future climate, and how to adapt PED to the future scenario. It will need more and more advanced machine learning and AI approaches to provide another level of analysis of the complex systems and component relationships that would be nearly impossible to recognize in real-world environment (Woods and Freas 2019).

In current digital PED twins, it lacks of a semantic model to standardize concept descriptions and data representations for interoperable interaction with different stakeholders and energy information communication/management. At the moment, semantic models and their applications are mostly designed to facilitate planning or analysis of urban energy systems through simulation or information representation and exchange, rather than facilitating energy-related operation and management or as part of a complex event processing system (Howell et al. 2017). Semantic heterogeneity between vocabularies and data representations is a common issue in existing digital twin models.

20.5.2 Business Models and Economic Analysis

There is a piece of very low information available on economic feasibility studies and business models of digital twin platforms. The concept of digital twin will transform the business of energy production and delivery amid the ongoing passage from centralized production (i.e. energy plants) to decentralized production (i.e. high penetration of renewables and PEDs) (Richter 2013). The renewable energy systems and the energy saving technologies that form a PED have an initial cost but also an energy production/saving potential during their lifetime. The business model of these technologies consists in exceeding the investment and maintenance costs with the savings (Qi et al. 2019) and revenues (IRENA 2019) from the energy saved or exported. The costs associated with the creation of a digital twin are already to be taken into account when designing a PED energy system as most of the infrastructure needed for the operation of a digital twin is necessary in a decentralized energy market, regardless of the ownership structure of the infrastructure itself. Given the profound inter-dependence of energy and monetary fluxes, it is paramount to have a detailed knowledge, hence a model, of the energy flows in a local grid (Roberts et al. 2019). In recent years some authors have started to study the interaction between prosumers within an energy producing district (Jing et al. 2020; Zhang et al. 2018), by proposing different business models, such as power purchase agreements (PPA), net-metering mechanism, and peer to peer (P2P) trading mechanism. If the energy producers have to be adequately remunerated by the energy consumers, the energy transaction within a local grid have to be adequately mapped. Once the sensors and the model of the local energy system are put in place, the use of a digital twin can bring a series of benefits from design to operation phase, which will facilitate the energy sharing and trading based on different business models.

Using a digital twin during design phase helps to predict the performance during operation phase. Continuous learning can improve the profitability of the energy investments and reduce the investment risk. The digital twin offers an insight into the aging of the infrastructure it represents by monitoring and re-calibration of certain parameters over time.

20.5.3 Data Security and Management

PED gathers dynamic energy and other information at a district level and generates big amount of data when it is digitalized, which enables cloud computing a viable strategy for modelling. However, when multiple organizations share the data, there is a risk of misusing the data (Rao and Selvamani 2015). Providing authentication, authorization and access control for data stored in cloud may increase the data security in terms of confidentiality, integrity and availability.

Confidentiality protects information from being accessed by unauthorized parties. It is an essential requirement to ensure the security of data in cloud storage and computing (Aloraini and Hammoudeh 2017). Applying data encryption can limit the access to stored data for PEDs.

PED is an integrated system where digital information for each subsystem is highly correlated. Any alteration of data may jeopardize the connections between systems. Thus, cloud service providers should check and maintain the data and computation regularly. But it is still a challenge to predict any future modification to the data based on historical performance. Another issue concerns to avoid inefficient download/upload of dynamic data.

PED comprises multiple agents. The subsystems may work independently and they need a system that is always available that is not under control of a single-cloud provider. Substantial efforts are needed for making the transition from single-cloud to multi-cloud computing.

20.6 Summary

This chapter presents a comprehensive review about digital twins for PED from aspects of concepts, working principles, tools/platform and applications, in order to address the issues of both 'how digital PED twin is made' and 'how digital PED twin optimises liveability'. A few available tools and platforms are reviewed for digital twin in built environment and PED context, such as ICL, Building minds, Predix, Ecodomus. Other platforms/tools either have a focus on industrial applications or are mainly developed based on existing GIS models of cities/districts, BIM models, data collection networks, as well as data communication and visualization platforms. Several successful application of digital PED twins are summarized, where lessons and observations are gained so that digital PED twins can be categorized into three tiers: (1) an enhanced version of BIM model only, (2) semantic platforms for data flow, and (3) AI-enabled agents for data analysis and feedback operation. Further challenges and opportunities lie in areas of data analysis and semantic interoperability, business models, data security and management.

References

Alonso R, Borras M, Koppelaar R, Lodigiani A, Loscos E, Yöntem E (2019) SPHERE: BIM digital twin platform. Proceedings 20 9. https://doi.org/10.3390/proceedings2019020009

Aloraini F, Hammoudeh M (2017) A survey on data confidentiality and privacy in cloud computing. In: ICFNDS '17, pp 1–7

Boje C, Guerriero A, Kubicki S, Rezgui Y (2020) Towards a semantic construction digital twin: directions for future research. Autom Constr 114:103179

BuildingMinds (2020) How digital building twins accelerate your sustainability—Buildingminds. Available at: https://buildingminds.com/articles/how-digital-building-twins-accelerate-your-sustainability/. Accessed 4 May 2020
CRUNCH (2018) CRUNCH. Available at: http://www.fwe-nexus.eu/workshops.php. Accessed 4 May 2020
Dasher360.com (2020) Dasher 360. Available at: https://dasher360.com/. Accessed 4 May 2020
Doyle S (2019) Digital urban planning: twins help make sense of smart cities. Available at: https://eandt.theiet.org/content/articles/2019/01/digital-urban-planning-twins-help-make-sense-of-smart-cities/. Accessed 31 Mar 2020
Ecodomus.com (2020) Ecodomus FM—Ecodomus. Available at: https://ecodomus.com/products/masonry-style/. Accessed 4 May 2020
Fischer R (2019) Digital transportation: the future of urban mobility. Available at: https://usgtima.org/perspectives/digital-transportation-the-future-of-urban-mobility/. Accessed 31 Mar 2020
Ge.com (2020) Digital Twin | GE Digital. Available at: https://www.ge.com/digital/applications/digital-twin. Accessed 4 May 2020
Heiskanen A (2019) Helsinki is building a digital twin of the city. Available at: https://aec-business.com/helsinki-is-building-a-digital-twin-of-the-city/. Accessed 31 Mar 2020
Howell S, Rezgui Y, Hippolyte JL, Jayan B, Li H (2017) Towards the next generation of smart grids: Semantic and holonic multiagent management of distributed energy resources. Renew Sustain Energy Rev 77:193–214
IESVE (2017) Dundalk Institute of Technology. Available at: https://www.iesve.com/icl/case-studies/2886/dundalk-institute-of-tecnhnology. Accessed 30 Mar 2020
IESVE (2020) The ICL Is Here! The new environmental digital twin for healthy sustainable communities | Discoveries | IES. Available at: https://www.iesve.com/discoveries/article/7734/the-icl-is-here. Accessed 4 May 2020
Institute for Manufacturing, University of Cambridge, (2020) Infrastructure Digital Twins. Available at: https://www.ifm.eng.cam.ac.uk/research/asset-management/research-projects/infrastructure-digital-twins/. Accessed 31 Mar 2020
IRENA (2019) Renewable power generation costs in 2018. International Renewable Energy Agency, Abu Dhabi. ISBN 978-92-9260-126-3
Jing R, Xie MN, Wang et al (2020) Fair P2P energy trading between residential and commercial multi-energy systems enabling integrated demand-side management. Appl Energy 262:114551
Khajavi SH, Motlagh NH, Jaribion A, Werner LC, Holmström J (2019) Digital twin: vision, benefits, boundaries, and creation for buildings. IEEE Access. https://doi.org/10.1109/ACCESS.2019.2946515
Macfarlane S (2020) SME blog—Ecodomus—the digital journey from an SME |Centre For Digital Built Britain. Available at: https://www.cdbb.cam.ac.uk/sme-blog-ecodomus-digital-journey-sme. Accessed 4 May 2020
Parrott A, Warshaw L (2017) Industry 4.0 and the digital twin-Manufacturing meets its match. s.l.: Deloitte University Press
Patrick B (2018) Meet Boston's digital twin. Available at: https://www.esri.com/about/newsroom/blog/3d-gis-boston-digital-twin/. Accessed 30 Mar 2020
Policy Lab and Spatial Services (2019) Twinning! Spatial services has created a digital twin of NSW. Available at: https://www.digital.nsw.gov.au/article/twinning-spatial-services-has-created-digital-twin-nsw. Accessed 30 Mar 2020
Poppe S (2016) Digital 3D modelling of Rennes Metropole. Available at: http://www.eurocities.eu/eurocities/documents/Digital-3D-modelling-of-Rennes-Metropole-WSPO-A8JDUM. Accessed 31 Mar 2020
Qin Q, Liang F, Li L et al (2017) Selection of energy performance contracting business models: a behavioural decision-making approach. Renew Sustain Energy Rev 72:422–433
Qi Q, Tao F, Hu T et al (2019) Enabling technologies and tools for digital twin. J Manuf Syst. https://doi.org/10.1016/j.jmsy.2019.10.001
Rao RV, Selvamani K (2015) Data security challenges and its solutions in cloud computing. In: International conference on intelligent computing, communication & convergence, pp 204–209
Richter M (2013) Business model innovation for sustainable energy: German utilities and renewable energy. Energy Policy 62:1226–1237
Roberts MB, Bruce A, MacGill I (2019) A comparison of arrangements for increasing self-consumption and maximizing the value of distributed photovoltaics on apartment buildings. Sol Energy 193:372–386
Research and Markets (2020) Digital twin cities, 2020: the global ecosystem, city-backed projects, main application sectors, and solution suppliers. s.l.: Research and Markets
Savian C (2020) Digital twins for the built environment: an introduction to the opportunities, benefits, challenges and risks. The Institution of Engineering and Technology
SET-Plan action 3.2 (2018) Implementation plan: Europe to become a global role model in integrated, innovative solution for the planning, deployment, and replication of Positive Energy Districts. https://setis.ec.europa.eu/system/files/setplan_smartcities_implementationplan.pdf. Accessed on 11 Mar 2020
Systèmes D (2019) Meet virtual Singapore, the city's 3D digital twin. Available at: https://www.smartcitylab.com/blog/digital-transformation/singapore-experiments-with-its-digital-twin-to-improve-city-life/. Accessed 30 Mar 2020
Theiet (2020) Built environment. Available at: https://www.theiet.org/impact-society/sectors/built-environment/?utm_source=redirect&utm_medium=legacyredirects&utm_campaign=2019relaunch. Accessed 4 May 2020
Urban Europe (2019) Framework definition for positive energy districts and neighbourhoods. https://jpi-

urbaneurope.eu/news/consultation-on-the-framework-definition-for-positive-energy-districts-and-neighbourhoods. Accessed 18 Feb 2020
Warner T (2018) Innovating the future with Canada'S digital technology supercluster | Safe Software. Available at: https://www.safe.com/blog/2018/10/innovating-future-canadas-digital-technology-supercluster/. Accessed 4 May 2020
Woods E, Freas B (2019) White paper of creating zero carbon communities: the role of digital twins, Commissioned by Integrated Environmental Solutions
Zhang C, Wu J, Zhou Y, Cheng M, Long C (2018) Peer-to-Peer energy trading in a Microgrid. Appl Energy 220:1–12

Digital Mapping of Techno-Economic Performance of a Water-Based Solar Photovoltaic/thermal (PVT) System for Buildings Over Large Geographical Cities

21

Santhan Reddy Penaka, Puneet Kumar Saini, and Xingxing Zhang

Abstract

Solar photovoltaic thermal (PVT) is an emerging technology, capable of producing electrical and thermal energy using a single collector. However, to achieve larger market penetration for this technology, it is imperative to have an understanding of the energetic performance for different climatic conditions and the economic performance under various financial scenarios. This chapter thus presents a techno-economic evaluation of a typical water based PVT system for electricity and domestic hot water applications in 85 locations worldwide. The simulations are performed using a validated tool with one-hour time step for output. The thermal performance of the collector is evaluated using energy utilization ratio and exergy efficiency as key performance indicators, which are further visualized by the digital mapping approach. The economic performance is assessed using net present value and payback period under two financial scenarios: (1) total system cost as a capital investment in the first year; (2) only 25% of total system cost is a capital investment and remaining 75% investment is considered with financing period with certain interest rate. The results show that such a PVT system has better energy and exergy performance for the locations with a low annual ambient temperature and vice versa. Furthermore, it is seen that the system boundaries, such as load profile, hot water storage volume, etc., can have a significant effect on the annual energy production of the system. Economic analysis indicates that the average net present values per unit collector area are 1800 € and 2200 € respectively among the 85 cities for financial model 1 and financial model 2. Nevertheless, from the payback period point of view, financial model 1 is recommended for the locations with high interest rate. The study is helpful to set an understanding of general factors influencing the techno-economic performance of PVT systems.

Keywords

PVT · Water-based PVT · Techno-economic analysis · Digital mapping

---

S. R. Penaka (✉) · P. K. Saini · X. Zhang
Department of Energy and Community Buildings, Dalarna University, 79188 Falun, Sweden
e-mail: pks@du.se

X. Zhang
e-mail: xza@du.se

X. Zhang (ed.), *Data-driven Analytics for Sustainable Buildings and Cities*, Sustainable Development Goals Series, https://doi.org/10.1007/978-981-16-2778-1_21

21.1 Introduction

21.1.1 Background and Existing Studies

The concept of "electrify everything" considers solar energy as a key renewable technology with an aim of de-carbonization of domestic heating demand (Jia et al. 2019). The rapid growth in Photovoltaic (PV) installation capacity from the last few years has further strengthen the importance of PV as the main driver of renewable transformation (Joshi and Dhoble 2018). PV remains an interesting subject area for many researchers, global leaders, and manufacturers because of its reliability, sustainability, ease of installation, and economic feasibility (Al-Waeli et al. 2017). However, the concurrence of heat/electricity demand and limited roof area in domestic dwellings does require technologies, which can generate energy efficiently in both thermal and electrical form. Therefore, there is a huge potential for well-designed systems by combining both solar PV and solar thermal technologies. A relatively new commercialized concept of solar photovoltaic/thermal (PVT) technology can achieve such a goal by generating both electrical and thermal energy together using a single panel (Gu et al. 2018). Realizing its importance, the Solar Heating and Cooling Program (SHC) of the International Energy Agency (IEA) has initiated the task 60 for PVT applications and solutions to HVAC systems in buildings (PVT systems IEA SHC 2019). The task is active from January 2018 and has built a huge knowledge base around PVT systems for its use in domestic and industrial applications.

PVT system can be categorized in several ways, however, the most common is based on heat-transfer medium (air based/liquid based) used in the PVT collector (Zhang et al. 2012). The liquid based types are dominating the current PVT market in terms of the number of installations due to high efficiency, and ease of integration in existing hydronic systems (Ramschak 2020). In a standard liquid based PVT collector, the heat carrier is usually water or brine mixture, which is allowed to circulate in a heat exchanger behind the PV cells. The circulation results in a heat transfer through the back sheet of the module, which raises the fluid temperature enough to use for various applications such as hot water, swimming pool heating etc. From a technical perspective, PVT technology is well developed and it can be coupled with various energy systems. For instance, it can go hand-in-hand with the emerging awareness of heat pump technology with/without borehole storage (IEA-SHC-Task60-Highlights 2020). However, the main barriers currently in PVT development and deployment are lack of testing standards, uncertain financial incentives, and business models across different regions in a niche market. Therefore, the business potential of PVT solution is not fully explored, although it can be a very efficient solution for domestic and industrial heating requirements.

There are several studies concerning the techno-economic analysis of PVT collectors with a focus on the component and system design (Gu et al. 2018; Buonomano et al. 2015; Riggs 2017; Wang et al. 2019; Kazem 2019). The most common way is to assess the energetic performance firstly and then carry out an economic evaluation based on dependent variables (Gu et al. 2018; Buonomano et al. 2015; Riggs 2017; Kazem 2019; Fudholi et al. 2014; Khelifa et al. 2015, 2016; Gagliano et al. 2019). The prevalent energy performance indexes are energy efficiency and exergy efficiencies, Zhang et al. (2012) while the most popular economic indicators are represented by Levelized Cost Of Energy (LCOE), Net Present Value (NPV), and payback period (Gu et al. 2018). To name a few studies for technical evaluation, Fudholi et al. (2014) investigated electrical and thermal performances on PVT water-based collectors by testing with specific inputs parameters ranging from 500 to 800 W/m^2 solar irradiance and mass flow rate of 0.011–0.041 kg/s. The test has recognized that absorber performed better at a mass flow rate of 0.041 kg/s and under 800 W/m^2 irradiance, with measured PV efficiency of

13.8%, the thermal efficiency of 54.6%, and overall collector efficiency of 68.4% (Fudholi et al. 2014). Shah and Srinivasa (2018) modelled a theoretical model using COMSOL multi-physics validation tool with Standard Test Conditions (STC) to measure the PV improved efficiency when it is integrated with hybrid PVT system. Another study performed by Buonomano (2016) has developed a numerical model to conduct the technical and economic analysis of PVT collectors and compared it with conventional PV collectors installed in Italy. The tool was validated using TRNSYS platform for the energetic and economic performance of systems integrated with PV and PVT collectors together. Yazdanpanahi (2015) presented a numerical simulation and experimental validation for evaluation of PVT exergy performance using one-dimensional steady thermal model and four-parameter current–voltage model for PVT water collector. In terms of economic studies, Gu et al. (2018) developed an analytical model on basis of combinations of Monte Carlo method to analyze techno-economic performances of solar PVT concentrator for Swedish climates, which considered several essential input uncertainties whereas economic variables were assessed initially. The obtained results for range of capital cost between 4482 and 5378 SEK/m^2 and system size of 10.37 m^2 during the system lifespan of 25 years. The results indicate LCOE of 1.27 SEK/kWh, and NPV of 18,812 SEK with a simple payback period of 10 years. It was concluded that the most important sensitivity factor is average daily solar irradiation followed by debt to equity ratio, capital price, regional heating price, and discount rate. Herrando et al. (2016) performed techno-economic analysis of hybrid PVT systems for electricity and domestic hot water (DHW) demand for a typical house in London and concluded that with such systems can meet 51% of electricity demand and 36% of DHW demand even during low solar Global Horizontal Irradiation (GHI) and ambient temperatures. In the economic aspect, it is also concluded that hybrid PVT technology has better energy yield per unit roof area, which can result in attractive NPV for investor while mitigating the CO_2 emissions. Riggs et al. (2017) developed a combined LCOE techno-economic model for different types of hybrid PVT systems applied for process heat application in United States. The sensitivity analysis of parameters affecting the Levelized Cost Of Heat (LCOH) was determined using technical, financial, and site-specific variables. Ahn et al. (2019) studied the importance of energy demands, solar energy resources and economic performances of hybrid PVT systems at different PV penetration levels using Monte Carlo method, whereas the study found that irrespective of PV penetration levels, the uncertainties in energy demands and solar irradiance can influence the energy performance of PVT systems. Heck et al. (2016) conducted Monte Carlo method for LCOE based on probability distribution, which concluded that this method provides more realistic information on risk/uncertainty, which triggers more scope of potential investment on electricity generation. However, this method is complex slightly than point values.

However, most of the existing studies focused on a single climate, with a straightforward economic-financial analysis. Furthermore, complicated procedures or individual software are used to estimate the performance of PVT collectors, where it lacks a comprehensive simulation of PVT techno-economic performance through a common tool over a large geographic area, aiming for application feasibility and business potentials. Besides, many studies have reported solar energy resource potential on buildings at different spatial scales using digital mapping methods, such as digital numerical maps (Jung et al. 2019), digital surface model (Oh and Park 2018), satellite imageries and geographic information systems (Mishra et al. 2020; Bremer et al. 2016), multi-scale uncertainty-aware ranking of different urban locations (Peronato et al. 2018), which provide direct evaluations for solar application, leading to robust planning decisions. Nevertheless, no study is found yet for mapping of techno-economic performance of PVT systems.

As a result, this chapter aims to fill these research gaps by utilizing a validated simulation tool to perform a comprehensive techno-economic performance simulation for a wide range of cities. The results are further analyzed and visualized using a digital numerical mapping approach to set a comparison among various regions.

21.1.2 Aim and Objectives

This study aims at simulation and mapping of the energetic and economic indicators of a typical PVT system over different regions, to establish a digital performance database for various key performance indicators (KPI). The economic feasibility of the PVT collector is obtained and compared under various financial scenario models. The data obtained from simulations are used to establish a simple correlation between variables affecting the PVT system.

The main objectives of this chapter are to:

- Assess the thermal and electrical performance of a typical PVT system in 85 locations across the World using a validated simulation tool.
- Evaluate the economic performance using NPV and payback period using two financial scenarios.
- Analysis and visualization of energy and economic performance.

The significance of this chapter lies in (1) understanding of typical PVT components behaviour at system level, (2) mapping of the collector energetic and economic performance at different climatic conditions across the world. This research results would reflect the concrete developments to this subject area and helps the promotion of the potential markets, e.g. discovering the economic feasibility of the PVT system, and feasible financial solutions to the PVT system in different regions. This chapter evaluates the related business benefits of a typical PVT system, which would help to develop a database as repository of PVT performances in different regions and contexts. The research results will be useful for researchers, planners, and policy-makers to further evaluate PVT potentials in a net-zero/positive energy district towards energy surplus and climate neutrality.

21.2 System Description and Research Methodology

21.2.1 Water-Based PVT Collector

Among different types of PVT technology, the water based PVT is the most common one that has great possibilities for system integration (Chow 2010). This PVT collector type is structured similarly to the typical flat-plate collector, as shown in Fig. 21.1. It is a sandwiched structure comprising several layers, including a glass cover placed on the top; a layer of PV cells or a commercial PV lamination laid beneath the cover with a small air gap in between; heat-exchanging tubes or flowing channels through the absorber and closely adhered to the PV layer; a thermally insulated layer located right below the flow channels. All the layers are fixed into a framed module using adequate clamps and connections. In the heat-exchanging tubes, water is the most commonly used heat carrier media due to high specific heat capacity and ease of availability. The glass cover is often optional depending on system design priority for type of output required (i.e. electricity or heat). The glass cover helps to reduce heat convection losses, but it also causes high solar reflectance losses and thus lowers optical efficiency. In many cases, glass cover is used when higher heat output is expected, while it is removed when the system is optimized for higher electrical output.

The electrical efficiency of PV cells increases when the pumped cooled water flows across the rigid series or parallel tubes. The flow control is an important factor to achieve overall high performance of the PVT collectors (Joshi and Tiwari 2007). In addition to electricity production, hot water is generated by absorbing extra heat from the PV layer, which can be used for several applications. The electrical and thermal efficiencies of PVT generally depend on PV cell type, fluid temperature, fluid flow rate, flow channel size/configuration, and ambient climatic

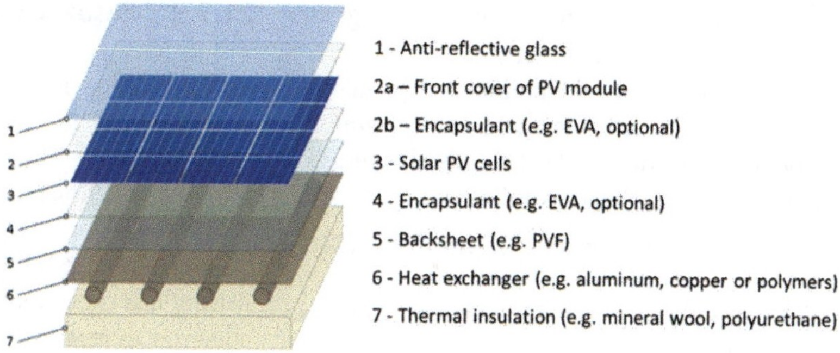

Fig. 21.1 Schematic cross-section of a covered flat-plate PVT collector (IEA SHC 2019)

condition. The collector energetic performance can be measured in terms of energy utilization ratio and exergy efficiency (Yazdanpanahi et al. 2015).

This chapter will focus on a typical PVT collector developed by a Spanish manufacturer named Abora solar. The collector is market available and more than 5700 m^2 of the gross collector is installed for a broad range of applications. The collector is a covered PVT type with an additional layer of glass on the top of the collector (in addition to a glass layer for PV cells) to reduce the heat convection losses. The rated power of the collector is 365 W at Standard Testing Conditions (STC), with a collector area of 1.96 m^2 consisting of 72 mono-crystalline cells. The main specifications and characteristics of analysed PVT collector are shown in Table 21.1.

21.2.2 Key Performance Indicators

The performance of such PVT collector is evaluated using standard key performance indicators. The performance of a collector over a specified period can be quantified using the energy utilization ratio (η_e)., which is defined as below (IEA SHC 2019):

$$\eta_e = \frac{Output\ energy_{electrical}}{GHI*collector\ area} + \frac{Output\ energy_{thermal}}{GHI*collector\ area} \quad (21.1)$$

where, GHI is Global horizontal Irradiation (kWh/m^2), and the collector aa is in m^2. However, the exergy value of both electricity and heat is different. Electricity can be regarded as pure exergy whereas heat contains some exergy value. To account for this, with the drawback of being somewhat less intuitive, is to replace "energy" by "exergy". The overall exergy efficiency takes

Table 21.1 Schematic design of the solar PV/LHP module-based heat-pump water heating system

Parameter	Description
Length * width * thickness	1970 * 995 * 107 mm
Gross collector area	1.96 m^2
Number of PV cells	72
Cell type	Mono-crystalline
Rated power	365 Wp
Electric efficiency at STC conditions	17%
Thermal efficiency at STC conditions	70%
Temperature coefficient of PV	−0.41%/°C
Thermal efficiency at zero mean temperature	0.7
Coefficient of thermal losses, a_1	5.98 W/m^2 K
Coefficient of thermal losses, a_2	0.021 W/m^2 K^2
Internal water volume	1.78 L

into account the difference of energy grades between heat and electricity and involves a conversion of low-grade thermal energy into the equivalent high-grade electrical energy using the theory of the Carnot cycle. The overall exergy of the PVT (ε_e) is defined as following expression,

$$\varepsilon_e = \eta_c \, \eta_{th} + \eta_{el}. \quad (21.2)$$

Carnot efficiency η_C (%), which is defined in the following,

$$\eta_C = 1 - \frac{T_{in}}{T_{out}} \quad (21.3)$$

where $\eta_{th}, \eta_{el}, T_{out}, T_{in}$ are thermal efficiency, electrical efficiency, outlet fluid temperature, and inlet fluid temperature respectively.

NPV is defined as a measurement of cumulative profit calculated by subtracting the present values of cash outflows (including initial cost) from the present values of cash inflows over the PVT collector's lifetime. In this chapter, we use NPV to evaluate a single investment to evaluate the acceptability of the project (Gu et al. 2018). A positive NPV indicates that the projected earnings, generated by a project or investment, exceed the anticipated costs. In general, an investment with a positive NPV will be a profitable one and the higher NPV means higher benefits. This concept is the basis for the NPV decision rule, which dictates that the only investments that should be made are those with positive NPV values. NPV is calculated using,

$$NPV = \sum_{t=0}^{n-1} \frac{CF_t}{(1+r)^t} - C_0 \quad (21.4)$$

where, CFt, r, n, t, C_0 are the cash flow of particular year (SEK), discount rate, number of years, year of NPV evaluation, and capital cost respectively.

The payback period is the time for a project to break even or recover its initial investment funds, where the cash flow starts to turn positive and can be given as in,

$$PP = T_{(CF_t > 0)} \quad (21.5)$$

21.2.3 Research Methodology

The simulation is carried using a validated tool developed by the manufacturer of the studied PVT collector. The "Abora hybrid simulation tool" (Abora Solar 2019) was used to map the performance across 85 cities shown in Fig. 21.2. The cities were chosen based on population density and geographical coordinates in different countries to represent a large market potential in these regions. A large number of selected locations for analysis are concentrated within Europe, with limited locations in India, USA, and Australia. The selection of locations is also restricted due to the availability of weather and GHI data in the simulation tool. The simulation tool accepts a wide range of design and financial input parameters, e.g. location and weather resources, electrical and thermal demands, local energy tariffs, specific storage volume, PVT panel and installation parameters, interest rate, and financing period, etc. The complete list of various inputs used is shown in. The performance model used in the tool for evaluation of PVT performance is validated in Oh and Park (2018), where a heat pump system integrated with 25 PVT modules was monitored, and measurements were also compared with the dynamic simulation model built in TRNSYS for Zaragoza, Spain. The thermal and electrical performance of collectors reasonably matches with measured data (4.2% deviation), however, a slightly higher deviation in heat pump performance was noted due to limitations in the black-box model of the heat pump in the studied energy system.

This chapter further applies the digital numerical map approach based on heat maps to visualize the performance of various indicators across simulated locations. The simulation results for all locations are exported to Microsoft Excel for calculations of energy and exergy efficiency (Microsoft Excel 2020). After then, the results are visualized using QGIS tool, which provides a heat map rendering to design a point layer data with a kernel density estimation processing algorithm (Welcome to the QGIS Project! 2020).

Fig. 21.2 Schematic simulated locations for techno-economic analysis

Table 21.2 Technical and economic input parameters

Technical parameters	Economic input parameters
Type of application (Domestic/industrial)	Type of mounting structure
Type of demand (hot water/space heating)	Type of inverter
Type of auxiliary system	Material profit margin
Number of bedrooms	Operation and maintenance margin
DHW temperature	Pricing of all system components
Dwellings occupancy	Annual maintenance cost
Number of collectors	Electricity price increment
Collector tilt	Auxiliary fuel price increment
Collector azimuth	Financing period models
Storage tank volume	Interest rate
Meteorological parameters (irradiation/ambient temperature/Albedo etc.)	Opening interest rate
Shadow loss percentage	
Number of additional PV panels	

Initially, a parametric study of components at the system level is considered according to the operation flow of the simulation tool indicated in the flow chart shown in Fig. 21.3. Then, the simulations are carried with defined boundary conditions and the results are represented subsequently as monthly electrical and thermal performances, energy savings, economic parameters, such as NPV, payback period.

This chapter also considers the economic performance of the collector in two different financial models, which are described below:

- Model 1: The total system cost is invested in the first year.
- Model 2: Only 25% of total system cost is a capital investment and the remaining 75% investment is considered with the financing period with a certain interest rate.

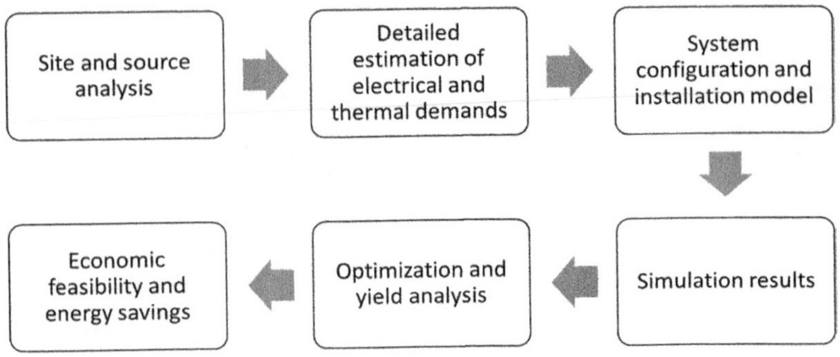

Fig. 21.3 Operation flow of the simulation tool

The economic analysis results highlight the economic parameters, such as NPV and payback period per unit collector area, for all the locations. Furthermore, the uncertainty and sensitivity parameters are discussed and the strategy in decision-making for investing in PVT technology is recommended. Digital mapping method is applied to compile and format the techno-economic performance data into a virtual image, which aims to produce a general map with KPIs of such a PVT system that gives appropriate representations of the dedicated areas.

21.3 Simulation Tool and Boundary Conditions

21.3.1 Location and Detailed Demand Analysis

The simulation tool considers the Meteonorm (2020) weather database to determine solar and meteorological resources, such as GHI, ambient temp, and wind speed. The thermal and electrical demands change with different categories of buildings, i.e. single and multifamily houses, tertiary buildings (such as hospitals, hotels, and gyms, etc.,) and can be selected individually within the tool interface. Specific key parameters, such as load profiles, the current auxiliary source of electricity, and energy system details, are included. The simulation engine assesses the total monthly and annual total demand depending on inputs for each application. The monthly energy load (L) needed to raise the temperature of supply water to the desired hot water temperature is calculated using,

$$L = m * C_p * N * (T_d - T_s) \qquad (21.6)$$

where the 'm' indicates the amount of hot water required per person in a day (in liters), 'C_p' is the specific heat capacity (J/kg K), 'N' is several days in a month (days), 'T_d' is desired water temperature (°C), and 'T_s' cold supply water temperature in (°C). The monthly demand can also be customized based on consumer utilization in that specific month. For a single-family house, the amount of DHW for one person in a day is considered as 28 L/person/day at 100% occupancy. The demand is kept constant to minimize the variables in the overall system, and thus to have a fair comparison of collector performance for various locations. The fraction of occupancy can be parameterized to meet the specific thermal demand for the individual location. For tertiary buildings (such as industrial applications), tools consider a different consumption depending on process characteristics.

This simulation tool offers to choose an auxiliary heating system to meet the load demand. This tool also accommodates that the total collector electricity generation can be utilized for self-consumption or if there is excess electrical energy, it can be sold to the electricity grid in the context of a positive energy building.

21.3.2 System Variables

This simulation tool consists of several PVT collectors and also recommends the number of collectors that would be required based on optimization of total demand and the storage tank capacity. The specific volume capacity (v/a), which is ratio of tank volume (liter) to collector gross area (m^2) can be changed depending on number of storage duration hours.

The shading loss fraction on PVT modules can be adjusted manually. There is the provision to integrate PV and PVT collectors in a scenario if the thermal demand is first fully met by PVT modules, and electrical demand is not covered fully.

21.3.3 Working Principle of the Simulation Tool

The simulation tool also optimizes the collector and installation parameters based on the demand, availability, and metrological conditions for a particular location. Simulation results highlight essential parameters such as GHI, irradiation on a tilted surface, thermal demand, thermal production, thermal solar coverage, electrical production, total electric and thermal savings, and environmental impact. The maximum power point Pm (in kW) generated by the PV cells is obtained by using Error! Reference source not found. depending on the global irradiation on the surface of the module G (W/m^2), ambient temperature Ta (°C), cell temperature Tc (°C), nominal power of photovoltaic collector Pn (kW), GSTC irradiance under STC conditions (W/m^2) i.e. 1000 W/m^2, the temperature variation coefficient of power (γ) (%/°C) (Al-Ghussain et al. 2019).

$$P_m = P_n * \frac{G}{G_{STC}}(1 - \gamma(T_c - 25)) \quad (21.7)$$

The cell temperature T_c is linked to the temperature of the absorber plate, which is dependent on the temperature of fluid going in and out of the module. Cell temperature is calculated for each simulation time step based on inlet and outlet temperatures, and electrical output is then calculated depending on the temperature coefficient of the module.

The instantaneous thermal efficiency of the collector is calculated based on,

$$\eta_{th} = \eta_o - a_1\left(\frac{T_m - T_a}{G}\right) - a_2\left(\frac{(T_m - T_a)^2}{G}\right) \quad (21.8)$$

where η_o is optical efficiency, a_1 is first order heat loss coefficient (W/m^2 K), a_2 is second order heat loss coefficient (W/m^2 K^2), T_m is average fluid temperature (°C), and T_a is ambient temperature (°C). The various characteristics of the simulated module are listed in Table 21.1, and are validated by real measurements as explained in Mishra et al. (2020).

The temperature leaving the PVT module T_o is determined using,

$$T_o = T_i + \left(\frac{m \cdot C_p}{G \cdot \eta_{th}}\right) \quad (21.9)$$

where T_o, m, C_p represents inlet temperature (° C), fluid mass flow rate (kg/s), fluid specific heat (kJ/kg K) respectively. Thermal solar coverage (Tsolar) is calculated using Eq. 21.10 in this simulation tool,

$$T_{solar}(\%) = \frac{\text{Total collector thermal production(kWh)}}{\text{Total thermal demand(kWh)}} \times 100 \quad (21.10)$$

21.3.4 System Pricing And Optimization

The detailed system cost of the PVT system is defined by customizing each component, such as flat or tilted mounting structure, single-phase or three-phase inverter, material marginal rate, electrical and combustible price escalation rate, annual maintenance cost, etc.

The simulation considers the appropriate dynamic inputs and generates the report of

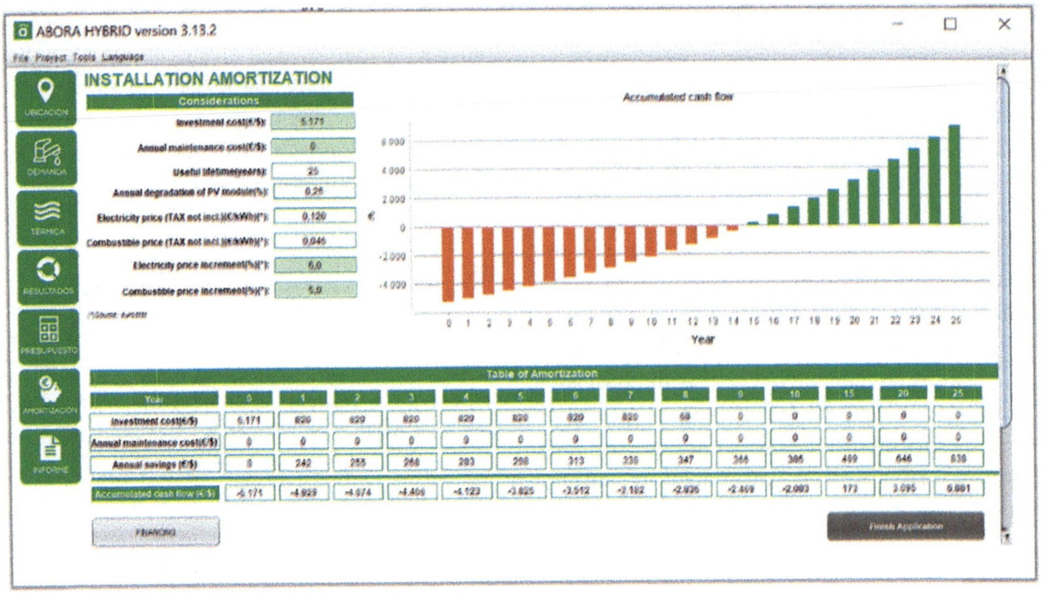

Fig. 21.4 Cost optimization of PVT system in the simulation tool

assessment on the key economic performance indicators i.e. lifetime cash flow with appropriate total annual savings, NPV, and payback period. This simulation tool allows collector economic performance with several financing options shown in Fig. 21.4. For instance,

The total system cost is invested in the first year as a capital investment.

- The 100% of total system cost can be invested in several years with monthly payment at a certain open and fixed interest rate.
- The 75% of total system cost can be invested in several years with monthly payment at a certain open and fixed interest rate and the remaining 25% of total system cost is to be invested initially as capital investment.

This simulation tool is also flexible in customizing several real-time scenarios, i.e. the number of payments in a single year, the total number of payments in the entire financing period. The early cancellation interest rate can be applied when the system is to be dismantled during the financing period.

21.3.5 Boundary Conditions

This section pre-determines the boundary conditions for the simulation as shown in Table 21.3.

Initially, the energy performance of the PVT system is simulated in 85 different locations using the simulation tool. In order to discover and compare the collector energy performance in different locations, the thermal demand is maintained the same in all selected locations. Therefore, the simulated system considers a single PVT collector (1.96 m^2), for a single-family house application with 5 people, for the same demand, and the same tank volume for all locations. These assumptions provide a common system boundary to understand the effect of climatic variables and financing parameters on collector performance. Two types of demands are considered as DHW and electricity use in the building. In the electricity model, no price difference in self-consumed and exported power to the grid is considered. In the thermal system configuration, the auxiliary source for the house is the electricity grid with appropriate energy prices for every location. The generated DHW by

Table 21.3 Boundary conditions for the simulation tool

Parameter	Description
Type of application	Single-family house
Type of demand	Electricity demand, and Thermal demand for DHW
Auxiliary system	Electrical heater
Auxiliary system energy price	This has been selected for the appropriate location
No. of people in house	5
DHW temperature	60 °C
Collector model	aH72SK
No. of collectors	1
Specific volume capacity	80 L/m^2
Inclination	These were selected optimally based on a parametric study for maximum energy production
Type of mounting structure	Tilted
Type of inverter	Single-phase inverter
Annual maintenance cost	Assumed that no maintenance is required for a single collector to reduce uncertainties
Electricity and combustible price increment	6% per year is assumed for all the locations
System lifetime	25 years
Interest rate	Selected appropriately for each location

the collector is utilized for household purposes using a storage tank connected to the auxiliary system which will deliver demand at the desired temperature of 60 °C, as shown in Fig. 21.5. For each location, the installed tilt and azimuth angles are taken optimally based on higher collector production. The specific volume capacity is assumed 80 L/m^2 for all the locations which is equivalent to total 150 L of storage tank capacity.

In the proposed simplified energy system, PVT collector is directly connected to the tank without any internal or external heat exchanger. The cold water from the tank enters the PVT module, exchange heat from the absorber, and hot water is fed to the top of the tank. The DHW cold water enters at bottom of the tank, and hot water leaves from top of the tank for DHW supply in building. The DHW distribution system and associated heat losses are not considered in the analysis. The maximum DHW supply temperature is set at 60 °C, and an electric auxiliary heater is provisioned in the tank for periods when the energy from PVT modules is not enough to meet the DHW load. Electric heater starts and stops at the determined dead band to optimize energy consumption, while maintaining the fixed supply DHW temperature. During the periods when tank temperature exceeds the set limit, the energy from PVT modules is fed to a heat sink (air/water heat exchanger), and this spilled energy from the collector is not counted as part of useful energy output.

In the electrical system configuration, the generated DC power will be converted to the AC power using an inverter. Then, it is utilized by household purposes and the remaining will be sent to the electricity grid, whereas the excess electricity demand is taken from the grid connection as shown in Fig. 21.5. As the tilt angle of the PVT collector is a key parameter that will also decide the collector production, a preliminary parametric study is carried for each location to determine the optimal tilt angle for maximum annual collector production.

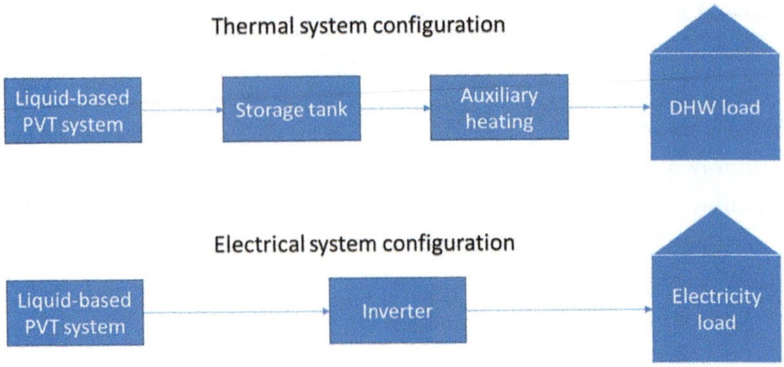

Fig. 21.5 Thermal and Electrical system configurations

The total system cost is determined using variables such a module cost, system components cost, annual operation, and maintenance cost. The electricity and auxiliary energy price escalation is assumed 6% per year for all the locations. Various parameters considered for economic analysis are shown in Table 21.4.

The payback time and NPV are estimated by considering a reference system using an electric heater. The price of electricity considered for various locations is shown in Fig. 21.6.

The economic performance of the collector in two different financial models is evaluated based:

- Model 1: The total system cost is invested in the first year,
- Model 2: The total system cost is paid for 7 years with a certain variable interest rate with every location.

21.4 Results and Discussion

This section details the simulation results using the digital mapping approach. Table 21.5 shows the inputs and results of key performance indicators for all selected locations, and the results are discussed.

Table 21.4 Parameters considered for economic analysis

Parameter	Value
Abora PVT collector	350 €
Cost for Connection kit	128 €
Tilted mounting structure	243 €
Storage tank	1553 €
Valve (servo meter)	127 €
Flowmeter	142 €
Copper tubes	19 €
Isolation tubes	14 €
Heat sink	474 €
Micro inverter	500 €
Legal regulations	377 €
Electricity price increment	6% annually
System life time	25 years
Electricity price	Variable based on each location

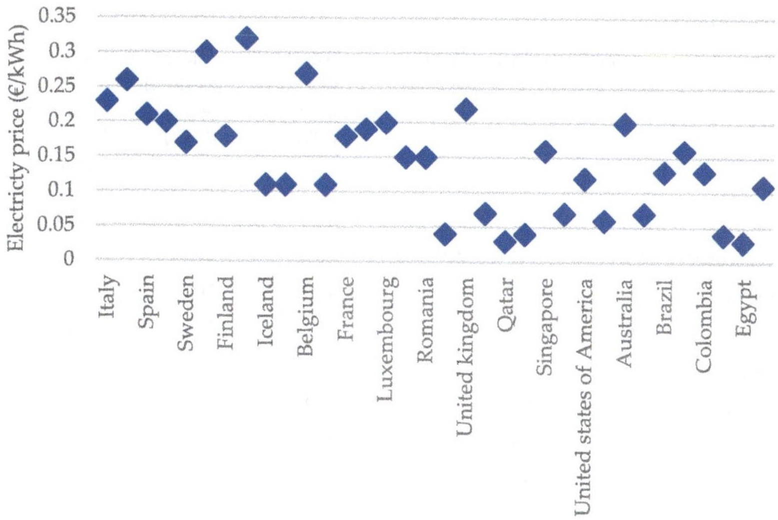

Fig. 21.6 Considered electricity prices in all countries [37]

Table 21.5 All simulated data of key performance indicators

Country	City	Latitude	Annual GHI (kWh)	Annual average temperature (°C)	Annual thermal production (kWh)	Annual electrical production (kWh)	NPV per unit collector area for financial model 1 (EUR)	NPV per unit collector area for financial model 2 (EUR)
Italy	Catania	38	1967	18	1790	487	5140	5541
	Florence	44	1632	16	1520	413	4039	4451
	Milan	45	1233	12	1153	317	2528	2955
	Rome	42	1585	17	1464	401	3797	4211
	Bari	41	1824	17	1679	458	4691	5096
Portugal	Lisbon	39	1939	18	1770	483	4766	5171
	Porto	41	1765	16	1640	447	4246	4657
	Setubal	39	1997	18	1823	495	4966	5368
Spain	Sevilla	37	2134	20	1882	520	4972	5361
	Valencia	39	2043	18	1831	505	4776	5167
	Zaragoza	42	2002	16	1795	498	4649	5041
	Barcelona	41	1904	18	1728	479	4387	4782
	Lugo	43	1567	13	1464	406	3393	3798
	Madrid	40	2019	15	1810	504	4709	5101
Switzerland	Bern	47	1335	10	1270	351	2576	3002
	Davos	47	1612	4	1562	426	2863	3286
	Lausanne	47	1408	12	1329	364	2108	2539
	Zurich	47	1249	10	1186	331	1648	1935

(continued)

Table 21.5 (continued)

Country	City	Latitude	Annual GHI (kWh)	Annual average temperature (°C)	Annual thermal production (kWh)	Annual electrical production (kWh)	NPV per unit collector area for financial model 1 (EUR)	NPV per unit collector area for financial model 2 (EUR)
Sweden	Gothenburg	58	1138	10	1073	305	1287	1726
	Linkoping	58	1132	8	1061	304	1257	1697
	Malmo	56	1183	9	1113	316	1424	1863
	Stockholm	59	1179	8	1105	317	1407	1846
	Uppsala	60	1099	8	1024	297	1142	1583
Denmark	Alborg	57	1116	8	1047	298	3041	3463
	Copenhagen	56	1144	10	1079	305	3195	3615
	Odense	55	1102	9	1040	295	2987	3409
Finland	Helsinki	60	1160	6	1086	312	1021	1464
	Oulu	65	1182	4	1112	321	1104	1545
Germany	Berlin	53	1194	10	1128	315	4582	4988
	Dortmund	52	1093	11	1037	291	4034	4446
	Frankfurt	50	1143	11	1078	302	4291	4701
	Hamburg	54	1146	11	1091	306	4363	4772
	Munich	48	1318	11	1257	345	5348	5747
Iceland	Reykjavik	64	968	6	932	266	−145	186
Norway	Bergen	60	926	9	875	253	−576	−163
	Oslo	60	1029	7	962	277	−408	3
	Trondheim	64	1166	7	1107	317	−136	273
Belgium	Brussels	51	1151	12	1094	306	3244	3664
Bulgaria	Sofia	43	1335	13	1264	348	364	813
France	Lyon	46	1422	14	1337	368	1899	2333
	Nantes	47	1408	13	1333	367	1889	2323
	Paris	49	1204	13	1134	315	1279	1718
	Toulouse	44	1522	15	1437	391	2197	2628
Greece	Athinai	38	1915	21	1731	474	3119	3540
Luxembourg	Luxembourg	50	1194	9	1128	318	1661	2096
Poland	Krakow	50	1191	10	1126	315	868	1267
	Warsaw	52	1213	10	1137	320	909	1307
Romania	Bucharest	44	1589	13	1482	406	1841	2153
	Cluj-Napoca	47	1443	11	1365	374	1516	1831
Ukraine	Kyiv	50	1330	10	1242	348	− 1287	− 1368
United Kingdom	Glasgow	56	1097	10	1045	294	2096	2527
	Liverpool	53	1013	11	965	273	1765	2199
	London	52	1107	13	1048	294	2109	2540

(continued)

Table 21.5 (continued)

Country	City	Latitude	Annual GHI (kWh)	Annual average temperature (°C)	Annual thermal production (kWh)	Annual electrical production (kWh)	NPV per unit collector area for financial model 1 (EUR)	NPV per unit collector area for financial model 2 (EUR)
China	Hong Kong	22	1338	24	1251	329	461	725
Qatar	Doha	25	1957	28	1715	462	−1468	−1168
Saudi Arabia	Medina	25	2349	29	1966	540	−828	−401
Singapore	Singapore	1	1618	27	1473	390	1461	1569
India	Bangalore	13	2093	25	1847	489	−12	178
	Bombay	19	1910	28	1687	445	−213	−21
	Hyderabad	17	2005	28	1765	466	−112	79
	Lucknow	27	1921	27	1717	453	−174	17
	New Delhi	29	2157	27	1878	505	35	224
	Surat	21	2168	28	1874	500	26	215
	Wadhwan	23	2159	28	1866	496	17	207
	Yavatmal	20	1938	28	1715	453	−179	13
USA	Chicago	42	1564	11	1475	402	987	1432
	Denver	40	1912	11	1796	483	1695	2133
	Houston	30	1720	21	1582	422	1211	1655
	Las Vegas	36	2278	21	1987	545	2136	2570
	Los Angeles	34	1973	20	1808	489	1722	2161
	New York	41	1597	14	1508	407	1052	1496
	Portland	46	1436	12	1361	374	732	1179
	San Francisco	38	1886	15	1757	478	1616	2056
	Washington	39	1602	15	1510	407	1053	1497
Mexico	Mexico City	20	1848	18	1727	451	−342	−224
Australia	Brisbane	−27	1898	21	1720	452	3940	4339
	Melbourne	−38	1528	15	1426	371	2872	3282
	Perth	−32	1930	19	1731	455	3990	4389
Argentina	Buenos Aires	−35	1703	18	1550	406	65	−2077
Brazil	Brasilia	−16	1928	22	1762	467	1985	2197
Chile	Santiago	−33	1732	15	1570	411	1785	2171
Colombia	Bogota	5	1560	14	1510	394	856	1107
Algeria	Algiers	37	2017	18	1835	495	−1027	−747
Egypt	Cairo	30	2009	22	1791	485	−1551	−1589
Morocco	Rabat	34	2094	18	1907	517	1616	1950

21.4.1 Energy Performance Evaluation of PVT Panel

21.4.1.1 Collector Thermal Production

The simulated results are visualised using geospatial maps, as they provide a clear indication for the understanding of regional trends for thermal and electrical output despite large data sets. Figure 21.7 shows the variation in the thermal output of the collector.

The general trend shows that thermal output is higher in countries with higher irradiation such as Saudi Arabia, Algeria, Morocco, Brazil, Mexico, India, etc., with annual thermal production above 1800 kWh (area-specific output 918 kWh/m^2) due to high GHI and ambient temperatures. The lower band of average collector production can be seen in Reykjavik, Iceland and for some locations in Norway with a specific output of 475 kWh/m^2 and 500 kWh/m^2 respectively. Similar thermal output is obtained for locations in counties such as Sweden, Finland, United Kingdom, Denmark, etc., with less than 510 kWh/m^2 annual production. The collector shows better performance in countries, such as Spain, Portugal, and Australia with collector production of above 1600 kWh (816 kWh/ m^2).

Figure 21.8 shows the correlation of collector thermal production with GHI and ambient temperature. All the simulated data points of these parameters are considered to define the possible trend. Results show that thermal output has a strong linear correlation with GHI with R^2 value close to 0.98. Thus, the location with higher GHI has higher thermal output. Also, thermal output shows a linear trend with ambient temperature for most of the data points, however, the correlation is not as strong as with GHI. Therefore, ambient temperature cannot be used as a sole indicator to estimate the collector output.

21.4.1.2 Collector Electrical Production

Figure 21.9 represents the electrical performance of the collector, which shows similar trends as thermal output. For locations in countries with high GHI such as Saudi Arabia, Algeria, Morocco, Brazil, India, etc. have generation above 500 kWh, and peak value in Saudi Arabia with 540 kWh. The electrical production is much less in Iceland with 266 kWh due to less available GHI, And the collector lower than 300 kWh in locations, such as Sweden, Finland, Denmark, Poland, United Kingdom, etc., The collector performed slightly better in Spain, Portugal, and Australia with more than 400 kWh annually. However, it shows there is no significant difference in thermal and electrical production trends. Furthermore, a correlation of collector electrical production with GHI and ambient temperature is developed based on all monthly points from all chosen locations and a positive correlation is realized as shown in Fig. 21.10. A large variation

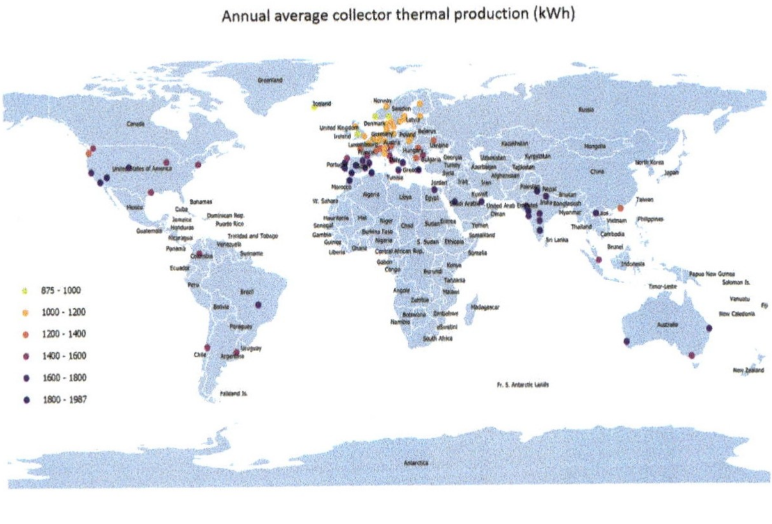

Fig. 21.7 Annual average collector thermal performance

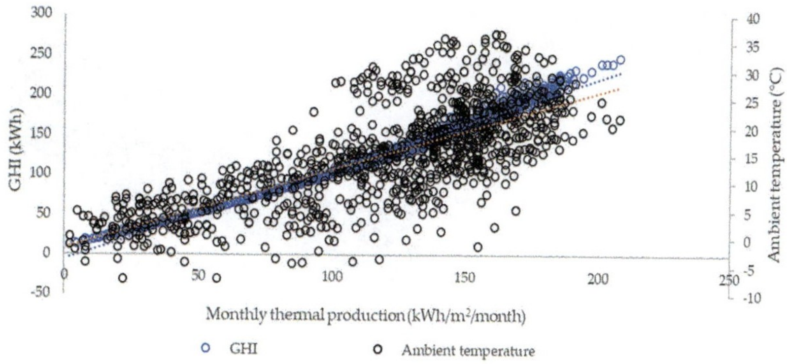

Fig. 21.8 Correlation of collector thermal production with Global Horizontal Irradiation (GHI) and ambient temperature

in electrical output for similar values of ambient temperature can be observed, which again shows that GHI is the critical parameter governing the electrical output of the collector.

A large variation in thermal and electrical output is seen for many counties and is reflected in Figs. 21.7 and 21.9. The range of collector output with a maximum and minimum value of thermal and electrical production is shown in Fig. 21.11.

The minimum thermal production in blue color represents minimum production for analyzed location, while the maximum thermal production is indicated with an orange color that represents the highest thermal production of a city in each country. The results show likely high variation in Italy, Spain, USA, and Australia, as many cities were simulated in those countries, and less variation is recorded in countries Denmark, Iceland, United Kingdom, etc., due to less simulated cities.

In general, PVT collector monthly production is an important key factor in the sizing of a solar system to match the monthly variation of energy consumption. Figures 21.12 and 21.13 show the variation in collector monthly thermal and

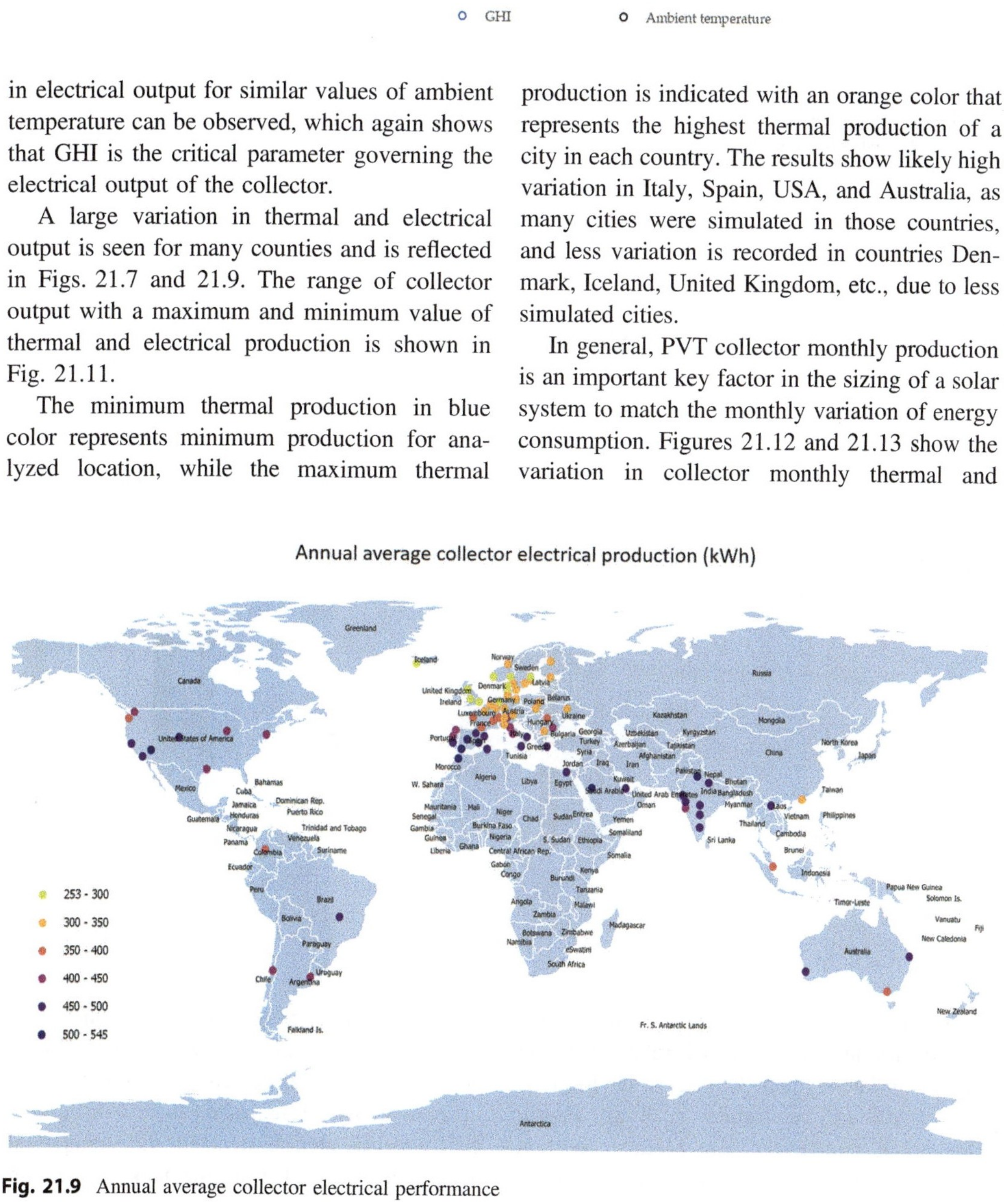

Fig. 21.9 Annual average collector electrical performance

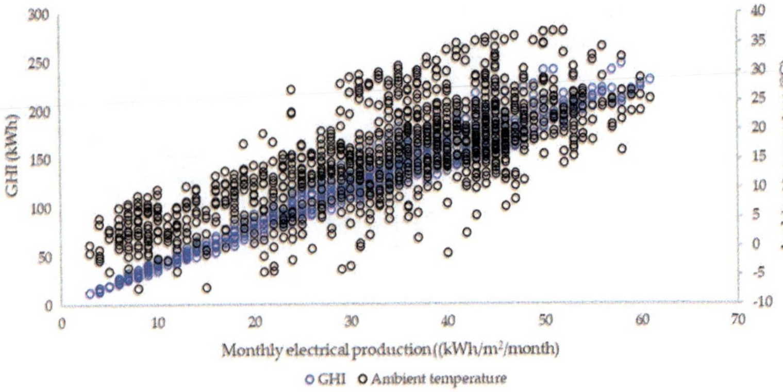

Fig. 21.10 Correlation of collector electrical production with Global Horizontal Irradiation (GHI) and ambient temperature

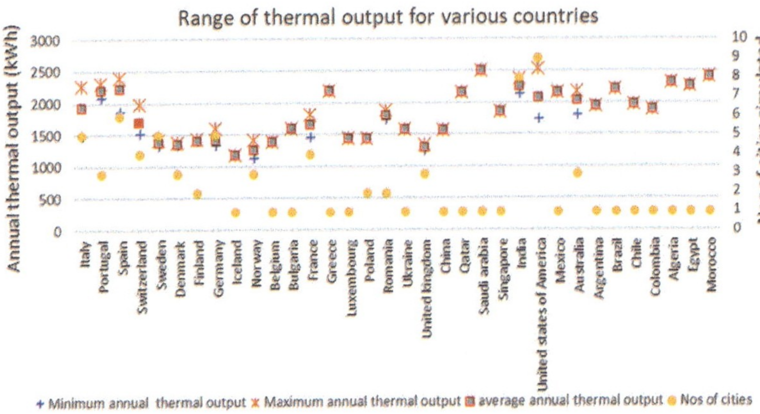

Fig. 21.11 Country-wise collector thermal performance uncertainty

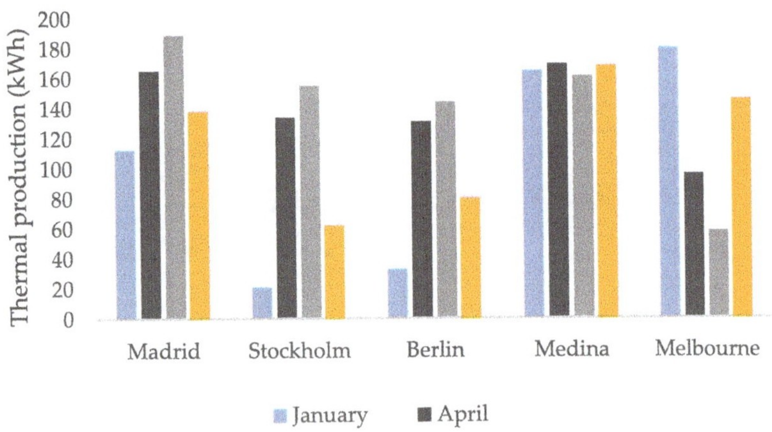

Fig. 21.12 Collector monthly thermal production variation

electrical production respectively. The thermal performance in April and July is relatively higher and less in January and October for the locations in the northern hemisphere such as Madrid, Stockholm, and Berlin. In Medina location, although GHI and ambient temperatures are higher in July, yet the thermal production in lower compared to October month. This is because the thermal demand in July is less than in October. Therefore, in July month, due to high GHI and less thermal demand, the storage tank losses will be higher as the tank temperature

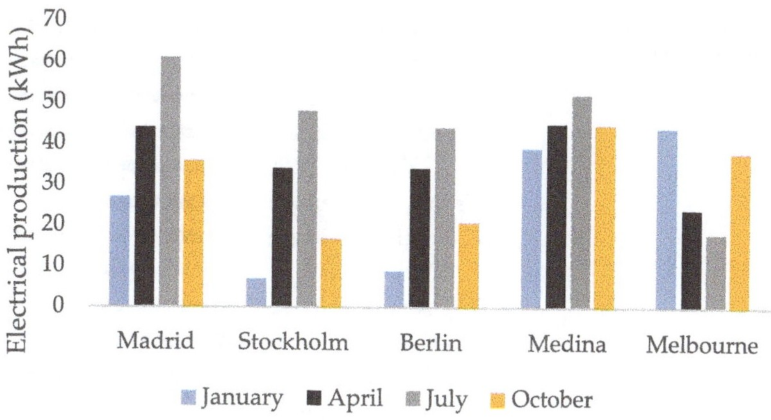

Fig. 21.13 Collector monthly electrical production variation

increases. Higher tank temperature results in lower thermal and electrical production of collector. As the GHI trend in the southern hemisphere is quite opposite to the northern hemisphere, the production in January and October is likely higher than April and July months. In Stockholm location, the variation between the months is significant because of seasonal variation in GHI, and the same is lower in Medina which results in more uniform monthly production.

The trends for monthly electrical production are slightly different than thermal output. For example, In Medina location, electrical production is higher in July than in October even though the ambient temperature is maximum in July. This is due to high GHI in July and is in line with findings that the major factor influencing the electrical production is GHI, rather than ambient temperature.

21.4.1.3 Collector Energy Utilization Ratio

The energy utilization ratio of the collector for various locations is shown in Fig. 21.14. The correlation trends between energy utilization ratio and annual average ambient temperature are shown in Fig. 21.15 with consideration of all selected 85 geographical locations to derive a possible trend between the parameters.

Some locations show interesting results of system boundaries on PVT collector performance. This can be realized by comparing the energy utilization ratio for Medina (high irradiation) and Davos (low irradiation location). The energy utilization for Davos (63%) is higher compare to Medina (52.5%), even though the absolute value of total energy output is higher for Medina (2506 kWh), compared to Davos (1988 kWh). This is because the load demand for Medina is comparably lower, while the other system design parameters remain the same (collector area, tank volume, etc.), which resulted in higher average tank temp, and thus lower collector efficiency for Medina. Results show that the total thermal demand for every location is varying depending on the ambient temperature as shown in Fig. 21.16. It is because of the temperature difference between the annual average ambient temperature of each location and desired water temperature (assumed 60 °C), which has to be covered by the collector thermal production.

21.4.1.4 Collector Exergy Efficiency

From the Carnot efficiency Error! Reference source not found., it can be noted that exergy efficiency is a function of inlet temperature and thermal output of the collector (assumed that the desired output temperature is fixed at 60 °C). So, it can be derived that locations with higher ambient temperature will result in less quality of exergy, and thus lower exergetic efficiency.

Figure 21.17 shows the correlation of exergetic efficiency with ambient temperature based on all selected 85 geographical locations to derive a possible trend between the parameters. The similar trends can be seen for some specific locations shown in Fig. 21.18. It can be seen that

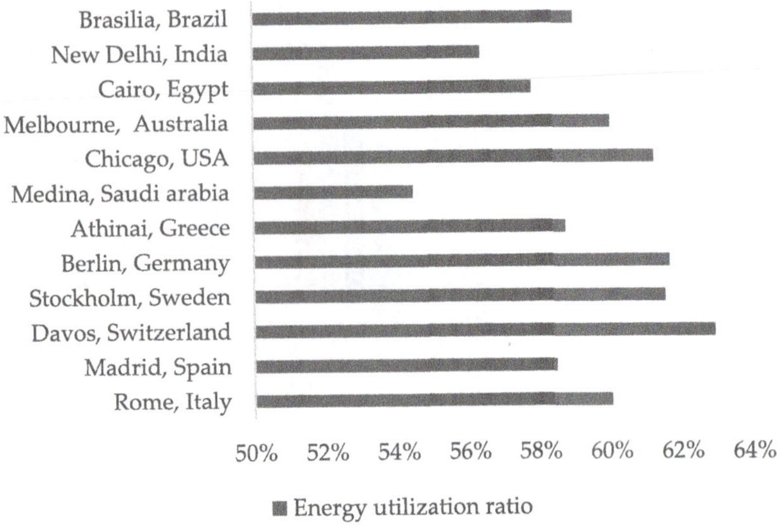

Fig. 21.14 Collector energy utilization ratio

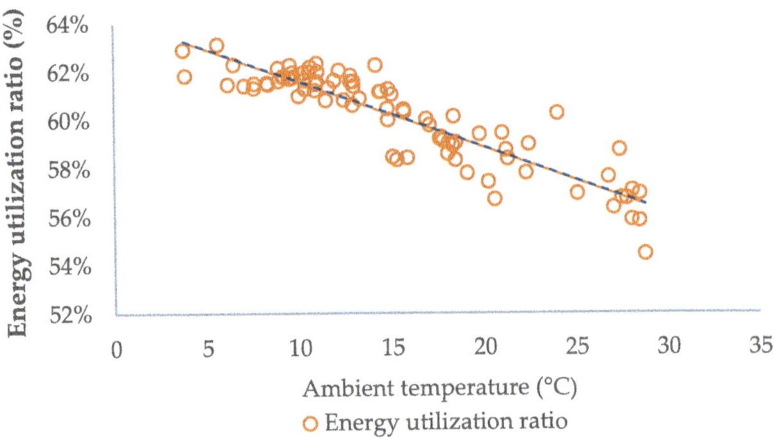

Fig. 21.15 Correlation of Energy utilization ratio with the annual average ambient temperature

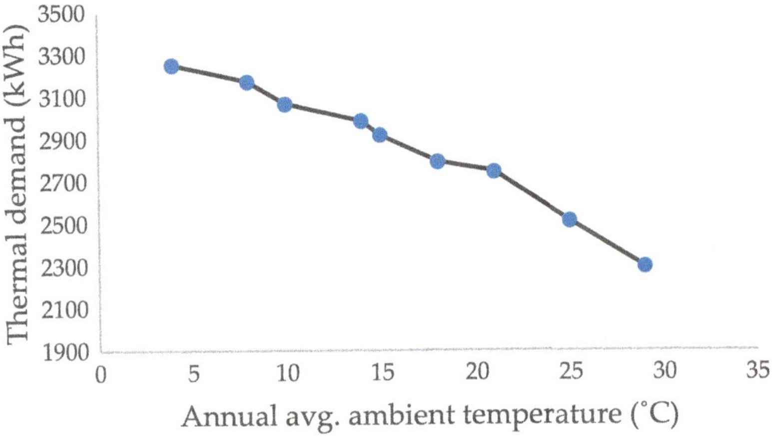

Fig. 21.16 Total thermal demand of single-family house relation with the average ambient temperature

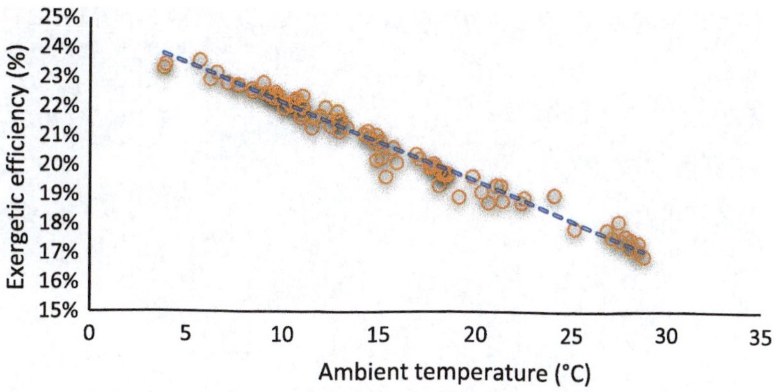

Fig. 21.17 Correlation of exergy efficiency with the annual average ambient temperature

even though the energy efficiency of Madrid is higher compared to the Davos, the exergy efficiency of the Davos is higher due to lower annual ambient temperature, and thus higher quality of heat is delivered to the user.

21.4.2 Economic Performance Evaluation of PVT Collector

Based on the above energy performance, the economic performance of such a PVT system is investigated in the 85 different locations. In this section, NPV per unit collector area is analyzed and represented.

21.4.2.1 Collector Economic Performance in Financing Model 1

This financing model scenario has assumed that the total cost of the system is invested in the first year of the system period. As the total system cost will be invested in the first year, no interest rate is not considered. Figure 21.19 is the digital representation of NPV potential per unit collector area with financial model 1 in all 85 geographical cities across the World and Fig. 21.20 shows the NPV potential per unit collector area in geographical cities in the European continent.

The cities with larger dots represent the high NPV potential and cities with smaller dot size represents the least NPV potential. The cities

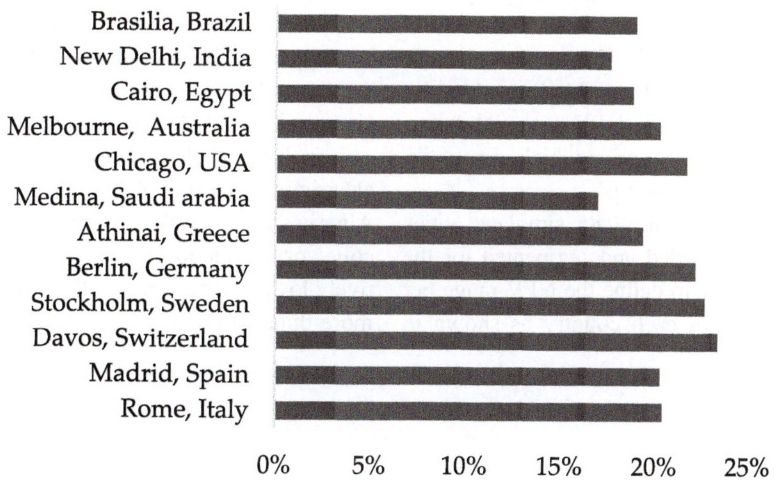

Fig. 21.18 Collector exegetic efficiency

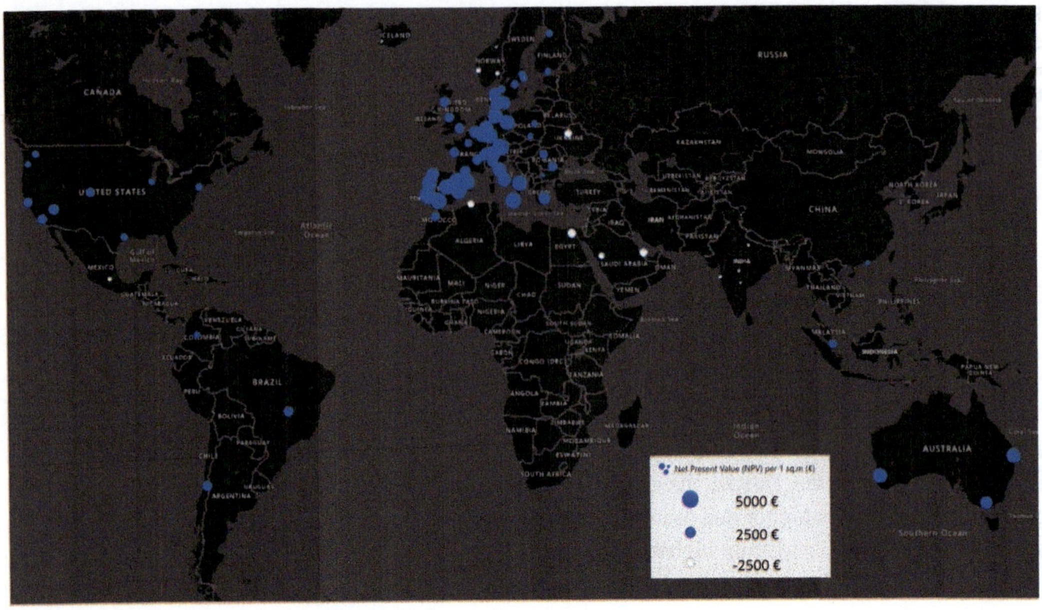

Fig. 21.19 NPV potential per unit collector area for financing model 1

Catania and Munich have the highest potential of 5140 € and 5348 € respectively, followed by Bari, Lisbon, Setubal, Sevilla, Valencia, Zaragoza, Madrid and Berlin cities has potentially more than 4500 € per unit collector area. This is due to their high available GHI and electricity grid price, so the energy savings are high in these locations which reflected in huge NPV potential for this system. Although cities, such as Oslo, Bergen, Reykjavik, etc., with relatively less electricity grid price, resulted in negative NPV due to lower available GHI. The cities with high collector production such as Medina, Algeria, Cairo have shown negative NPV potential due to a very less electricity grid price which eventually showed fewer energy savings.

The NPV potential in all 85 simulated cities has been selected divided and segmented for the appropriate countries to define the NPV range per unit collector area of each country as shown in Fig. 21.21. A large variation in NPV can be seen in few countries, such as Italy, Portugal, due to variability in GHI for simulated locations. However, a smaller variation is identified in countries such as China, Argentina, Brazil, etc., this is because only one city has been simulated in this chapter, which is part of the key uncertainty.

Figure 21.22 shows the payback period of this PVT system for a single-family house of 5 people in several countries based on financial model 1. The results show that the total system cost will be returned in the first 10 years in countries, such as Australia, Belgium, Denmark, Germany, Greece, Italy, Portugal, Spain, Switzerland, etc. This is due to high collector production and high electricity grid price. Although countries such as Algeria, Saudi Arabia, Egypt have the highest collector production, the grid price is comparatively lower, which reflects the payback period of more than 20 years.

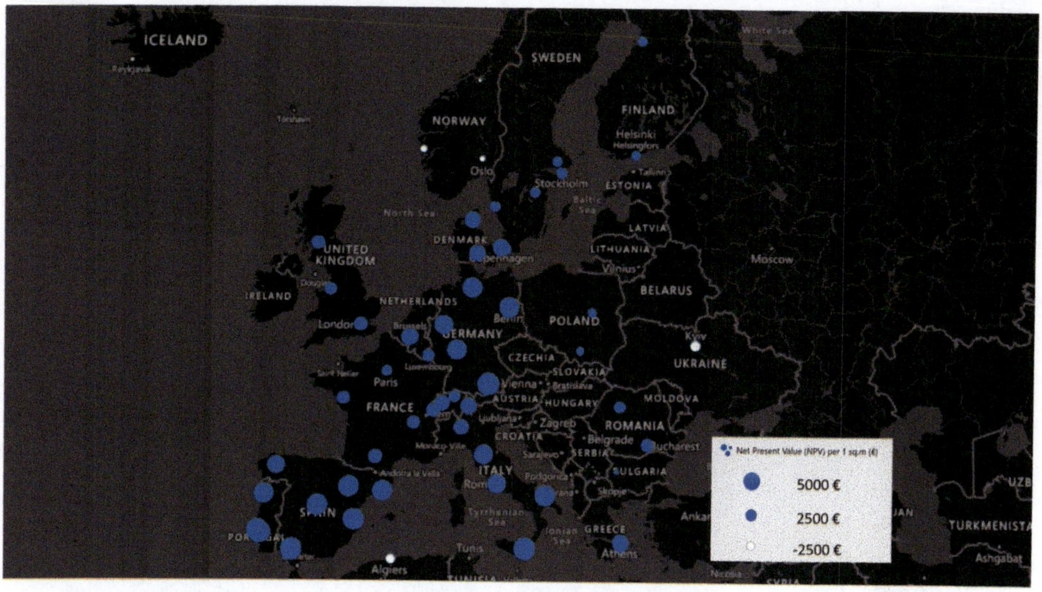

Fig. 21.20 NPV potential per unit collector area in Europe for financing model 1

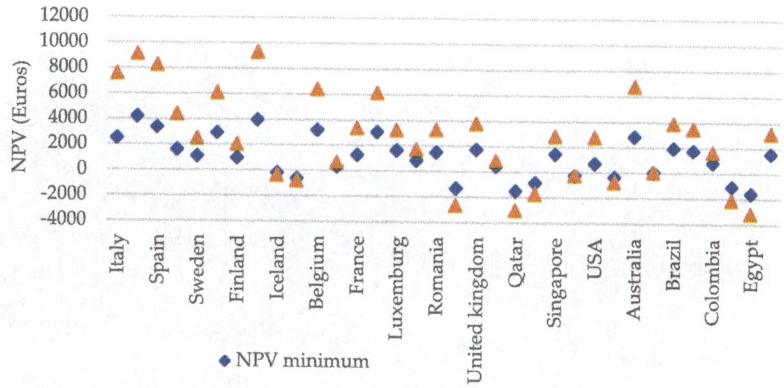

Fig. 21.21 Country-wise NPV potential per unit collector area for financial model 1

21.4.2.2 Collector Economic Performance in Financing Model 2

This financing model has been analyzed by assuming that 75% of total system cost is paid within a financing period of 7 years with a certain interest rate and the remaining 25% of total system cost is invested in the first year without any interest rate. The NPV potential per unit collector area with financing model 2 in 85 geographical cities across the world is shown in Fig. 21.23 and NPV potential per unit collector area in a specific European continent is shown in Fig. 21.24.

The cities with larger dots represent the high NPV potential and cities with smaller dots represent the lower NPV potential. The cities performed high NPV potential in financing model 1, such as Catania and Munich, which has shown improved NPV of 5140 € and 5348 € respectively because of Zero interest rates in those countries. This is because if the interest rate is zero, the user needs to pay the part of system cost in later years, and the present value of this investment will be lower due to the time value of money. This will reduce the accumulated investment and thus higher NPV. However, if the interest rate is high, the extra amount paid due to

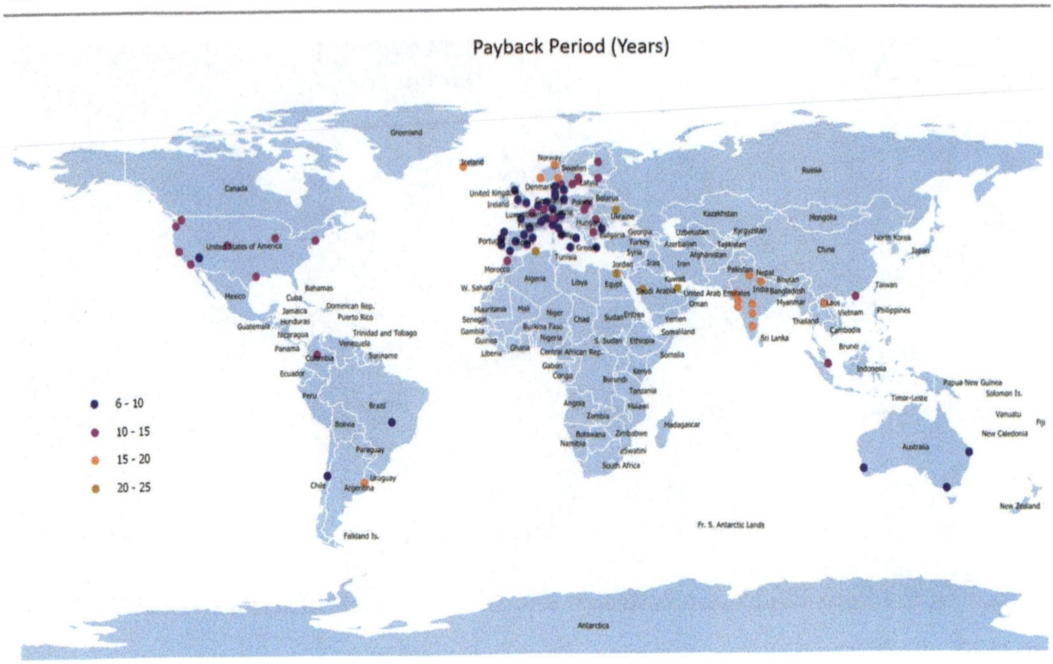

Fig. 21.22 Country-wise average payback period of the PVT collector system

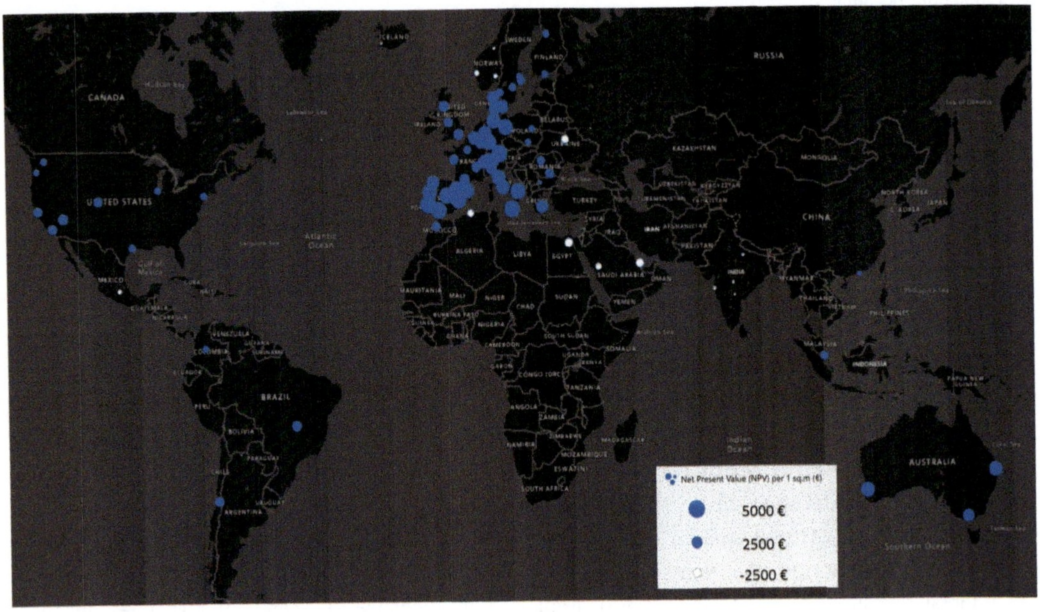

Fig. 21.23 NPV potential per unit collector area for financing model 2

high interest in later years, which will overweigh the advantage due to the time value of money and it will decrease the overall NPV. Therefore, Financial model 1 is recommended for countries with high interest rate to maximize the NPV, and minimize the payback. Whereas, financial model 2 is recommended for counties with zero or lower interest rates to maximize the NPV.

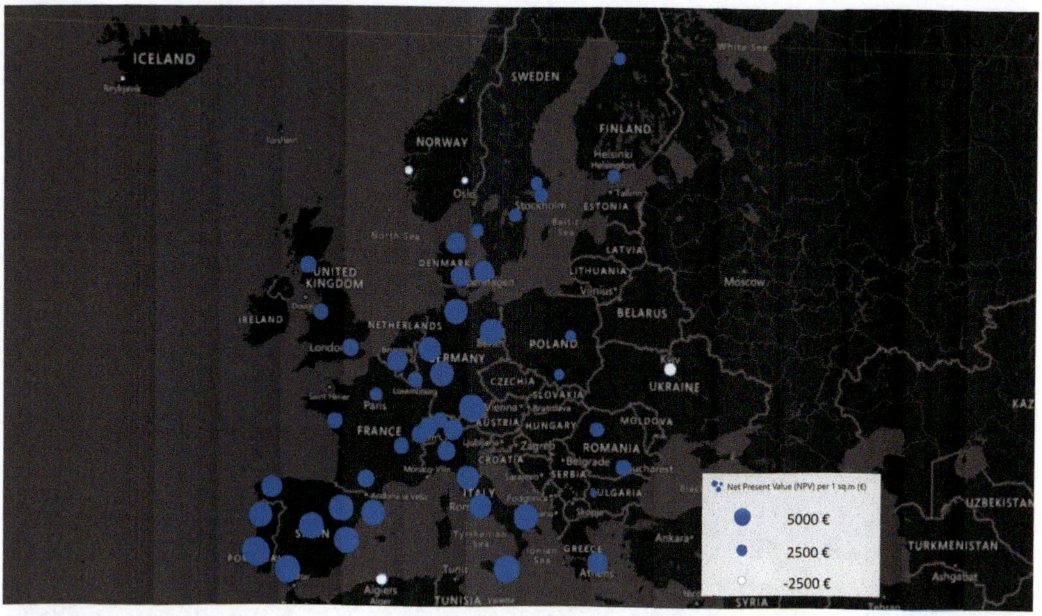

Fig. 21.24 NPV potential per unit collector area in Europe for financing model 2

Figure 21.25 shows the NPV potential per unit collector area in each country for the financing model 2. As compared with financing model 1, there is slightly better performance in NPV in most of the countries. Thus, there is not much variation has been identified in this model 2 compared with model 1.

The effect of NPV change due to financial model 2 compare to model 1 is shown in Fig. 21.26. As expected, the countries with high interest rate have shown a negative effect on NPV and countries with less and zero interest rates has shown better NPV potential, such as USA, Australia and most of the European countries. However, due to high interest rate of 38% in Argentina, a huge negative impact is identified with this financing model 2. Furthermore, a correlation is derived between NPV variation with an interest rate of a specific location in Fig. 21.27.

21.4.2.3 Uncertainties

In this chapter, the authors acknowledge the possible uncertainties in energy performance analysis. For instance, the delivery water temperature are assumed 60 °C and 28 L DHW

Fig. 21.25 Country-wise NPV potential per unit collector area for financing model 2

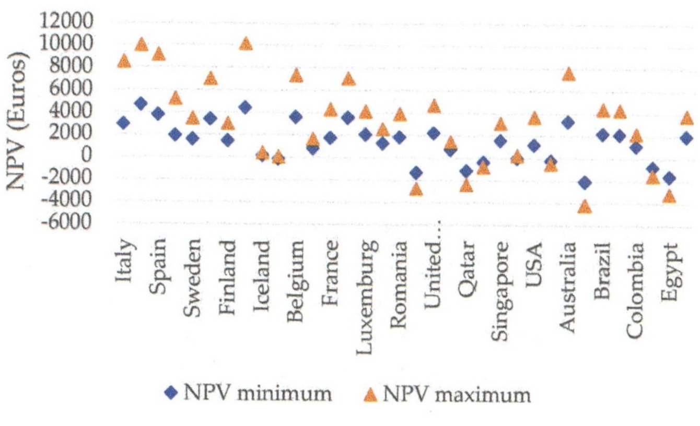

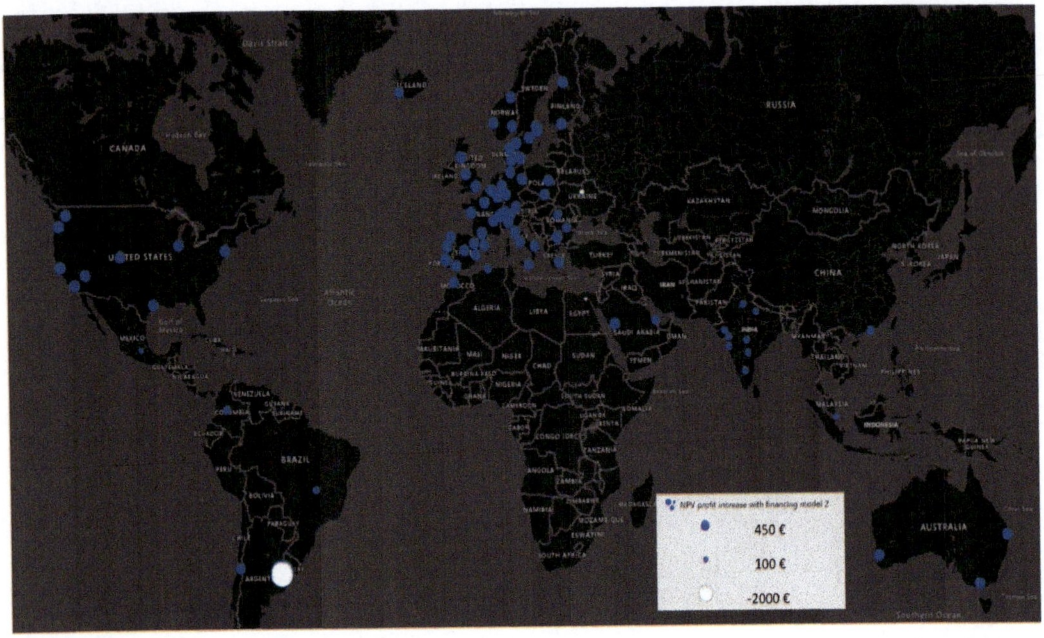

Fig. 21.26 NPV profit increase with financing model 2

Fig. 21.27 Correlation of NPV potential variation with an interest rate

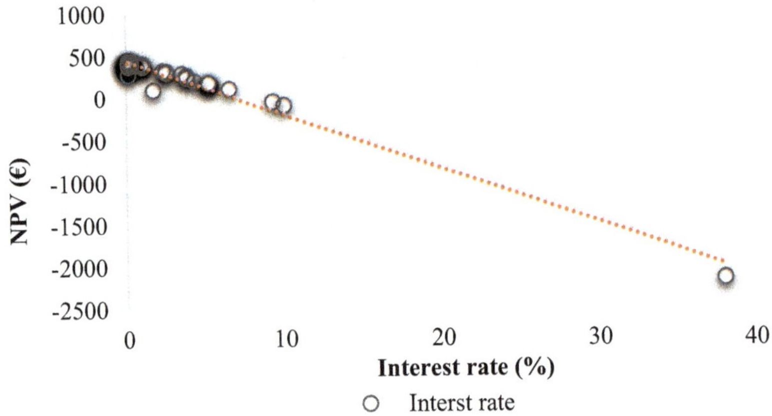

demand per person for all the locations across all the cities. Also, the specific volume ratio (v/a) has been assumed as 80 L/m^2 for all the locations but since it may vary depending on the location and type of application, the resulted collector production would be slightly different in real-time but this approach has been assumed to achieve the goals of this chapter.

Furthermore, as the grid price is a key parameter of the total system energy savings, the auxiliary energy price is taken as the generalized price for every specific country, whereas in real-time case the energy price would be different for every state/city/municipality depending on localized energy policy. It has been considered because of the unavailability of precise data, which may not be significantly higher. The interest rate is chosen for each country for deriving the NPV potential difference between financing model 1 and model 2. But only few countries, which have negative and zero interest rate, have been assumed as 0.1% due to the simulation tool incapability of accepting negative or null values. However, it has also realized that uncertainty of difference between the negative interest rates and assumed interest rates has not been less than 1% which is not significantly affecting the NPV potential difference. Hence, the assumptions have been considered to achieve the aims in possible optimistic and realistic approaches irrespective of the uncertainties.

21.5 Conclusions

The performance of an energy system consists of a PVT collector performance and storage tank is evaluated for 85 locations across a large cities. The optimal tilt angle of PVT collector, load demand, and electricity prices are chosen appropriately for each simulated location. The results show that the major parameter influencing the PVT performance is GHI, and results derived a strong linear correlation between collector output and GHI. The other factor influencing energetic performance is ambient temperature, source, and load water temperatures. The energetic utilization ratio is dependent on total thermal demand and specific volume ratio (v/a ratio), as it can have a major influence on the fluid temperature in the storage tank, and thus collector total production. The electrical production by PVT collector is higher in high ambient temperature locations. The highest and lowest energy utilization ratio of the collector is recorded in Reykjavik, Iceland (63%), and Medina, Saudi Arabia (54%) respectively. The highest and lowest exergetic efficiency of the collector has been recorded in Reykjavik, Iceland (23%), and Medina, Saudi Arabia (17%) respectively. Most importantly, the results show that the higher energetic output does not guarantee high economic feasibility. There are several factors such as electricity price, interest rate, and selection of financial model which can highly affect the economic feasibility of PVT collector. The average NPV per unit collector area of 85 geographical cities for financial model 1 and financial model 2 is 1886 € and 2221 € respectively. The NPV and payback period analysis of the PVT system has shown positive results for the cities, which have high collector production and high electricity grid price reflecting high energy savings. However, the financing model 1 is highly recommended for the locations with high interest rates and financial model 2 is beneficial for the locations with less interest rates. This chapter offers potential insight into the promotion of the PVT market in different regions.

References

Al-Ghussain L, Taylan O, Baker DK (2019) An investigation of optimum PV and wind energy system capacities for alternate short and long-term energy storage sizing methodologies. Int J Energy Res 43 (1):204–218. https://doi.org/10.1002/er.4251

Ahn H, Rim D, Pavlak GS, Freihaut JD (2019) Uncertainty analysis of energy and economic performances of hybrid solar photovoltaic and combined cooling, heating, and power (CCHP + PV) systems using a Monte-Carlo method. Appl Energy 255:113753. https://doi.org/10.1016/j.apenergy.2019.113753

Al-Waeli AHA, Sopian K, Kazem HA, Chaichan MT (2017) Photovoltaic/Thermal (PV/T) systems: status and future prospects. Renew Sustain Energy Rev 77:109–130. https://doi.org/10.1016/j.rser.2017.03.126

Bremer M, Mayr A, Wichmann V, Schmidtner K, Rutzinger M (2016) A new multi-scale 3D-GIS-approach for the assessment and dissemination of solar income of digital city models. Comput Environ Urban Syst 57:144–154. https://doi.org/10.1016/j.compenvurbsys.2016.02.007

Buonomano A, De Luca G, Figaj RD, Vanoli L (2015) Dynamic simulation and thermo-economic analysis of a PhotoVoltaic/Thermal collector heating system for an indoor–outdoor swimming pool. Energy Convers Manage 99:176–192. https://doi.org/10.1016/j.enconman.2015.04.022

Buonomano A, Calise F, Vicidomini M (2016) Design, simulation and experimental investigation of a solar system based on PV panels and PVT collectors. Energies; Basel 9(7):497. https://doi.org/10.3390/en9070497

Chow TT (2010) A review on photovoltaic/thermal hybrid solar technology. Appl Energy 87(2):365–379. https://doi.org/10.1016/j.apenergy.2009.06.037

'Contact', Abora Solar. https://abora-solar.com/en/contact/. Accessed 29 Nov 2019

Fudholi A, Sopian K, Yazdi MH, Ruslan MH, Ibrahim A, Kazem HA (2014) Performance analysis of photovoltaic thermal (PVT) water collectors. Energy Convers Manage 78:641–651. https://doi.org/10.1016/j.enconman.2013.11.017

Gagliano A, Tina GM, Nocera F, Grasso AD, Aneli S (2019) Description and performance analysis of a flexible photovoltaic/thermal (PV/T) solar system. Renew Energy 137:144–156. https://doi.org/10.1016/j.renene.2018.04.057

Gu Y, Zhang X, Are Myhren J, Han M, Chen X, Yuan Y (2018) Techno-economic analysis of a solar photovoltaic/thermal (PV/T) concentrator for building application in Sweden using Monte Carlo method. Energy Convers Manage 165:8–24. https://doi.org/10.1016/j.enconman.2018.03.043

Heck N, Smith C, Hittinger E (2016) A Monte Carlo approach to integrating uncertainty into the levelized cost of electricity. Electr J 29(3):21–30. https://doi.org/10.1016/j.tej.2016.04.001

Herrando M, Markides CN (2016) Hybrid PV and solar-thermal systems for domestic heat and power provision in the UK: techno-economic considerations. Appl Energy 161:512–532. https://doi.org/10.1016/j.apenergy.2015.09.025

IEA-SHC-Task60-Highlights-2019.pdf. Accessed: 10 Jun 2020. Available: https://task60.iea-shc.org/Data/Sites/1/publications/IEA-SHC-Task60-Highlights-2019.pdf

IEA SHC || Task 60 || IEA SHC || Task 60. http://task60.iea-shc.org/. Accessed 21 Nov 2019

'Intro', Meteonorm (de). https://meteonorm.com/. Accessed 22 Jul 2020

Jia Y, Alva G, Fang G (2019) Development and applications of photovoltaic–thermal systems: a review. Renew Sustain Energy Rev 102:249–265. https://doi.org/10.1016/j.rser.2018.12.030

Jung J, Han S, Kim B (2019) Digital numerical map-oriented estimation of solar energy potential for site selection of photovoltaic solar panels on national highway slopes. Appl Energy 242:57–68. https://doi.org/10.1016/j.apenergy.2019.03.101

Joshi SS, Dhoble AS (2018) Photovoltaic-Thermal systems (PVT): technology review and future trends. Renew Sustain Energy Rev 92:848–882. https://doi.org/10.1016/j.rser.2018.04.067

Joshi AS, Tiwari A (2007) Energy and exergy efficiencies of a hybrid photovoltaic–thermal (PV/T) air collector. Renewable Energy 32(13):2223–2241. https://doi.org/10.1016/j.renene.2006.11.013

Kazem HA (2019) Evaluation and analysis of water-based photovoltaic/thermal (PV/T) system. Case Stud Thermal Eng 13:100401. https://doi.org/10.1016/j.csite.2019.100401

Khelifa A, Touafek K, Moussa HB, Tabet I, Haloui H (2015) Analysis of a Hybrid Solar Collector Photovoltaic Thermal (PVT). Energy Proc 74:835–843. https://doi.org/10.1016/j.egypro.2015.07.819

Khelifa A, Touafek K, Ben Moussa H, Tabet I (2016) Modeling and detailed study of hybrid photovoltaic thermal (PV/T) solar collector. Solar Energy 135:169–176. https://doi.org/10.1016/j.solener.2016.05.048

Microsoft Excel, Spreadsheet Software, Excel Free Trial. https://www.microsoft.com/en-ww/microsoft-365/excel. Accessed 29 Jul 2020

Mishra T, Rabha A, Kumar U, Arunanchalam K, Sridhar V (2020) Assessment of solar power potential in a hill state of India using remote sensing and Geographic Information System. Remote Sens Appl: Soc Environ p 100370. https://doi.org/10.1016/j.rsase.2020.100370

Oh M, Park H-D (2018) A new algorithm using a pyramid dataset for calculating shadowing in solar potential mapping. Renew Energy 126:465–474. https://doi.org/10.1016/j.renene.2018.03.068

Peronato G, Rastogi P, Rey E, Andersen M (2018) A toolkit for multi-scale mapping of the solar energy-generation potential of buildings in urban environments under uncertainty. Sol Energy 173:861–874. https://doi.org/10.1016/j.solener.2018.08.017

PVT systems IEA SHC 60 - Annex 180504.pdf. Accessed: Nov 25, 2019. Available: http://task60.iea-shc.org/Data/Sites/1/publications/PVT%20systems%20IEA%20SHC%2060%20-%20Annex%20180504.pdf

Ramschak V (2020) /ieashc-task60-2020-0001. IEA SHC Task 60. https://doi.org/10.18777/ieashc-task60-2020-0001

Riggs BC et al (2017) Techno-economic analysis of hybrid PV/T systems for process heat using electricity to subsidize the cost of heat. Appl Energy 208:1370–1378. https://doi.org/10.1016/j.apenergy.2017.09.018

Shah R, Srinivasan P (2018) Hybrid Photovoltaic and Solar Thermal Systems (PVT): performance simulation and experimental validation. Mater Today: Proc 5(11), Part 2, 22998–23006. https://doi.org/10.1016/j.matpr.2018.11.028

Wang K, Herrando M, Pantaleo AM, Markides CN (2019) Technoeconomic assessments of hybrid photovoltaic-thermal vs. conventional solar-energy systems: case studies in heat and power provision to sports centres. Appl Energy 254:113657. https://doi.org/10.1016/j.apenergy.2019.113657

Welcome to the QGIS project! https://qgis.org/en/site/. Accessed 29 Jul 2020

Yazdanpanahi J, Sarhaddi F, Mahdavi Adeli M (2015) Experimental investigation of exergy efficiency of a solar photovoltaic thermal (PVT) water collector based on exergy losses. Solar Energy 118:197–208. https://doi.org/10.1016/j.solener.2015.04.038

Zhang X, Zhao X, Smith S, Xu J, Yu X (2012) Review of R&D progress and practical application of the solar photovoltaic/thermal (PV/T) technologies. Renew Sustain Energy Rev 16(1):599–617. https://doi.org/10.1016/j.rser.2011.08.026

MIX
Papier aus verantwortungsvollen Quellen
Paper from responsible sources
FSC® C105338

Printed by Books on Demand, Germany